Characterizations, Mechanical Properties and Constitutive Modeling of Advanced Materials

Characterizations, Mechanical Properties and Constitutive Modeling of Advanced Materials

Editors

Madhav Baral
Charles Lu

MDPI

Basel • Beijing • Wuhan • Barcelona • Belgrade • Novi Sad • Cluj • Manchester

Editors

Madhav Baral
Department of Mechanical
and Aerospace Engineering
University of Kentucky
Lexington, KY
USA

Charles Lu
Department of Mechanical
and Aerospace Engineering
University of Kentucky
Lexington, KY
USA

Editorial Office
MDPI AG
Grosspeteranlage 5
4052 Basel, Switzerland

This is a reprint of articles from the Special Issue published online in the open access journal *Materials* (ISSN 1996-1944) (available at: https://www.mdpi.com/journal/materials/special_issues/3O7B9269X0).

For citation purposes, cite each article independently as indicated on the article page online and as indicated below:

Lastname, A.A.; Lastname, B.B. Article Title. *Journal Name* **Year**, *Volume Number*, Page Range.

ISBN 978-3-7258-1916-4 (Hbk)
ISBN 978-3-7258-1915-7 (PDF)
doi.org/10.3390/books978-3-7258-1915-7

Contents

Preface

This Special Issue aims to highlight the latest advancements in the critical areas of characterizations, mechanical properties, and constitutive modeling of advanced materials, integrating both experimental and theoretical perspectives. This collection of works spans a diverse range of topics, from the characterization of cutting-edge materials like nanomaterials and biomaterials to the detailed investigation of their mechanical properties under various conditions. Additionally, it features the development of sophisticated constitutive models that enhance our understanding and prediction of material behavior. By bringing together pioneering research and insights, this Special Issue aspires to contribute significantly to the advancement of materials science and engineering.

Madhav Baral and Charles Lu
Editors

materials

MDPI

Article

An Algorithm for Modeling Thermoplastic Spherulite Growth Using Crystallization Kinetics

Jamal F. Husseini [1], Evan J. Pineda [2,*] and Scott E. Stapleton [1]

1 Department of Mechanical Engineering, University of Massachusetts Lowell, Lowell, MA 01854, USA; jamal_husseini@student.uml.edu (J.F.H.); scott_stapleton@uml.edu (S.E.S.)
2 Multiscale and Multiphysics Modeling Branch, NASA Glenn Research Center, Cleveland, OH 44135, USA
* Correspondence: evan.j.pineda@nasa.gov

Abstract: Crystallization kinetics were used to develop a spherulite growth model, which can determine local crystalline distributions through an optimization algorithm. Kinetics were used to simulate spherulite homogeneous nucleation, growth, and heterogeneous nucleation in a domain discretized into voxels. From this, an overall crystallinity was found, and an algorithm was used to find crystallinities of individual spherulites based on volume. Then, local crystallinities within the spherulites were found based on distance relative to the nucleus. Results show validation of this model to differential scanning calorimetry data for polyether ether ketone at different cooldown rates, and to experimental microscopic images of spherulite morphologies. Application of this model to various cooldown rates and the effect on crystalline distributions are also shown. This model serves as a tool for predicting the resulting semi-crystalline microstructures of polymers for different manufacturing methods. These can then be directly converted into a multiscale thermomechanical model.

Keywords: thermoplastics; crystallization; modeling; spherulites

Citation: Husseini, J.F.; Pineda, E.J.; Stapleton, S.E. An Algorithm for Modeling Thermoplastic Spherulite Growth Using Crystallization Kinetics. *Materials* **2024**, *17*, 3411. https://doi.org/10.3390/ma17143411

Academic Editors: Madhav Baral and Charles Lu

Received: 12 June 2024
Revised: 1 July 2024
Accepted: 3 July 2024
Published: 10 July 2024

1. Introduction

High performance semi-crystalline thermoplastic resins such as polyether-ether-ketone (PEEK) and polyether-ketone-ketone (PEKK) are gaining popularity for use in aerospace applications due to desirable material properties, manufacturability, and repairability while in service. The thermal history during manufacturing ultimately influences the growth and formation of microscale spherulites, which may impact the thermomechanical properties of these materials. For thermoplastics to be used for novel applications, understanding the relationship between the microscale crystalline morphology and bulk properties is imperative. Using this understanding, computational models can be developed that predict thermomechanical properties and residual stresses based on the material microstructure and spherulite growth derived from processing conditions.

At the microscale, semicrystalline thermoplastics contain structures referred to as spherulites, which consist of crystalline and amorphous phases with distinct mechanical properties. During thermal cooldown, spherulites grow radially from a nucleation site until they impinge on neighboring spherulites. As such, the material is not fully crystalline and there are gradients in the crystallinity of the spherulites themselves, resulting in a higher crystallinity at the core than the perimeter of the spherulites. Thus, the different spherulite morphologies and local crystalline distributions have an impact on the thermoelastic properties, residual stresses, and failure.

Multiple studies have tried to examine spherulite fracture, but there is still not a complete consensus on this phenomenon [1–3]. Studies such as [1] have shown scanning electron microscope (SEM) images of a fracture surface where there is nuclei "pull-out" within the spherulite, whereas studies such as [2] indicate that cracks tend to propagate through the spherulite nucleus depending on its orientation. These studies highlight that understanding the crystalline microstructure of these materials is necessary to predict mechanical properties and failure. Understanding these phenomena is often difficult using experiments alone, as different manufacturing techniques such as ultrasonic welding, additive manufacturing, automated fiber placement (for composites), and stamp forming may produce unknown microstructures [4–8]. Computational simulation techniques can be used to establish structure–property relationships for thermoplastic materials and supplement experimental data.

Computational models have been developed to predict spherulite nucleation, growth, and crystallinity based on thermal history. These models typically use the Avrami equation [9] to model crystallization, which has shown good empirical agreement [10–13]. More modern approaches utilize crystallization kinetics, due originally to Lauritzen and Hoffman [14], to model crystalline growth [15]. From these methods, models have been developed which can describe the isothermal and non-isothermal crystallization of thermoplastic polymers [16–22]. Choe and Lee [10] reported an approach adopted from [12] to model the non-isothermal crystallization of PEEK. Their results showed that for different cooldown rates, the proposed model predicted the non-isothermal crystallization well as compared to experimental differential scanning calorimetry (DSC) values. Guan and Pitchuami [23] adapted this model to predict 2D PEEK spherulite growth around fibers based on processing parameters of a tow placement process. Saber [24] used crystallization kinetics to model spherulite growth in 2D and 3D domains with neat resin and with fibers, using scans of microstructures for fiber locations. These models were validated against cross-polarized microscopy images of in situ spherulite growth and relative crystallinity as a function of time in neat resin from DSC data, and then used in more complex model geometries. While these models serve as important tools in modeling spherulite growth, they do not address the distribution of crystalline and amorphous phases locally within the spherulites. If the ultimate goal is a thermomechanical model for predicting properties and failure, this additional level of detail is necessary.

In this study, a crystallization kinetic model was developed from [10] using constants for PEEK and used to simulate different spherulite morphologies. Ultimately, the intent of the current development is to couple the output of the crystallization simulations to a multiscale thermomechanical model [25–27]. Therefore, the model domain was discretized into voxels, where, after the spherulite growth, each voxel was assigned crystallinity through an optimization scheme that used the overall crystallinity of the polymer as the objective. This optimization algorithm was used to determine the crystallinities of each spherulite and the local crystalline distributions within a spherulite. Results of this model were first validated against DSC data from the literature, as well as microscopic images of spherulites taken under cross-polarized light from [5]. Finally, different cooldown rates were simulated using this algorithm and the predictions presented. Coupling with multiscale thermomechanical models will be presented in subsequent publications.

2. Materials and Methods

2.1. Spherulite Crystallization Growth Simulation

A simulation was developed that can model the growth of thermoplastic spherulites based on non-isothermal cooldown. The governing kinetic equations were introduced in [10] and applied in simulations by [23] for non-isothermal crystallization of PEEK. A flow chart showing the growth simulation process can be found in Figure 1.

The simulation works by discretizing a volume with lengths $l_X \times l_Y \times l_Z$ (or $l_X \times l_Y$ for a 2D simulation) into a grid of sub-volume elements (voxels) referred to as subcells for this study. The subcell length, which controls the coarseness of the subcell discretization, is defined by user input as l_s. The number of subcells in each coordinate direction is l_X/l_s, l_Y/l_s, and l_Z/l_s, respectively. The growth simulation assumes that the thermoplastic starts from a complete melt with no "memory" of previous nucleation, meaning that all nucleation events are random and not dependent on time history. This simulation also assumes that the entire volume considered is heated uniformly at each time increment. Once all input parameters defined by the user are preprocessed, the voxelated volume consisting of the subcells is constructed. If fibers are present in the simulation, the centers are placed on the

$x - y$ plane and are assumed to continue straight in the z-direction. Subcells within the fiber radius are identified and designated as regions where crystallization cannot occur. At $t = 0$ s of the simulation, an instantaneous nucleation occurs and is governed by

$$N = \frac{k_1}{4\pi v_0^3} \tag{1}$$

where N is the number of nuclei per unit volume, and k_1 and v_o are kinetic constants for PEEK and can be found in Table 1. If there were fibers present, a fiber nucleation factor, N_F, was used and from the literature was assumed to be $N_F = 4$ nuclei/μm^2. N_F only controls nucleation on the surface of the fiber. This is outlined to demonstrate that the algorithm can handle fiber surface nucleation, but henceforth this study will only present results without fibers. As the growth simulation continues, $t > 0$ s, it was assumed that the spherulites are allowed to grow radially from each nucleation site. Note that the code could be easily altered to accommodate different growth patterns, such as parabolic growth from fiber surfaces, if deemed necessary. The growth rate of the spherulites, G, is described as

$$G = v_0 e^{\left(\frac{-E_d}{RT}\right)} e^{\left(\frac{-\psi_1 T_m^0}{T(T_m^0 - T)}\right)} \tag{2}$$

where E_d, R, ψ_1, and T_m^0 are kinetic parameters for PEEK outlined in Table 1. T is the current temperature based on the user prescribed cooldown rate and time step, highlighting that the simulation is both time- and temperature-dependent. When a subcell's center was found to be within the growth radius of a spherulite nucleus, that subcell was "captured" as part of the spherulite. During the simulation, new nuclei were allowed to form based on the relation

$$I_{nuc} = \frac{k_2}{4\pi v_0^3} e^{\left(\frac{-E_d}{RT}\right)} e^{\left(\frac{-\psi_2 T_m^0}{T(T_m^0 - T)}\right)} \tag{3}$$

where the kinetic constants k_2 and ψ_2 are also outlined in Table 1. I_{nuc} is in units of $\left[\text{Nuclei}/\mu\text{m}^3/\text{s}\right]$, where, for a given time increment, the nuclei per unit volume or $\left[\text{Nuclei}/\mu\text{m}^3\right]$ is found and multiplied by the volume of amorphous subcells, or subcells which have not been captured by spherulite growth. If the number of nuclei is above one for that time increment, a new nucleus is added randomly to the domain.

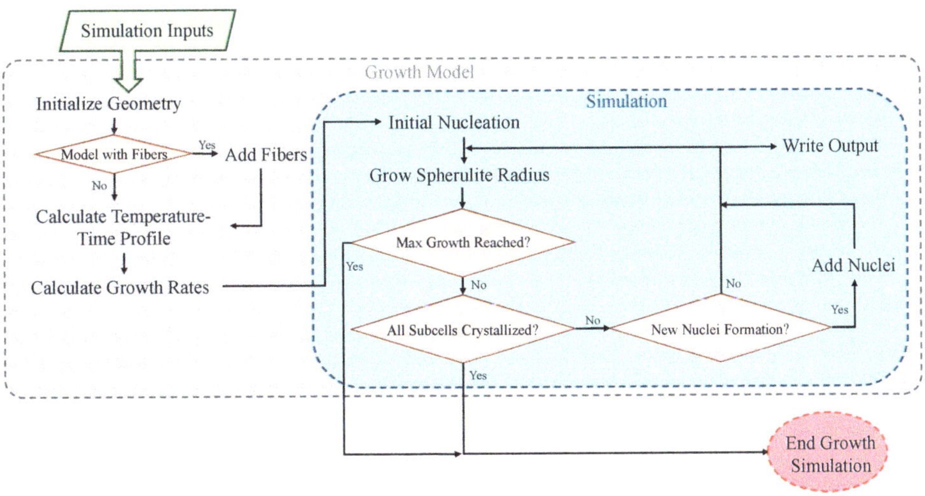

Figure 1. Flow chart describing process of thermoplastic spherulite growth simulation.

Table 1. List of PEEK crystallization kinetic values used as inputs for spherulite growth simulation.

Symbol	Value	Units
v_o	7.50×10^8	$[\mu m/s]$
R	1.986	[cal/molK]
k_1	9.03×10^{24}	$[s^{-3}]$
k_2	9.32×10^{38}	$[s^{-4}]$
T_m^0	385–395	[°C]
E_d	1.52×10^4	[cal/mol]
ψ_1	529	[°C]
ψ_2	1517	[°C]

The growth simulation continues until the end of the cooldown or until there are no longer any subcells to capture within a spherulite. Periodicity was also enforced, where all fibers and spherulite growths were properly reflected over the simulation domain boundaries.

2.2. Local Crystallinity Distribution

A spherulite is composed of lamella stacks that grow radially outwards from the nucleation center during the non-isothermal cool-down from melt [28,29]. Amorphous material occupies the volume in between crystalline lamella stacks. Due to the branch-like nature of lamella growing away from the nucleus, there is a local distribution of crystallinity within the spherulites that is important to capture in computational models when investigating the mechanical behavior of these materials [2].

From the growth simulation, relative crystallinity was calculated as the ratio of the number of subcells that were enveloped during the radial growth of the spherulites versus the total number of subcells. The relative crystallinity varies from zero, a pure amorphous material, to one, a material where spherulites have reached a maximum growth and occupy 100% of the volume of the material.

For PEEK, a sample with a relative crystallinity of one may still only have an overall degree of crystallinity, χ, of 20–30% due to the inter-amorphous regions with spherulites between lamellae. To model this, each subcell within a spherulite was assigned a local crystallinity, v_c. This local crystallinity was optimized with the constraint that the crystallinity must decrease radially from the respective nucleus, and the objective function yielded the assumed maximum PEEK χ of 30% [30]. To this regard, if there were still uncaptured subcells once the simulation ended, χ was found by multiplying the relative crystallinity by the assumed maximum crystallinity, χ_{max}, of 30%.

The optimization process started by calculating an overall degree of crystallinity, χ, as

$$\chi = \chi_V \times \chi_{max} \tag{4}$$

where χ_V is the relative degree of crystallinity defined by

$$\chi_V = \frac{V_{sph}}{V} \tag{5}$$

and V is the total volume of the domain, V_{sph} is the volume of spherulites, and χ_{max} is the assumed maximum degree of crystallinity.

Once χ was calculated, the crystallinity within each spherulite was found. It was assumed that smaller spherulites had a higher crystallinity than larger ones due to denser crystallization near the nucleus, which also composes more of the volume in small spherulites as compared to large ones. The crystallinity of each spherulite was found through minimizing an objective function

$$min\left|\boldsymbol{\chi}_{sph} \cdot \mathbf{v}_{sph} - \chi_V\right| \tag{6}$$

where χ_{sph} is a vector containing the crystallinities of individual spherulites ordered from lowest to highest and $\mathbf{v}_{sph}$ is a vector containing the associated volumes. The optimization enforced a linear inequality such that

$$\mathbf{A}\,\chi_{sph} \leq \mathbf{b} \tag{7}$$

where **A** and **b** are defined as

$$\mathbf{A} = \begin{bmatrix} -1 & 1 & 0 & \ldots \\ 0 & -1 & 1 & \ldots \\ 0 & 0 & -1 & 1 \\ \ldots & \ldots & \ldots & -1 \end{bmatrix} \text{ and } \mathbf{b} = \begin{bmatrix} 0 \\ 0 \\ \ldots \end{bmatrix}. \tag{8}$$

The optimization was used to find χ_{sph}, which achieved a target χ.

After the crystallinities of each spherulite were found in χ_{sph}, the crystallinities of each subcell within a spherulite were determined using the same optimization process with a different objective function and linear inequality. For each spherulite i, the objective function is defined by

$$\min\left|\frac{\chi_{sub}\cdot\mathbf{n}^{i}_{sub}}{\sum\mathbf{n}^{i}_{sub}} - \chi^{i}_{sph}\right| \tag{9}$$

where χ_{sub} is a vector containing the crystallinities of subcells within a spherulite, $\mathbf{n}^{i}_{sub}$ is a vector containing the number of subcells with centroids located at particular radial distances from the nucleus where the jth entry of $\mathbf{n}^{i}_{sub}$ is the number of subcells within a radius, r_j. A schematic of this is shown in Figure 2 depicting how $\mathbf{n}^{i}_{sub}$ is determined for an arbitrary spherulite demonstrated in a 2D cross section. The subcells that fall within a given radius are all assigned the same crystallinity determined from the optimization. The summation of $\mathbf{n}^{i}_{sub}$ is the total number of subcells within a spherulite. It is assumed that the subcells at the nucleus have a fixed crystallinity of 85%, which is approximately the maximum possible crystallinity [26]. The linear inequality for this optimization is defined as

$$\mathbf{C}\,\chi_{sub} \leq \mathbf{d} \tag{10}$$

where **C** and **d** are defined as

$$\mathbf{C} = \begin{bmatrix} -1 & 1 & 0 & \ldots \\ 0 & -1 & 1 & \ldots \\ 0 & 0 & -1 & 1 \\ \ldots & \ldots & \ldots & -1 \end{bmatrix} \text{ and } \mathbf{d} = \begin{bmatrix} 0 \\ 0 \\ \ldots \end{bmatrix}. \tag{11}$$

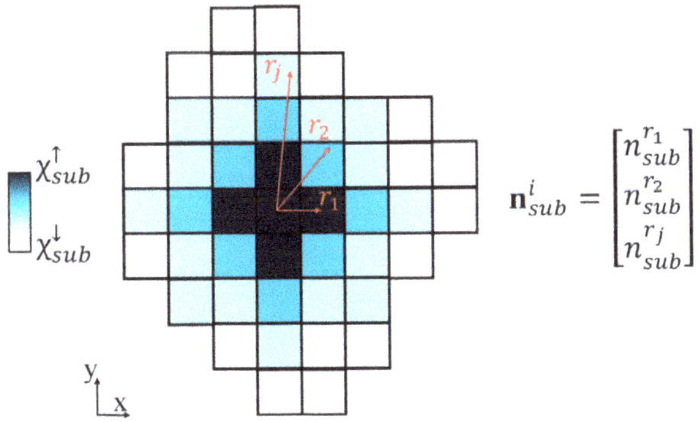

Figure 2. 2D Representation of how crystallinity decreases radially from the spherulite nucleus where $\mathbf{n}^{i}_{sub}$ is a vector containing the number of subcells assigned a specific crystallinity.

3. Results and Discussion

3.1. Parameter Sensitivity and Validation

A parameter convergence study was conducted to determine the maximum subcell refinement required to capture the crystallinity evolution accurately. A cube domain was used where all lengths were equal (i.e., $l_X = l_Y = l_Z$), the subcell dimension, l_s, was varied, and the crystallinity outputs were measured. The domain size was normalized by subcell length $\left(\frac{l_X}{l_s}\right)$, where the minimum subcell length could be determined based on the desired domain. To examine this, a thermal cooldown of 10 °C/min was applied to a domain where $l_X = 15$ μm and the subcell length was refined from 1 μm to 0.1 μm and the relative crystallinity, χ_V, was recorded. The domain size was also varied from 10 μm to 30 μm keeping $l_s = 1$ constant. The results are shown in Figure 3.

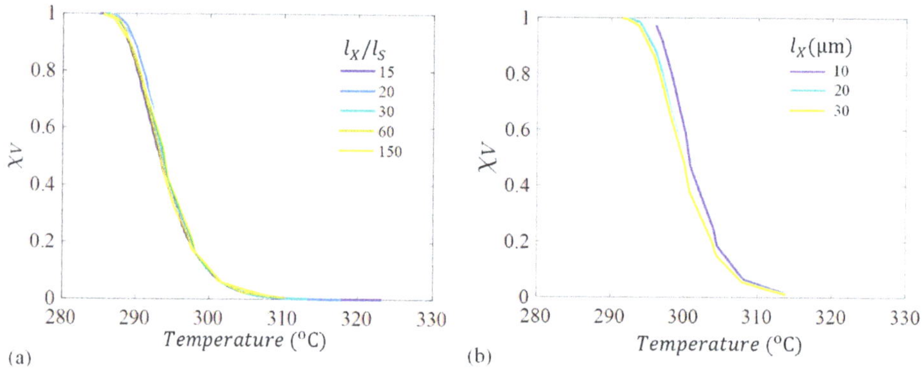

Figure 3. Relative crystallinity, χ_V, versus temperature for a simulation with a 10 °C/min cooldown for (**a**) increasing subcell refinement and (**b**) increasing domain size.

Shown in Figure 3a, there is minimal dependency on subcell refinement for the crystal growth, but Figure 3b shows that domain size does affect the spherulite volume fraction. This domain size dependency is likely due to homogeneous nucleation because a larger overall volume means that more nuclei can form during cooldown. This is why the curves start at the same temperature at $\chi_V = 0$, but the new nuclei that form in the larger domains cause the curves to differ from smaller domains. For the remainder of the results, a standard domain size of 30 μm and a subcell refinement of 1 μm were used.

A validation of the crystallization kinetic mode was performed, where different non-isothermal cooldown rates were applied and χ_V was recorded. The model predictions were compared to differential scanning calorimetry (DSC) experiments from the literature for PEEK [10]. Figure 4 shows that there is very good agreement between the experimental and simulation results, which implies that the kinetic constants used (Table 1) can predict the spherulite growth for PEEK for different non-isothermal cooldown rates. There are slight differences at high χ_V, which is possibly due to the domain being discretized into voxels, causing jumps in χ_V when a new subcell is captured, or nucleates.

It was noted in this study that changing equilibrium temperature, T_m^0, which for PEEK is commonly reported as 395 °C [31] but was measured in [10] as 385 °C, resulted in a shift of relative crystallinity. The effect of these two bounds on recorded χ_V was compared to the experimental results [10] at a 10 °C/min cooldown and shown in Figure 5a. Figure 5a shows that χ_V shifts by approximately 10 °C when T_m^0 changed, and this value had to be chosen carefully to represent experimental results. The shaded area in between the two bounds shows where results may fall if an intermediate T_m^0 value is chosen. For this study, a $T_m^0 = 392$ °C was chosen for thermal cooldowns 10 °C/min and below. By contrast, for higher thermal cooldowns, $T_m^0 = 395$ °C was used, which better represented the experimental results. This shift in relative crystallinity, which is almost equal to the difference in T_m^0 bounds, suggests that relative crystallinity is nearly constant with respect to the temperature to the equilibrium melting point, shown in Figure 5b. Figure 5b shows that the curves line up closely when adjusted based on respective T_m^0. Moreover, this aspect can be beneficial for future simulations with new materials where T_m^0 may not be known.

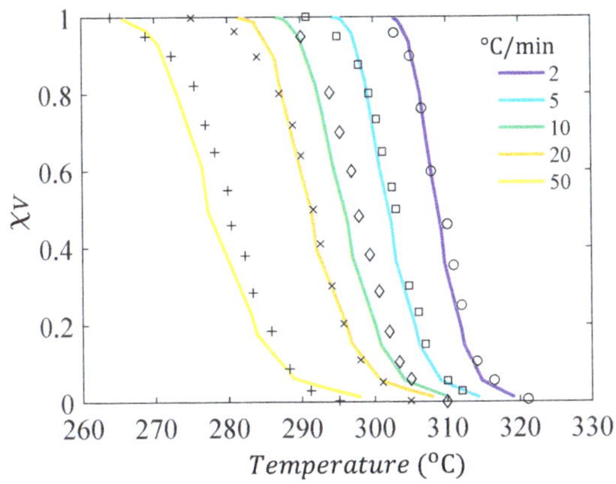

Figure 4. Validation of crystallization kinetic model with experimental DSC results [10] for thermal cooldowns of 2(+), 5(×), 10(◊), 20(□), and 50(o) °C/min.

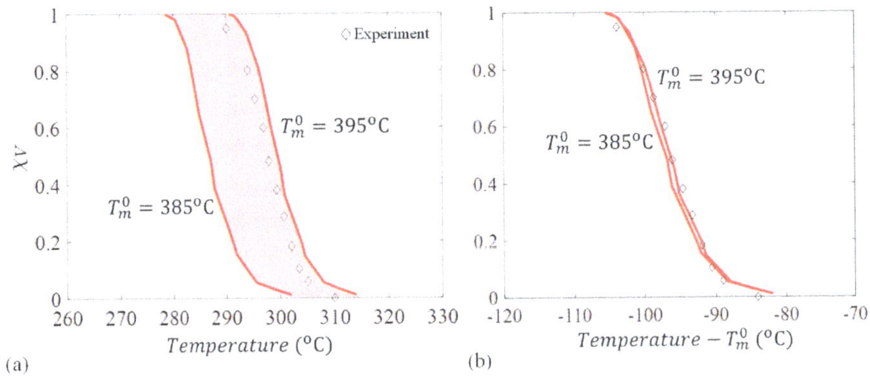

Figure 5. (**a**) Sensitivity of relative crystallinity, χ_V, based on reported equilibrium melting temperatures, T_m^0, and (**b**) adjusted with respect to T_m^0 for PEEK.

3.2. Spherulite Morphology Validation

Once the model was validated to ensure the spherulite growth was representative of experimental DSC results, it was used to compare simulated morphologies to experimental scans. A study by [5] was simulated where a film of PEEK was cooled at 20 °C/min. For the simulation, a domain size of 100 × 100 × 1 μm was used with a subcell length of 1 μm. A side-by-side comparison is shown in Figure 6.

The results reported by [5] showed that there was a distribution in spherulite sizes of approximately 15–30 μm. The simulation shows similar results with spherulite measurements depicted in Figure 7.

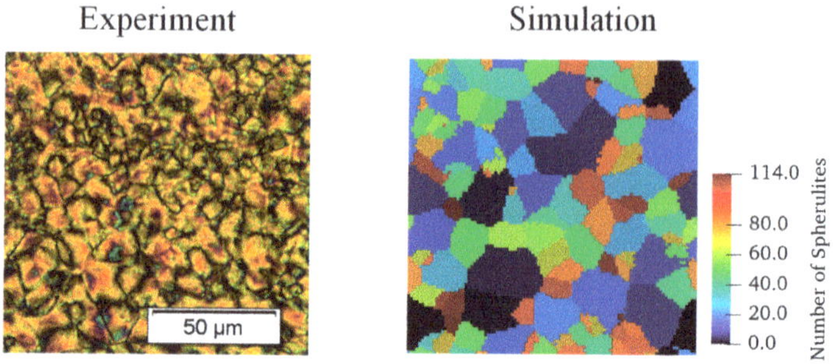

Figure 6. Spherulite morphologic comparison between a thin PEEK sample cooled at 20 °C/min [5] and simulation.

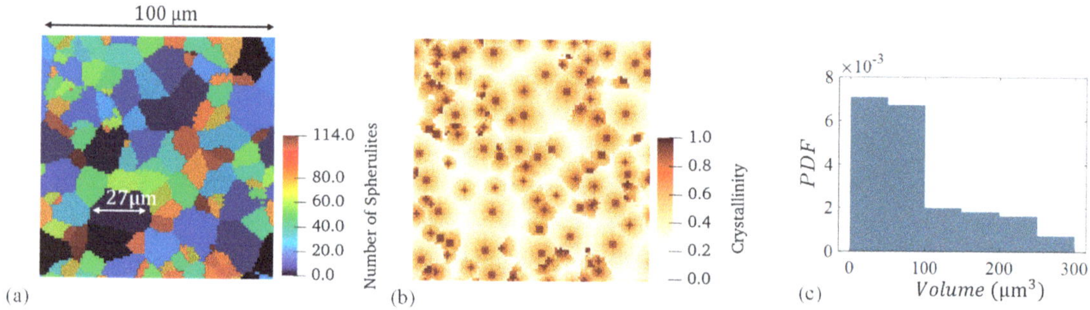

Figure 7. (**a**) Spherulite morphology from a sample cooled at 20 °C/min with the largest spherulite diameter labelled, (**b**) the corresponding local crystallinity distribution, and (**c**) a probability density function (PDF) of spherulite volumes.

In Figure 7a, the growth simulation showed similar spherulite measurements compared to experiments where the largest spherulite had a diameter in the upper end of the reported range. Figure 7c shows the probability distribution function (PDF) of all spherulite volumes. Spherulites are not perfectly circular, and their volumes do not directly correspond to their diameter, which is typically reported for these experimental measurements. Figure 7b shows the calculated crystallinity distribution for this sample, where the nuclei are clearly visible. The results show the simulation is producing expected spherulite sizes compared to experimental results. More advanced image analysis techniques are needed to measure and validate the crystallinity distribution within spherulites in these types of experimental results.

Typical cooldown rates may vary depending on manufacturing and application. For example, tow placement processes such as automated fiber placement can have a cooldown rate on the order of 2000 °C/min [32]. To examine this effect on spherulite growth and morphology, simulations with different cooldown rates (20, 200, 500 and 2000 °C/min) were conducted using a 30 × 30 × 30 μm domain size and compared in Figure 8.

Figure 8 shows the nucleation, spherulite evolution, and resulting local crystallinity for different cooldown rates. Figures of the growth were taken at initial nucleation, halfway through the cool down, and at the final time step. Both 20 and 200 °C/min reached a state with fully grown spherulites that impinged with one another, where the cooldown of 500 °C/min had some spherulite growth but not full impingement and 2000 °C/min was quick enough to not let the spherulites grow past initial nucleation or form new nuclei. As a result, the crystallinity of each nucleus in the 2000 °C/min simulation was equal to 85% where the surrounding is pure amorphous. It can also be seen that the 20 °C/min simulation allowed for spherulites to grow significantly larger than the 200 °C/min simulation, but more new nuclei formed in the latter case. The result shows that a slower cooldown rate promotes spherulite growth, whereas a higher cooldown rate promotes more nucleation with smaller spherulites, which is consistent with experimental results shown in [33,34]. This relationship between growth and cooldown rates can be seen in Table 2, except for 2000 °C/min, because there was no spherulite growth. A distribution of spherulite relative volumes, the volume of each spherulite divided by its respective average, can be seen in Figure 9. The crystallinity in Figure 8 also represents this finding and shows that larger spherulites have a gradual crystalline transition from the nucleus outwards. The relationship between thermal cooldown, growth radius, and homogeneous nucleation rate for these cases is shown in Figure 10.

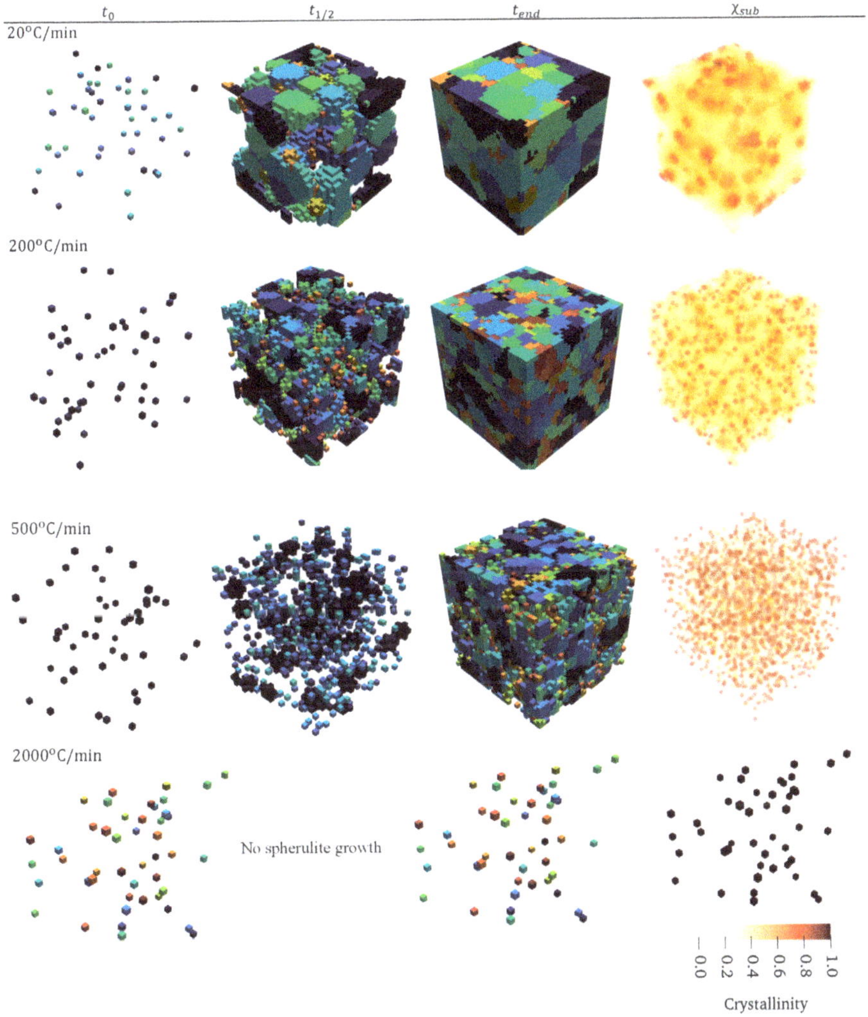

Figure 8. Spherulite nucleation, growth, and crystallinity for cooldown rates of 20, 200, 500, and 2000 °C/min compared at the initial time step, halfway through the simulation, and the final time step, along with the final subcell crystallinity distribution. The spherulite colors in the first three columns represent spherulite numbers.

Table 2. Minimum, maximum, average, and standard deviation of spherulite volumes from three different thermal cooldowns.

Cooldown (°C/min)	Minimum Volume (μm^3)	Maximum Volume (μm^3)	Average (μm^3)	Standard Deviation
20	4	767	221.31	212.91
200	1	213	32.88	35.61
500	1	106	8.60	12.81

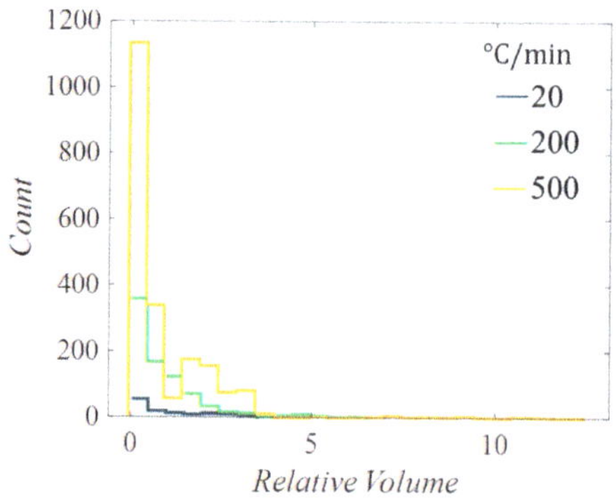

Figure 9. Number of spherulites versus relative volume for simulations of three different cooldown rates.

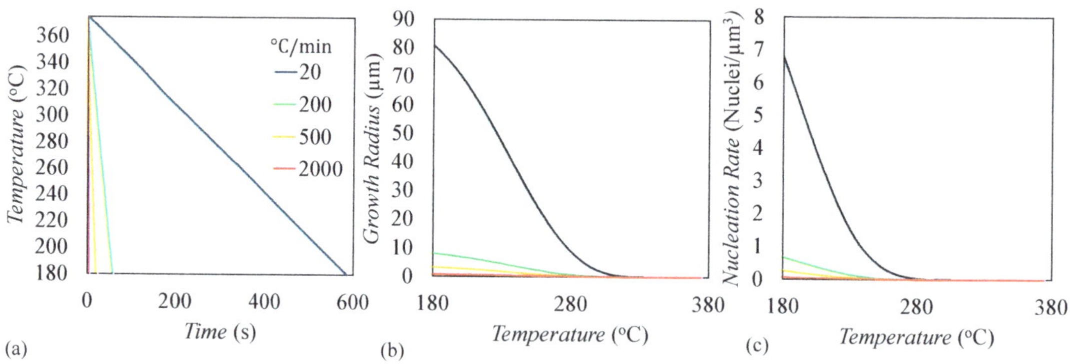

Figure 10. (**a**) Temperature versus time, (**b**) nuclei growth radius, and (**c**) nucleation rate for four different simulations with increasing thermal cooldown rates.

Table 2 shows that the minimum spherulite volume for the 200 °C/min and 500 °C/min cooldown was 1 μm^3, meaning it was only a nucleus that did not grow because the subcell dimension was $l_s = 1 \times 1 \times 1$ μm. This can occur when a new nucleus is placed in between surrounding spherulites and is not given any room to grow, or near the end of a fast cooldown where growth rate is very low, and the final temperature is reached before growth to a neighboring subcell is reached. This can be changed by increasing the subcell refinement, which will decrease the probability that a nucleus will be fully surrounded by spherulites. Further, lower growth rates could be captured because the subcell-to-subcell distance would be smaller. While this may be desirable, increasing the discretization will increase computation time, and further studies are needed to understand required refinement to achieve sufficient fidelity. The 20 °C/min simulation has the tightest distribution where bins are similar heights, meaning the spherulites are similar volumes. As the cooldown rate increases, so does the number of small spherulites, as well as the distribution spread, meaning that there is a large range of spherulite sizes relative to the average.

Figure 10a shows the temperature versus time profile for the four examined cooldown rates. Figure 10b shows the corresponding growth radius for all cooldown rates, which is calculated by integrating Equation (2) with respect to time. This shows that the slower the thermal cooldown, the larger the growth radius can form by the end of the simulation. Lastly, Figure 10c shows the homogeneous nucleation rate as a function of temperature, which is calculated by integrating Equation (3) with respect to time and shows similar relationships of the growth radius. While this shows that slower cooldowns should form more nuclei per unit volume, this is opposite of what is displayed in Figure 7. This is because as the growth radius increases, the amount of subcells not captured by a spherulite decreases, so in slower cooldowns, there is significantly less volume for new nucleation. By contrast, in fast cooldowns, the growth radius is not very large and there is a large amount of remaining amorphous volume, so when the nucleation rate is multiplied by the available amorphous volume to determine how many nuclei form, that value is higher than slower cooldowns shown in Figure 11.

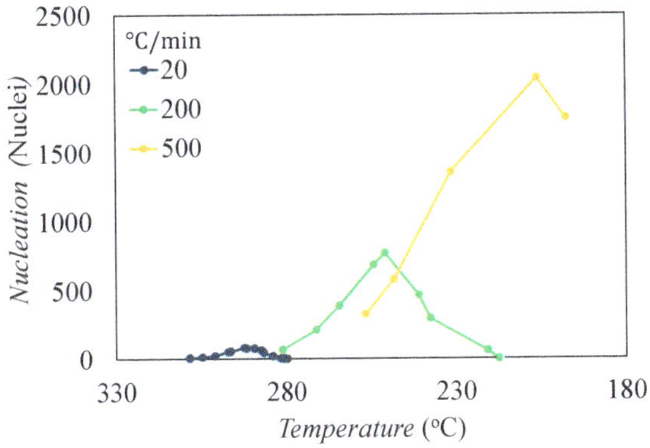

Figure 11. Nucleation versus temperature for three different thermal cooldowns.

Figure 11 shows the number of nuclei that form during the thermal cooldown rates. This is found by multiplying nucleation rate (Figure 10) by the amorphous volume at a given time step. Seen in Figure 11, nucleation happens at different temperature ranges for each of the cooldowns. At the temperature range of the 20 °C/min simulation, all three cooldown rates have very similar nucleation rates (Figure 10c), but the spherulite growth radius is much higher, meaning there is not much amorphous volume, resulting in low nuclei formation. This shows how the 20 °C/min cooldown has fewer, larger spherulites, and at ~280 °C there is full impingement. The 200 °C/min nucleation primarily happens in a temperature range of 280 °C–230 °C where the spherulite growth radius for this cooldown rate is low and has a higher nucleation rate than the 20 °C/min cooldown. Due to the spherulites not growing as large, there is more amorphous volume available, and more nucleation can occur. The 500 °C/min cooldown has the lowest spherulite growth radius of the three cases, and a lower nucleation rate than the 200 °C/min cooldown. But, because there is less growth, there is significantly more amorphous volume available, and the nucleation is much higher. The 2000 °C/min case had no new nucleation or growth.

4. Conclusions

In this study, a model was developed that uses crystallization kinetics to model the growth of semi-crystalline spherulites within a domain. Optimization was used, with assumed constraints on spherulite geometries, to determine inter-spherulitic crystallinities. The kinetic constants and equations used in this model were obtained from the previous literature for PEEK. The equations were applied to a model domain that was discretized into voxelated regions called subcells. After growing the spherulites in this domain, overall crystallinity was determined, and optimization schemes were used to determine crystallinities for each spherulite and then crystallinities for each subcell. Results of this study investigated the model validation against DSC data for PEEK from literature, and then sensitivity of the model to domain size, subcell discretization, and certain kinetic constants. Then, a simulated microstructure was compared to an experimental image from PEEK after a cooldown where spherulite sizes were compared. Finally, the effect of cooldown rate was examined by simulating three rates of 20, 200, 500 and 2000 °C/min. The results of these simulations showed that the fastest cooldown rate resulted in no spherulite growth, just nucleation, the rate of 20 °C/min showed fewer but larger spherulites grown, 200 °C/min showed more but smaller spherulites, and 500 °C/min showed some spherulite growth but not full impingement. These models were then processed through the optimization schemes and the final crystalline distributions were shown. The crystalline distributions show the nucleus of the spherulites having higher crystallinities that decrease radially outward.

Previous studies have shown that thermomechanical modeling of thermoplastics is a multiscale problem due to the local distribution and orientation of the lamellae, and the current model was developed to directly convert into a geometry for NASA's Multiscale Analysis Tool (NASMAT), which relies on subcell discretization for its semi-analytical multiscale recursive micromechanics methods [25,35–37]. Knowing the crystallinity for each individual subcell from this analysis allows for lower length scales, such as individual lamella surrounded by amorphous material, to be simulated and integrated into the mesoscale containing the morphology of the spherulites. Coupling with NASMAT will also allow for linear and nonlinear thermomechanical analysis, where stiffness, strength, fracture toughness, thermal conductivity, etc., can be simulated for different morphologies resulting from different manufacturing processes. Future work on this model will include studying the effect of these properties with the inclusion of fibers, where fiber morphologies [38] and different spherulite geometries from spherulite nucleation on the fiber surface can impact mechanical results. Finally, an anisotropic core for the spherulites will be considered as this impacts the fracture behavior [2].

Author Contributions: Conceptualization, J.F.H. and E.J.P.; methodology, J.F.H.; software, J.F.H.; validation, J.F.H.; formal analysis, J.F.H.; investigation, J.F.H.; resources, J.F.H.; data curation, J.F.H.; writing—original draft preparation, J.F.H., E.J.P. and S.E.S.; writing—review and editing, J.F.H., E.J.P. and S.E.S.; visualization, J.F.H.; supervision, E.J.P. and S.E.S.; project administration, E.J.P. and S.E.S.; funding acquisition, J.F.H., E.J.P. and S.E.S. All authors have read and agreed to the published version of the manuscript.

Funding: This work was supported by a NASA Space Technology Graduate Research Opportunity Fellowship under grant number 80NSSC21K1285. The authors would also like to acknowledge the NASA Thermoplastics Development for Exploration Applications Project and the Office of Naval Research for their support.

Institutional Review Board Statement: Not applicable.

Informed Consent Statement: Not applicable.

Data Availability Statement: Dataset available on request from the authors.

Acknowledgments: The authors would like to acknowledge Navid Zobeiry and Mathew Wynn from the University of Washington for experimental images and data.

Conflicts of Interest: The authors declare no conflicts of interest.

References

1. Chu, J.-N.; Schultz, J.M. The Influence of Microstructure on the Failure Behaviour of PEEK. *J. Mater. Sci.* **1990**, *25*, 3746–3752. [CrossRef]
2. Marinosci, V.; Chen, K.; Helthuis, N.G.J.; Grouve, W.J.B.; de Vries, E.G.; Bao, N.; Akkerman, R.; de Rooij, M.B.; Chu, L. Direct Observation of the Fracture Behavior of the Polyether Ketone Ketone (PEKK) Spherulites. *J. Appl. Polym. Sci.* **2024**, *141*, e54764. [CrossRef]

3. Barocio, E.; Brenken, B.; Favaloro, A.; Pipes, R.B. Interlayer Fusion Bonding of Semi-Crystalline Polymer Composites in Extrusion Deposition Additive Manufacturing. *Compos. Sci. Technol.* **2022**, *230*, 109334. [CrossRef]
4. Bhudolia, S.K.; Gohel, G.; Leong, K.F.; Islam, A. Advances in Ultrasonic Welding of Thermoplastic Composites: A Review. *Materials* **2020**, *13*, 1284. [CrossRef] [PubMed]
5. Lee, A.; Wynn, M.; Quigley, L.; Salviato, M.; Zobeiry, N. Effect of Temperature History during Additive Manufacturing on Crystalline Morphology of PEEK. *Adv. Ind. Manuf. Eng.* **2022**, *4*, 100085. [CrossRef]
6. Oromiehie, E.; Gain, A.K.; Donough, M.J.; Prusty, B.G. Fracture Toughness Assessment of CF-PEEK Composites Consolidated Using Hot Gas Torch Assisted Automated Fibre Placement. *Compos. Struct.* **2022**, *279*, 114762. [CrossRef]
7. Pourahmadi, E.; Ganesan, R.; Shadmehri, F. Micromechanical Characterization of Carbon/PEEK Thermoplastic Composite Material in-Situ Consolidated by Automated Fiber Placement: Stiffness Prediction. *Compos. Sci. Technol.* **2024**, *246*, 110390. [CrossRef]
8. Yarlagadda, S.; Advani, S.; Deitzel, J.; Heider, D.; Molligan, D.; Roseman, D.; Simacek, P.; Tierney, J.; Gillespie, J., Jr. Formability of TUFF Composite Blanks. In Proceedings of the SAMPE 2019, Charlotte, NC, USA, 20–23 May 2019.
9. Avrami, M. Kinetics of Phase Change. I General Theory. *J. Chem. Phys.* **1939**, *7*, 1103–1112. [CrossRef]
10. Choe, C.R.; Lee, K.H. Nonisothermal Crystallization Kinetics of Poly(Etheretherketone) (PEEK). *Polym. Eng. Sci.* **1989**, *29*, 801–805. [CrossRef]
11. Malkin, A.Y.; Beghishev, V.P.; Keapin, I.A.; Bolgov, S.A. General Treatment of Polymer Crystallization Kinetics—Part 1. A New Macrokinetic Equation and Its Experimental Verification. *Polym. Eng. Sci.* **1984**, *24*, 1396–1401. [CrossRef]
12. Tobin, M.C. Theory of Phase Transition Kinetics with Growth Site Impingement. I. Homogeneous Nucleation. *J. Polym. Sci. Polym. Phys. Ed.* **1974**, *12*, 399–406. [CrossRef]
13. Kelly, C.A.; Jenkins, M.J. Modeling the Crystallization Kinetics of Polymers Displaying High Levels of Secondary Crystallization. *Polym. J.* **2022**, *54*, 249–257. [CrossRef]
14. Treatise on Solid State Chemistry. Available online: https://www.springer.com/series/10833 (accessed on 7 March 2024).
15. Fundamentals of Polymer Science: An Introductory Text [Second Edition] 9780203755211, 0203755219, 9781351446396, 1351446398. Available online: https://dokumen.pub/fundamentals-of-polymer-science-an-introductory-text-second-edition-9780203755211-0203755219-9781351446396-1351446398.html (accessed on 7 March 2024).
16. Müller, A.J.; Michell, R.M.; Lorenzo, A.T. Isothermal Crystallization Kinetics of Polymers. In *Polymer Morphology*; John Wiley & Sons, Ltd.: Hoboken, NJ, USA, 2016; pp. 181–203. ISBN 978-1-118-89275-6.
17. Bessard, E.; De Almeida, O.; Bernhart, G. Unified Isothermal and Non-Isothermal Modelling of Neat PEEK Crystallization. *J. Therm. Anal. Calorim.* **2014**, *115*, 1669–1678. [CrossRef]
18. Choupin, T.; Fayolle, B.; Régnier, G.; Paris, C.; Cinquin, J.; Brulé, B. Isothermal Crystallization Kinetic Modeling of Poly(Etherketoneketone) (PEKK) Copolymer. *Polymer* **2017**, *111*, 73–82. [CrossRef]
19. Godara, A.; Raabe, D.; Green, S. The Influence of Sterilization Processes on the Micromechanical Properties of Carbon Fiber-Reinforced PEEK Composites for Bone Implant Applications. *Acta Biomater.* **2007**, *3*, 209–220. [CrossRef]
20. Tierney, J.J.; Gillespie, J.W., Jr. Crystallization Kinetics Behavior of PEEK Based Composites Exposed to High Heating and Cooling Rates. *Compos. Part A Appl. Sci. Manuf.* **2004**, *35*, 547–558. [CrossRef]
21. Velisaris, C.N.; Seferis, J.C. Crystallization Kinetics of Polyetheretherketone (Peek) Matrices. *Polym. Eng. Sci.* **1986**, *26*, 1574–1581. [CrossRef]
22. Choupin, T. Mechanical Performances of PEKK Thermoplastic Composites Linked to Their Processing Parameters. Ph.D. Thesis, Ecole Nationale Supérieure D'arts et Métiers; ENSAM: Paris, France, 2017.
23. Guan, X.; Pitchumani, R. Modeling of Spherulitic Crystallization in Thermoplastic Tow-Placement Process: Spherulitic Microstructure Evolution. *Compos. Sci. Technol.* **2004**, *64*, 1363–1374. [CrossRef]
24. Saber, C. Fonctionnalisation de Composites C/PEKK Pour Application Aérospatiale: Caractérisation, Modélisation et Influence Sur Les Propriétés Du Composite. Doctoral Dissertation, Université Paul Sabatier-Toulouse III, Toulouse, France, 2018.
25. Pisani, W.A.; Radue, M.S.; Chinkanjanarot, S.; Bednarcyk, B.A.; Pineda, E.J.; Waters, K.; Pandey, R.; King, J.A.; Odegard, G.M. Multiscale Modeling of PEEK Using Reactive Molecular Dynamics Modeling and Micromechanics. *Polymer* **2019**, *163*, 96–105. [CrossRef]
26. Pisani, W.A.; Newman, J.K.; Shukla, M.K. Multiscale Modeling of Polyamide 6 Using Molecular Dynamics and Micromechanics. *Ind. Eng. Chem. Res.* **2021**, *60*, 13604–13613. [CrossRef]
27. Pineda, E.J.; Husseini, J.; Kemppainen, J.; Odegard, G.M.; Bednarcyk, B.A.; Pisani, W.; Stapleton, S.E. Multiscale Modeling of Thermoplastics Using Atomistic-Informed Micromechanics. In Proceedings of the AIAA SCITECH 2023 Forum, National Harbor, MD, USA, 23–27 January 2023; p. 0140.
28. Chan, C.-M.; Li, L. Direct Observation of the Growth of Lamellae and Spherulites by AFM. In *Intrinsic Molecular Mobility and Toughness of Polymers II*; Kausch, H.-H., Ed.; Springer: Berlin/Heidelberg, Germany, 2005; pp. 1–41. ISBN 978-3-540-31601-5.
29. Gránásy, L.; Pusztai, T.; Tegze, G.; Warren, J.A.; Douglas, J.F. Growth and Form of Spherulites. *Phys. Rev. E* **2005**, *72*, 011605. [CrossRef]
30. Doumeng, M.; Makhlouf, L.; Berthet, F.; Marsan, O.; Delbé, K.; Denape, J.; Chabert, F. A Comparative Study of the Crystallinity of Polyetheretherketone by Using Density, DSC, XRD, and Raman Spectroscopy Techniques. *Polym. Test.* **2021**, *93*, 106878. [CrossRef]

31. Blundell, D.J.; Osborn, B.N. The Morphology of Poly(Aryl-Ether-Ether-Ketone). *Polymer* **1983**, *24*, 953–958. [CrossRef]
32. Modeling of Heat Transfer and Crystallization in Thermoplastic Composite Tape Placement Process—Fazil, O. Sonmez, H. Thomas Hahn. 1997. Available online: https://journals.sagepub.com/doi/abs/10.1177/089270579701000301 (accessed on 5 June 2024).
33. Li, X.; Zhang, T.; Li, S.; Liu, H.; Zhao, Y.; Wang, K. The Effect of Cooling Rate on Resistance-Welded CF/PEEK Joints. *J. Mater. Res. Technol.* **2021**, *12*, 53–62. [CrossRef]
34. Gao, S.-L.; Kim, J.-K. Cooling Rate Influences in Carbon Fibre/PEEK Composites. Part 1. Crystallinity and Interface Adhesion. *Compos. Part A Appl. Sci. Manuf.* **2000**, *31*, 517–530. [CrossRef]
35. Kashmari, K.; Deshpande, P.; Patil, S.; Maiaru, M.; Odegard, G. A Multiscale Approach to Investigate the Effect of Temperature and Crystallinity on the Development of Residual Stresses in Semicrystalline PEEK. In Proceedings of the American Society for Composites—37th Technical Conference, ASC 2022, Tucson, AZ, USA, 19–21 September 2022. [CrossRef]
36. Bednarcyk, B.A.; Aboudi, J.; Arnold, S.M. Micromechanics of Composite Materials Governed by Vector Constitutive Laws. *Int. J. Solids Struct.* **2017**, *110–111*, 137–151. [CrossRef]
37. Pineda, E.J.; Bednarcyk, B.A.; Ricks, T.M.; Arnold, S.M.; Henson, G. Efficient multiscale recursive micromechanics of composites for engineering applications. *Int. J. Multiscale Comput. Eng.* **2021**, *19*, 77–105. [CrossRef]
38. Schey, M.J.; Beke, T.; Appel, L.; Zabler, S.; Shah, S.; Hu, J.; Liu, F.; Maiaru, M.; Stapleton, S. Identification and Quantification of 3D Fiber Clusters in Fiber-Reinforced Composite Materials. *JOM* **2021**, *73*, 2129–2142. [CrossRef]

materials

MDPI

Article

Modeling of Eyld2000-2d Anisotropic Yield Criterion Considering Strength Differential Effect and Analysis of Optimal Calibration Strategy

Kai Du [1,2,*], Li Dong [1], Hao Zhang [1], Zhenkai Mu [3], Hongrui Dong [4], Haibo Wang [5], Yanqiang Ren [1], Liang Sun [1], Liang Zhang [1] and Xiaoguang Yuan [1,2,*]

1 School of Materials Science and Engineering, Shenyang University of Technology, Shenyang 110870, China
2 State Key Laboratory of Light Alloy Casting Technology for High-End Equipment, Shenyang 110022, China
3 Hebei Key Laboratory of Material Near-Net Forming Technology, Hebei University of Science and Technology, Shijiazhuang 050018, China; muzhenk1229@163.com
4 School of Materials Science and Engineering, Beihang University, Beijing 100191, China; donghongrui@buaa.edu.cn
5 School of Mechanical and Materials Engineering, North China University of Technology, Beijing 100144, China; wanghaibo@ncut.edu.cn
* Correspondence: dukai@sut.edu.cn (K.D.); yuanxg@sut.edu.cn (X.Y.)

Abstract: Sheet metals usually experience various loading paths such as uniaxial tension, uniaxial compression, biaxial tension, and simple shear during the forming process. However, the existing constitutive models cannot always accurately describe blanks' anisotropic yield and plastic flow behavior of blanks under all typical stress states. Given this, this paper improves the Eyld2000-2d yield criterion by introducing hydrostatic pressure to the A-Eyld2000-2d yield criterion that can describe the strength differential effect of materials. Meanwhile, to control the curvature of the yield surface more effectively, the near-plane strain yield stresses were added in the parameter identification process to calibrate the exponent m, so that the exponent is no longer considered as a constant value. Taking the widely used AA6016-T4, AA5754-O, DP980, and QP980 blanks in the automotive stamping industry as an example, the effectiveness of the new model and different parameter identification methods was verified by predicting experimental data under various simple and complex loading paths. Subsequently, the new model employing the optimal parameter identification strategy was compared with four widely used asymmetric yield criteria under associated and non-associated flow rules, including CPB06, LHY2013, S-Y2004, and Hu & Yoon2021, to further verify the accuracy of the proposed constitutive model. The results indicate that parameter identification strategy with variable exponent can significantly improve the flexibility of the yield criterion in describing the plastic anisotropy of blanks. Compared to the other yield criteria examined in this work, the new model provides the best prediction accuracy for the yield stresses and plastic flows of all blanks, especially in the near-plane strain and simple shear stress states. Modeling under the concept of anisotropic hardening can more accurately capture the evolving plastic behavior of blanks than isotropic hardening.

Keywords: sheet metal; yield criterion; strength differential effect; plastic anisotropy; evolving plastic behavior

Citation: Du, K.; Dong, L.; Zhang, H.; Mu, Z.; Dong, H.; Wang, H.; Ren, Y.; Sun, L.; Zhang, L.; Yuan, X. Modeling of Eyld2000-2d Anisotropic Yield Criterion Considering Strength Differential Effect and Analysis of Optimal Calibration Strategy. *Materials* **2023**, *16*, 6445. https://doi.org/10.3390/ma16196445

Academic Editors: Tomasz Trzepieciński, Madhav Baral and Charles Lu

Received: 31 August 2023
Revised: 19 September 2023
Accepted: 22 September 2023
Published: 27 September 2023

1. Introduction

With increasingly strict requirements for lightweight and crashworthiness in the automotive manufacturing industry, an increasing number of aluminum alloy and advanced high-strength steel blanks are widely used [1–6]. However, sheet metals usually exhibit serious anisotropic behavior during the forming process and are subjected to various stress states such as uniaxial tension (UT), uniaxial compression (UC), equi-biaxial tension (EBT), near-plane strain (NPS), and simple shear (SS) [7,8]. These complex plastic deformation behaviors pose a grand challenge in the development of high-fidelity constitutive models. Nevertheless, establishing a high-precision constitutive model that can characterize typical stress states and anisotropic behavior is still a main research interest in the field of plastic forming.

To date, researchers have already proposed various anisotropic constitutive models. Hill [9] proposed the famous quadratic anisotropic yield criterion based on the von Mises isotropic yield criterion, which has become one of the most widely used yield criteria due to its simple expression and ease of calculation [10]. Considering the poor ability of the secondary yield criterion to describe the plastic deformation behavior of sheet metals, especially aluminum alloys, Barlat and Lian [11] developed the Barlat89 yield criterion. With the demand for accurate predictions of more mechanical properties, Barlat et al. [12,13] have successively established the yield criteria of Balat94 and Balat96. However, the convexity of the yield criterion cannot be guaranteed, which limits their application. To solve the above issue, Barlat et al. [14] proposed the Yld2000-2d yield criterion applicable to the plane stress state by introducing a fourth-order linear operator to the Cauchy stress tensor. Because Yld2000-2d can describe the anisotropic behavior of blanks more accurately, it has become one of the most widely used advanced yielding criteria in industry and academia. Subsequently, Barlat et al. [15] proposed the Yld2004-18p yield criterion applicable to three-dimensional stress states, which effectively predicted the phenomenon of six or eight earings appearing in deep drawing tests of cylindrical cups for blanks with strong anisotropy. However, due to the large amount of experimental data and complex calculation process required for calibrating parameters, they have not been widely used in industry. Distinguishing from the linear transformation approach, Banabic et al. [16] developed the BBC2005 yield criterion by adding anisotropy parameters in Hershey1954, which can accurately predict the shape of the yield surface. Cazacu and Barlat [17] constructed an orthotropy yield criterion based on the J_2- and J_3-based Drucker frameworks using the theory of the representation of tensor functions, which can accurately describe the plastic anisotropic behaviors of AA6016-T4 and AA2093-T3. Another anisotropic form of the Drucker yield function was introduced through linear transformation tensor [18], which is implemented into Ansys LS-DYNA as *Mat_263 with four ductile fracture criteria developed by the same authors. Meanwhile, to reduce the input of experimental data, Khalfallah et al. [19,20] further proposed a simplified calibration program for the CB2001 yield criterion and verified the effectiveness of the newly developed parameter identification strategy through simulation analysis of cross-shaped deep-drawn cup and tube hydroforming experiments. Lou et al. [21] introduced a reduced Yld2004 function under associate flow rule to model anisotropic plastic behavior both in strength and plastic deformation for spatial and plane stress loading conditions. Recently, Lee et al. [22] coupled quadratic S-Y2009 with non-quadratic Hosford72 and proposed the CQN2017 yield criterion, which can describe the yield stresses anisotropy of blanks under the non-associated flow rule (non-AFR). Inspired by CQN2017, Hu et al. [23] further coupled the fourth-order polynomial yield criterion with the Hosford isotropic yield criterion under the associated flow rule (AFR), which can accurately describe the anisotropic behavior of materials; even for blanks with strong plastic flow anisotropy, it can provide accurate prediction levels. Chen et al. [24] proposed another form of the CQN function by coupling the quadratic S-Y2009 function with the non-quadratic Drucker function to achieve higher computation efficiency with similar accuracy. Hou et al. [25] further replaced the coupling function based on stress components with the coupling function based on stress invariants.

However, most of the yield criteria mentioned above cannot describe the asymmetric yield behavior under tension and compression of materials. Spitzig et al. [26] and Spitzig and Richmond [27] found that the UT yield behavior of aluminum alloys and steels was influenced by superimposed hydrostatic pressure. Therefore, Stoughton and Yoon [28] developed a quadratic asymmetric yield criterion to describe the strength differential (SD) effect of AA2008-T4 and AA2090-T3 aluminum alloys. To seek a yield criterion suitable for describing the anisotropic behavior of HCP structure, Cazacu and Barlat [29] modified the even-form Drucker1949 yield criterion to an odd-form one and proposed the CB2004 yield criterion that uses the invariants of the stress deviator to characterize the asymmetric yield behavior of materials. Subsequently, Cazacu et al. [30] further developed the CPB06 yield criterion containing a fourth-order linear operator. To improve the flexibility of CPB06 in describing plastic anisotropy, Plunkett et al. [31] and Li et al. [32] added additional linear transformation tensors to the CPB06 yield criterion and established CPB06ex2 and M_CPB06 yield criteria. Khan et al. [33] proposed a new method to describe the SD effect by using the Lode angle parameter, which can describe the tension–compression asymmetry individually. Lou et al. [34] modified the Yld2000-2d yield criterion to accurately describe the asymmetric yield behavior of aluminum alloys by introducing the first stress invariant. Yoon et al. [35] and Lou et al. [36] developed two asymmetric yield functions in a form of three stress invariants. Furthermore, Hou et al. [37–39] improved the KB93 and Min2016 yield criteria under the non-AFR, which can describe the tension–compression asymmetry of yield stresses and r-values, respectively. Hu and Yoon [40] simplified the expression of the Yoon2014 yield criterion by analyzing the transformation tenors on the deviatoric stress invariants, which achieved the analytical calibration of anisotropic parameters while retaining the traditional model's ability to describe the SD effect. Hu et al. [41] established a more flexible asymmetric yield criterion by reconstructing LHY2013 and verified its accuracy based on FCC and HCP materials. Recently, Lou and Yoon [42] proposed a Lode-dependent asymmetric–anisotropic (LAA) framework by analyzing the correspondence between stress triaxiality and normalized third deviatoric stress invariant under uniaxial and equi-biaxial stress states, which can accurately predict the asymmetric behavior of yield stresses and r-values with a 45° increment under uniaxial loading paths.

To accurately describe the dominant plasticity behaviors under SS and NPS stress states, Vegter et al. [43] optimized the exponents of the Yld2000-2d and Yld2004-18p yield criteria based on interpolation methods, which filled the gap in traditional calibration methods for predicting the mechanical properties of SS and NPS stress states. Similarly, Du et al. [44] accurately predicted the normal and diagonal planes yield loci of AA6016-T4, AA5182-O, MP980, and DP490 blanks by incorporating NPS yield stresses along the 0°, 45°, and 90° directions into the calibration of anisotropy parameters and exponents of BBC2008. To address the issue of insufficient accuracy in describing plastic flow under NPS loading paths, Hou et al. [45] further introduced the directions of plastic strain rate along the 0°, 45°, and 90° directions in the NAFR-Poly4 yield criterion to calibrate the anisotropy parameters of the plastic potential function. In addition, it is worth noting that the Yld2000-2d yield criterion can accurately predict the plastic deformation behavior under corresponding stress states when identifying anisotropic parameters through NPS or SS mechanical properties, but it cannot describe both stress states simultaneously [46]. Therefore, He et al. [47] improved the Yld2000-2d yield criterion by introducing a shear-related additional term to enhance the flexibility of the Yld2000-2d, while ensuring the accuracy of describing biaxial tensile (BT) stress states, it can effectively predict anisotropic yield and plastic flow near the SS stress states. Recently, Hu et al. [48] proposed a more flexible Analytical Poly6-18p yield function based on the Analytical Poly6-16p yield criterion, which not only achieved accurate modeling of SS stress along the 45° direction but also incorporated SS stress along the 0° direction into the modeling category.

In summary, considering that existing constitutive models usually cannot accurately describe the anisotropic behavior of blanks under various typical loading conditions such as UT, UC, EBT, NPS, and SS, in this work, the Eyld2000-2d yield criterion was improved to a new model, i.e., the A-Eyld2000-2d yield criterion, that can describe the SD effect by introducing hydrostatic pressure. Meanwhile, to more effectively control the yield locus of the sheet metals, the mathematical constraint that the exponent m is a constant value was removed by increasing NPS yield stresses during the parameter calibration process. Through experimental data measured under UT, UC, SS, and BT stress states, the differences in describing the anisotropic yield and plastic flow of AA6016-T4, AA5754-O,

DP980, and QP980 using the new model with different parameter identification strategies were systematically evaluated. Subsequently, the new model employing the best parameter identification strategy was compared with four commonly used asymmetric yield criteria, i.e., CPB06, LHY2013, S-Y2004, and Hu & Yoon2021, to further verify the validity and applicability of the new model. Finally, the influence of different hardening concepts on the prediction accuracy of yield criteria in describing the evolving plastic behavior of sheet metals was discussed.

2. Modeling Strategies for Describing the Plastic Anisotropic Behavior of Materials under Various Typical Stress States

2.1. Asymmetric Eyld2000-2d Yield Criterion

This section improves the Eyld2000-2d yield criterion proposed by He et al. [47] under AFR to describe the SD effect of blanks. The equivalent stress of the newly proposed asymmetric Eyld2000-2d yield criterion (A-Eyld2000-2d) is defined as:

$$\varphi = \overline{\sigma} = \widetilde{I} + f^{1/m} \tag{1}$$

$$\widetilde{I} = h_x \sigma_{11} + h_y \sigma_{22} \tag{2}$$

$$f = \frac{1}{2}(f_1 + f_2) \tag{3}$$

where φ is the yield function, $\overline{\sigma}$ is the equivalent stress (in this work the rolling direction (RD) was chosen as the reference direction), $\widetilde{I}$ is the hydrostatic pressure, m is the exponent of the yield criterion, f is the sum of the two isotropic functions f_1 and f_2, where f_1 and f_2 are expressed as

$$f_1 = \frac{1}{2}(|X'_1 - X'_2|^m + |2X'_1 + X'_2|^m + |X'_1 + 2X'_2|^m) \tag{4}$$

$$f_2 = \frac{1}{2}(|2X''_1 + X''_2|^m + |X''_1 + 2X''_2|^m) \tag{5}$$

where X'_i and X''_j $(i,j = 1,2)$ are the eigenvalues of the stress deviators $\boldsymbol{X}'$ and $\boldsymbol{X}''$, i.e.,

$$X'_{1,2} = \frac{1}{2}(X'_{11} + X'_{22} \pm \sqrt{(X'_{11} - X'_{22})^2 + 4X'^{2}_{12}}) \tag{6}$$

$$X''_{1,2} = \frac{1}{2}(X''_{11} + X''_{22} \pm \sqrt{(X''_{11} - X''_{22})^2 + 4X''^{2}_{12}}) \tag{7}$$

where X'_{ij} and X''_{ij} $(i,j = 1,2)$ are obtained through linear transformation of Cauchy stress, there are

$$\boldsymbol{X}' = \begin{bmatrix} X'_{11} \\ X'_{22} \\ X'_{12} \end{bmatrix} = \begin{bmatrix} \frac{2}{3}C'_{11} - \frac{1}{3}C'_{12} & -\frac{1}{3}C'_{11} + \frac{2}{3}C'_{12} & 0 \\ \frac{2}{3}C'_{21} - \frac{1}{3}C'_{22} & -\frac{1}{3}C'_{21} + \frac{2}{3}C'_{22} & 0 \\ 0 & 0 & C'_{66} \end{bmatrix} \begin{bmatrix} \sigma_{11} \\ \sigma_{22} \\ \sigma_{12} \end{bmatrix} \tag{8}$$

$$\boldsymbol{X}'' = \begin{bmatrix} X''_{11} \\ X''_{22} \\ X''_{12} \end{bmatrix} = \begin{bmatrix} \frac{2}{3}C''_{11} - \frac{1}{3}C''_{12} & -\frac{1}{3}C''_{11} + \frac{2}{3}C''_{12} & 0 \\ \frac{2}{3}C''_{21} - \frac{1}{3}C''_{22} & -\frac{1}{3}C''_{21} + \frac{2}{3}C''_{22} & 0 \\ 0 & 0 & C''_{66} \end{bmatrix} \begin{bmatrix} \sigma_{11} \\ \sigma_{22} \\ \sigma_{12} \end{bmatrix} \tag{9}$$

where C'_{ij} and C''_{ij} $(i,j = 1,2 \text{ or } i = j = 6)$ are the anisotropy parameters of the yield criterion, and σ_{ij} $(i,j = 1,2)$ is the in-plane stress component of the Cauchy stress tensor.

2.2. Exponent *m* Study of A-Eyld2000-2d

To quantitatively clarify the potential regulative effects of the exponent m on the curvature of yield locus, the normalized yield loci of the A-Eyld2000-2d yield criterion are calculated using isotropic mechanical properties and a specific set of anisotropic parameters by adjusting the value of m, as shown in Figure 1. In general, the uniaxial and equi-biaxial tensile/compressive yield stresses are not affected by changes in the exponent, but the yield points near the NPS and SS stress states exhibit strong exponent sensitivity. When using isotropic mechanical properties, as the exponent increases, the boundary near the SS stress state shrinks first and then expands, while the region corresponding to the NPS stress state continues to expand outward, as shown in Figure 1a. In contrast, when the anisotropy parameters remain constant, as the exponent increases, the normalized yield loci of the NPS and SS stress state regions gradually shrink inward, as shown schematically in Figure 1b. Furthermore, to further understand the prediction capability of the A-Eyld2000-2d yield criterion on plastic anisotropy behavior within the investigated exponent range of $m \in [2, 20]$, the ratios PTR and STR of NPS and SS yield stresses to UT yield stress, as defined by Hou et al. [49], are used to quantify the maximum and minimum boundaries described in Figure 1. After calculation, the values of PTR and STR in Figure 1a are 1.2879 and 0.6496, while the values of PTR and STR are 1.0528 and 0.5359 in Figure 1b, respectively, which are sufficient to accurately describe the yield behavior of the vast majority of conventional blanks. The above characteristics provide a theoretical basis for calibrating the exponent and anisotropy parameters of the A-Eyld2000-2d yield criterion using NPS and SS mechanical properties.

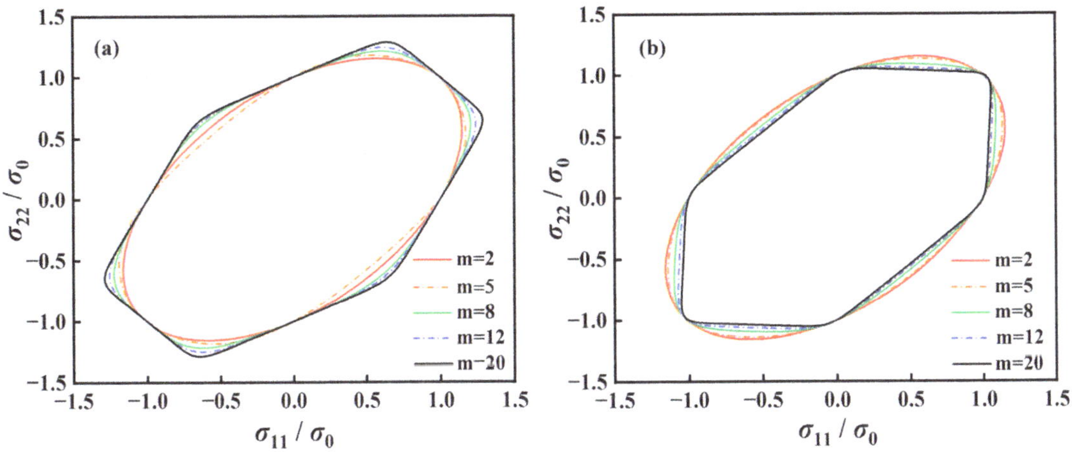

Figure 1. Normalized yield loci of the A-Eyld2000-2d yield criterion calculated using (**a**) isotropic mechanical properties and (**b**) a specific set of anisotropic parameters at different exponents m.

2.3. Parameter Calibration Programs for A-Eyld2000-2d Yield Criterion

The description ability of the yield criterion is not only related to the modeling expression but also influenced by the calibration methods. Therefore, four different parameter identification strategies are designed for A-Eyld2000-2d in this section to illustrate the regulative effects of the selected mechanical properties and exponent m on the curvature of the yield locus. Table 1 lists the mechanical properties employed by different calibration methods, where "$\surd$" and "$\times$" represent selected and unselected, respectively. Method #1 is the same as the original Eyld2000-2d yield criterion and does not consider the asymmetric yield behavior of blanks; Method #2 additionally introduces σ_{C0} and σ_{C90} to calculate h_x and h_y based on Method #1; Method #3 further adds the NPS yield stresses to identify anisotropy parameters; Method #4 removes the mathematical constraint that the exponent m is an integer compared with Method #3. Note that when $m \in N^*$, i.e., Methods #1, #2, and #3, m is set to 6 for BCC materials and 8 for FCC materials. In addition, it is usually difficult to measure the plastic strain rate r_s of blanks under SS stress state. Therefore, to avoid the formation of underdetermined equations during calibration process, which leads to non-uniqueness of parameters, in this work, the anisotropy parameters C''_{12} and C''_{21} related to the normal stress component are made equal.

Table 1. Mechanical properties required for different parameter identification strategies of the A-Eyld2000-2d yield criterion.

Mechanical Properties		**Parameter Identification Strategies**							
σ	r	**#1 m∈N***		**#2 m∈N***		**#3 m∈N***		**#4 m∈N**	
σ_{T0}	r_{T0}	√	√	√	√	√	√	√	√
σ_{T45}	r_{T45}	√	√	√	√	√	√	√	√
σ_{T90}	r_{T90}	√	√	√	√	√	√	√	√
σ_b	r_b	√	√	√	√	√	√	√	√
σ_{C0}	-	×	-	√	-	√	-	√	-
σ_{C90}	-	×	-	√	-	√	-	√	-
σ_{S45}	-	√	-	√	-	√	-	√	-
σ_{PS}^{0}	-	×	-	×	-	√	-	√	-
σ_{PS}^{90}	-	×	-	×	-	√	-	√	-

For different parameter identification strategies, the following minimization objective function is defined to compute the anisotropy parameters.

$$
\begin{aligned}
&F\left[C'_{11}, C'_{12}, C'_{21}, C'_{22}, C'_{66}, C''_{11}, C''_{12}, C''_{22}, C''_{66}, h_x, h_y, m\right]_{\mathrm{Min}} = \\
&\lambda_1\left[\sum_{t=1}^{3}\left(\frac{\sigma_{Ti}^{\mathrm{cal}}}{\sigma_{Ti}^{\mathrm{exp}}}-1\right)^2\right]+\lambda_2\left[\sum_{t=1}^{3}\left(\frac{r_{Ti}^{\mathrm{cal}}}{r_{Ti}^{\mathrm{exp}}}-1\right)^2\right]+\lambda_3\left(\frac{\sigma_{S45}^{\mathrm{cal}}}{\sigma_{S45}^{\mathrm{exp}}}-1\right)^2 \\
&+\lambda_4\left[\left(\frac{\sigma_{b}^{\mathrm{cal}}}{\sigma_{b}^{\mathrm{exp}}}-1\right)^2+\left(\frac{r_{b}^{\mathrm{cal}}}{r_{b}^{\mathrm{exp}}}-1\right)^2\right] \\
&+\lambda_5\left[\left(\frac{\sigma_{C0}^{\mathrm{cal}}}{\sigma_{C0}^{\mathrm{exp}}}-1\right)^2+\left(\frac{\sigma_{C90}^{\mathrm{cal}}}{\sigma_{C90}^{\mathrm{exp}}}-1\right)^2\right] \\
&+\lambda_6\left\{\left[\frac{\left(\sigma_{PS}^{0}\right)^{\mathrm{cal}}}{\left(\sigma_{PS}^{0}\right)^{\mathrm{exp}}}-1\right]^2+\left[\frac{\left(\sigma_{PS}^{90}\right)^{\mathrm{cal}}}{\left(\sigma_{PS}^{90}\right)^{\mathrm{exp}}}-1\right]^2\right\}
\end{aligned}
\tag{10}
$$

where i = 0, 45, and 90, "cal" and "exp" represent the theoretical calculation values and experimental measurement values, respectively. $\lambda_j (j = 1, 2, \ldots, 6)$ is the weighting factor, when the corresponding mechanical properties are used in the parameter identification processes $\lambda = 1$; conversely, $\lambda = 0$. In this work, the Particle Swarm Optimization function in MATLAB is used to calculate the anisotropy parameters in different calibration methods [37]. To demonstrate the differences in describing plastic anisotropy among several parameter identification strategies, they are applied to AA6016-T4, AA5754-O, DP980, and QP980 blanks, which are widely used in the automotive stamping industry, wherein the mechanical properties of AA6016-T4 are obtained from the reports of Du et al. [50,51], and the mechanical properties of AA5754-O, DP980, and QP980 are extracted from the research of Hou et al. [38].

3. Results and Discussions

3.1. Prediction Results of Different Parameter Identification Strategies

Figure 2 shows the normalized yield loci of AA6016-T4, AA5754-O, DP980, and QP980 blanks predicted by the A-Eyld2000-2d yield criterion with different parameter identification strategies and compared with the experimental data. Note that the equivalent plastic strains (EPSs) selected for calibrating the anisotropy parameters of four blanks are 0.056, 0.074, 0.050, and 0.060, respectively. These are the maximum EPSs that can be achieved by obtaining experimental data within a uniform deformation range in all loading paths investigated, i.e., UT, UC, EBT, NPS, and SS stress states and different sampling directions. The anisotropy parameters calculated by different parameter identification strategies are summarized in Table A1 in Appendix A. It can be observed that the four parameter identification strategies can reasonably describe the yield loci of four blanks. Compared to Method #1, the other three methods can accurately predict the UC yield stresses, and the ability to describe the SD effect has been improved. This can be attributed to the fact that Methods #2, #3, and #4 employed UC yield stresses during the parameter calibration process, as shown in Table 1. It can also be observed that Methods #1 and #2 exhibit slight deviations in characterizing the NPS stress states, especially for QP980, as shown in Figure 2d. This is because these two methods do not use NPS yield stresses to calibrate the anisotropy parameters, resulting in the inability to more accurately control the curvature of the yield loci. In addition, the four methods can accurately predict the SS yield stresses of AA5754-O, DP980, and QP980 blanks, as shown in Figure 2b–d. However, only Method #4 provides the best prediction accuracy for SS yield stress of AA6016-T4, as shown in Figure 2a. This can be attributed to the exponent of Method #4 having more flexible adjustment space. To further evaluate the ability of the A-Eyld2000-2d yield criterion with different parameter identification strategies in describing yield loci, the errors are calculated through the analytical indicator used by Du et al. [52]:

$$\Delta_{\mathrm{yl}} = \sqrt{\frac{1}{N}\sum_{t=1}^{N}\left\{\frac{f\left[(\sigma_{ij})^{\exp}(t)\right]}{\overline{\sigma}} - 1\right\}^2} \tag{11}$$

where σ_{ij} $(i, j = 1, 2)$ is the experimental stress component, N is the number of loading paths, t is the tth loading path.

Figure 3 shows the prediction errors of the normalized yield loci calculated by the A-Eyld2000-2d yield criterion with different parameter identification strategies for AA6016-T4, AA5754-O, DP980, and QP980 blanks. It can be observed that Method #4 provides the best prediction accuracy for these four blanks, while Method #1 provides the highest level of prediction errors. Compared with Method #1, the errors of AA6016-T4, AA5754-O, DP980, and QP980 blanks predicted by Method #4 decreased by 51.57%, 23.94%, 46.32%, and 50.43%, respectively. Methods #3 and #4 are compared to further illustrate the contribution of exponent m to improving the accuracy of yield criteria in describing the yield locus of blanks, which have great accuracy and insignificant difference in describing the mechanical properties of DP980 and QP980 blanks under SS and NPS stress states, as shown in Figure 2c,d. This can be attributed to Method #4 removing the restriction of the integer exponent of the yield criterion, and thereby enhancing the ability of the yield criterion to adjust the curvature of the yield locus. Meanwhile, the quantitative analysis of errors showed that compared with Method #3, the errors of AA6016-T4, AA5754-O, DP980, and QP980 blanks predicted by Method #4 decreased by 46.27%, 2.05%, 1.35%, and 17.86%, respectively, as shown in Figure 3. This proves that changing the exponent from integer to non-integer has a positive effect on improving the prediction accuracy of the yield criterion. Note that DP980 always has the lowest prediction errors for the four engineering materials investigated. This is because there is no phenomenon of residual austenite transforming into martensite in DP980 compared with QP980 [41]. Meanwhile, the shapes of the yield loci of DP980 are more rounded than that of aluminum alloy blanks. Therefore, regardless of the parameter calibration method used, it will always provide the best prediction accuracy for the DP980.

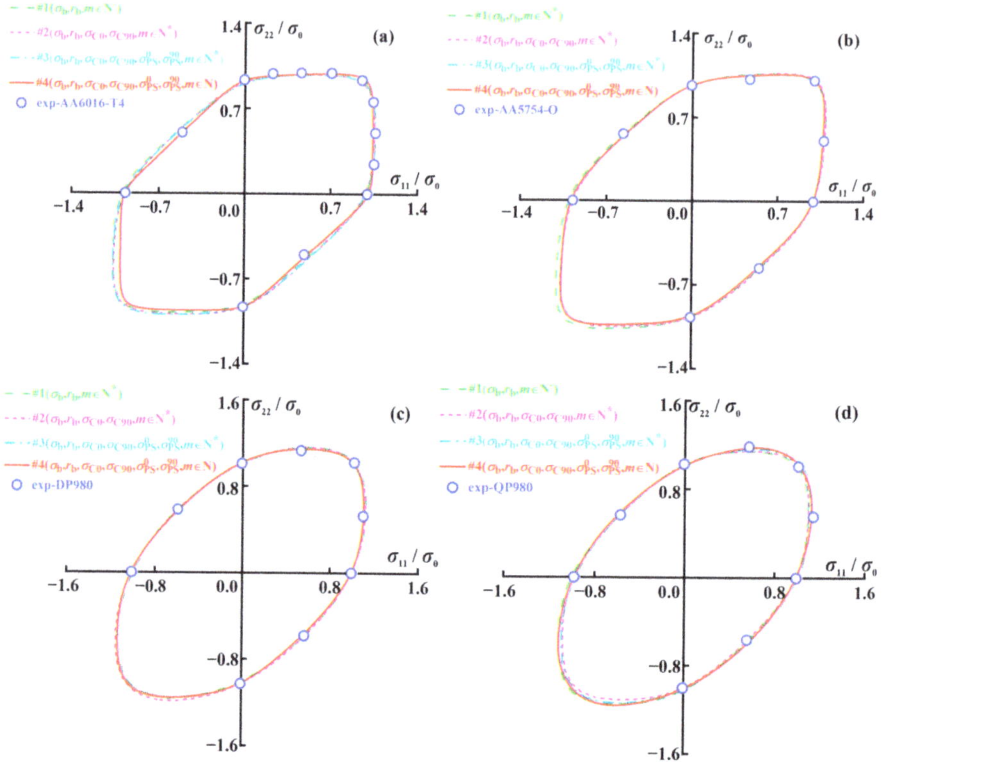

Figure 2. Normalized yield loci of (**a**) AA6016-T4, (**b**) AA5754-O, (**c**) DP980, and (**d**) QP980 blanks predicted by A-Eyld2000-2d yield criterion with different parameter identification strategies.

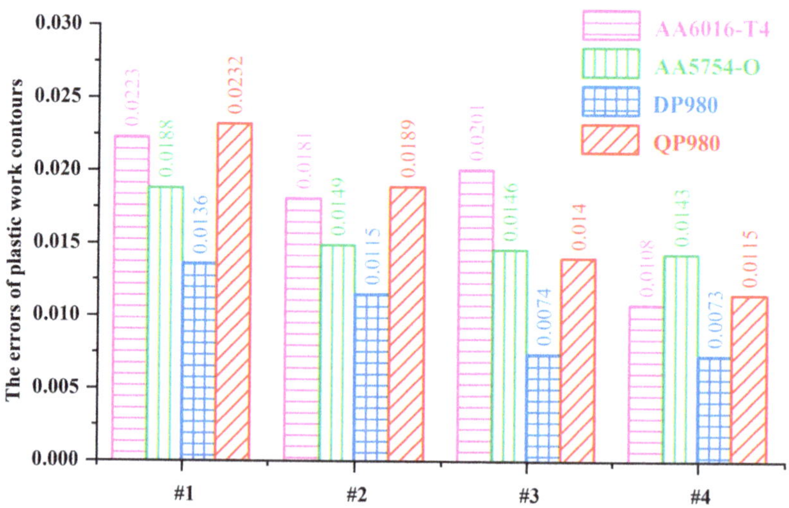

Figure 3. Prediction errors of normalized yield loci of AA6016-T4, AA5754-O, DP980, and QP980 blanks calculated by the A-Eyld2000-2d yield criterion with different parameter identification strategies.

Figures 4 and 5 show the normalized UT yield stresses and r-values of AA6016-T4, AA5754-O, DP980, and QP980 blanks predicted by the A-Eyld2000-2d yield criteria with different parameter identification strategies, respectively. It can be observed that the four parameter identification strategies can accurately predict the normalized UT yield stresses and r-values along the RD, diagonal direction (DD), and transverse direction (TD), but there are certain deviations when describing the normalized UT yield stresses and r-values at 15°, 30°, 60°, and 75° directions along the RD. This can be attributed to the fact that these four methods employ mechanical properties along the RD, DD, and TD to identify anisotropy parameters. Therefore, when predicting normalized yield stress and r-values along the RD, DD, and TD, it is more accurate than other angles. However, it is worth noting that Method #2 has obvious errors in describing the UT yield stresses of blanks, especially for AA6016-T4 and QP980, as shown in Figure 4a,d. Meanwhile, Method #4 always provides the best prediction accuracy for in-plane anisotropic behavior of r-values, which is the most evident in AA6016-T4 and AA5754-O, i.e., the prediction results provided by Methods #1, #2, and #3 deviate significantly from the experimental values, as shown in Figure 5a,b. To further evaluate the ability of the A-Eyld2000-2d yield criterion with different parameter identification strategies to describe normalized UT yield stresses and r-values, the analysis indicators proposed by Stoughton and Yoon [53] are used to calculate the prediction errors Δ_{UT} and Δ_r:

$$\Delta_{UT} = \sqrt{\frac{1}{12}\left[\begin{array}{l} \left(\frac{\sigma_0^{cal}}{\sigma_0^{exp}}-1\right)^2+2\left(\frac{\sigma_{15}^{cal}}{\sigma_{15}^{exp}}-1\right)^2+ \\ 2\left(\frac{\sigma_{30}^{cal}}{\sigma_{30}^{exp}}-1\right)^2+2\left(\frac{\sigma_{45}^{cal}}{\sigma_{45}^{exp}}-1\right)^2+2\left(\frac{\sigma_{60}^{cal}}{\sigma_{60}^{exp}}-1\right)^2 \\ +2\left(\frac{\sigma_{75}^{cal}}{\sigma_{75}^{exp}}-1\right)^2+\left(\frac{\sigma_{90}^{cal}}{\sigma_{90}^{exp}}-1\right)^2 \end{array}\right]} \tag{12}$$

$$\Delta_r = \sqrt{\frac{1}{12}\left[\begin{array}{l} \left(\frac{r_0^{cal}}{r_0^{exp}}-1\right)^2+2\left(\frac{r_{15}^{cal}}{r_{15}^{exp}}-1\right)^2+ \\ 2\left(\frac{r_{30}^{cal}}{r_{30}^{exp}}-1\right)^2+2\left(\frac{r_{45}^{cal}}{r_{45}^{exp}}-1\right)^2+2\left(\frac{r_{60}^{cal}}{r_{60}^{exp}}-1\right)^2 \\ +2\left(\frac{r_{75}^{cal}}{r_{75}^{exp}}-1\right)^2+\left(\frac{r_{90}^{cal}}{r_{90}^{exp}}-1\right)^2 \end{array}\right]} \tag{13}$$

Due to the significant anisotropic behavior in the 105°, 120°, 135°, 150°, and 165° directions along the RD, the weights of yield stresses and r-values under the UT stress states in the 15°, 30°, 45°, 60°, and 75° directions are increased.

Table 2 lists the errors of the normalized UT yield stress and r-values predicted by the A-Eyld2000-2d yield criterion with different parameter identification strategies. The average errors of r-values provided by the four methods for all engineering materials investigated are 0.0911, 0.1229, 0.1294, and 0.0293, respectively. Obviously, compared with Methods #1, #2, and #3, Method #4 provides the minimum Δ_r for these four blanks, and the average errors are reduced by 67.87%, 76.19%, and 77.40%, respectively. The average errors of the four methods in predicting UT yield stresses are 0.0114, 0.0265, 0.0147, and 0.0121, respectively. This indicates that except for Method #2, the prediction accuracy of the other methods is not significantly different. In summary, it can be concluded that among all the calibration methods investigated, Method #4 is the best and can accurately describe the plastic anisotropy behavior of blanks under UT, UC, SS, NPS, and EBT stress states, especially for the plastic flow anisotropic behavior.

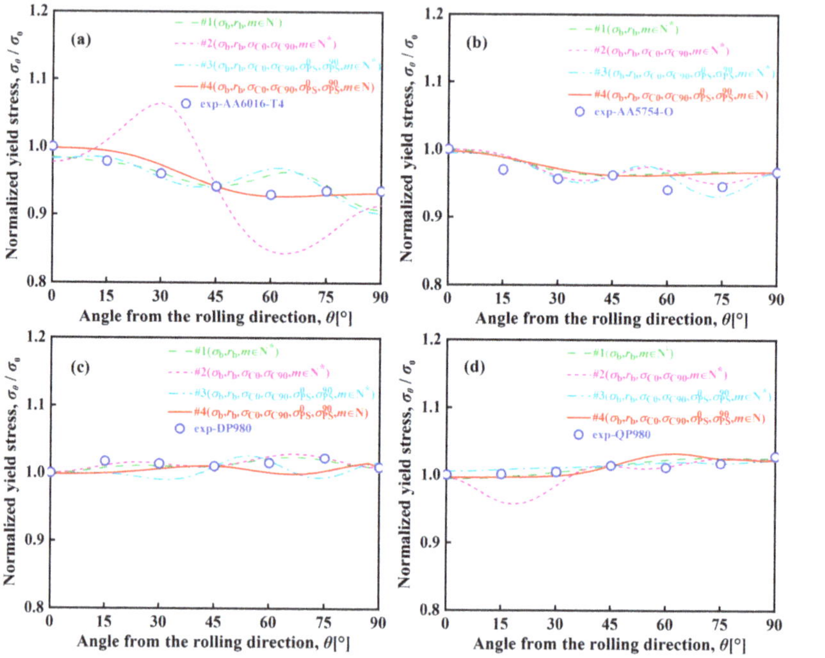

Figure 4. Normalized UT yield stresses of (**a**) AA6016-T4, (**b**) AA5754-O, (**c**) DP980, and (**d**) QP980 blanks predicted by the A-Eyld2000-2d yield criterion with different parameter identification strategies.

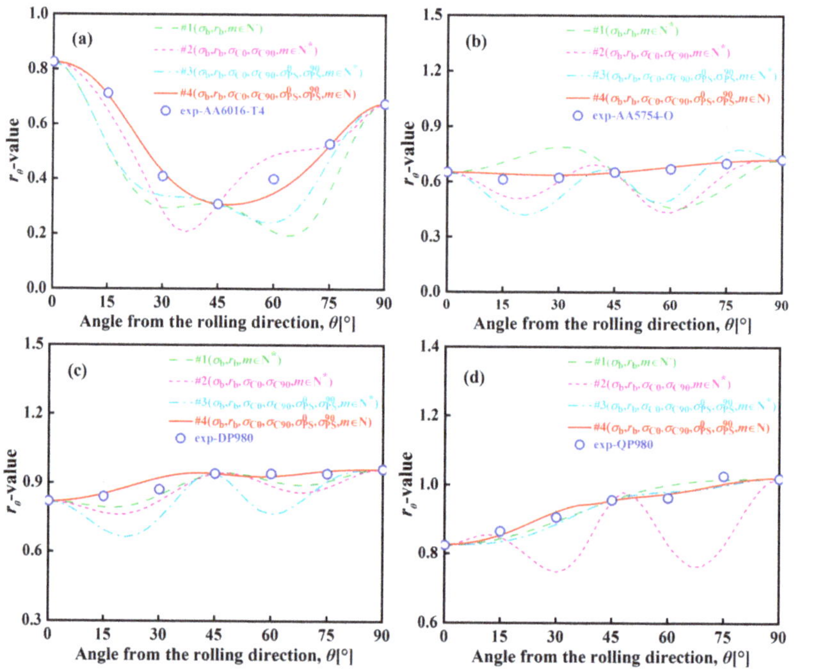

Figure 5. *r*-values of (**a**) AA6016-T4, (**b**) AA5754-O, (**c**) DP980, and (**d**) QP980 blanks predicted by the A-Eyld2000-2d yield criterion with different parameter identification strategies.

Table 2. Errors of the normalized UT yield stress and r-values predicted by the A-Eyld2000-2d yield criterion with different parameter identification strategies.

Methods	AA6016-T4		AA5754-O		DP980		QP980	
	Δ_{UT}	Δ_r	Δ_{UT}	Δ_r	Δ_{UT}	Δ_r	Δ_{UT}	Δ_r
#1	0.0170	0.2865	0.0172	0.0254	0.0051	0.0324	0.0064	0.0199
#2	0.0662	0.1590	0.0160	0.1573	0.0047	0.0542	0.0190	0.1209
#3	0.0211	0.2133	0.0145	0.1642	0.0166	0.1196	0.0066	0.0206
#4	0.0089	0.0553	0.0165	0.0247	0.0124	0.0255	0.0104	0.0115

3.2. Prediction Results of Several Asymmetric Yield Criteria

To further present the advancement of the new model, the A-Eyld2000-2d yield criterion employing the best parameter identification strategy is compared with several asymmetric yield criteria, CPB06 [30], LHY2013 [34], S-Y2004 [28], and Hu & Yoon2021 [40], which are widely used under the AFR and non-AFR. Combining quantitative analysis of errors, a systematic evaluation is made for the describing ability of different asymmetric yield criteria to the anisotropic behavior of four blanks.

Table 3 lists the mechanical properties required to calibrate anisotropic parameters for several asymmetric yield criteria. Note that the yield models S-Y2004 and Hu & Yoon2021 under the non-AFR use S-Y2009 as the plastic potential function (PPF). The anisotropy parameters calculated by several asymmetric yield criteria are summarized in Tables A2 and A3 in Appendix B.

Table 3. Mechanical properties required to calibrate anisotropic parameters for several asymmetric yield criteria.

Mechanical Properties		Asymmetric Yield Criteria									
		AFR						NAFR			
σ	r	CPB06		LHY2013		A-Eyld2000-2d		S-Y2004		Hu &Yoon2021	
σ_{T0}	r_{T0}	√	√	√	√	√	√	√	×	√	×
σ_{T45}	r_{T45}	×	×	√	√	√	√	√	×	√	×
σ_{T90}	r_{T90}	√	√	√	√	√	√	√	×	√	×
σ_b	r_b	√	×	√	√	√	√	√	×	√	×
σ_{C0}	-	√	-	√	-	√	-	√	-	√	-
σ_{C45}	-	×	-	×	-	×	-	×	-	√	-
σ_{C90}	-	√	-	√	-	√	-	√	-	√	-
σ_{Cb}	-	×	-	×	-	×	-	×	-	√	-
σ_{S45}	-	×	-	×	-	√	-	×	-	×	-
σ_{PS}^{0}	-	×	-	×	-	√	-	×	-	×	-
σ_{PS}^{90}	-	×	-	×	-	√	-	×	-	×	-

Figure 6 shows the normalized yield locus of AA6016-T4, AA5754-O, DP980, and QP980 blanks predicted by several asymmetric yield criteria. CPB06 shows significant deviation when describing the yield loci of AA6016-T4 and AA5754-O, as shown in Figure 6a,b. However, for DP980 and QP980, the errors of normalized yield loci predicted by the five asymmetric yield criteria are relatively small, and the differences are mainly reflected in the SS and NPS stress states, as shown in Figure 6c,d. In addition, it can also be observed that S-Y2004 and Hu & Yoon2021 have very similar abilities to describe the normalized yield loci of four blanks, and their prediction accuracy is always better than CPB06 under NPS and SS stress states. To further quantitatively evaluate the ability of five asymmetric yield criteria to describe the normalized yield loci of four blanks, the errors between the normalized yield loci and experimental data are calculated by Equation (11).

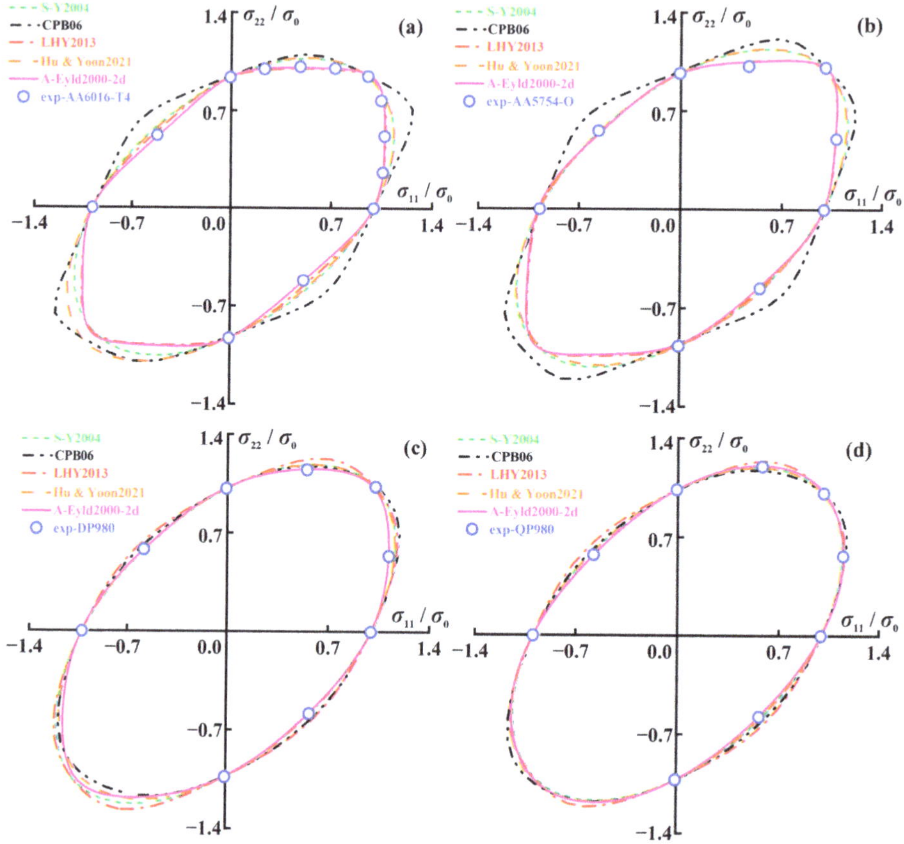

Figure 6. Normalized yield loci of (**a**) AA6016-T4, (**b**) AA5754-O, (**c**) DP980, and (**d**) QP980 blanks predicted by several asymmetric yield criteria.

Figure 7 shows the errors of normalized yield loci of AA6016-T4, AA5754-O, DP980, and QP980 blanks calculated by several asymmetric yield criteria. The average errors of the five yield criteria for the four engineering materials investigated are 0.0271, 0.0508, 0.0215, 0.0265, and 0.0069, respectively. It can be observed that A-Eyld2000-2d can provide the minimum prediction errors for the normalized yield loci of these four blanks, especially for NPS and SS stress states, which can be attributed to the new model using more comprehensive stress state data to calibrate the anisotropy parameters and exponent m. In addition, CPB06 provides the highest errors for AA6016-T4 and AA5754-O. Compared to A-Eyld2000-2d, the errors of both blanks have increased by 92.72% and 82.71%, respectively, while LHY2013 always has lower errors than S-Y2004 and Hu & Yoon2021, and the average errors are reduced by 20.66% and 18.87%, respectively. However, LHY2013 provides the worst prediction accuracy for DP980 and QP980, which is inconsistent with conventional cognition that increasing the number of yield stresses to identify anisotropy parameters usually helps to improve the prediction accuracy of yield criteria in describing yield loci directly related to stress. This can be attributed to the high dependence on blanks, i.e., among the four engineering materials investigated in this study, the CPB06 yield criterion is more suitable for describing the yield loci of BCC materials, while the LHY2013 yield criterion shows higher applicability to FCC materials.

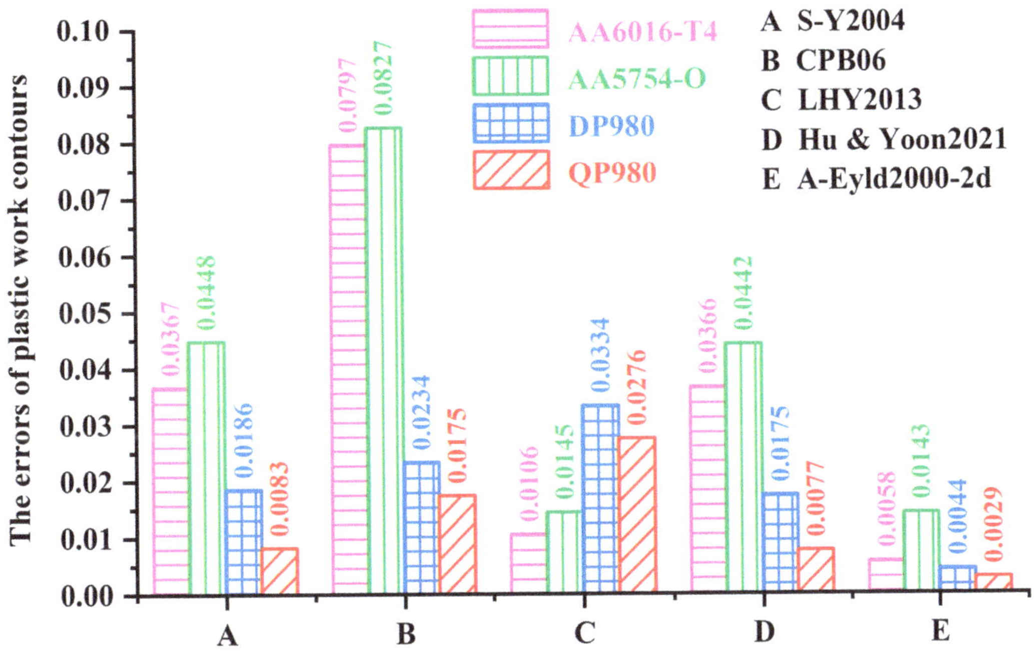

Figure 7. Prediction errors of normalized yield loci of AA6016-T4, AA5754-O, DP980, and QP980 calculated by several asymmetric yield criteria.

Figures 8 and 9 show the normalized UT yield stresses and *r*-values of AA6016-T4, AA5754-O, DP980, and QP980 blanks predicted by several asymmetric yield criteria, respectively. It can be observed that the curves of CPB06 in characterizing the in-plane anisotropic behaviors of four blanks show extremely obvious and strong upward convex or downward concave trends. The curves deviate significantly from the experimental values of UT yield stresses for these four blanks. Because the CPB06 yield criterion does not use the mechanical properties along the DD to identify anisotropy parameters, it exhibits drastic fluctuations in predicting the normalized UT yield stresses, as shown in Figures 8 and 9. This indicates that the prerequisite for reasonably predicting the UT mechanical properties of blanks through the yield criterion is to ensure the priority of calibration along the RD, DD, and TD. However, LHY2013 has large errors in predicting the normalized UT yield stresses and *r*-values of DP980 and QP980, and the curves show a relatively obvious fluctuation. Except for CPB06 and LHY2013, the other three yield criteria can predict the in-plane anisotropic behaviors of four blanks more reasonably. The prediction curves of S-Y2004 and Hu & Yoon2021 have the smallest variations and show the weakest fluctuations. Furthermore, although S-Y2004, Hu & Yoon2021, and A-Eyld2000-2d show different trends in predicting the normalized UT yield stresses, they all show good accuracy in predicting the normalized UT yield stresses and *r*-values of four blanks along the RD, DD, and TD. To further quantitatively evaluate the ability of the five yield criteria to predict the normalized UT yield stresses and *r*-values of four blanks, the errors Δ_{UT} and Δ_r of several asymmetric yield criteria to predict the normalized UT yield stresses and *r*-values are calculated by Equations (12) and (13), respectively.

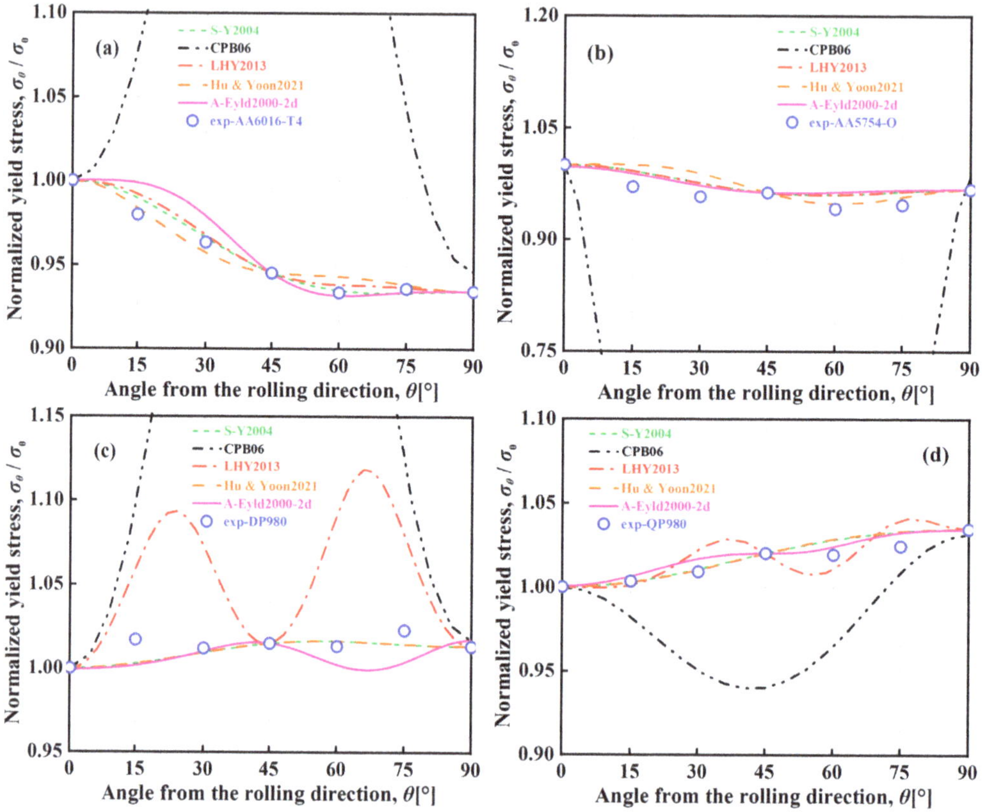

Figure 8. Normalized UT yield stresses of (**a**) AA6016-T4, (**b**) AA5754-O, (**c**) DP980, and (**d**) QP980 predicted by several asymmetric yield criteria.

Table 4 shows the errors Δ_{UT} and Δ_r for predicting the normalized UT yield stresses of four blanks based on several asymmetric yield criteria. The average errors of the normalized UT yield stresses predicted by the five yield criteria for the four engineering materials investigated are 0.0082, 0.3323, 0.0201, 0.0092, and 0.0107, respectively, and the average errors of r-values are 0.0339, 1.6855, 0.1331, 0.0339, and 0.0504, respectively. In general, CPB06 always provides the highest errors. Compared with the other four yield criteria, the average errors of CPB06 in describing normalized UT yield stresses increased by 97.53%, 93.95%, 97.23%, and 96.78%, respectively. The average errors in describing r-values increased by 97.99%, 92.10%, 97.99%, and 97.01%, respectively, while the quadratic yield function S-Y2004 achieves the best returns in predicting normalized UT yield stresses. Due to the application of the same PPF in S-Y2004 and Hu & Yoon2021, the calculated Δ_r are completely equal. In summary, compared with other asymmetric yield criteria, A-Eyld2000-2d can always make the most accurate prediction of the normalized yield loci of four blanks, especially under NPS and SS stress states. Considering that the blanks are mainly subjected to multi-axial stress states in the actual forming process, and none of the yield criteria can accurately predict the plastic anisotropy behavior under all loading paths, although A-Eyld2000-2d is not stable enough and there exists a certain deviation to describing the normalized UT yield stresses and r-values, as shown in Figures 8a–c and 9c,d, it is still the best asymmetric yield criteria investigated.

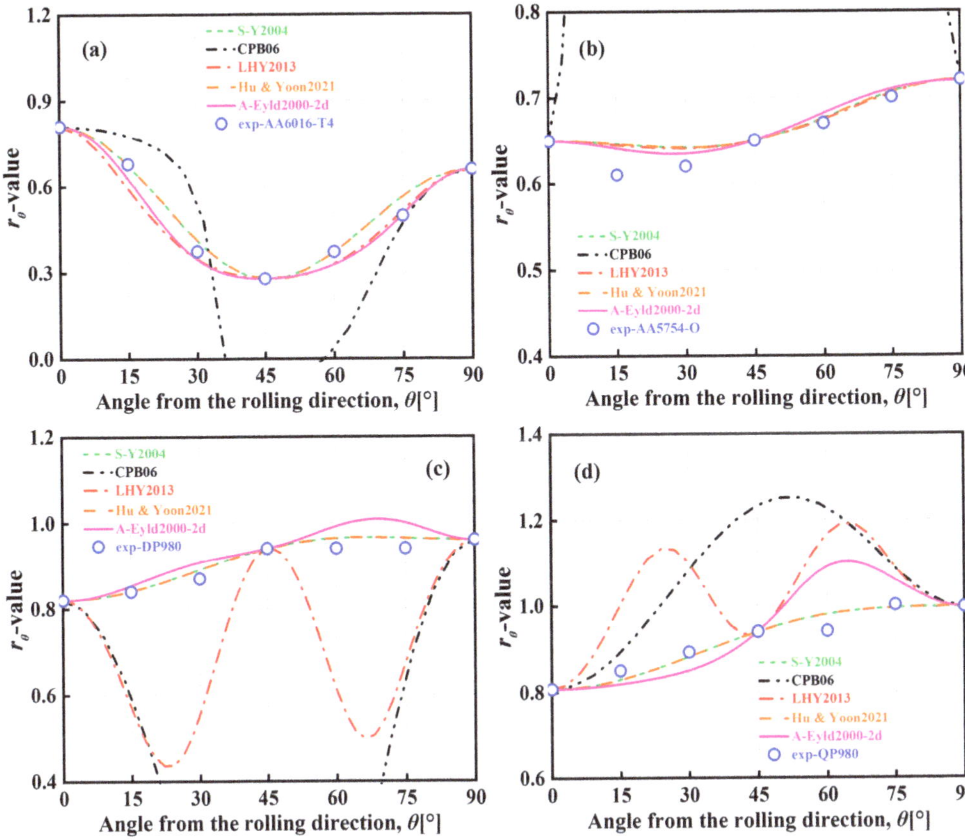

Figure 9. r-values of (**a**) AA6016-T4, (**b**) AA5754-O, (**c**) DP980, and (**d**) QP980 predicted by several asymmetric yield criteria.

Table 4. Errors of normalized UT yield stresses and r-values predicted by several asymmetric yield criteria.

Yield Criteria	AA6016-T4		AA5754-O		DP980		QP980	
	Δ_{UT}	Δ_r	Δ_{UT}	Δ_r	Δ_{UT}	Δ_r	Δ_{UT}	Δ_r
S-Y2004	0.0044	0.0689	0.0166	0.0287	0.0067	0.0178	0.0051	0.0201
CPB06	0.3354	0.8146	0.5127	4.8851	0.4343	0.8401	0.0466	0.2023
LHY2013	0.0058	0.0769	0.0166	0.0275	0.0490	0.2692	0.0090	0.1588
Hu & Yoon2021	0.0053	0.0689	0.0193	0.0287	0.0068	0.0178	0.0052	0.0201
A-Eyld2000-2d	0.0103	0.0645	0.0165	0.0247	0.0107	0.0393	0.0051	0.0730

3.3. Evolving Plastic Behavior

The deformation characteristics of sheet metals under a specific level of EPS cannot fully reflect the anisotropic behavior during the forming process [54], i.e., the materials exhibit an anisotropic hardening (AH) phenomenon. Therefore, this section further describes the yield loci, UT yield stresses, and r-values of four blanks at different EPSs and compared with the predicted results of the isotropic hardening (IH) concept.

Figure 10 shows the yield loci of AA6016-T4, AA5754-O, DP980, and QP980 blanks predicted by A-Eyld2000-2d based on the IH and AH concepts at various EPSs. Whether applying IH or AH concepts, both can accurately predict the initial yield loci of four blanks at EPS = 0.002. However, the IH concept exhibits obvious defects in describing the yield loci of four blanks with the increase of EPS. Overall, the SS yield stress and UC yield stress along the RD predicted by the IH concept for these four blanks have significant errors compared to experimental data, especially in the stage of large plastic strain. In addition, the UC yield stresses along the TD of AA5754-O, DP980, and QP980 predicted by the IH concept also have obvious errors, as shown in Figure 10b–d. Furthermore, it can be observed that the IH concept is difficult to provide accurate prediction results for AA5754-O and QP980 under NPS and EBT stress states, as shown in Figure 10b,d. In contrast, the AH concept can always accurately predict the yield loci of four blanks with changes in EPS. This can be attributed to the AH concept using instantaneous mechanical properties to identify the anisotropy parameters of the yield criterion. In other words, the IH concept cannot accurately predict the yield loci under all EPS conditions, which is most significant in AA6016-T4 and QP980, as shown in Figure 10a,d.

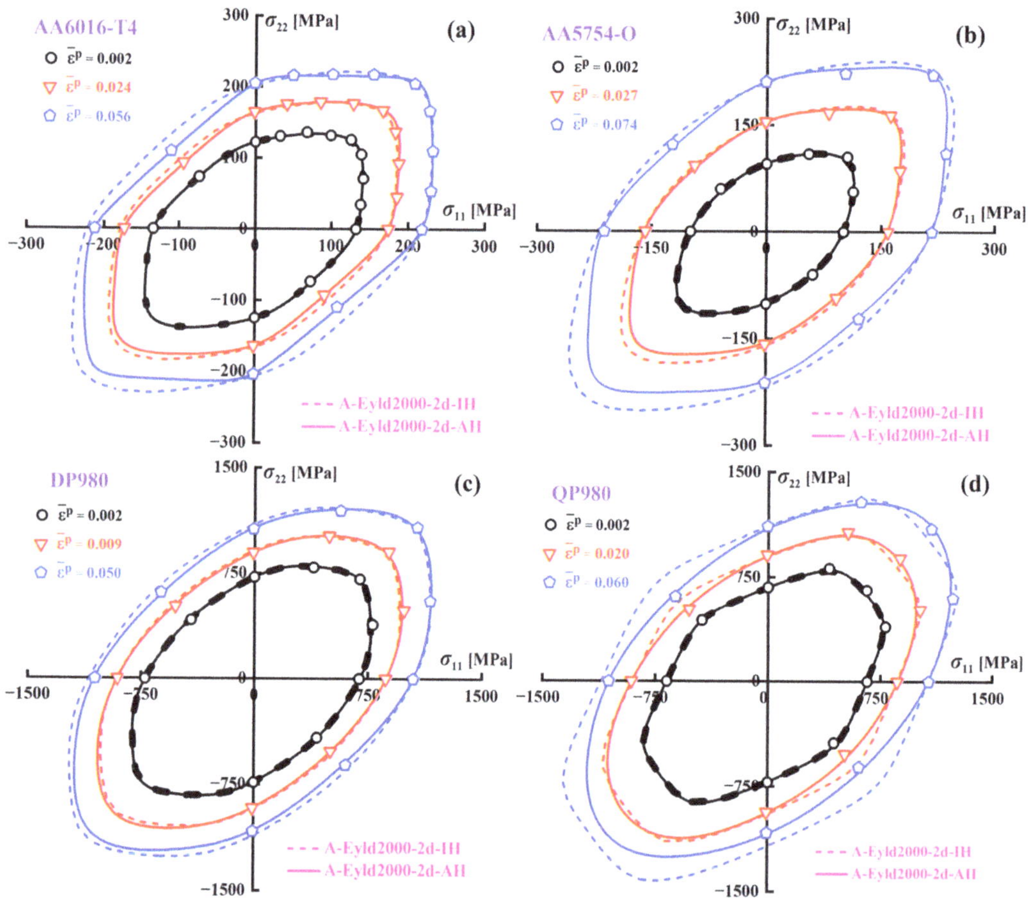

Figure 10. Yield loci of (**a**) AA6016-T4, (**b**) AA5754-O, (**c**) DP980, and (**d**) QP980 blanks predicted by IH and AH concepts at various EPSs.

Figure 11 shows the UT yield stresses of AA6016-T4, AA5754-O, DP980, and QP980 predicted by the A-Eyld2000-2d yield criterion based on the IH and AH concepts at various EPSs. For AA6016-T4 and QP980, because the curve shapes of UT yield stress predicted by the IH concept are calculated through the anisotropic parameters at the initial yield point, the prediction accuracy gradually deteriorates with the increase of equivalent plastic strain, while the accuracy of UT curves predicted based on the AH concept gradually improves, as shown in Figure 11a,d. Meanwhile, it can also be observed that the curve obtained based on the AH concept is more stable with the continuous increase of EPS, which makes the prediction accuracy of UT yield stresses at 15° intervals along the RD much higher than that of the IH concept. This indicates that the prediction effect of the A-Eyld2000-2d yield criterion based on the AH concept on the UT yield stresses of four blanks become increasingly excellent.

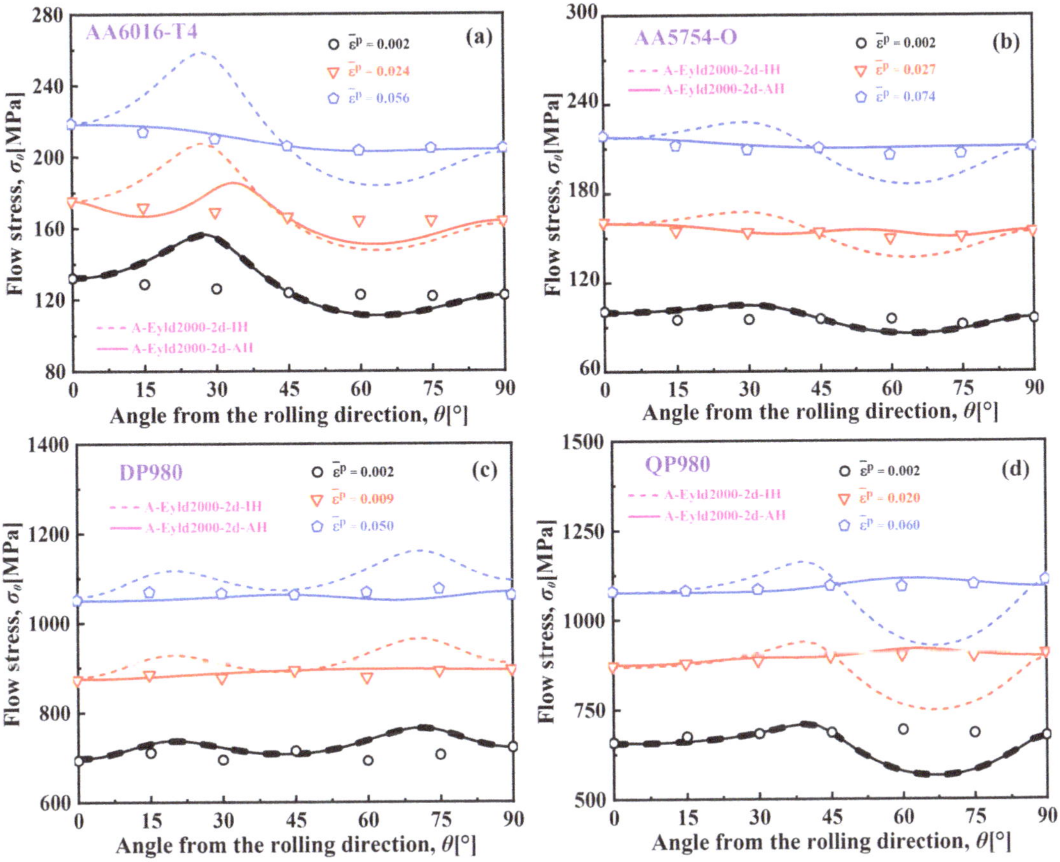

Figure 11. UT yield stresses of (**a**) AA6016-T4, (**b**) AA5754-O, (**c**) DP980, and (**d**) QP980 predicted by IH and AH concepts at various EPSs.

Figure 12 shows the *r*-values of AA6016-T4, AA5754-O, DP980, and QP980 predicted by the A-Eyld2000-2d yield criterion based on the IH and AH concepts at various EPSs. These two different concepts have the same ability to describe the *r*-values at the initial yield point for these four blanks, and can accurately predict *r*-values along the RD, DD, and TD. However, the IH concept is difficult to provide reasonable predictions for the *r*-values of AA6016-T4 and QP980 with the EPS increases. In contrast, the AH concept can effectively describe the *r*-values along the RD, DD, and TD, and the prediction accuracy of *r*-values in other directions has also become increasingly accurate, as shown in Figure 12a–d. In addition, although there is no evolving flow behavior in the *r*-values of AA5754-O and DP980, the ability of the AH concept to describe *r*-values at 15° intervals along the RD is also becoming stronger with the increase of EPS, as shown in Figure 12b,c. This indicates that it is necessary to describe the evolving plastic behavior of blanks through the AH concept.

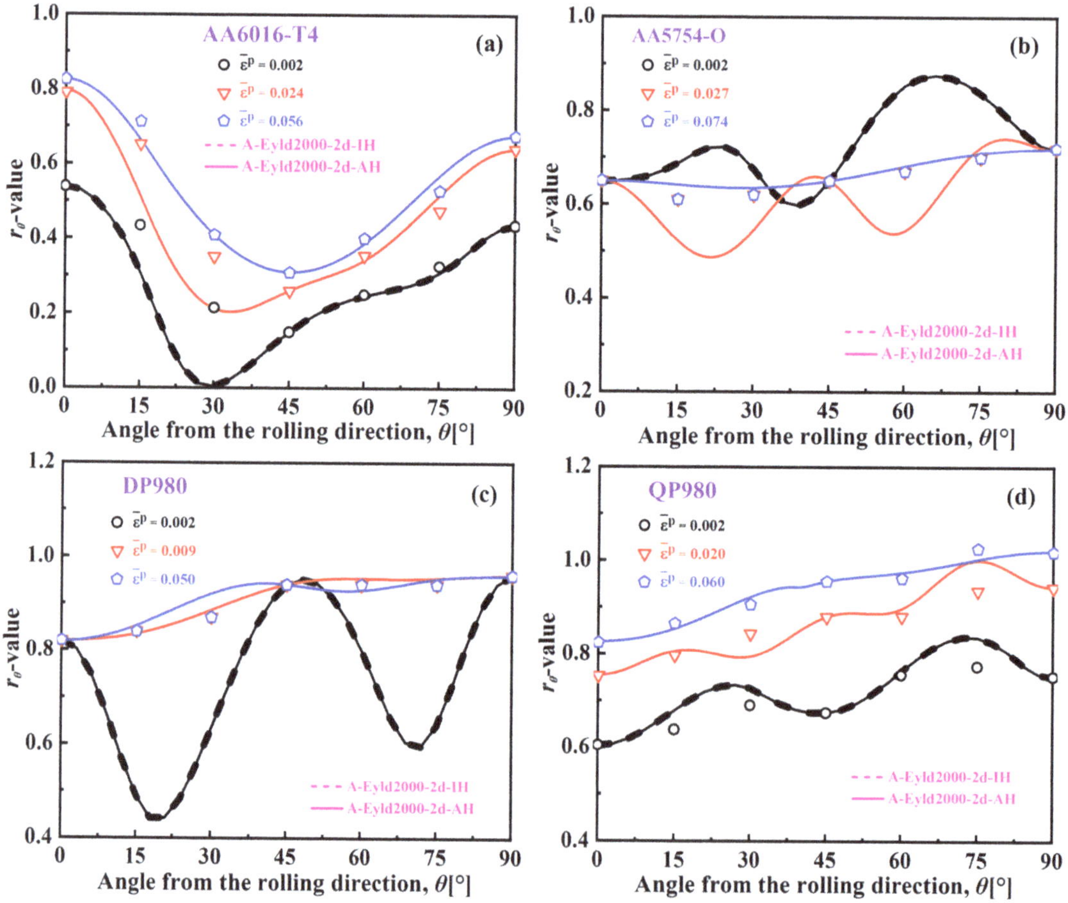

Figure 12. r-values of (**a**) AA6016-T4, (**b**) AA5754-O, (**c**) DP980, and (**d**) QP980 blanks predicted by IH and AH concepts at various EPSs.

4. Conclusions

In this work, the Eyld2000-2d yield criterion was improved by introducing hydrostatic pressure and proposed an A-Eyld2000-2d asymmetric yield criterion that can describe the SD effect and anisotropic hardening. Four different parameter identification strategies were designed for the new model, and their validity and applicability in describing plastic anisotropy were verified through AA6016-T4, AA5754-O, DP980, and QP980 blanks. Meanwhile, the A-Eyld2000-2d employing the optimal parameter identification strategy was compared with several existing pressure-sensitive asymmetric yield criteria under AFR and non-AFR. Finally, the influence of different hardening concepts on the evolving plastic behavior of blanks was analyzed, and the following conclusions can be drawn:

(1) The A-Eyld2000-2d yield criterion proposed by coupling hydrostatic pressure method has higher prediction accuracy for the tensile and compressive asymmetric yield behavior of sheet metals compared to the traditional Eyld2000-2d yield criterion.

(2) It is feasible to introduce additional NPS yield stresses to calibrate the exponent m of the new model. The parameter identification strategy with a non-integer exponent can significantly improve the prediction ability of the yield criterion for plastic anisotropic behavior, especially for in-plane plastic flow.

(3) The A-Eyld2000-2d yield criterion can provide the best prediction accuracy for the yield loci of four blanks, compared to the S-Y2004, CPB06, LHY2013, and Hu & Yoon2021 yield criteria. The ability to describe UT yield stresses and r-values also shows reasonable prediction accuracy, which is most suitable for characterizing the plastic anisotropy behavior of these four blanks.

(4) The deformation characteristics of the initial yield point cannot fully reflect the plastic anisotropic behavior of blanks during the entire deformation stage, considering that the anisotropic hardening concept can significantly improve the ability of the yield criterion to describe the evolving yield behavior and provide sufficient power for the accurate characterization of anisotropic behavior at various EPSs.

Author Contributions: K.D.: conceptualization, validation, formal analysis, writing—original draft, writing—review and editing, funding acquisition; L.D.: methodology, visualization, investigation; H.Z.: data curation; Z.M.: supervision, investigation, funding acquisition; H.D.: review and editing; H.W.: supervision, validation; Y.R.: validation; L.S.: software, review and editing; L.Z.: investigation; X.Y.: review and editing, funding acquisition, project administration. All authors have read and agreed to the published version of the manuscript.

Funding: The authors appreciate the financial support from the National Natural Science Foundation of China (Grant Nos. 52371116, 52305396, 52205353, and 52205326), the Doctoral Start-up Foundation of Liaoning Province (Grant No. 2023-BS-125), and the Beijing Municipal Education Commission and Beijing Municipal Natural Science Foundation (Grant No. KZ200010009041).

Institutional Review Board Statement: Not applicable.

Informed Consent Statement: Not applicable.

Data Availability Statement: The data that support the findings of this study are available from the corresponding authors upon reasonable request.

Conflicts of Interest: The authors declare that they have no known competing financial interest or personal relationships that could have appeared to influence the work reported in this paper.

Appendix A. Anisotropy Parameters Calculated by Different Parameter Identification Strategies

Table A1. Anisotropy parameters calculation results of AA6016-T4, AA5754-O, DP980, and QP980 blanks.

Materials	Methods	Anisotropy Parameters											
		C'_{11}	C'_{12}	C'_{21}	C'_{22}	C'_{66}	C''_{11}	C''_{12}	C''_{22}	C''_{66}	h_x	h_y	m
AA6016-T4	#1	0.0639	1.2217	−1.0993	−1.1564	−0.6901	0.0240	−0.9873	0.0670	−1.5581	-	-	8
	#2	0.0540	1.1636	1.0995	−0.0462	0.9001	0.1775	0.3873	0.8453	−1.5309	−0.0076	0.0116	8
	#3	−0.0491	−1.2111	−1.0499	0.0423	−0.8238	0.0056	−0.9575	0.1264	1.5649	−0.0066	0.0136	8
	#4	1.0829	1.0797	−0.0457	−1.1160	0.9356	0.0202	−0.8239	−0.3891	1.2707	−0.0235	0.0021	17.3657
AA5754-O	#1	−0.9159	0.0657	0.0588	−1.0801	1.0802	−0.0689	−1.0264	0.0650	−1.2148	-	-	8
	#2	0.9543	−0.1929	−0.9290	−0.7634	−1.0551	−1.0760	0.0305	−1.1241	1.2151	−0.0171	0.0021	8
	#3	−0.9956	−1.0216	0.9945	−0.0689	−1.0386	−0.0486	1.0663	−0.0795	1.2319	−0.0088	−0.0046	8
	#4	−0.9972	0.0456	−0.0092	−1.1091	1.0798	−1.0682	0.0736	−1.0055	1.2479	−0.0101	−0.0049	8.4664
DP980	#1	−0.0499	−1.0020	−1.0562	−0.0830	1.0854	0.6179	−1.1072	0.5978	−1.2492	-	-	6
	#2	−1.0467	−0.0945	−0.0527	−0.9848	1.0894	−0.6309	1.1125	−0.6240	1.1477	0.0019	0.0092	6
	#3	1.1080	−0.0161	−1.1077	−1.0329	−1.0674	0.1468	−0.7867	−0.1393	−0.9006	0.0046	0.0027	6
	#4	−0.0183	1.0318	1.1271	0.0186	1.0985	0.1808	−0.5419	−0.4794	−0.8793	0.0047	0.0028	6.0340
QP980	#1	−1.1328	−0.0153	0.0074	−1.0673	−1.1022	0.2161	−0.7535	0.0491	0.6944	-	-	6
	#2	0.0104	−1.1038	1.0940	1.1144	1.0678	−0.7402	−0.2993	−0.1210	1.0231	−0.0109	−0.0110	6
	#3	−1.1117	−0.0749	−0.0059	−1.0243	−1.0820	−0.5555	1.0464	−0.5353	1.3200	−0.0188	−0.0014	6
	#4	−1.0697	0.2114	−0.2727	−0.8849	1.2262	0.3140	−0.9775	−0.2876	−0.8423	−0.0123	−0.0019	2.8502

Appendix B. Anisotropy Parameters Calculated by Different Yield Criteria

Table A2. Anisotropy parameters of AA6016-T4 and AA5754-O calculated by different yield criteria.

Yield Criteria	AA6016-T4		AA5754-O	
S-Y2004	α_1	-7×10^{-5}	α_1	-8×10^{-5}
	α_2	-3×10^{-6}	α_2	4×10^{-5}
	λ_{y1}	1.0274	λ_{y1}	1.0279
	λ_{y2}	1.1466	λ_{y2}	1.0605
	ν_y	0.5199	ν_y	0.5586
	ρ_y	1.7016	ρ_y	1.7021
CPB06	C_{11}	0.0089	C_{11}	-3×10^{-6}
	C_{22}	−0.0012	C_{22}	0.0003
	C_{33}	−0.0012	C_{33}	0.0075
	C_{12}	0.0006	C_{12}	0.0085
	C_{13}	3×10^{-6}	C_{13}	−0.0010
	C_{23}	0.0085	C_{23}	−0.0010
	C_{66}	0.0042	C_{66}	−0.0483
LHY2013	a_1	0.9230	a_1	−1.7729
	a_2	1.1087	a_2	0.2303
	a_3	1.1223	a_3	−1.5438
	a_4	1.0992	a_4	−0.8316
	a_5	1.0218	a_5	−0.0958
	a_6	0.9753	a_6	−1.7788
	a_7	0.9160	a_7	0.9472
	a_8	1.2183	a_8	−1.2773
	h_x	−0.0134	h_x	−0.0125
	h_y	−0.0006	h_y	0.0096
Hu & Yoon2021	μ	0.0047	μ	−0.0097
	C_{11}	−0.0267	C_{11}	−0.0057
	C_{22}	−0.0087	C_{22}	0.0263
	C_{66}	−24.6485	C_{66}	14.3624
	E	1.1158	E	1.0325
	F	1.0323	F	0.9847
	G	1.0055	G	1.0560
A-Eyld2000-2d	C'_{11}	0.0626	C'_{11}	−0.9972
	C'_{12}	1.1393	C'_{12}	0.0456
	C'_{21}	1.0457	C'_{21}	−0.0092
	C'_{22}	−0.0548	C'_{22}	−1.1091
	C'_{66}	0.2116	C'_{66}	1.0798
	C''_{11}	0.0605	C''_{11}	1.0682
	C''_{12}	−0.7675	C''_{12}	0.0736
	C''_{22}	−0.4126	C''_{22}	−1.0055
	C''_{66}	1.4650	C''_{66}	1.2479
	h_x	−0.0150	h_x	−0.0101
	h_y	0.0010	h_y	−0.0049
	m	11.2927	m	8.4664

Table A3. Anisotropy parameters of DP980 and QP980 calculated by different yield criteria.

Yield Criteria	DP980		QP980	
S-Y2004	α_1	7×10^{-6}	α_1	-5×10^{-6}
	α_2	1×10^{-5}	α_2	-6×10^{-6}
	λ_{y1}	0.9860	λ_{y1}	1.0091
	λ_{y2}	0.9549	λ_{y2}	0.9458
	v_y	0.5138	v_y	0.4805
	ρ_y	1.4520	ρ_y	1.4455
CPB06	C_{11}	0.0019	C_{11}	−0.0005
	C_{22}	−0.0036	C_{22}	0.0028
	C_{33}	−0.0029	C_{33}	0.0238
	C_{12}	−0.0027	C_{12}	0.0241
	C_{13}	−0.0034	C_{13}	0.0037
	C_{23}	0.0016	C_{23}	−0.0017
	C_{66}	−0.0013	C_{66}	−0.0257
LHY2013	a_1	−1.6343	a_1	0.0185
	a_2	1.5577	a_2	−1.6411
	a_3	1.6999	a_3	−1.7365
	a_4	0.0370	a_4	−0.0551
	a_5	0.0027	a_5	0.8565
	a_6	1.6425	a_6	1.6198
	a_7	0.9453	a_7	0.6292
	a_8	0.8113	a_8	1.6660
	h_x	0.0071	h_x	−0.0045
	h_y	0.0098	h_y	−0.0060
Hu & Yoon2021	μ	0.0028	μ	−0.0032
	C_{11}	0.0065	C_{11}	−0.0019
	C_{22}	0.0103	C_{22}	−0.0039
	C_{66}	8.3327	C_{66}	21.4960
	E	0.9684	E	0.9373
	F	0.9594	F	0.9729
	G	0.9687	G	0.9686
A-Eyld2000-2d	C'_{11}	0.0030	C'_{11}	−0.8073
	C'_{12}	−1.0565	C'_{12}	0.6162
	C'_{21}	1.1262	C'_{21}	1.2326
	C'_{22}	−0.0031	C'_{22}	0.6194
	C'_{66}	1.1000	C'_{66}	1.2455
	C''_{11}	−0.0219	C''_{11}	0.2957
	C''_{12}	−0.1449	C''_{12}	0.7318
	C''_{22}	−0.7115	C''_{22}	0.1844
	C''_{66}	−0.9697	C''_{66}	−0.1656
	h_x	0.0084	h_x	−0.0047
	h_y	0.0061	h_y	−0.0060
	m	5.4945	m	3.4011

References

1. Hou, Y.; Myung, D.; Park, J.K.; Min, J.; Lee, H.-R.; El-Aty, A.A.; Lee, M.-G. A review of characterization and modelling approaches for sheet metal forming of lightweight metallic materials. *Materials* **2023**, *16*, 836. [CrossRef] [PubMed]
2. Wang, J.; Han, M.; Zhang, C.; Rayhan, H.M.A.; Li, X.; Lou, Y. Anisotropic hardening of TRIP780 steel sheet: Experiments and analytical modeling. *Materials* **2023**, *16*, 1414. [CrossRef] [PubMed]
3. Han, G.; He, J.; Li, S. Simple shear deformation of sheet metals: Finite strain perturbation analysis and high-resolution quasi-in-situ strain measurement. *Int. J. Plast.* **2022**, *150*, 103194. [CrossRef]
4. Liu, X.; He, J.; Huang, S.Y. Mechanistically informed artificial neural network model for discovering anisotropic path-dependent plasticity of metals. *Mater. Des.* **2023**, *226*, 111697. [CrossRef]
5. He, J.; Guo, C.; Li, W.K. Understanding the helicoidal damage behavior of bio-inspired laminates by conducting multiscale concurrent simulation and experimental analysis. *Compos. Struct.* **2023**, *314*, 116972. [CrossRef]
6. Guo, C.; He, J. Multi-scale concurrent analysis for bio-inspired helicoidal CFRP laminates and experimental investigation. *Compos. Struct.* **2022**, *296*, 115886. [CrossRef]

7. Brosius, A.; Küsters, N.; Lenzen, M. New method for stress determination based on digital image correlation data. *CIRP Ann. -Manuf. Technol.* **2018**, *67*, 269–272. [CrossRef]
8. Lou, Y.; Zhang, C.; Zhang, S.; Yoon, J.W. A general yield function with differential and anisotropic hardening for strength modelling under various stress states with non-associated flow rule. *Int. J. Plast.* **2022**, *158*, 103414. [CrossRef]
9. Hill, R. A theory of the yielding and plastic flow of anisotropic metals. *Proc. R. Soc. London. Ser. A Math. Phys. Sci.* **1948**, *193*, 281–297.
10. Yoshida, F.; Hamasaki, H.; Uemori, T. A user-friendly 3D yield function to describe anisotropy of steel sheets. *Int. J. Plast.* **2013**, *45*, 119–139. [CrossRef]
11. Barlat, F.; Lian, K. Plastic behavior and stretchability of sheet metals. Part I: A yield function for orthotropic sheets under plane stress conditions. *Int. J. Plast.* **1989**, *5*, 51–66. [CrossRef]
12. Barlat, F.; Becker, R.C.; Hayashida, Y.; Maeda, Y.; Yanagawa, M.; Chung, K.; Brem, J.C.; Lege, D.J.; Matsui, K.; Murtha, S.J.; et al. Yielding description for solution strengthened aluminum alloys. *Int. J. Plast.* **1997**, *13*, 385–401. [CrossRef]
13. Barlat, F.; Maeda, Y.; Chung, K.; Yanagawa, M.; Brem, J.C.; Hayashida, Y.; Lege, D.J.; Matsui, K.; Murtha, S.J.; Hattori, S.; et al. Yield function development for aluminum alloy sheets. *J. Mech. Phys. Solids.* **1997**, *45*, 1727–1763. [CrossRef]
14. Barlat, F.; Brem, J.C.; Yoon, J.W.; Chung, K.; Dick, R.E.; Lege, D.J.; Pourboghrat, F.; Choi, S.-H.; Chu, E. Plane stress yield function for aluminum alloy sheets-part 1: Theory. *Int. J. Plast.* **2003**, *19*, 1297–1319. [CrossRef]
15. Barlat, F.; Aretz, H.; Yoon, J.W.; Karabin, M.E.; Brem, J.C.; Dick, R.E. Linear transfomation-based anisotropic yield functions. *Int. J. Plast.* **2005**, *21*, 1009–1039. [CrossRef]
16. Banabic, D.; Comsa, D.S.; Sester, M.; Selig, M.; Kubil, W.; Mattiasson, K.; Sigvant, M. Influence of constitutive equations on the accuracy of prediction in sheet metal forming simulation. In Proceedings of the NUMISHEET, Interlaken, Switzerland, 1–5 September 2008; pp. 37–42.
17. Cazacu, O.; Barlat, F. Generalization of Drucker's yield criterion to orthotropy. *Math. Mech. Solids* **2001**, *6*, 613. [CrossRef]
18. Lou, Y.; Yoon, J.W. Anisotropic yield function based on stress invariants for BCC and FCC metals and its extension to ductile fracture criterion. *Int. J. Plast.* **2018**, *101*, 125–155. [CrossRef]
19. Khalfallah, A.; Alves, J.L.; Oliveira, M.C.; Menezes, L.F. Influence of the characteristics of the experimental data set used to identify anisotropy parameters. *Simul. Model. Pract. Th.* **2015**, *53*, 15–44. [CrossRef]
20. Khalfallah, A.; Oliveira, M.C.; Alves, J.L.; Menezes, L.F. Constitutive parameter identification of CB2001 yield function and its experimental verification using tube hydroforming tests. *Int. J. Mech. Sci.* **2020**, *185*, 105868. [CrossRef]
21. Lou, Y.; Zhang, S.; Yoon, J.W. A reduced Yld2004 function for modeling of anisotropic plastic deformation of metals under triaxial loading. *Int. J. Mech. Sci.* **2019**, *161*, 103997. [CrossRef]
22. Lee, E.-H.; Stoughton, T.B.; Yoon, J.W. A yield criterion through coupling of quadratic and non-quadratic functions for anisotropic hardening with non-associated flow rule. *Int. J. Plast.* **2017**, *99*, 120–143. [CrossRef]
23. Hu, Q.; Yoon, J.W.; Manopulo, N.; Hora, P. A coupled yield criterion for anisotropic hardening with analytical description under associated flow rule: Modeling and validation. *Int. J. Plast.* **2021**, *136*, 102882. [CrossRef]
24. Chen, Z.; Wang, Y.; Lou, Y. User-friendly anisotropic hardening function with non-associated flow rule under the proportional loadings for BCC and FCC metals. *Mech. Mater.* **2022**, *165*, 104190. [CrossRef]
25. Hou, Y.; Min, J.; Lin, J.; Lee, M.-G. Modeling stress anisotropy, strength differential, and anisotropic hardening by coupling quadratic and stress-invariant-based yield functions under non-associated flow rule. *Mech Mater.* **2022**, *174*, 104458. [CrossRef]
26. Spitzig, W.A.; Sober, R.J.; Richmond, O. Pressure dependence of yielding and associated volume expansion in tempered martensite. *Acta. Metall.* **1975**, *23*, 885–893. [CrossRef]
27. Spitzig, W.A.; Richmond, O. The effect of pressure on the flow stress of metals. *Acta. Metall.* **1984**, *32*, 457–463. [CrossRef]
28. Stoughton, T.B.; Yoon, J.W. A pressure-sensitive yield criterion under a non-associated flow rule for sheet metal forming. *Int. J. Plast.* **2004**, *20*, 705–731. [CrossRef]
29. Cazacu, O.; Barlat, F. A criterion for description of anisotropy and yield differential effects in pressure-insensitive metals. *Int. J. Plast.* **2004**, *20*, 2027–2045. [CrossRef]
30. Cazacu, O.; Plunkett, B.; Barlat, F. Orthotropic yield criterion for hexagonal closed packed metals. *Int. J. Plast.* **2006**, *22*, 1171–1194. [CrossRef]
31. Plunkett, B.; Cazacu, O.; Barlat, F. Orthotropic yield criteria for description of the anisotropy in tension and compression of sheet metals. *Int. J. Plast.* **2008**, *24*, 847–866. [CrossRef]
32. Li, Z.; Yang, H.; Liu, J.; Liu, F. An improved yield criterion characterizing the anisotropic and tension-compression asymmetric behavior of magnesium alloy. *J. Magnes. Alloys* **2022**, *10*, 569–584. [CrossRef]
33. Khan, A.S.; Yu, S.; Liu, H. Deformation induced anisotropic responses of Ti-6Al-4V alloy Part II: A strain rate and temperature dependent anisotropic yield criterion. *Int. J. Plast.* **2012**, *38*, 14–26. [CrossRef]
34. Lou, Y.; Huh, H.; Yoon, J.W. Consideration of strength differential effect in sheet metals with symmetric yield functions. *Int. J. Mech. Sci.* **2013**, *66*, 214–223. [CrossRef]
35. Yoon, J.W.; Lou, Y.; Yoon, J.; Glazoff, M.V. Asymmetric yield function based on the stress invariants for pressure sensitive metals. *Int. J. Plast.* **2014**, *56*, 184–202. [CrossRef]
36. Lou, Y.; Zhang, S.; Yoon, J.W. Strength modeling of sheet metals from shear to plane strain tension. *Int. J. Plast.* **2020**, *134*, 102813. [CrossRef]

37. Hou, Y.; Min, J.; Guo, N.; Shen, Y.; Lin, J. Evolving asymmetric yield surfaces of quenching and partitioning steels: Characterization and modeling. *J. Mater. Process. Technol.* **2021**, *290*, 116979. [CrossRef]
38. Hou, Y.; Min, J.; Stoughton, T.B.; Lin, J.; Carsley, J.E.; Carlson, B.E. A non-quadratic pressure-sensitive constitutive model under non-associated flow rule with anisotropic hardening: Modeling and validation. *Int. J. Plast.* **2020**, *135*, 102808. [CrossRef]
39. Hou, Y.; Min, J.; Lee, M.-G. Non-associated and non-quadratic characteristics in plastic anisotropy of automotive lightweight sheet metals. *Automot. Innov.* **2023**, *6*, 364–378. [CrossRef]
40. Hu, Q.; Yoon, J.W. Analytical description of an asymmetric yield function (Yoon2014) by considering anisotropic hardening under non-associated flow rule. *Int. J. Plast.* **2021**, *140*, 102978. [CrossRef]
41. Hu, Q.; Chen, J.; Yoon, J.W. A new asymmetric yield criterion based on Yld2000-2d under both associated and non-associated flow rules: Modeling and validation. *Mech. Mater.* **2022**, *167*, 104245. [CrossRef]
42. Lou, Y.; Yoon, J.W. Lode-dependent anisotropic-asymmetric yield function for isotropic and anisotropic hardening of pressure-insensitive materials. Part I: Quadratic function under non-associated flow rule. *Int. J. Plast.* **2023**, *166*, 103647. [CrossRef]
43. Vegter, H.; Abspoel, M.; Mulder, J. A plane stress yield surface using Bézier curve interpolation in two directions. *IOP Conf. Ser. Mater. Sci. Eng.* **2019**, *651*, 012052. [CrossRef]
44. Du, K.; Huang, S.; Shi, M.; Li, L.; Huang, H.; Zhang, S.; Zheng, W.; Yuan, X. Effects of biaxial tensile mechanical properties and non-integer exponent on description accuracy of anisotropic yield behavior. *Mater. Des.* **2021**, *212*, 110210. [CrossRef]
45. Hou, Y.; Du, K.; El-Aty, A.A.; Lee, M.-G.; Min, J. Plastic anisotropy of sheet metals under plane strain loading: A novel non-associated constitutive model based on fourth-order polynomial functions. *Mater. Des.* **2022**, *223*, 111187. [CrossRef]
46. Zhang, K.; He, Z.; Zheng, K.; Yuan, S. Experimental verification of anisotropic constitutive models under tension-tension and tension-compression stress states. *Int. J. Mech. Sci.* **2020**, *178*, 105618. [CrossRef]
47. He, Z.; Zhang, K.; Zhu, H.; Lin, Y.; Fu, M.W.; Yuan, S. An anisotropic constitutive model for forming of aluminum tubes under both biaxial tension and pure shear stress states. *Int. J. Plast.* **2022**, *152*, 103259. [CrossRef]
48. Hu, Q.; Yoon, J.W.; Chen, J. Analytically described polynomial yield criterion by considering both plane strain and pure shear states. *Int. J. Plast.* **2023**, *162*, 103514. [CrossRef]
49. Hou, Y.; Min, J.; El-Aty, A.A.; Han, H.N.; Lee, M.-G. A new anisotropic-asymmetric yield criterion covering wider stress states in sheet metal forming. *Int. J. Plast.* **2023**, *166*, 103653. [CrossRef]
50. Du, K.; Huang, S.; Hou, Y.; Wang, H.; Wang, Y.; Zheng, W.; Yuan, X. Characterization of the asymmetric evolving yield and flow of 6016-T4 aluminum alloy and DP490 steel. *J. Mater. Sci. Technol.* **2023**, *133*, 209–229. [CrossRef]
51. Du, K.; Huang, S.; Li, X.; Wang, H.; Zheng, W.; Yuan, X. Evolution of yield behavior for AA6016-T4 and DP490—Towards a systematic evaluation strategy for material models. *Int. J. Plast.* **2022**, *154*, 103302. [CrossRef]
52. Du, K.; Ren, Y.; Hou, Y.; Li, X.; Chen, S.; Sun, L.; Zuo, X.; Yuan, X. Parameters identification strategy of yield criterion for accurately predicting anisotropic behavior under near plane strain loading. *Chin. Mech. Eng.* **2023**, *34*, 2381–2394. [CrossRef]
53. Stoughton, T.B.; Yoon, J.W. Anisotropic hardening and non-associated flow in proportional loading of sheet metals. *Int. J. Plast.* **2009**, *25*, 1777–1817. [CrossRef]
54. Du, K.; Huang, S.; Wang, H.; Yu, F.; Pan, L.; Huang, H.; Zheng, W.; Yuan, X. Effect of different yield criteria and material parameter identification methods on the description accuracy of the anisotropic behavior of 5182-O aluminum alloy. *J. Mater. Eng. Perform.* **2022**, *31*, 1077–1095. [CrossRef]

materials

MDPI

Article

Anisotropic Hardening and Plastic Evolution Characterization on the Pressure-Coupled Drucker Yield Function of ZK61M Magnesium Alloy

Jianwei You [1,*], Jiangnan Liu [1], Can Zhou [2], Wei Gao [1] and Yuhong Yao [1,*]

[1] School of Materials Science and Chemical Engineering, Xi'an Technological University, Xi'an 710021, China; liujiangnan@xpu.edu.cn (J.L.); gaowei@xatu.edu.cn (W.G.)

[2] School of Mechanical Engineering, Xi'an Jiaotong University, Xi'an 710021, China; 779560229@stu.xjtu.edu.cn

* Correspondence: youjianwei@st.xatu.edu.cn (J.Y.); yaoyuhong@xatu.edu.cn (Y.Y.)

Abstract: This paper studies the plastic behavior of the ZK61M magnesium alloy through a combination method of experiments and theoretical models. Based on a dog-bone specimen under different loading directions, mechanical tests under uniaxial tension were carried out, and the hardening behavior was characterized by the Swift–Voce hardening law. The von Mises yield function and the pressure-coupled Drucker yield function were used to predict the load–displacement curves of the ZK61M magnesium alloy under various conditions, respectively, where the material parameters were calibrated by using inverse engineering. The experimental results show that the hardening behavior of the ZK61M magnesium alloy has obvious anisotropy, but the effect of the stress state is more important on the strain hardening behavior of the alloy. Compared with the von Mises yield function, the pressure-coupled Drucker yield function is more accurate when characterizing the plastic behavior and strain hardening in different stress states of shear, uniaxial tension, and plane strain tension for the ZK61M alloy.

Keywords: ZK61M magnesium alloy; hardening behavior; anisotropy; yield function; plastic evolution

Citation: You, J.; Liu, J.; Zhou, C.; Gao, W.; Yao, Y. Anisotropic Hardening and Plastic Evolution Characterization on the Pressure-Coupled Drucker Yield Function of ZK61M Magnesium Alloy. *Materials* **2024**, *17*, 1150. https://doi.org/10.3390/ma17051150

Academic Editors: Madhav Baral and Charles Lu

Received: 22 December 2023
Revised: 8 February 2024
Accepted: 19 February 2024
Published: 1 March 2024

1. Introduction

ZK61M magnesium alloy is extensively utilized in the domestic military industry to manufacture components that are subjected to significant mechanical stress, such as aircraft skins, panels, and interior components, as well as complex die forging parts. The incorporation of a zirconium element into magnesium alloys offers numerous benefits, for example, enhanced mechanical properties and robust overall performance. As a hexagonal dense packing (HCP) metal, the mechanical behavior of the ZK61M magnesium alloy is very complex. Selecting an appropriate yield function based on the hardening characteristics of the ZK61M magnesium alloy is advantageous in enhancing the dependability of numerical simulations for steel forming and fulfilling the requirements of real-world applications.

To study the yield behavior of metals, scholars have carried out a lot of research. The Hill48 yield criterion [1] was the first attempt to analyze anisotropy and is now one of the most widely used quadratic yield functions, providing accurate prediction of these hardening curves under uniaxial tension along different directions and under equibiaxial tension. However, due to the quadratic form of the Hill48 function, it is impossible to distinguish the difference in the yield surface of metals with different crystal structures. Hosford et al. [2,3] used an exponential form to combine the yield surfaces of body-centered cubic (BCC) and face-centered cubic (FCC) metals. Barlat et al. [4] extended isotropic functions to anisotropy by introducing anisotropic coefficients through linear transformation tensors. Barlat et al. [5,6] proposed the Yld2000-2d and Yld2004-18p yield criteria to characterize the anisotropic behavior of metals under plane and spatial stress states, respectively. Cazacu et al. [7] introduced an orthogonal anisotropic yield criterion in the form of the principal value of stress deviation to capture the anisotropy and tension–compression asymmetry. Some yield functions were established based on stress invariants. Yoon et al. [8] proposed an asymmetric yield function with the first invariant to accurately characterize the tension–compression asymmetry. Lou and Yoon [9] effectively distinguished the anisotropic behavior of metals with BCC and FCC structures by correcting for the effect of the third invariant, which was used to model the deep drawing of the 6K21 aluminum alloy [10]. Lou et al. [11] proposed a pressure-coupled Drucker yield function to simulate the strength of sheet metal between shear and plane strain, which takes into account the influence of three stress invariants: pressure sensitivity, Lode dependence, and strength difference.

In recent years, some anisotropic hardening models were proposed to capture the evolution of the yield surface through the analytical description of the anisotropy coefficient. Stoughton and Yoon [12] established the S-Y2009 yield function, and the numerical result was consistent with the hardening behavior in reality. Lee et al. [13] proposed a CQN model by the coupling of the above quadratic yield function and the Hershey–Hosford yield function. Chen et al. [14] coupled the S-Y2009 function with the Drucker function to explain the difference in the yield behavior of metals with BCC and FCC structures, which was used for the plastic evolution characterization for 304 stainless steel [15]. Hou et al. [16] proposed an anisotropic hardening model by coupling the asymmetric Hill48 function with the isotropic stress-invariant yield function. Hu et al. [17] coupled the yield criterion of fourth-order polynomials with the non-quadratic yield function under the associated flow rule to analytically describe the evolution process of anisotropic yield behavior. Wu et al. [18] established a Cazacu2004 yield function [19] with the temperature variable to describe the tension–compression asymmetry of a Mg-Gd-Y alloy under various temperatures. Lou et al. [20] proposed a stress invariant-based yield function to accurately simulate the strain hardening behavior of metals with BCC, FCC, and HCP structures under different stress states, and the convexity was analyzed by a GINCA method [21]. Hou et al. [22] proposed a constitutive model to accurately describe the anisotropy behavior of sheet metal. Zhang and Lou [23] characterized the evolving plastic behavior of BCC and FCC metals by coupling the enhanced pressure-coupled Drucker yield function and the S-Y2009 model. The yield criterion proposed by Hu et al. considers the pure shear stress along various directions to simultaneously predict the mechanical behavior under both pure shear and uniaxial tension [24]. Lou and Yoon [25] constructed an anisotropic asymmetric hardening model by coupling two Hill48 functions with Lode correlation weight functions.

Recently, Bassini et al. [26] investigated the effect of cold rolling on the microstructural and mechanical properties of a dual-phase steel. Baral et al. [27] analyzed plastic evolution and its modeling of an Al-Si-Mg die-cast alloy. Ha et al. [28] characterized the plastic anisotropy of annealed, commercially pure aluminum through experiments and modeling. Allen et al. [29] studied anisotropic hardening and texture evolution due to dislocation transmutation in twining using crystal plasticity modeling. Imandoust et al. [30] reviewed the effect of rare-earth elements on texture evolution and anisotropic hardening during the processing of magnesium alloys. Ha et al. [31] investigated the plastic anisotropy of a bake-hardening AA6013 aluminum sheet. Knysh and Korkolis [32] identified the post-necking hardening response of rate- and temperature-dependent metals. Proust et al. [33] modeled the texture, twinning, and hardening evolution during deformation of hexagonal materials. Dick and Korkolis [34] investigated the anisotropic hardening of thin-walled tubes using a new experimental method for combined tension and shear loading. Generally speaking, the more material parameters in the yield function, the more accurately the anisotropic hardening behavior can be characterized. Previous research mostly considered the anisotropic and strength differential effect, but did not consider the effect of stress states on the flow curves. The stress-state effect on the different strain hardening behaviors and their modeling is the focus of this study.

This investigation aimed to characterize the distinct hardening behavior of ZK61M in different stress states, from shear to plane strain tension, using experiments and analytical and numerical modeling. The mechanical experiments were conducted under various stress states and loading directions. The hardening behavior was analyzed by using the Swift–Voce hardening equation. The inverse engineering method was used to optimize the prediction result. A comparative analysis was conducted to assess the precision of the pressure-coupled Drucker yield function and the von Mises yield function in characterizing the plastic deformation behavior of the ZK61M magnesium alloy.

2. Experiment

Materials and Experiments

The parent material of the ZK61M magnesium alloy is a sheet metal with 2 mm thickness, which was prepared by using the rolling process. The size of these experimental specimens is shown in Figure 1, including dog bone, R20 notched, R5 notched, and shear specimens. To investigate the anisotropy of the ZK61M magnesium alloy, four specimens were cut along three different angles, 0°, 45°, and 90°, namely, RD, DD, and TD. Three replicates were conducted for each specimen in order to validate the precision and reliability of the collected data.

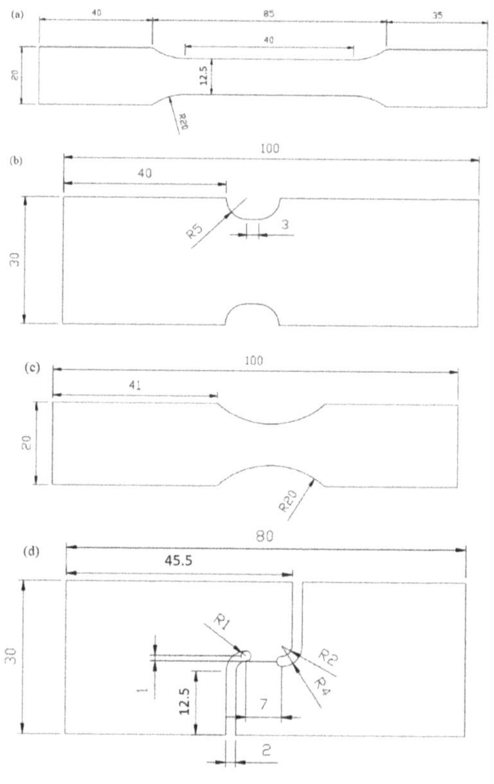

Figure 1. Size structure of (**a**) dog-bone specimen; (**b**) R20 notched specimen; (**c**) R5 notched specimen; and (**d**) shear specimen. (Unit: mm).

The experimental equipment is shown in Figure 2. A universal mechanical experimental machine manufactured by WANCE, Shenzhen, China and XTOP digital image correlation system in Xi'an, China were utilized to conduct tests and obtain the deformation process, respectively. The experimental apparatus is capable of preserving the symmetry and stability of clamping, hence enabling the acquisition of precise experimental load and displacement measurements. In the experimental procedure, it is important to determine the tensile velocity of these specimens. Given the quasi-static strain rate (0.001/s), the tensile velocity of the dog bone, R20 notched, R5 notched, and shear specimens was determined to be 3.6, 0.5, 0.5, and 0.5 mm/min, respectively. A 30 mm gauge was established for the dog bone and R20 notched specimens, and a 20 mm gauge was created for the R5 notched and shear specimens, as shown in Figure 1. The load–displacement curves of all tests and the longitudinal strain–width strain curves of the dog-bone specimen were obtained by the analysis of XTOP and GOM systems, as presented in Figures 3 and 4.

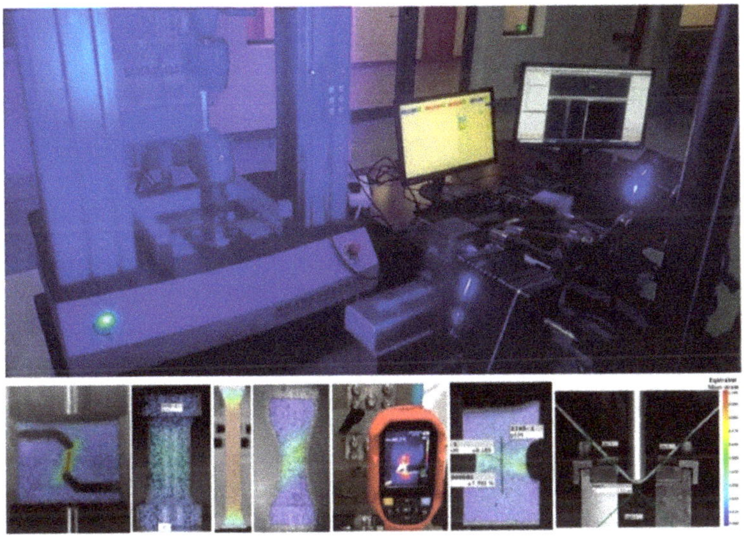

Figure 2. Universal mechanical experimental machine and XTOP digital image correlation system.

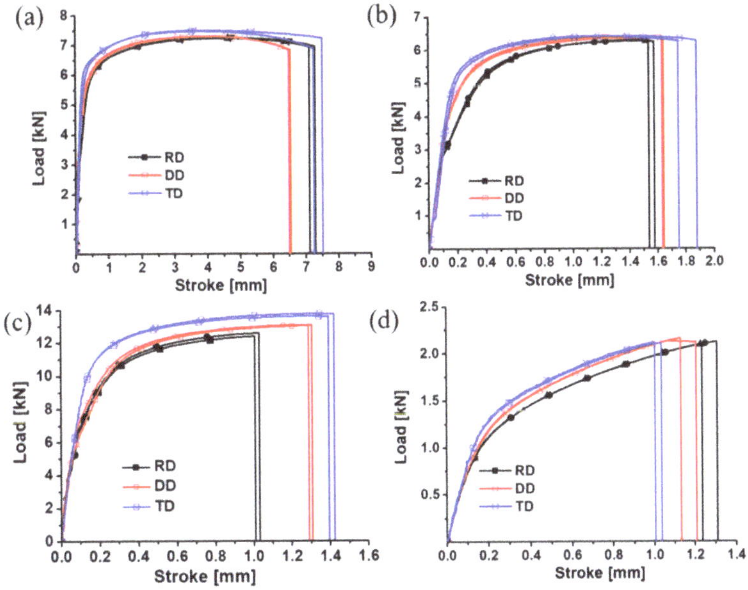

Figure 3. Load–displacement curves of four different samples of ZK61M: (**a**) dog-bone specimen; (**b**) R20 notched specimen; (**c**) R5 notched specimen; (**d**) shear specimen.

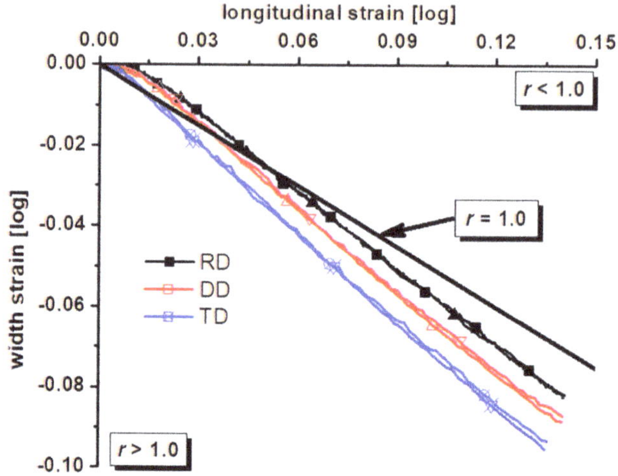

Figure 4. Longitudinal strain–width strain curves of the dog-bone specimen for the ZK61M magnesium alloy.

From the load–displacement curves, the majority of the specimens have a notable degree of repeatability. It is evident that the load–displacement curve of each specimen exhibits a significant anisotropy. The dog-bone specimen exhibits a maximum strength difference of approximately 3.8%. Similarly, the R20 notched specimen demonstrates a maximum strength difference of approximately 2.2%, while the R5 notched specimen displays a maximum strength difference of approximately 9.5%. The shear specimen exhibits a maximum strength difference of approximately 3.9%. The conspicuousness of the maximum strength disparity among the four types of samples is evident. This indicates that the ZK61M magnesium alloy exhibits clear anisotropy under uniaxial tension, plane strain tension, and shear strength. From the longitudinal strain–width strain curves, the r-value also presents some anisotropy. The result indicates that the ZK61M magnesium alloy possesses the anisotropic plastic flow phenomenon.

The experimental true stress–true plastic strain curves of the dog-bone specimen along RD are shown in Figure 5 and are described by the Swift and Voce hardening laws, respectively. The parameters in the Swift and Voce hardening laws are calibrated by the simplex method in the Origin 8.5.1 software. The calibrated coefficients are presented in Table 1.

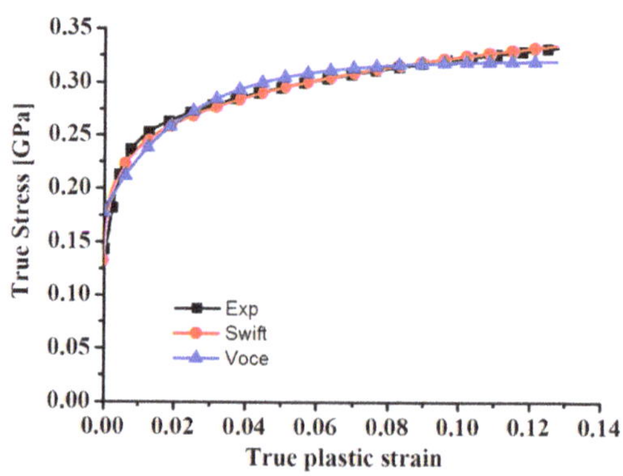

Figure 5. Comparison of the fitting hardening curves based on Swift and Voce hardening laws with the experimental true stress–true plastic strain curve of the dog-bone specimen along RD.

Table 1. Swift and Voce hardening coefficients of the dog-bone specimen along RD for ZK61M magnesium alloy.

Swift Coefficients in $\sigma_{Swift}(\varepsilon) = k(e_0 + \varepsilon)^n$			Voce Coefficients in $\sigma_{Voce}(\varepsilon) = A - (A - B)^{-C\varepsilon}$		
k	e_0	n	A	B	C
0.443 GPa	1.406×10^{-4}	0.136	0.321 GPa	0.1777 GPa	43.079

3. Model and Method

3.1. Pressure-Coupled Drucker Yield Function

The precise modeling of sheet metal strength between shear and plane strain is of significance in ensuring the reliability of the analysis of the sheet metal forming process. In the case of nearly isotropic metals, the normalized third stress invariant exhibits equivalence between uniaxial tension and plane strain tension. It is imperative to consider the influence of pressure in order to accurately model the disparity in strength between shear and plane strain tension. Therefore, by coupling the pressure effect with the Drucker function, the following function is proposed:

$$f(\sigma_{ij}) = a[3b\eta + \frac{1}{3}(27 - 4c\xi^2)^{1/6}]\bar{\sigma}_{VM} \quad (1)$$

where $\bar{\sigma}_{VM}$ is the von Mises equivalent stress, η is the stress triaxiality, ξ is the normalized Lode parameter, and a, b, and c are the material parameters. Here, parameter a is calculated as

$$a = \frac{1}{b + \frac{1}{3}(27 - 4c)^{1/6}} \quad (2)$$

The value of parameter a depends on the stress–strain curve calculated by the experiment. Based on the conclusion that the material strength is linearly affected by the hydrostatic pressure, parameter b is introduced as a pressure-sensitive parameter, and parameter c is used to simulate the dependence of the yield on the third invariant. The value of parameter c ranges from −3.375 to 2.25 to guarantee the convexity of the above function, which is the same as that in the Drucker function. The effect of the third stress invariant on the yield surface is shown in Figure 6. When $c = 0$, the third stress invariant has no effect on the yield, and the yield surface is reduced to the von Mises yield surface. When $c < 0$, the Drucker yield surface around the normalized plane strain tension is greater than the von Mises yield surface, which means that higher stress is required to activate the plastic deformation around the normalized plane strain tension. On the contrary when $c > 0$, the Drucker yield surface is smaller than the von Mises yield surface, indicating that the stress required for the material to plastically deform under plane strain tension is lower.

The above typical yield surface is shown in Figure 7, where the stress–strain curve is assumed to be measured by the uniaxial tensile test. Parameter b is equal to 0.05 to consider the pressure sensitivity, and parameter c is equal to 2 to couple the third stress invariant effect. Here, parameter a is equal to 1.6821, which is calculated from Equation (2).

The pressure sensitivity of Equation (1) enables it to model the difference between shear and plane strain tension strength, while the Lode dependence gives the yield function flexibility to characterize the different strength ratios of uniaxial and plane strain tension. Therefore, the pressure-coupled Drucker yield function can accurately predict sheet metal strength under shear, uniaxial tension, and plane strain tension.

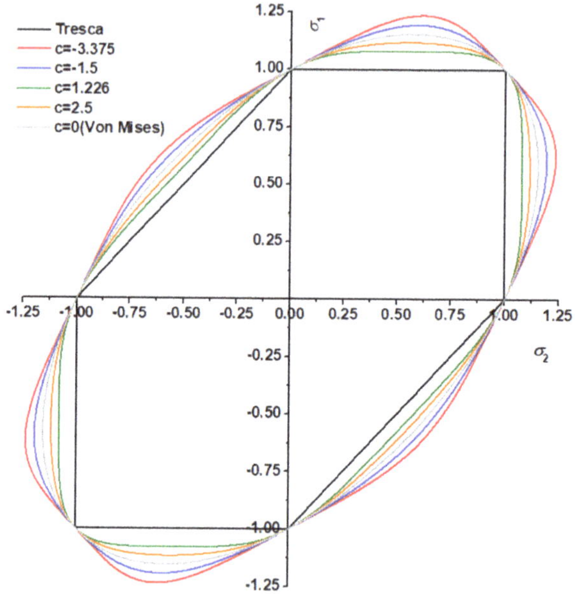

Figure 6. Effect of the third stress invariant on the yield surface under biaxial loading.

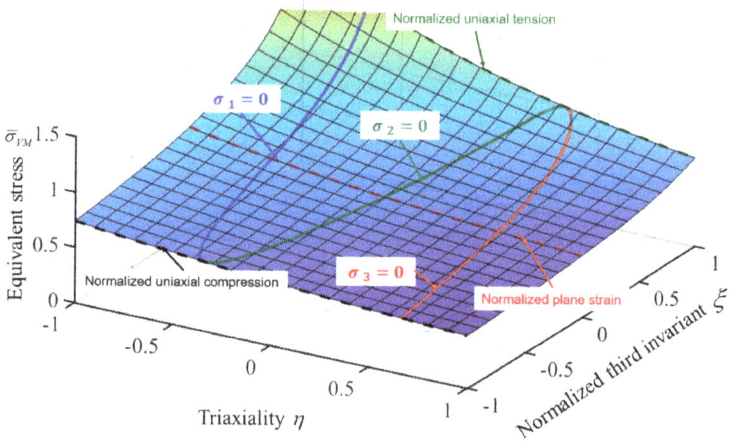

Figure 7. A typical yield surface of the pressure-coupled Drucker yield function with b = 0.05 and c = 2 in $(\eta, \xi, \bar{\sigma}_{VM})$ space.

3.2. Inverse Engineering

The flow chart of inverse engineering is depicted in Figure 8, which is applied to references [11,20,35–37]. Initially, an appropriate value is assigned as the initial value for the optimization parameter. Subsequently, the parameter is numerically simulated, and the numerical data are extracted and compared to the corresponding experimental value. The discrepancy between the numerical and the experimental values is quantified by utilizing the error function to assess the precision of the prediction. To obtain the most optimal optimization settings, it is important to minimize the error function. If the value of the error function exceeds the expected value, a simulation is conducted to predict the next optimization parameter. The updated prediction value is then compared to the experimental value once again, ensuring that the error function is less than or equal to the expected value. This process is repeated until the optimal parameter is obtained.

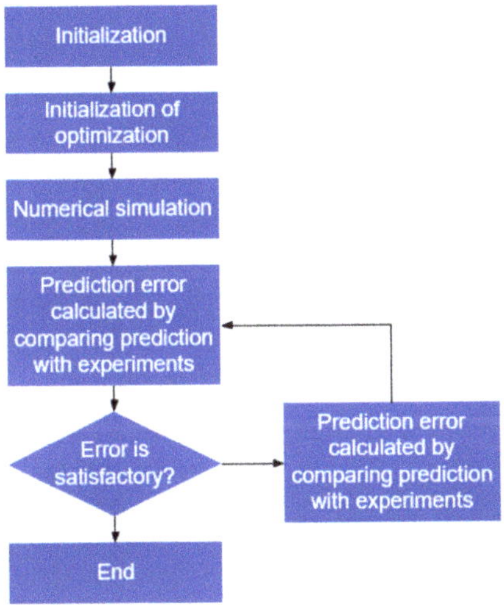

Figure 8. Flow chart of the inverse engineering method.

4. Comparison of the Predicted Load–Displacement Curves between Different Yield Functions

4.1. Mesh Sensitivity Analysis

The load–displacement curve of R20 notched specimens was numerically predicted by using ABAQUS/Explicit 6.14 with different element sizes to study the element size effect. The 1/8 finite element model was used to study the influence of six different element sizes on the numerical result of the R20 notched specimen, as shown in Figure 9. Three symmetric boundary constraints were applied to the numerical model accordingly because the 1/8 FE model was adopted to reduce computation time. The element used was a type of brick element with reduced integration called C3D8R. The element numbers were 220, 440, 660, 880, 1100, and 4400 for the six different meshing results in the figure, respectively. From model 1 to model 5, the number of units in the thickness direction gradually increased from one layer to five layers, and the number of units in each layer was consistent. The mesh in the thickness direction of model 6 was five layers, but the overall mesh size was smaller than that of model 5. Here, the elastic modulus in the material card was set to 45 GPa, the Poisson's ratio was set to 0.33, and the density was set to 1.8×10^{-6} kg/mm^3. The strain hardening of all models was modeled using the Swift–Voce hardening law as

$$\bar{\sigma} = \frac{1}{2}[k(e_0 + \varepsilon)^n + (A - (A - B)exp(- C\varepsilon)] \tag{3}$$

where k, e_0, n, A, B, and C are strain hardening parameters. This can be seen from the coefficients in Table 2. The flow chart of the numerical simulation is referenced in reference [38].

Table 2. Swift–Voce hardening coefficients of the dog-bone specimen along RD.

$\sigma_{Swift-Voce}(\varepsilon) = \alpha\left[k(e_0 + \varepsilon)^n\right] + (1 - \alpha)[A - (A - B)^{-C\varepsilon}]$						
k	e_0	n	A	B	C	α
0.443	1.406×10^{-4}	0.136	0.321	0.178	43.079	0.5

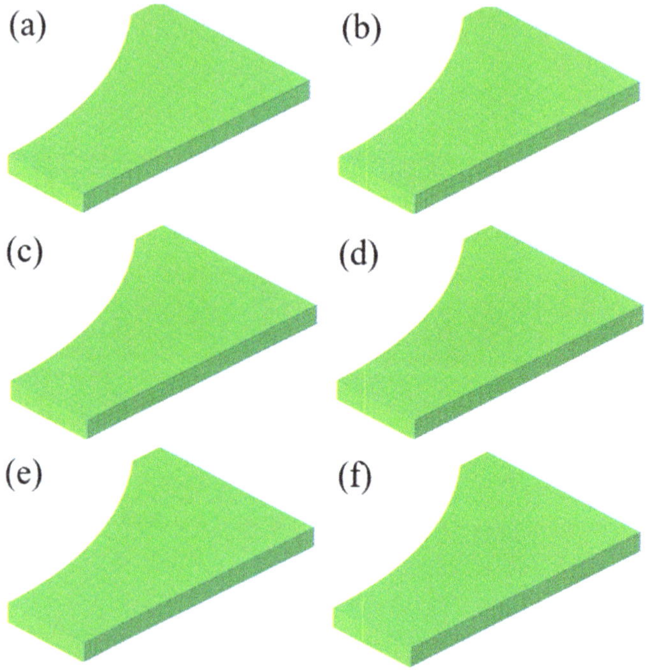

Figure 9. Finite element models with different mesh sizes for R20 notched specimen with different element numbers: (**a**) model #1: 220 elements; (**b**) model #2: 440 elements; (**c**) model #3: 660 elements; (**d**) model #4: 880 elements; (**e**) model #5: 1100 elements; and (**f**) model #6: 4400 elements.

Figure 10 shows the predicted load–displacement curves of the six finite element models with different element sizes in Figure 9. Through a comparison with the experimental result, it is found that the element size has little influence on the numerical results of the R20 notched specimen when it is smaller than 0.5 mm in the edge length of the C3D8R elements. Therefore, numerical models with a similar element size to #1 were adopted for the simulation of all the specimens. In addition, the yield strength in the numerical simulation is different from that in the experiments, as shown in Figure 10. This is because the stress–strain curve is calibrated from the dog-bone specimens for which the stress state is uniaxial tension, while the stress state of the notched R20 specimens is between uniaxial tension and plane strain tension. The elastic regime of the numerical and experimental curves is observed to be different after the force is higher than 3.2 kN. This is because, in experiments, small plastic deformation takes place at the grain boundary, thereby reducing the slope of the force–stroke curves. The change in elastic modulus in the simulation at about 3.2 kN is expected to be caused by a numerical error in the computation of the onset of plastic deformation under small plastic deformation by the return-mapping algorithm. Last, the predicted force–stroke curve at large displacement is observed to decrease, while the experimental results do not drop. This is because the Swift–Voce hardening law calibrated by the stress–strain curve from the dog-bone specimen underestimates the strain hardening behavior under large plastic deformation. Therefore, the strain hardening behavior under large plastic strain is calibrated by the inverse engineering approach in Sections 4.2 and 4.3 of this study.

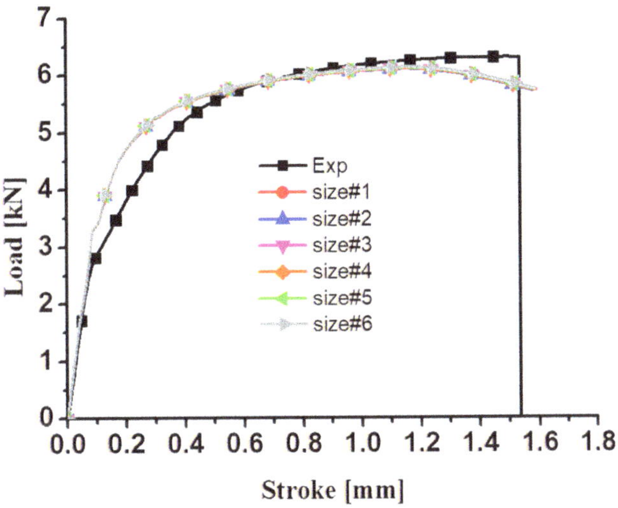

Figure 10. Comparison between the predicted load–displacement curves of six finite element models with different element sizes with the experimental results for the R20 notched specimen.

4.2. Von Mises Yield Function

The numerical simulation of R20 notched, R5 notched, and shear specimens was performed, where the yield function was the von Mises criterion. The numerical load–displacement curves were compared with the experimental results, as presented in Figure 11. Table 2 displays the parameters associated with the anticipated Swift–Voce hardening law. Clearly, the prediction error of the R20 notched specimen is approximately 20% at the elastic deformation stage, while that is approximately 5%. The error magnitude observed during the elastic stage of the R5 notched specimen is around 25%, while the error magnitude during the plastic stage is approximately 2%. The displacement increment is observed to have a significant effect on the error value of the shear specimen. Specifically, the error value initially rises, followed by a subsequent decline. It is noteworthy that the total error value is greater in magnitude. The numerical load exhibits an approximate increase of 4% compared to the experimental load of the shear specimen, and a 6% increase compared to the experimental load of the notched specimen. This indicates that the von Mises function has a tendency to overestimate the plane strain tensile strength of ZK61M by approximately 6% and the shear strength by approximately 4%.

To enhance the prediction performance of the above numerical simulation, the inverse engineering method was used to optimize the Swift–Voce hardening parameters. The preliminary guess of the material parameters in Equation (3) is based on the material parameters calibrated by the true stress–true plastic strain curve of the dog-bone specimen in Table 2. The optimized material parameters are shown in Table 3. Using the optimized hardening parameters, the R20 notched, R5 notched, and shear specimens were numerically simulated, as shown in Figure 12. The comparison shows that the Swift–Voce strain hardening coefficients optimized by the inverse engineering method can more accurately characterize the strain hardening behavior of the alloy. This proves that the inverse engineering method is a promising method to characterize the strain hardening of metals under various loading conditions.

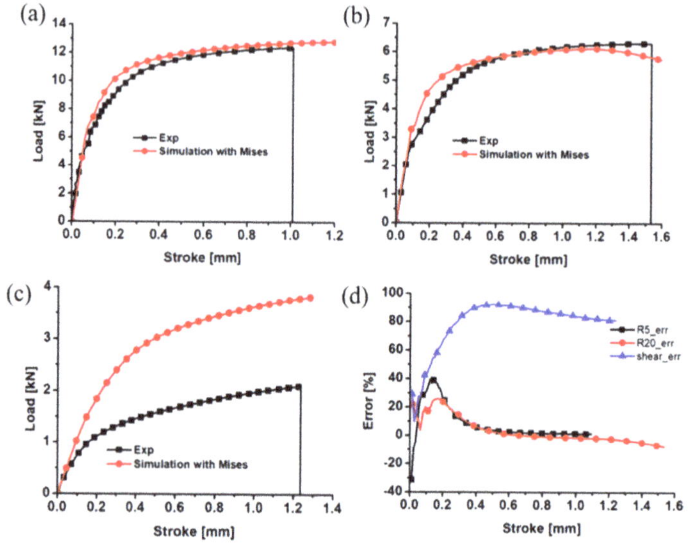

Figure 11. Comparison of the von Mises load–displacement curves and the experimental results for the (**a**) R5 notched specimen; (**b**) R20 notched specimen; (**c**) shear sample; and (**d**) prediction error.

Table 3. Swift–Voce hardening coefficients optimized by the inverse engineering method under the von Mises yield function.

$\sigma_{Swift-Voce}(\varepsilon) = \alpha\left[k(e_0+\varepsilon)^n\right]+(1-\alpha)[A-(A-B)^{-C\varepsilon}]$						
k	e_0	n	A	B	C	α
0.443	1.406×10^{-4}	0.136	0.321	0.178	43.079	0.584

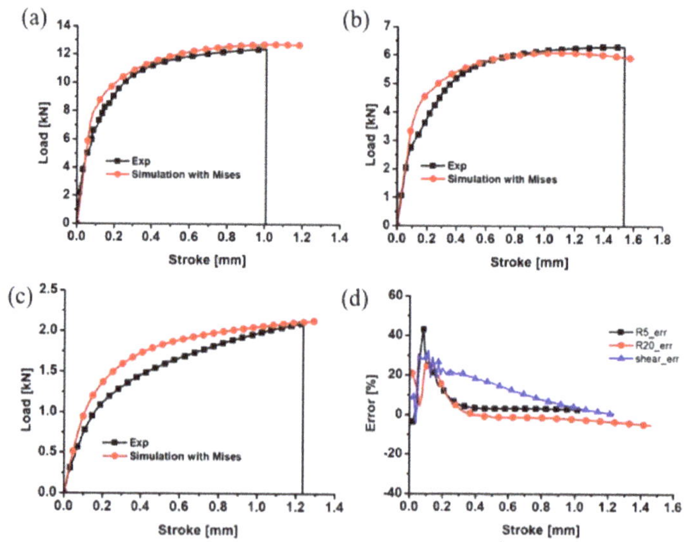

Figure 12. Comparison of the von Mises load–displacement curves with the inverse engineering method and the experimental results for the (**a**) R5 notched specimen; (**b**) R20 notched specimen; (**c**) shear sample; and (**d**) prediction error.

4.3. Pressure-Coupled Drucker Yield Function

The evidence derived from the comparison between Figures 6 and 7 demonstrates that the pressure-coupled Drucker function exhibits a satisfactory level of adaptability in the context of strength modeling for metals subjected to diverse loading situations. To ensure the precision of the numerical simulation, the pressure-coupled Drucker function was employed to estimate the load–displacement curves of material constants b and c, as well as the Swift–Voce hardening law. In order to streamline the procedure, the numerical simulation made the assumption of isotropic hardening. This decision was based on the fact that the tests conducted in this study involved loads that were approximately proportionate, and the changes in stress states were deemed to be negligible. The strain hardening model incorporates the combined model of the Swift–Voce law in Equation (3) and the hardening coefficient specified in Table 3. The inverse engineering method was employed to enhance the prediction performance of the R20 notched, R5 notched, and shear specimens. The simulation employed the associated flow rule and maintained a constant volume during plastic deformation. The optimized Swift–Voce parameters are listed in Table 4, b = −0.0268, and c = 2.2496. The yield surface in $(\eta, \xi, \bar{\sigma}_{VM})$ space is shown in Figure 13.

Table 4. Swift–Voce hardening calibrated by the inverse engineering method under the pressure-coupled Drucker yield function.

$\sigma_{Swift-Voce}(\varepsilon) = \alpha\left[k(e_0+\varepsilon)^n\right] + (1-\alpha)[A-(A-B)^{-C\varepsilon}]$						
k	e_0	n	A	B	C	α
0.443	1.406×10^{-4}	0.136	0.321	0.178	43.079	1.608

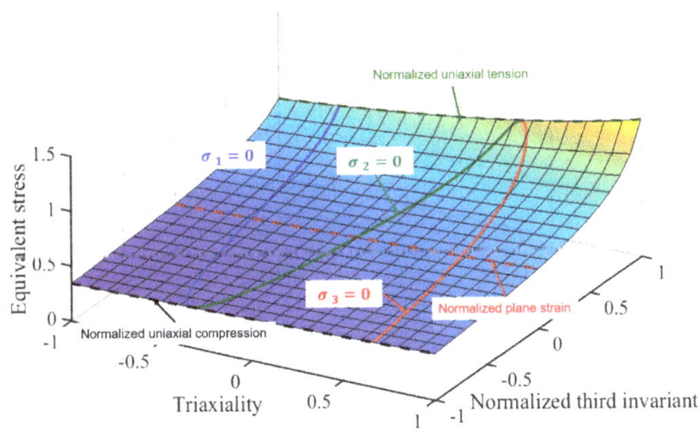

Figure 13. The pressure-coupled Drucker yield surface in $(\eta, \xi, \bar{\sigma}_{VM})$ space.

In Figure 14a, a comparison is made between the numerical load–displacement curve and the experimental curve. The result indicates that the pressure-coupled Drucker function accurately predicts the load–displacement curve of the R5 notched specimen prior to fracture. However, when comparing the prediction performance of the R5 notched specimen using the pressure-coupled Drucker yield function to that of the von Mises yield function, no advantage is observed.

For the prediction case of the R20 notched specimen in Figure 14b, the numerical load of the pressure-coupled Drucker function is higher than the experimental value, but it reasonably predicts the strength at the beginning of yield. After yield, the strength predicted by the pressure-coupled Drucker function is still about 3% lower than the experimental result. The pressure-coupled Drucker function can predict the trend more accurately than the von Mises function.

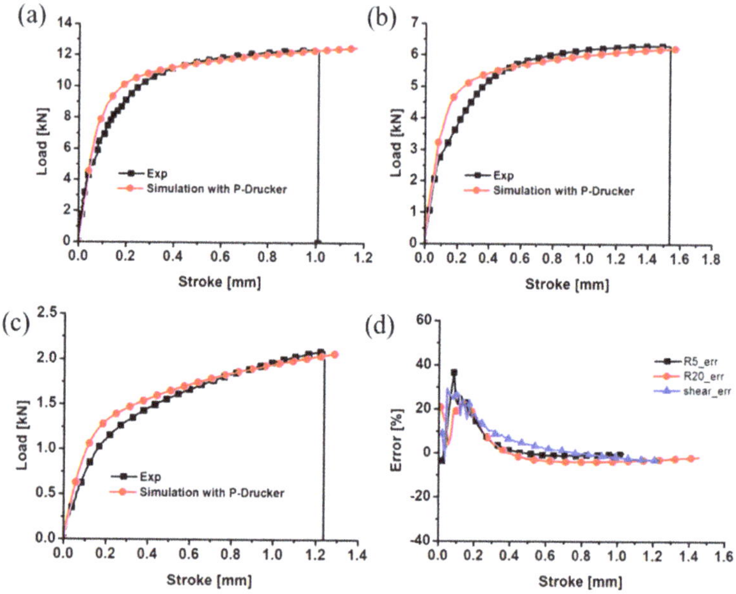

Figure 14. Comparison of the pressure-coupled Drucker load–displacement curves and the experimental results for the (**a**) R5 notched specimen; (**b**) R20 notched specimen; (**c**) shear sample; and (**d**) prediction error.

For the prediction case of the shear specimen in Figure 14c, the pressure-coupled Drucker function reasonably describes the trend from yield to fracture, and the predicted error value is within 3%. In addition, the maximum load predicted by the pressure-coupled Drucker function is basically consistent with the experimental value. Compared with the von Mises yield function, the pressure-coupled Drucker function was more accurate.

5. Conclusions

In this paper, the plastic behavior of ZK61M is characterized from shear to plane strain tension using experiments, analytical modeling, and numerical simulation. The result shows that the plastic behavior of the metal is very complicated and cannot be properly modeled by the von Mises yield function. The pDrucker yield function can significantly improve the prediction accuracy of the load–stroke curves for shear, notched, and plane strain tension specimens. The pDrucker yield function and the combined Swift–Voce hardening law were calibrated by the inverse engineering approach and the advanced calibration approach further reduced the prediction error to less than 10% for most parts of the stroke before fracture for these three specimens. Numerical simulation showed that there is little effect of element size on the force–stroke curves predicted in numerical simulation. According to the result, the pDrucker function and the Swift–Voce hardening law are suggested to model complicated plastic behaviors of metals under wide stress states from shear to plane strain tension. The inverse engineering approach is recommended to calibrate parameters in the constitutive model to improve the prediction accuracy of numerical simulation.

Author Contributions: J.Y.: Writing—original draft, Investigation, Data curation, Formal analysis; J.L.: Supervision, Writing—review and editing; C.Z.: Software, Visualization; W.G.: Data curation, Investigation; Y.Y.: Conceptualization, Project administration. All authors have read and agreed to the published version of the manuscript.

Funding: This research was funded by the National Key Research and Development Program of China (Grant No. 2022YFE0122900), the National Natural Science Foundation of China (Grant No. 52075423 & U2141214), and the Fundamental Research Funds for the Central Universities (Grant No. xtr012019004 & zrzd2017027).

Institutional Review Board Statement: Not applicable.

Informed Consent Statement: Not applicable.

Data Availability Statement: Data are contained within the article.

Conflicts of Interest: The authors declare no conflicts of interest.

References

1. Hill, R. A theory of the yielding and plastic flow of anisotropic metals. *Proc. R. Soc. A-Math Phys. Eng. Sci.* **1948**, *193*, 281–297.
2. Logan, R.W.; Hosford, W.F. Upper-bound anisotropic yield locus calculations assuming <111>-pencil glide. *Int. J. Mech. Sci.* **1980**, *22*, 419–430.
3. Hosford, W.F. Comments on anisotropic yield criteria. *Int. J. Mech. Sci.* **1985**, *27*, 423–427. [CrossRef]
4. Barlat, F.; Lege, D.J.; Brem, J.C. A six-component yield function for anisotropic materials. *Int. J. Plast.* **1991**, *7*, 693–712. [CrossRef]
5. Barlat, F.; Brem, J.C.; Yoon, J.W.; Chung, K.; Dick, R.E.; Lege, D.J.; Pourboghrat, F.; Choi, S.H.; Chu, E. Plane stress yield function for aluminum alloy sheets-part 1: Theory. *Int. J. Plast.* **2003**, *19*, 1297–1319. [CrossRef]
6. Barlat, F.; Aretz, H.; Yoon, J.W.; Karabin, M.E.; Brem, J.C.; Dick, R.E. Linear transfomation-based anisotropic yield functions. *Int. J. Plast.* **2005**, *21*, 1009–1039. [CrossRef]
7. Cazacu, O.; Plunkett, B.; Barlat, F. Orthotropic yield criterion for hexagonal closed packed metals. *Int. J. Plast.* **2006**, *22*, 1171–1194. [CrossRef]
8. Yoon, J.W.; Lou, Y.S.; Yoon, J.; Glazoff, M.V. Asymmetric yield function based on the stress invariants for pressure sensitive metals. *Int. J. Plast.* **2014**, *56*, 184–202. [CrossRef]
9. Lou, Y.S.; Yoon, J.W. Anisotropic yield function based on stress invariants for BCC and FCC metals and its extension to ductile fracture criterion. *Int. J. Plast.* **2018**, *101*, 125–155. [CrossRef]
10. Wang, S.C.; Shang, H.C.; Zhang, Z.; Lou, Y. Multi-scale numerical investigation of deep drawing of 6K21 aluminum alloy by crystal plasticity and a stress-invariant based anisotropic yield function under non-associated flow rule. *J. Manuf. Process.* **2023**, *102*, 736–755. [CrossRef]
11. Lou, Y.S.; Zhang, S.; Yoon, J.W. Strength modeling of sheet metals from shear to plane strain tension. *Int. J. Plast.* **2020**, *134*, 102813. [CrossRef]
12. Stoughton, T.B.; Yoon, J.W. Anisotropic hardening and non-associated flow in proportional loading of sheet metals. *Int. J. Plast.* **2009**, *25*, 1777–1817. [CrossRef]
13. Lee, E.-H.; Stoughton, T.B.; Yoon, J.W. A yield criterion through coupling of quadratic and non-quadratic functions for anisotropic hardening with non-associated flow rule. *Int. J. Plast.* **2017**, *99*, 120–143. [CrossRef]
14. Chen, Z.; Wang, Y.; Lou, Y.S. User-friendly anisotropic hardening function with non-associated flow rule under the proportional loadings for BCC and FCC metals. *Mech. Mater.* **2022**, *165*, 104190. [CrossRef]
15. Gao, X.; Wang, S.; XU, Z.; Zhou, J.; Wan, X.; Rayhan, H.M.A.; Lou, Y.S. Plastic Evolution Characterization for 304 Stainless Steel by CQN ChenModel under the Proportional Loading. *Materials* **2023**, *16*, 6828. [CrossRef]
16. Hou, Y.; Min, J.Y.; Lin, J.P.; Lee, M.G. Modeling stress anisotropy, strength differential, and anisotropic hardening by coupling quadratic and stress-invariant-based yield functions under non-associated flow rule. *Mech. Mater.* **2022**, *174*, 104458. [CrossRef]
17. Hu, Q.; Yoon, J.W.; Manopulo, N.; Hora, P. A coupled yield criterion for anisotropic hardening with analytical description under associated flow rule: Modeling and validation. *Int. J. Plast.* **2021**, *36*, 102882. [CrossRef]
18. Wu, P.; Lou, Y.S.; Chen, Q.; Ning, H. Modeling of temperature- and stress state-dependent yield and fracture behaviors for Mg-Gd-Y alloy. *Int. J. Mech. Sci.* **2022**, *229*, 107506. [CrossRef]
19. Cazacu, O.; Barlat, F. A criterion for description of anisotropy and yield differential effects in pressure-insensitive metals. *Int. J. Plast.* **2004**, *20*, 2027–2045. [CrossRef]
20. Lou, Y.S.; Zhang, C.; Zhang, S.; Yoon, J.W. A general yield function with differential and anisotropic hardening for strength modelling under various stress states with non-associated flow rule. *Int. J. Plast.* **2022**, *158*, 103414. [CrossRef]
21. Lou, Y.S.; Zhang, C.; Wu, P.; Yoon, J.W. New geometry-inspired numerical convex analysis method for yield functions under isotropic and anisotropic hardenings. *Int. J. Solids Struct.* **2024**, *286–287*, 112582. [CrossRef]
22. Hou, Y.; Du, K.; El-Aty, A.A.; Lee, M.G.; Min, J.Y. Plastic anisotropy of sheet metals under plane strain loading: A novel non-associated constitutive model based on fourth-order polynomial functions. *Mater. Des.* **2022**, *223*, 111187. [CrossRef]
23. Zhang, C.; Lou, Y.S. Characterization and modelling of evolving plasticity behaviour up to fracture for FCC and BCC metals. *J. Mater. Process. Technol.* **2023**, *317*, 117997. [CrossRef]
24. Hu, Q.; Yoon, J.W.; Chen, J. Analytically described polynomial yield criterion by considering both plane strain and pure shear states. *Int. J. Plast.* **2023**, *162*, 103514. [CrossRef]

25. Lou, Y.S.; Yoon, J.W. Lode-dependent anisotropic-asymmetric yield function for isotropic and anisotropic hardening of pressure-insensitive materials. Part I: Quadratic function under non-associated flos invariant-based anw rule. *Int. J. Plast.* **2023**, *166*, 103647. [CrossRef]
26. Bassini, E.; Marchese, G.; Sivo, A.; Martelli, P.A.; Gullino, A.; Ugues, D. Effect of Cold Rolling on Microstructural and Mechanical Properties of a Dual-Phase Steel for Automotive Field. *Materials* **2022**, *15*, 7482. [CrossRef]
27. Baral, M.; Ha, J.; Korkolis, Y.P. Plasticity and ductile fracture modeling of an Al–Si–Mg die-cast alloy. *Int. J. Fract.* **2019**, *216*, 101–121. [CrossRef]
28. Ha, J.; Fones, J.; Kinsey, B.L.; Korkolis, Y.P. Plasticity and Formability of Annealed, Commercially-Pure Aluminum: Experiments and Modeling. *Materials* **2020**, *13*, 4285. [CrossRef]
29. Allen, R.M.; Toth, L.S.; Oppedal, A.L.; Kadiri, H.E. Crystal Plasticity Modeling of Anisotropic Hardening and Texture Due to Dislocation Transmutation in Twinning. *Materials* **2018**, *11*, 1855. [CrossRef]
30. Imandoust, A.; Barrett, C.D.; Al-Samman, T.; Inal, K.A.; Kadiri, H.E. A review on the effect of rare-earth elements on texture evolution during processing of magnesium alloys. *J. Mater. Sci.* **2017**, *52*, 1–29. [CrossRef]
31. Ha, J.; Baral, M.; Korkolis, Y.P. Plastic anisotropy and ductile fracture of bake-hardened AA6013 aluminum sheet. *Int. J. Solids Struct.* **2019**, *155*, 123–139. [CrossRef]
32. Knysh, P.; Korkolis, Y.P. Identification of the post-necking hardening response of rate- and temperature-dependent metals. *Int. J. Solids Struct.* **2017**, *115–116*, 149–160. [CrossRef]
33. Proust, G.; Tomé, C.; Kaschner, G. Modeling texture, twinning and hardening evolution during deformation of hexagonal materials. *Acta Mater.* **2007**, *55*, 2137–2148. [CrossRef]
34. Dick, C.P.; Korkolis, Y.P. Anisotropy of thin-walled tubes by a new method of combined tension and shear loading. *Int. J. Plast.* **2015**, *71*, 87–112. [CrossRef]
35. Lou, Y.S.; Wu, P.; Zhang, C.; Wang, J.Z.; Li, X.Y.; Chai, R.G.; Yoon, J.W. A stress-based shear fracture criterion considering the effect of stress triaxiality and Lode parameter. *Int. J. Solids Struct.* **2022**, *256*, 111993. [CrossRef]
36. Wu, P.; Zhang, C.; Lou, Y.S. Two-component DF2016 criterion to characterize the fracture behavior of magnesium rare-earth alloys. *Theor. Appl. Fract. Mech.* **2023**, *127*, 103677. [CrossRef]
37. Shang, H.C.; Wang, S.C.; Zhou, L.C.; Lou, Y.S. Neural network-based ductile fracture model for 5182-O aluminum alloy considering electroplastic effect in electrically-assisted processing. *Eng. Fract. Mech.* **2023**, *290*, 109476. [CrossRef]
38. Wu, P.; Zhang, C.; Lou, Y.S.; Chen, Q.; Ning, H.Q. Constitutive relationship and characterization of fracture behavior for WE43 alloy under various stress states. *Trans. Nonferr. Met. Soc.* **2023**, *33*, 438–453. [CrossRef]

materials MDPI

Article

A Modified DF2016 Criterion for the Fracture Modeling from Shear to Equibiaxial Tension

Xiaona Xu, Ruqiang Yan * and Xucheng Fang

School of Mechanical Engineering, Xi'an Jiao Tong University, 28 Xianning West Road, Xi'an 710049, China; xnxu18@xjtu.edu.cn (X.X.); xucheng.fang@stu.xjtu.edu.cn (X.F.)
* Correspondence: yanruqiang@xjtu.edu.cn; Tel.: +86-135-8405-4760

Abstract: This study introduces a modified DF2016 criterion to model a ductile fracture of sheet metals from shear to equibiaxial tension. The DF2016 criterion is modified so that a material constant is equal to the fracture strain at equibiaxial tension, which can be easily measured by the bulging experiments. To evaluate the performance of the modified DF2016 criterion, experiments are conducted for QP980 with five different specimens with stress states from shear to equibiaxial tension. The plasticity of the steel is characterized by the Swift–Voce hardening law and the pDrucker function, which is calibrated with the inverse engineering approach. A fracture strain is measured by the XTOP digital image correlation system for all the specimens, including the bulging test. The modified DF2016 criterion is also calibrated with the inverse engineering approach. The predicted force–stroke curves are compared with experimental results to evaluate the performance of the modified DF2016 criterion on the fracture prediction from shear to equibiaxial tension. The comparison shows that the modified DF2016 criterion can model the onset of the ductile fracture with high accuracy in wide stress states from shear to plane strain tension. Moreover, the calibration of the modified DF2016 criterion is comparatively easier than the original DF2016 criterion.

Keywords: ductile fracture; DF2016 criterion; stress triaxiality; Lode parameter; advanced high-strength steel; sheet metal forming

Citation: Xu, X.; Yan, R.; Fang, X. A Modified DF2016 Criterion for the Fracture Modeling from Shear to Equibiaxial Tension. *Materials* **2024**, *17*, 958. https://doi.org/10.3390/ma17040958

Academic Editors: Madhav Baral and Charles Lu

Received: 25 December 2023
Revised: 31 January 2024
Accepted: 17 February 2024
Published: 19 February 2024

1. Introduction

With the continuous development of the aerospace and automotive industry, people are no longer satisfied with basic safety or strength requirements but hope that the materials can meet the requirements of weight reduction and energy conservation while meeting the strength standards. Advanced high-strength steel, aluminum alloys, and other metal materials have excellent material properties, especially in terms of strength and plasticity, making it possible to reduce weight, save energy, and meet safety standards. Therefore, they have shown excellent application prospects. However, a fracture that may occur during deformation processes, such as stamping and deep drawing, can pose a serious threat to the safety of practical applications. Therefore, it is necessary to study the deformation behavior of metals and to accurately predict the occurrence of fractures.

Researchers have developed many yield criteria to mathematically characterize the yield behavior of metals. First, many isotropic yield functions were developed to improve the modeling accuracy of yielding at different stress states of compression, shear, tension, etc., by considering the effect of pressure and the third stress invariant. These yield functions include the Tresca, von Mises, Drucker, Drucker–Prager, etc. For sheet metals, texture is formed during rolling, and sheet metals show dependence of plastic behavior on loading directions. Accordingly, many anisotropic yield functions were proposed. The Hill48 yield criterion [1] is one of the most representative research results, which accurately predicts the uniaxial and equibiaxial tensile strength along the rolling direction (RD), transverse direction (TD) and normal direction (ND) by introducing four anisotropic parameters based on the Huber–von Mises yield function. On this basis, the yield criteria after Hill48 increase the number of anisotropic parameters through linear transformation of stress tensors to improve the accuracy of the yield equation in characterizing plastic deformation. Barlat et al. [2] put forward the plane stress non-quadratic yield criterion to describe anisotropic metal sheets, such as aluminum alloy sheets. Barlat et al. [3–5] developed more accurate anisotropic yield functions based on a similar method to more accurately characterize the anisotropic behavior of metals and alloys. Other popular anisotropic yield functions were also proposed by Banabic et al. [6], Aretz and Barlat [7], Cazacu et al. [8], Cazacu [9], Yoshida et al. [10], Lou and Yoon [11], etc. Anisotropic hardening was extensively analyzed in the last 15 years by Stoughton and Yoon [12], Lee et al. [13], Park et al. [14], Hou et al. [15–18], Hu et al. [19–23], Du et al. [24], etc. Plastic behavior under various stress states was modeled recently by Hu et al. [25], Lou et al. [26,27], etc. These advances dramatically improve the plasticity modeling accuracy of metals under different loading directions and wide stress states.

Ductile fracture is increasingly investigated in the last 20 years since the 15 fracture experiments of AA2024-T351 by Bao and Wierzbicki [28]. Thereafter, many ductile fracture criteria were developed, including the modified Mohr–Coulomb criterion [29], the DF2012 [30], DF2016 [31], Mu [32], Ganjiani–Homayounfard [33], Hu–Chen [34], Zheng [35], Zhang [36], Quach [37], Shang et al. [38], etc. These criteria are expressed in a form of mixed stress and strain and based on micromechanisms of ductile fractures by nucleation, growth, and the coalescence of voids [39,40]. Stress-based ductile fracture criteria were proposed by Khan and Liu [41], Stoughton and Yoon [42], Mohr and Marcadet [43], sDF2016 [44], etc. An anisotropic ductile fracture was also studied in the last 10 years. Modeling approaches of anisotropic ductile fracture were proposed by Beese et al. [45], Luo et al. [46], Jia and Bai [47], and Lou and Yoon [48]. Park et al. [49] numerically studied ductile fracture modeling in pre-cracked tensile tests of SUS304L stainless steel. Baral et al. [50] modelled plasticity and ductile fracture of an Al-Si-Mg die-cast alloy. Bidadi et al. [51] investigated the effects of model mixity and the loading rate on the fracture behavior of cracked thin-walled 304L stainless steel sheets with large non-linear plastic deformations. Khan and Liu [52] proposed a ductile fracture criterion to consider strain rate and temperature effect. Wcislik and Lipies [53] reviewed the numerical modeling of void development in metals to investigate the mechanism of a ductile fracture during plastic deformation. Baral and Korkolis [54] investigated ductile fracture under proportional and non-proportional multiaxial loading. Alrasheedi et al. [55] investigated the tensile deformation and fracture of unreinforced AZ91 and reinforced AZ91-C at temperatures up to 300 °C. Ha et al. [56] characterized the ductile fracture of an aluminum sheet under proportional loading. Egidio et al. [57] analyzed the influence of microstructure on fracture mechanisms of the heat-treated AlSi10Mg alloy produced by laser-based powder bed fusion. Korkolis and Kyriakides [58] investigated the effect of the strain path on the failure of inflated aluminum tubes. Torabi et al. [59] investigated the fracture behavior of AA7075-AA6061 and AA7075-Cu friction-stir welded joint. Roth and Mohr [60] characterized the effect of the strain rate on the fracture of advanced high-strength steel.

In this study, the DF2016 fracture criterion is modified to model the fracture behavior of advanced metals from shear to equibiaxial tension. In the modified DF2016 criterion, the material constant C_3 is equal to the fracture strain at equibiaxial tension, which can be easily measured by the bulging experiments. Therefore, the material calibration of the modified DF2016 criterion is relatively simple. The modified DF2016 criterion is applied to model the fracture behavior of an advanced high-strength steel of QP980. Five different specimens are tested to characterize plasticity and fracture behaviors from shear to equibiaxial tension with the strain measurement by the XTOP digital image correlation system. Plasticity is characterized by the Swift–Voce hardening law and the pDrucker function. The modified DF2016 criterion is calibrated with an inverse engineering approach. The predicted load–stroke curves with fractures are compared with the experimental results to evaluate the performance of the modified DF2016 criterion from shear to equibiaxial tension.

2. A Modified DF2016 Fracture Criterion

The DF2016 criterion is proposed to characterize fracture onset for sheet metals from shear to plane strain tension. It is expressed as below:

$$\left(\frac{2\tau_{max}}{\overline{\sigma}_{VM}}\right)^{C_1}\left(\langle\frac{f(\eta,L,C)}{f(1/3,-1,C)}\rangle\right)^{C_2}\overline{\varepsilon}_f^p = C_3 \quad \langle x\rangle = \begin{cases} x & if\ x\geq 0 \\ 0 & if\ x<0 \end{cases} \tag{1}$$

with

$$f(\eta,L,C) = \eta + C_4\frac{(3-L)}{3\sqrt{L^2+3}} + C \tag{2}$$

where η is the stress triaxiality, L is the Lode parameter, τ_{max} is the maximum shear stress, $\overline{\sigma}_{VM}$ is the von Mises equivalent strain, and $\overline{\varepsilon}_f^p$ is the equivalent plastic strain at fracture. There are five fracture parameters of C_1, C_2, C_3, C_4, and C. The DF2016 criterion reduces to the DF2014 criterion when $C_4 = 1$ and the DF2012 criterion by setting $C_4 = 0$ and $C = 1/3$. The DF2016 criterion can be reformulated in a form of the Lode parameter and stress triaxiality because the maximum shear stress normalized by the von Mises effective stress is solely a function of the Lode parameter as below:

$$\left(\frac{2}{\sqrt{L^2+3}}\right)^{C_1}\left(\langle\frac{f(\eta,L,C)}{f(1/3,-1,C)}\rangle\right)^{C_2}\overline{\varepsilon}_f^p = C_3 \tag{3}$$

In the DF2016 criterion, the material constant C_3 is equal to the equivalent plastic strain to fracture at uniaxial tension. The fracture strain at equibiaxial tension can be easily predicted by the bulging test with DIC technique. Therefore, the DF2016 criterion is modified so that C_3 is equal to the equivalent plastic strain at equibiaxial tension by bulging tests as below:

$$\left(\frac{2}{\sqrt{L^2+3}}\right)^{C_1}\left(\langle\frac{f(\eta,L,C)}{f(2/3,1,C)}\rangle\right)^{C_2}\overline{\varepsilon}_f^p = C_3 \tag{4}$$

In the numerical application of the ductile fracture criterion above, the von Mises equivalent stress and strain are computed again based on the stress components updated based on the yield function, which is used to describe the plastic deformation of metals. After that, the stress triaxiality and Lode parameter are then computed according to their definitions based on the von Mises equivalent stress to compute damage and fractures during plastic deformation.

3. Materials and Experiments

This part aims to collect mechanical behavior data under uniaxial tension, hole tension, plane strain tension, shear tension, and equibiaxial tension to assess the plastic behaviors of the QP980 steel. The material is manufactured by BAOSTEEL in Shanghai, China. The chemical composition of QP980 in weight percent is 0.2% C, 1.49% Si, 1.82% Mn, 0.011% P, 0.0043% S and 0.046% Al. The thickness of the steel was 1.0 mm. Five specimens were cut as shown in Figure 1 to characterize the mechanical properties of QP980 steels. These five specimens included the dogbone specimens, the specimens with a central hole, the notched specimens, the in-plane shear specimens, and the circular specimens. The first four specimens were tested with a universal material testing machine, and the deformation was measured with the XTOP DIC method. The circular specimens were used to conduct the bulging test. The dogbone specimens were used to characterize plasticity at uniaxial tension along three different loading directions. The specimens with a central hole were used to characterize fracture behavior under uniaxial tension, the notched specimens were used to characterize the fracture behavior under plane strain tension, the in-plane shear specimens were used to characterize fracture strain under shear, and the circular specimens for bulging tests were used to measure the fracture strain of the steel under equibiaxial tension. The dimensional information of the specimens was designed as shown in Figure 1, including the initial gauge length.

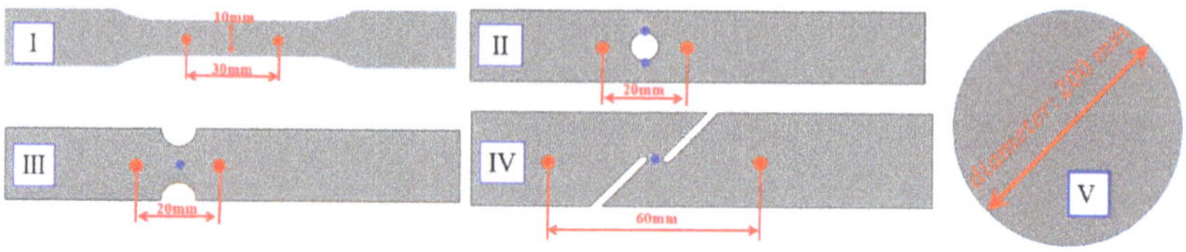

Figure 1. Four types of specimens [61]: (**I**) dogbone specimens; (**II**) specimens with central hole; (**III**) notched specimens; (**IV**) in-plane shear specimens; and (**V**) bulging specimens.

The universal mechanical testing system in Figure 2 was used to load the first four specimens. The loading velocity was 3.6 mm/min for the dogbone specimens to ensure that the strain rate during the tests was about 0.001/s. Deformation processes during experiments were recorded using the XTOP digital image correlation. Force during experiments was measured with a load cell. The measured force–stroke curves for the dogbone specimens were compared in Figure 3 along three directions of RD, DD, and TD. The evolution of plastic strain along the longitudinal and width directions is shown in Figure 4 to evaluate the anisotropic plastic deformation along the three directions. The comparison shows that the anisotropy in strength and plastic deformation is negligible. Therefore, the material was assumed to be isotropic in this study.

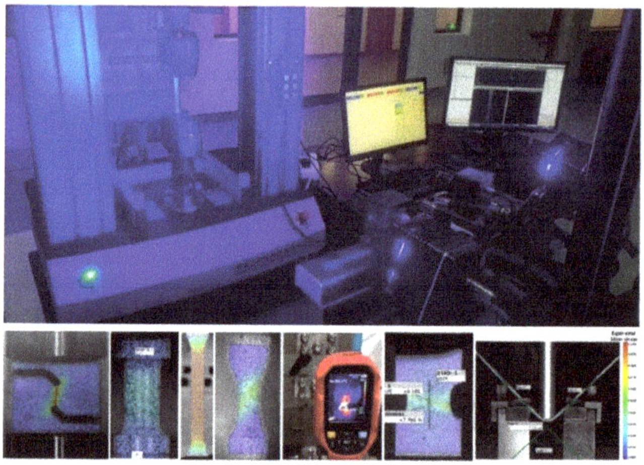

Figure 2. Universal mechanical testing system and the XTOP digital image correlation [62].

The load–stroke curves were also measured for the specimens with a central hole, notched specimens, and shear specimens along the rolling direction since the material was assumed to be isotropic. The loading velocity was set as 0.5 mm/min for the specimens with a central hole, notched specimens, and the in-plane shear specimens so that the strain rate during the tests were about 0.001/s. The measured load–stroke curves were compared in Figure 5 for the specimens with a central hole, Figure 6 for notched specimens, and Figure 7 for the shear specimens. It was obvious that the repeatability of the tests was reliable regarding the hardening behavior of the material. However, the stroke at failure was not as repeatable as the hardening behavior, especially for the shear test. The poor repeatability in the stroke at failure may be due to manufacturing error, inhomogeneous microstructure, etc. In this study, the most repeatable experiments with mean stroke at failure were selected to represent the experimental results for different specimens. Therefore, test #1 was selected for the further analysis of the specimen with central hole and #2 for the notched and shear specimens.

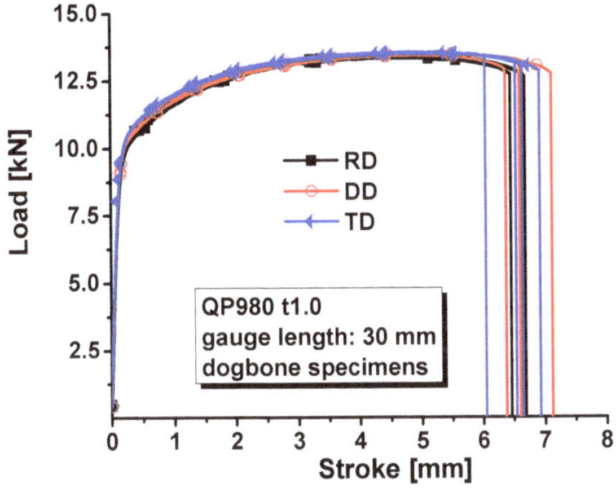

Figure 3. Load–stroke curves of QP980 for dogbone specimens.

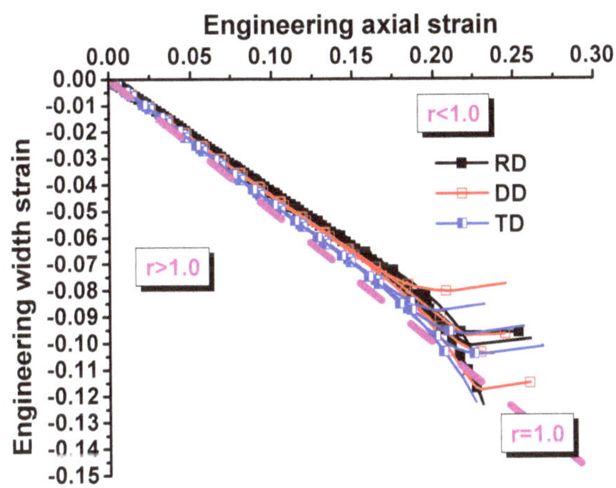

Figure 4. Relations between axial and width strain evolution of QP980 for dogbone specimens.

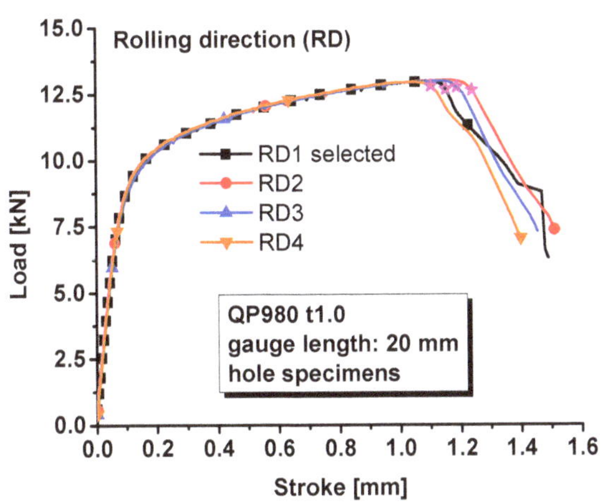

Figure 5. Load–stroke curves of QP980 for hole specimens.

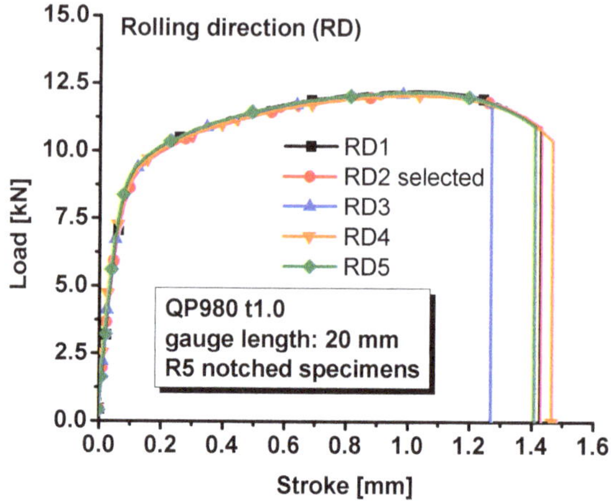

Figure 6. Load–stroke curves of QP980 for notched R5 specimens.

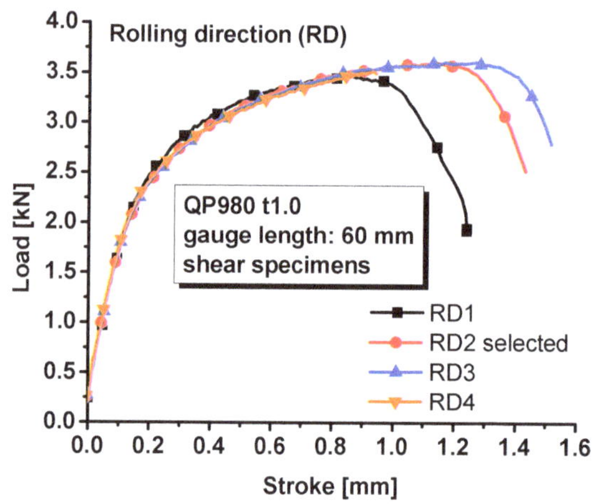

Figure 7. Load–stroke curves of QP980 for shear specimens.

Bulging tests were conducted for the QP980 steel with the specimen V in Figure 1. The punch velocity was 5 mm/min. Three bulging tests were carried out to ensure the repeatability of the experiments. The pressure and dome height are plotted in Figure 8. The evolution of equivalent strain is also shown against the dome height in the figure. It was measured that the fracture strain at the dome was about 0.5361 for QP980. There were two fracture strains shown in the figure, and the smaller one was selected so that fracture prediction was somewhat conservative. All the strains from DIC were the von Mises equivalent strain. To be consistent, the von Mises equivalent strain and its increment were computed by the plastic strain increments to compute the damage and fracture during simulation of plastic deformation. According to the modified DF2016 criterion, this fracture strain was equal to the parameter C_3 in the modified DF2016 criterion in Equation (4).

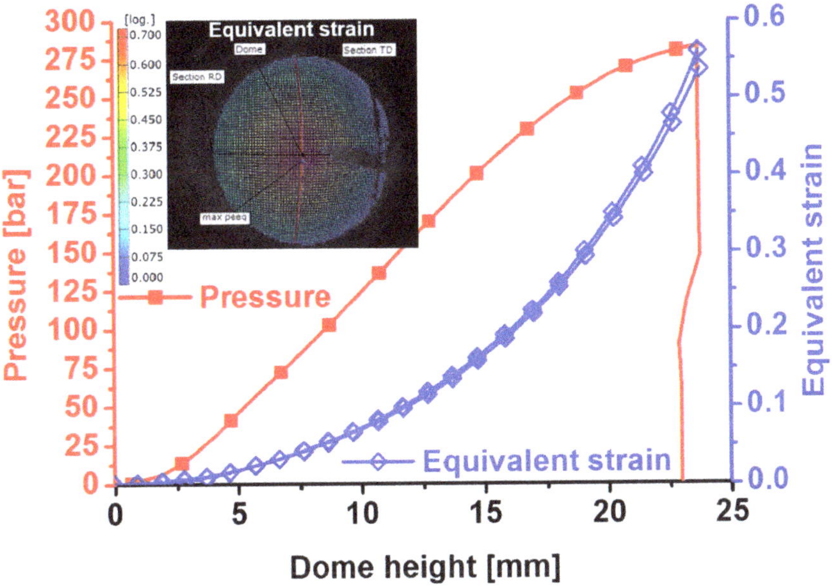

Figure 8. Bulging experimental results of QP980.

4. Plasticity Modeling

The strain hardening behavior is characterized by the dogbone specimens and fitted with the Swift–Voce hardening law below in Equation (5). The fitted parameters are summarized in Table 1 for the Swift, Voce, and Swift–Voce hardening laws. The fitted hardening laws are used to predict the strain hardening of QP980 and compared with experimental results, as shown in Figure 9. In the finite element analysis, small elements with an edge size of about 0.1 mm are adopted for the severe plastic zones of these specimens. All the simulations are conducted with C3D8R brick elements. All the other settings in the simulation are identical with the experimental conditions for all three specimens. The Swift and Swift–Voce hardening law are almost overlapped. The Voce model predicts the worst flow curve for QP980. The comparison demonstrates that the Swift–Voce hardening law fits the experimental result with the best agreement compared to the Swift and Voce hardening laws.

$$\overline{\sigma} = \alpha K(e_0 + \overline{\varepsilon}^p)^n + (1-\alpha)(A - (A-B)exp(-C\overline{\varepsilon}^p)) \quad (5)$$

Table 1. Coefficients of Swift–Voce function.

	K [GPa]	e_0	n	A [GPa]	B [GPa]	C	α
Swift	1.6562	0.0014	0.1451	\	\	\	\
Voce	\	\	\	1.2543	0.7393	21.5435	\
Swift–Voce	1.6562	0.0014	0.1451	1.2543	0.7393	21.5435	1.0262

Then the fitted Swift–Voce law is used to predict the load–stroke responses of specimens with a central hole, notched R5 specimens, and the in-plane shear specimens. The numerical simulation is conducted with Abaqus/Explicit 6.14. The minimum element size is about 0.5 mm. The predicted results are compared with experimental results in Figure 10. The prediction errors by numerical simulation are also computed with respect to stroke increase. It is observed that the predicted force is about 2% larger than the experimental results for the specimens with a central hole, 1% higher than the experimental results for the notched R5 specimens, and 4%~9% higher than the in-plane shear specimen results. The simulation error is too big, especially for the in-plane shear specimens, and not acceptable. The big error is due to the fact that the von Mises yield function cannot take the effect of the stress state into account on yielding and plastic deformation. The Lode parameter is 0.0 for shear and plane strain tension, but the strain hardening behavior is characterized by uniaxial tension of dogbone specimens whose Lode parameter is −1.

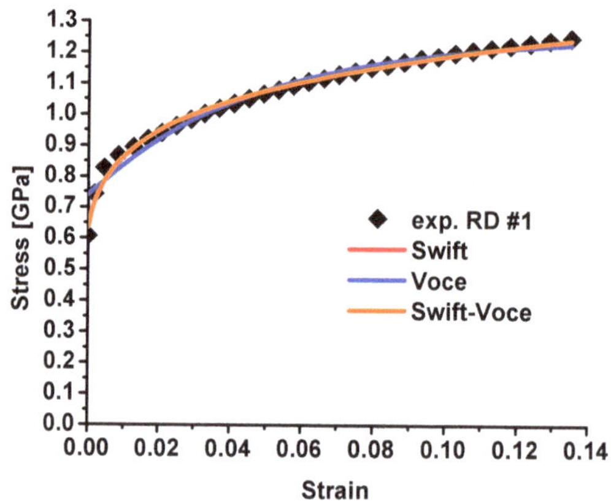

Figure 9. Comparison of the fitted hardening laws with experimental results for QP980.

The big load–stroke prediction error is due to the fact that the effect of stress states is not considered for the strength modeling from shear to plane strain tension. Therefore, the pDrucker yield function [26] is applied for QP980 to consider the effect of stress states on strength and expressed as follows:

$$\overline{\sigma}(\sigma_{ij}) = a\left(bI_1 + \left(J_2^3 - cJ_3^2\right)^{1/6}\right) \tag{6}$$

with

$$a = \frac{1}{b + \frac{1}{3}(27 - 4c)^{1/6}} \tag{7}$$

where I_1, J_2, and J_3 are the three stress invariants and a, b, and c are material parameters to adjust the yield surface. The computation of the parameter a in Equation (7) is based on the assumption that the strain hardening is characterized by uniaxial tensile tests by dogbone specimens. The a parameter can also be computed in a different form if the strain hardening curve is characterized at equibiaxial tension by the bulging tests. Details are suggested in the publication [26]. In the pDrucker function, there are three parameters to model the effect of stress states on the strength of QP980 steel sheets from shear to plane strain tension. These three parameters are calibrated with the flow curves under shear, uniaxial tension, and equibiaxial tension. By introducing the pressure effect and the dependence of the third stress invariant, the pDrucker function can predict different yield stresses in shear, uniaxial tension, and plane strain tension. This difference cannot be modeled by the von Mises yield function because the von Mises function only considers the effect of the second stress invariant on yielding.

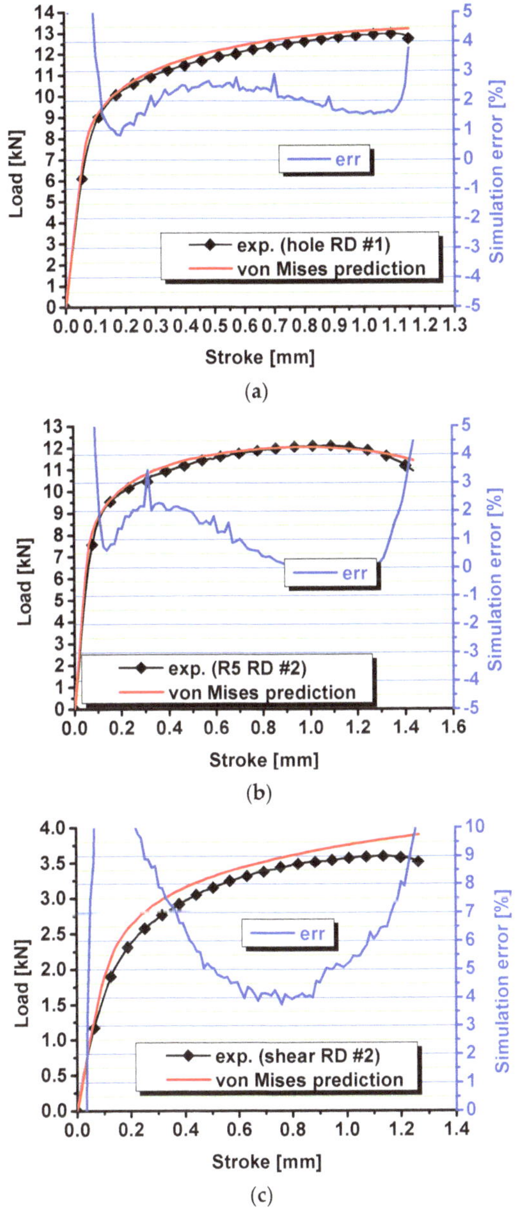

Figure 10. Comparison of the predicted load–stroke curves via von Mises function with experimental results for (**a**) specimens with a central hole; (**b**) notched specimens; and (**c**) shear specimens.

The pDrucker function and the Swift–Voce hardening law are calibrated with the inverse engineering approach [26] with the calibrated parameters in Table 2. Then, the load–stroke curves are predicted by the Abaqus/Explicit and compared with the experimental results in Figure 11. The prediction error is also computed and compared in the figure. The comparison shows that the error ranges from about −2% to 2% for the specimens with a central hole, from −1% to 1% for the notched specimens, and from −1% to 3% for the shear specimens. Compared with the prediction with the von Mises yield function, the prediction error for the load–stroke curves is significantly reduced by the pDrucker function, especially for the in-plane shear specimens, for which the error is reduced from 4%~9% to −1%~3%. Therefore, the Swift–Voce hardening law and the pDrucker function calibrated in Table 2 are used to model the plastic behavior of QP980 for these three specimens with stress states from the shear to plane strain tension. The significant reduction in the prediction error of the load–stroke curves is because the pDrucker yield function adjusts the relative strength of shear, uniaxial tension, and plane strain tension by optimizing the parameters of a, b, and c during inverse engineering approach, as shown in Figure 12, for the comparison of the von Mises and pDrucker yield surfaces. The error in the force–stroke curve for shear specimens is significantly reduced by the pDrucker yield function because the difference between the von Mises and pDrucker yield surfaces is very apparent, as shown in Figure 12. The yield surface difference under uniaxial and plane strain tension is not as obvious as that around shear. Therefore, the prediction accuracy improvement in the force–stroke curves is not obvious for specimens with a central hole and notched specimens. The error even increases slightly for the specimens with a central hole, which is due to the fact that the inverse engineering approach minimizes the total error in the load–stroke prediction for the three specimens. In the studied case, the total error is reduced, but the method sacrifices the prediction accuracy for the specimens with a central hole.

Table 2. Coefficients of the Swift–Voce function and the pDrucker function calibrated with the inverse engineering approach.

pDrucker			Swift–Voce Hardening Law						
a	b	c	K [GPa]	e_0	n	A [GPa]	B [GPa]	C	α
1.8769	−0.02486	1.2692	1.796	0.0080	0.1862	1.4050	0.6993	15.2893	0.5

The hardening law was fitted for low strain values before necking based on the dogbone specimens, and the fitted strain hardening parameters are summarized in Table 1. However, the fitted Swift–Voce hardening law cannot accurately predict the reaction forces for shear specimens, specimens with a central hole, and notched specimens, as shown in Figure 10. The force–stroke curves predicted by the flow curve calibrated with the inverse engineering approach matches with the experimental results with higher accuracy, as shown in Figure 11. To further improve the prediction accuracy of the force–stroke curves for different specimens, the evolution of yield surfaces is suggested to be considered during plastic deformation at different stress states.

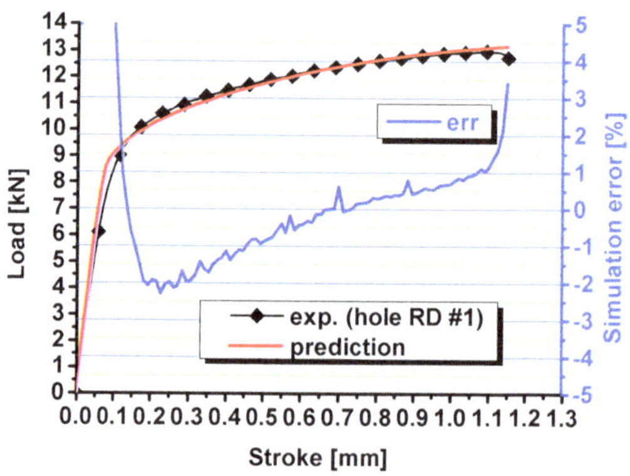

(a)

Figure 11. *Cont.*

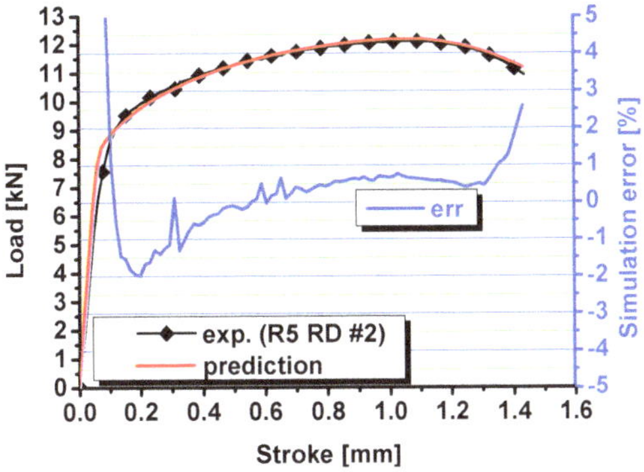

(b)

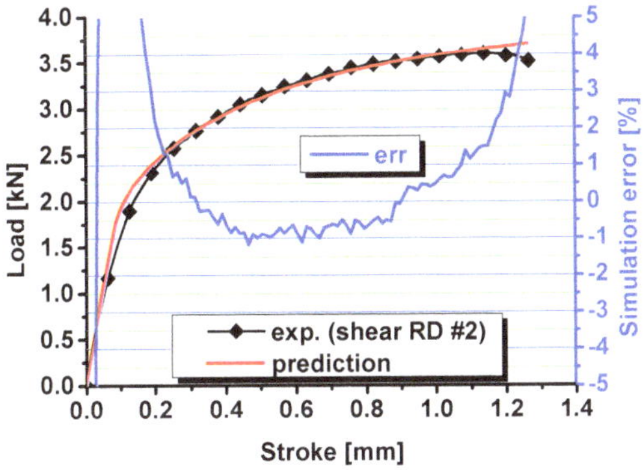

(c)

Figure 11. Comparison of the predicted load–stroke curves via the pDrucker function with experimental results for (**a**) specimens with a central hole; (**b**) notched specimens; and (**c**) shear specimens.

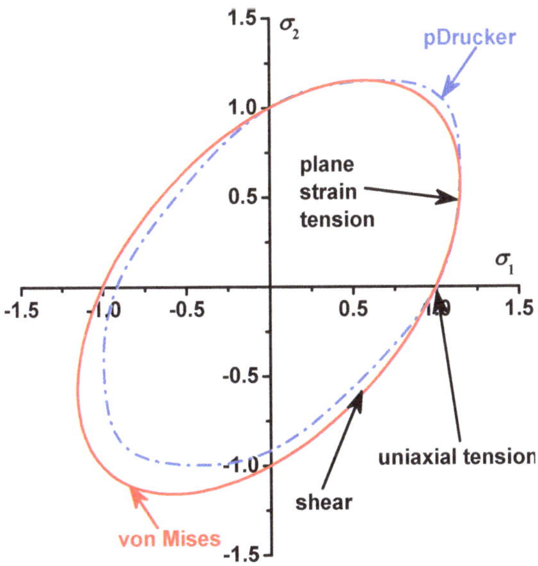

Figure 12. Comparison of the von Mises and pDrucker yield surfaces.

5. Fracture Modeling with the Modified DF2016 Criterion

The modified DF2016 criterion in Equation (4) is used to model the fracture behavior of QP980 from shear to equibiaxial tension. The material constant C_3 is equal to the fracture strain at equibiaxial tension, which is measured with the bulging test to be 0.5361 with the help of the XTOP digital image correlation method. The other fracture parameters in the modified DF2016 criterion are calibrated with the inverse engineering approach and summarized in Table 3. The inverse calibration of fracture parameters is conducted by minimizing the error between the predicted fracture stroke and experimental results for shear, specimens with a central hole, and plane strain tension of notched specimens. The fracture stroke during the simulation is determined by the sharp drop of load–stroke curves during simulation.

Table 3. Fracture parameters of the modified DF2016 criterion calibrated with the inverse engineering approach.

C_1	C_2	C_3	C_4	C
0.0654	1.1221	0.5361	4.542	3.0

In the implementation of the modified DF2016 criterion to the finite element simulation, the pDrucker yield function is used to describe the plastic deformation of the metal. The corresponding pDrucker equivalent strain is computed based on the pDrucker yield function to describe the strain hardening of the metal. The plastic strain increment components are obtained at the end of each integration and then used to compute the von Mises equivalent strain increment. The computed von Mises equivalent strain is used to compute damage and fractures based on the modified DF2016 criterion. All the equivalent strain used in damage and fracture computation is the von Mises equivalent strain because the calibration of the fracture criterion is based on the von Mises strain computed by DIC in different experiments.

The modified DF2016 criterion is implemented into ABAQUS/Explicit to predict the onset of the fracture for QP980 under various stress states from shear to plane strain tension. The load–stroke curves with an element deletion from the modified DF2016 criterion are predicted and compared with the experimental results of the specimens with a central hole, notched specimens, and shear specimens in Figure 13. For the specimens with a central hole, the predicted fracture stroke is 1.24 mm, while the experimental result is 1.15 mm. The difference between experiments and prediction is 0.09 mm, and the error is about 7.8% for the specimens with a central hole. For the notched specimens, the numerical prediction of the fracture stroke is 1.44 mm, and the experimental results is 1.43. The predicted fracture stroke is 0.01 mm longer than the experimental results, and the error is 0.7%. For the shear specimens, the predicted fracture stroke is 1.15 mm, while the experimental result is 1.26 mm. The experimental result is 0.09 mm higher than the prediction, and the error is 8.7%. Based on the comparison between prediction and experimental results in Figure 13, the prediction of the fracture stroke is all less than 10%. Considering the difficulty of the fracture prediction under complicated stress states, the prediction accuracy of less than 10% is definitely acceptable for engineering applications.

In addition, oscillation behaviors are observed for the force evolution in the simulation for specimens with a central hole, notched specimens, and shear. This is because all the numerical simulations are conducted with explicit formulation via ABAQUS/Explicit. The oscillation can be reduced by decreasing the mass scaling factor during the simulation or removed by the simulation with an implicit scheme via ABAQUS/Standard. However, simulations with ABAQUS/Standard cannot remove elements after a fracture.

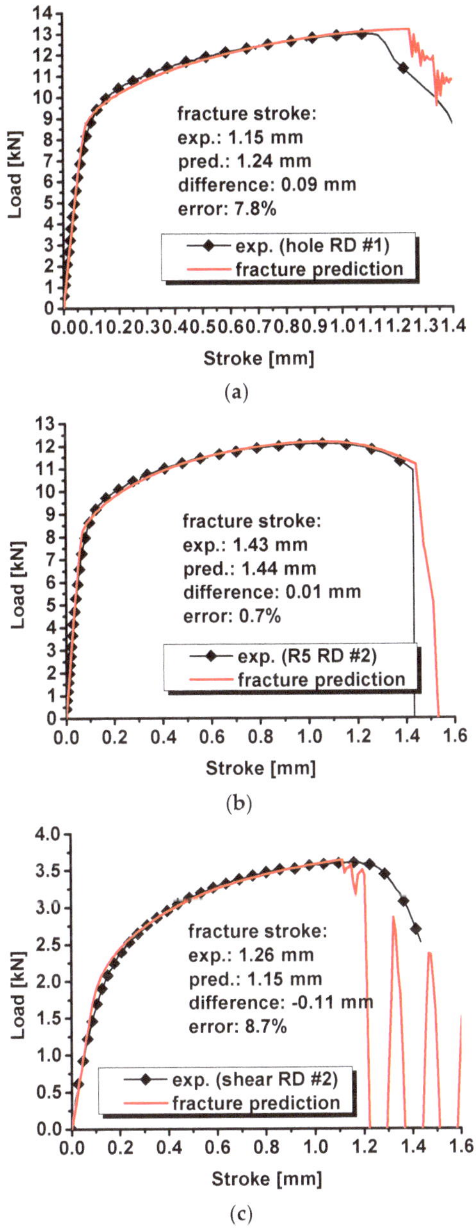

Figure 13. Comparison of the predicted load–stroke curves with the onset of ductile fracture via the modified DF2016 criterion with experimental results for (**a**) specimens with a central hole; (**b**) notched specimens; and (**c**) shear specimens.

The purpose of this study is to introduce the modified DF2016 criterion, which is relatively simple in parameter calibration compared to the original DF2016 criterion. Fracture prediction is not conducted in this study with the von Mises yield function because a simulation with the von Mises yield function results in a big error in the force–stroke curve prediction for the shear specimens. However, a fracture can be predicted with the modified

DF2016 criterion with the von Mises yield function, but the fracture parameters of the modified DF2016 criterion need to be calibrated again to obtain good fracture prediction results. The key problem is that the predicted reaction force with the von Mises function is not as accurate as that with the pDrucker function.

6. Conclusions

This study proposed a modified DF2016 criterion to model a ductile fracture from shear to equibiaxial tension for sheet metals. The modified DF2016 criterion is applied to describe the onset of a fracture for four specimens of QP980 steel. The result shows that the modified DF2016 criterion predicts the ductile fracture with acceptable accuracy. In addition, the modified DF2016 criterion is user-friendly since the fracture parameter C_3 is equal to the fracture strain at equibiaxial tension, which can be measured directly with the bulging test. According to the high accuracy and user-friendliness, the modified DF2016 criterion is suggested to be applied to model the fracture behavior of sheet metals from shear to equibiaxial tension.

Author Contributions: Conceptualization, X.X. and R.Y.; Methodology, X.X. and R.Y.; Software, X.X.; Validation, X.X.; Formal analysis, X.X.; Investigation, X.X. and X.F.; Resources, R.Y.; Writing—original draft, X.X.; Writing—review & editing, R.Y.; Visualization, X.X.; Supervision, R.Y.; Project administration, R.Y.; Funding acquisition, R.Y. All authors have read and agreed to the published version of the manuscript.

Funding: This research received no external funding.

Institutional Review Board Statement: Not applicable.

Informed Consent Statement: Not applicable.

Data Availability Statement: Data are contained within the article.

Conflicts of Interest: The authors declare no conflict of interest.

References

1. Hill, R. A theory of the yielding and plastic flow of anisotropic metals. *Proc. R. Soc. Lond. Ser. A Math. Phys. Sci.* **1948**, *193*, 281–297.
2. Barlat, F.; Lian, K. Plastic behavior and stretchability of sheet metals. Part I: A yield function for orthotropic sheets under plane stress conditions. *Int. J. Plast.* **1989**, *5*, 51–66. [CrossRef]
3. Barlat, F.; Lege, D.J.; Brem, J.C. A six-component yield function for anisotropic materials. *Int. J. Plast.* **1991**, *7*, 693–712. [CrossRef]
4. Barlat, F.; Brem, J.C.; Yoon, J.W.; Chung, K.; Dick, R.; Lege, D.J.; Pourboghrat, F.; Choi, S.-H.; Chu, E. Plane stress yield function for aluminum alloy sheets—Part 1: Theory. *Int. J. Plast.* **2003**, *19*, 1297–1319. [CrossRef]
5. Barlat, F.; Aretz, H.; Yoon, J.W.; Karabin, M.E.; Brem, J.C.; Dick, R.E. Linear transfomation-based anisotropic yield functions. *Int. J. Plast.* **2005**, *21*, 1009–1039. [CrossRef]
6. Banabic, D.; Aretz, H.; Comşa, D.S.; Părăianu, L. An improved analytical description of orthotropy in metallic sheets. *Int. J. Plast.* **2005**, *21*, 493–512. [CrossRef]
7. Aretz, H.; Barlat, F. New convex yield functions for orthotropic metal plasticity. *Int. J. Non-Linear Mech.* **2013**, *51*, 97–111. [CrossRef]
8. Cazacu, O.; Plunkett, B.; Barlat, F. Orthotropic yield criterion for hexagonal closed packed metals. *Int. J. Plast.* **2006**, *22*, 1171–1194. [CrossRef]
9. Cazacu, O. New yield criteria for isotropic and textured metallic materials. *Int. J. Solids Struct.* **2018**, *139–140*, 200–210. [CrossRef]
10. Yoshida, F.; Hamasaki, H.; Uemori, T. A user-friendly 3D yield function to describe anisotropy of steel sheets. *Int. J. Plast.* **2013**, *45*, 119–139. [CrossRef]
11. Lou, Y.; Yoon, J.W. Anisotropic yield function based on stress invariants for BCC and FCC metals and its extension to ductile fracture criterion. *Int. J. Plast.* **2018**, *101*, 125–155. [CrossRef]
12. Stoughton, T.B.; Yoon, J.W. Anisotropic hardening and non-associated flow in proportional loading of sheet metals. *Int. J. Plast.* **2009**, *25*, 1777–1817. [CrossRef]
13. Lee, E.-H.; Stoughton, T.B.; Yoon, J.W. A yield criterion through coupling of quadratic and non-quadratic functions for anisotropic hardening with non-associated flow rule. *Int. J. Plast.* **2017**, *99*, 120–143. [CrossRef]
14. Park, N.S.; Stoughton, T.B.; Yoon, J.W. A criterion for general description of anisotropic hardening considering strength differential effect with non-associated flow rule. *Int. J. Plast.* **2019**, *121*, 76–100. [CrossRef]

15. Hou, Y.; Min, J.; Stoughton, T.B.; Lin, J.; Carsley, J.E.; Carlson, B.E. A non-quadratic pressure-sensitive constitutive model under non-associated flow rule with anisotropic hardening: Modeling and validation. *Int. J. Plast.* **2020**, *135*, 102808. [CrossRef]
16. Hou, Y.; Min, J.; Guo, N.; Shen, Y.; Lin, J. Evolving asymmetric yield surfaces of quenching and partitioning steels: Characterization and modeling. *J. Mater. Process. Technol.* **2021**, *290*, 116979. [CrossRef]
17. Hou, Y.; Du, K.; El-Aty, A.A.; Lee, M.-G.; Min, J. Plastic anisotropy of sheet metals under plane strain loading: A novel non-associated constitutive model based on fourth-order polynomial functions. *Mater. Des.* **2022**, *223*, 111187. [CrossRef]
18. Hou, Y.; Min, J.; Lin, J.; Lee, M.-G. Modeling stress anisotropy, strength differential, and anisotropic hardening by coupling quadratic and stress-invariant-based yield functions under non-associated flow rule. *Mech. Mater.* **2022**, *174*, 104458. [CrossRef]
19. Hu, Q.; Yoon, J.W.; Manopulo, N.; Hora, P. A coupled yield criterion for anisotropic hardening with analytical description under associated flow rule: Modeling and validation. *Int. J. Plast.* **2021**, *136*, 102882. [CrossRef]
20. Hu, Q.; Yoon, J.W. Analytical description of an asymmetric yield function (Yoon2014) by considering anisotropic hardening under non-associated flow rule. *Int. J. Plast.* **2021**, *140*, 102978. [CrossRef]
21. Hu, Q.; Yoon, J.W.; Stoughton, T.B. Analytical determination of anisotropic parameters for Poly6 yield function. *Int. J. Mech. Sci.* **2021**, *201*, 106467. [CrossRef]
22. Hu, Q.; Chen, J.; Yoon, J.W. A new asymmetric yield criterion based on Yld 2000-2d under both associated and non-associated flow rules: Modeling and validation. *Mech. Mater.* **2022**, *167*, 104245. [CrossRef]
23. Hu, Q.; Yoon, J.W.; Chen, J. Analytically described polynomial yield criterion by considering both plane strain and pure shear states. *Int. J. Plast.* **2023**, *162*, 103514. [CrossRef]
24. Du, K.; Huang, S.; Hou, Y.; Wang, H.; Wang, Y.; Zheng, W.; Yuan, X. Characterization of the asymmetric evolving yield and flow of 6016-T4 aluminum alloy and DP490 steel. *J. Mater. Sci. Technol.* **2023**, *133*, 209–229. [CrossRef]
25. Hu, Q.; Li, X.; Han, X.; Li, H.; Chen, J. A normalized stress invariant-based yield criterion: Modeling and validation. *Int. J. Plast.* **2017**, *99*, 248–273. [CrossRef]
26. Lou, Y.; Zhang, C.; Zhang, S.; Yoon, J.W. A general yield function with differential and anisotropic hardening for strength modelling under various stress states with non-associated flow rule. *Int. J. Plast.* **2022**, *158*, 103414. [CrossRef]
27. Lou, Y.; Yoon, J.W. Lode-dependent anisotropic-asymmetric yield function for isotropic and anisotropic hardening of pressure-insensitive materials. Part I: Quadratic function under non-associated flow rule. *Int. J. Plast.* **2023**, *166*, 103647. [CrossRef]
28. Bao, Y.; Wierzbicki, T. On fracture locus in the equivalent strain and stress triaxiality space. *Int. J. Mech. Sci.* **2004**, *46*, 81–98. [CrossRef]
29. Bai, Y.; Wierzbicki, T. Application of extended Mohr–Coulomb criterion to ductile fracture. *Int. J. Fract.* **2010**, *161*, 1–20. [CrossRef]
30. Lou, Y.; Huh, H.; Lim, S.; Pack, K. New ductile fracture criterion for prediction of fracture forming limit diagrams of sheet metals. *Int. J. Solids Struct.* **2012**, *49*, 3605–3615. [CrossRef]
31. Lou, Y.; Chen, L.; Clausmeyer, T.; Tekkaya, A.E.; Yoon, J.W. Modeling of ductile fracture from shear to balanced biaxial tension for sheet metals. *Int. J. Solids Struct.* **2017**, *112*, 169–184. [CrossRef]
32. Mu, L.; Jia, Z.; Ma, Z.; Shen, F.; Sun, Y.; Zang, Y. A theoretical prediction framework for the construction of a fracture forming limit curve accounting for fracture pattern transition. *Int. J. Plast.* **2020**, *129*, 102706. [CrossRef]
33. Ganjiani, M.; Homayounfard, M. Development of a ductile failure model sensitive to stress triaxiality and Lode angle. *Int. J. Solids Struct.* **2021**, *225*, 111066. [CrossRef]
34. Hu, Q.; Li, X.; Han, X.; Chen, J. A new shear and tension based ductile fracture criterion: Modeling and validation. *Eur. J. Mech. A-Solid.* **2017**, *66*, 370–386. [CrossRef]
35. Zheng, L.; Wang, K.; Jiang, Y.; Wan, M.; Meng, B. A new ductile failure criterion for micro/meso scale forming limit prediction of metal foils considering size effect and free surface roughening. *Int. J. Plast.* **2022**, *157*, 103406. [CrossRef]
36. Zhang, T.; Yuan, H.; Yang, S. Fracture energy and tensile strength depending on stress triaxiality along a running crack front in three-dimensional cohesive modeling. *Eng. Fract. Mech.* **2020**, *227*, 106919. [CrossRef]
37. Quach, H.; Kim, J.-J.; Nguyen, D.-T.; Kim, Y.-S. Uncoupled ductile fracture criterion considering secondary void band behaviors for failure prediction in sheet metal forming. *Int. J. Mech. Sci.* **2020**, *169*, 105297. [CrossRef]
38. Shang, X.; Cui, Z.; Fu, M.W. A ductile fracture model considering stress state and Zener–Hollomon parameter for hot deformation of metallic materials. *Int. J. Mech. Sci.* **2018**, *144*, 800–812. [CrossRef]
39. Weck, A.; Wilkinson, D. Experimental investigation of void coalescence in metallic sheets containing laser drilled holes. *Acta Mater.* **2008**, *56*, 1774–1784. [CrossRef]
40. Lou, Y.; Yoon, J.W.; Huh, H.; Chao, Q.; Song, J.H. Correlation of the maximum shear stress with micro-mechanisms of ductile fracture for metals with high strength-to-weight ratio. *Int. J. Mech. Sci.* **2018**, *146*, 583–601. [CrossRef]
41. Khan, A.S.; Liu, H. A new approach for ductile fracture prediction on Al 2024-T351 alloy. *Int. J. Plast.* **2012**, *35*, 1–12. [CrossRef]
42. Stoughton, T.B.; Yoon, J.W. A new approach for failure criterion for sheet metals. *Int. J. Plast.* **2011**, *27*, 440–459. [CrossRef]
43. Mohr, D.; Marcadet, S.J. Micromechanically-motivated phenomenological Hosford–Coulomb model for predicting ductile fracturev initiation at low stress triaxialities. *Int. J. Solids Struct.* **2015**, *67–68*, 40–55. [CrossRef]
44. Lou, Y.; Wu, P.; Zhang, C.; Wang, J.; Li, X.; Chai, R.; Yoon, J.W. A stress-based shear fracture criterion considering the effect of stress triaxiality and Lode parameter. *Int. J. Solids Struct.* **2022**, *256*, 111993. [CrossRef]
45. Beese, A.M.; Luo, M.; Li, Y.; Bai, Y.; Wierzbicki, T. Partially coupled anisotropic fracture model for aluminum sheets. *Eng. Fract. Mech.* **2010**, *77*, 1128–1152. [CrossRef]

46. Luo, M.; Dunand, M.; Mohr, D. Experiments and modeling of anisotropic aluminum extrusions under multi-axial loading–Part II: Ductile fracture. *Int. J. Plast.* **2012**, *32*, 36–58. [CrossRef]
47. Jia, Y.; Bai, Y. Ductile fracture prediction for metal sheets using all-strain-based anisotropic eMMC model. *Int. J. Mech. Sci.* **2016**, *115*, 516–531. [CrossRef]
48. Lou, Y.; Yoon, J.W. Anisotropic ductile fracture criterion based on linear transformation. *Int. J. Plast.* **2017**, *93*, 3–25. [CrossRef]
49. Park, S.-J.; Lee, K.S.; Nam, W.S.; Kim, K.H.; Park, B.J. Numerical Study on a Ductile Fracture Model in Pre-Cracked Tension Tests of SUS304L. *Materials* **2024**, *17*, 276. [CrossRef]
50. Baral, M.; Ha, J.; Korkolis, Y.P. Plasticity and ductile fracture modeling of an Al–Si–Mg die-cast alloy. *Int. J. Fract.* **2019**, *216*, 101–121. [CrossRef]
51. Bidadi, J.; Googarchin, H.S.; Akhavan-Safar, A.; Silva, L.F.M. Effects of Mode Mixity and Loading Rate on Fracture Behavior of Cracked Thin-Walled 304L Stainless Steel Sheets with Large Non-Linear Plastic Deformation. *Materials* **2023**, *16*, 7690. [CrossRef] [PubMed]
52. Khan, A.S.; Liu, H. Strain rate and temperature dependent fracture criteria for isotropic and anisotropic metals. *Int. J. Plast.* **2012**, *37*, 1–15. [CrossRef]
53. Wcisli, W.; Lipiec, S. Voids development in metals: Numerical modelling. *Materials* **2023**, *16*, 4998. [CrossRef] [PubMed]
54. Baral, M.; Korkolis, Y.P. Ductile fracture under proportional and non-proportional multiaxial loading. *Int. J. Solids Struct.* **2021**, *210–211*, 88–108. [CrossRef]
55. Alrasheedi, N.H.; Ataya, S.; Seleman, M.M.E.-S.; Ahmed, M.M.Z. Tensile Deformation and Fracture of Unreinforced AZ91 and Reinforced AZ91-C at Temperatures up to 300 °C. *Materials* **2023**, *16*, 4785. [CrossRef] [PubMed]
56. Ha, J.; Baral, M.; Korkolis, Y.P. Ductile fracture of an aluminum sheet under proportional loading. *J. Mech. Phys. Solids* **2019**, *132*, 103685. [CrossRef]
57. Egidio, G.D.; Martini, C.; Borjesson, J.; Ghassemali, E.; Ceschini, L.; Morri, A. Influence of Microstructure on Fracture Mechanisms of the Heat-Treated AlSi10Mg Alloy Produced by Laser-Based Powder Bed Fusion. *Materials* **2023**, *16*, 2006. [CrossRef] [PubMed]
58. Korkolis, Y.P.; Kyriakides, S. Path-dependent failure of inflated aluminum tubes. *Int. J. Plast.* **2009**, *25*, 2059–2080. [CrossRef]
59. Torabi, A.R.; Mirzavand, M.; Saboori, B.; Cicero, S. Fracture Behavior of AA7075-AA6061 and AA7075-Cu Friction-Stir Welded Joints Containing Blunt V-Notches under Opening-Mode Loading. *Materials* **2023**, *16*, 1757. [CrossRef]
60. Roth, C.C.; Mohr, D. Effect of strain rate on ductile fracture initiation in advanced high strength steel sheets: Experiments and modeling. *Int. J. Plast.* **2014**, *56*, 19–44. [CrossRef]
61. Zhang, C.; Wang, Y.; Chen, Z.; Yang, N.; Lou, Y.; Clausmeyer, T.; Tekkaya, A.E.; Zhang, Q. Charaterization of plasticity and fracture of an QP1180 steel sheet. *Procedia Manuf.* **2020**, *50*, 529–534. [CrossRef]
62. Lou, Y.; Chen, Z.; Zhang, C.; Jiang, N.; Yang, G. Charaterization of anisotropic hardening of AA3003-O. *IOP Conf. Ser. Mater. Sci. Eng.* **2022**, *1270*, 012120. [CrossRef]

materials

MDPI

Article

Study of Dynamic Failure Behavior of a Type of PC/ABS Composite

Jiayu Zhou [1], Zhaodong Xia [1], Dongfang Ma [2,*] and Huanran Wang [1]

[1] Key Laboratory of Impact and Safety Engineering, Ministry of Education, Ningbo University, Ningbo 315211, China; zsetelle5@gmail.com (J.Z.); xiazhaodong2024@163.com (Z.X.); wanghuanran@nbu.edu.cn (H.W.)

[2] College of Science & Technology, Ningbo University, Ningbo 315300, China

* Correspondence: madongfang@nbu.edu.cn; Tel.: +86-(0)-158-057-461-00

Abstract: PC/ABS composites are commonly used in airbag covers. In this paper, uniaxial tensile experiments of a PC/ABS composite at different temperatures and strain rates were conducted. The results showed that the temperature and loading rate affect the mechanical properties of the PC/ABS composite. As the temperature increases, the yield stress decreases and the strain at the moment of fracture increases, but the strain rate at the same temperature has a relatively small effect on the mechanical properties, which are similar to ductile materials. The experimental results were applied to the Abaqus model which considered thermal effects and the exact Johnson–Cook constitutive parameters were calculated by applying the inverse method. Based on the constitutive model and the failure analysis findings acquired by DIC, the uniaxial tensile test at the room temperature and varied strain rates were simulated and compared to the test results, which accurately reproduced the test process. The experiment on target plate intrusion of the PC/ABS composite was designed, and a finite-element model was established to simulate the experimental process. The results were compared with the experiments, which showed that the constitutive and the failure fracture strains were valid.

Keywords: failure analysis; polymers and plastics; Johnson–Cook constitutive; numerical simulation; inversion method

Citation: Zhou, J.; Xia, Z.; Ma, D.; Wang, H. Study of Dynamic Failure Behavior of a Type of PC/ABS Composite. *Materials* **2024**, *17*, 1728. https://doi.org/10.3390/ma17081728

Academic Editors: Madhav Baral and Charles Lu

Received: 8 March 2024
Revised: 2 April 2024
Accepted: 7 April 2024
Published: 10 April 2024

1. Introduction

In recent decades, polymers have become increasingly popular in various industries due to their superior properties, such as high specific strength and low cost. ABS (acrylonitrile-butadiene-styrene) has a good durability and rigidity, containing rubber particles which allow it to suffer more plastic deformation under impact loads [1]. PC (polycarbonate) is often used in structural support due to its wide temperature and impact resistance [2]. TC-45M is a composite material made of a mixture of PC and ABS, which exhibits better impact toughness and tensile strength compared to pure polymer [3–5]. It has gradually replaced PC materials and is extensively used in the automotive industry for airbag covers. The study of its failure behavior can provide effective theoretical support for the study of vehicle safety performance.

During the last few decades, uniaxial tensile or compressive testing has represented a widely used method for studying polymer mechanical properties at different temperatures and strain rates Zheng et al. performed uniaxial stretching of Poly-Ether-Ether-Ketone at elevated temperatures and simulated its deformation behavior using a phenomenological model named DSGZ [6]. A study on the uniaxial compressive deformation behavior of PC/ABS blends at different rates and temperatures was conducted by Wang et al. [7]. They utilized a modified DSGZ model to characterize the deformation after unloading and reloading. Louche et al. conducted uniaxial tensile experiments on ABS polymer materials at various strain rates and temperatures to investigate their performance under impact loading, simulated the experimental process using the J-C constitutive (Johnson–Cook constitutive) to simulate the experimental process, and finally compared the experimental and numerical results to prove the correctness of the model [8]. Based on uniaxial tensile tests and single-edge-notch-tension (SENT) tests, two rubber-toughened thermoplastic polymer blends with different volume fractions of PC and ABS were analyzed experimentally, as well as by constitutive models and finite-element simulations with regard to their large strain deformation and fracture behavior by Hund et al. [9]. The impact behavior and modeling of ABS and polybutylene-terephthalates (PBT) were obtained as a function of impact velocity and temperature from a multiaxial impact test by Duan et al., and the deformation and failure of polymers were analyzed using a combination of experiments and finite-element analysis [10,11].

When a car is subjected to a violent impact force, airbag ejection leads to an impact loading, and the cover plant temperature increases [12]. However, there are limited thermodynamic constitutive models available for impact loading and the viscoplasticity model which considers temperature, suffers from the issue of too many parameters. In this paper, a uniaxial tensile test of the PC/ABS composite was conducted. Based on the experimental results of middle and low strain rates, the initial parameters of the J-C constitutive were obtained by MATLAB fitting. It should be noted that the parameters determined in this way may have some inaccuracy [13–15]. Subsequently, the initial parameters were substituted into Abaqus 2020 to simulate the experimental process, and the parameters were gradually adjusted to invert the modified constitutive model, and the results have a high accuracy [16–18], and apply to the high strain rate case. Based on the determined constitutive model and fracture parameters, numerical simulations of high-speed tensile testing of PC/ABS were conducted and compared with experimental results. Finally, a target plate impact experiment for the PC/ABS composite was designed, and a finite-element model was established to simulate the experimental process. The results were compared with the experiments, indicating the validity of the constitutive model and the failure fracture strain.

2. Materials and Methods

2.1. Material

The material used in the experiments of this paper is the PC/ABS composite. The PC/ABS composite material can combine the excellent properties of PC and ABS, improve the heat resistance, impact resistance and tensile strength of ABS, reduce the cost of PC and the viscosity of the melt, improve the processing performance, and reduce the internal stress of the product and the sensitivity of the impact strength to the thickness of the product. In addition, it also has a low price, low density, and other characteristics. The PC/ABS can be used as a structural material. It has been widely used in the car tool industry as a car cover.

A composite is not a simple blend of many materials. Therefore, it is not possible to derive accurate results of PC/ABS failure behavior from a single ABS or PC [19]. To accurately simulate the failure behavior of automotive covers during vehicle impacts, the material behavior at different temperatures and strain rates is investigated in this paper.

2.2. Uniaxial Tensile Test of the PC/ABS Composite Material

The PC/ABS composite model TC-45M (Dongguan Xinrui polymer material Technology Co., Ltd., Dongguan, China) was used in this study, and a specimen of the PC/ABS composite was designed for testing in the MTS-810 dynamic and static materials testing machine (MTS Systems Corporation, Eden Prairie, MN, USA), as shown in Figure 1a. The MTS-810 has a measured strain rate range of 10^{-4}–$10\ s^{-1}$. The low and medium strain rate tensile test at various temperatures can be realized with the temperature chamber. The influence of specimen size on the force balance error should be considered in the case of high-speed drawing [20]. The specific dimensions of the specimen are shown in Figure 1b. The specimen length of 15 mm can greatly reduce the influence of specimen size on the results of high-speed tensile experiments and can be used in quasi-static experiments too [21].

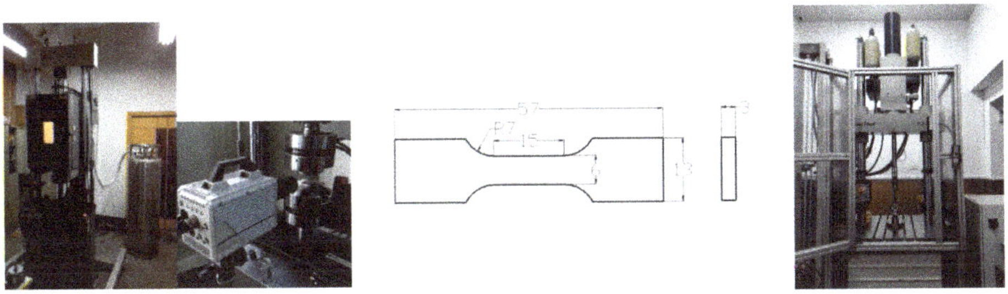

(a) (b) (c)

Figure 1. Design of uniaxial tensile experiment: (**a**) MTS-810 dynamic and static materials testing machine; (**b**) Size of test material (unit: mm); (**c**) ZwickRoell-5020 high-speed hydraulic tensile testing machine system.

The MTS-810, shown in Figure 1a, was used to perform uniaxial tensile tests on the PC/ABS composite at various ambient temperatures and strain rates. The ambient temperature was accurately controlled by a temperature chamber (238.15 K was achieved by continuously passing liquid nitrogen into the chamber). The PC/ABS composite is mostly used in automobile manufacturing, so its failure behavior under a high strain rate is also the focus of this paper. The ZwickRoell-5020 high-speed hydraulic tensile testing machine system (ZwickRoell GmbH & Co. KG, Ulm, Germany) was selected, shown in Figure 1c, which can also cooperate with the temperature chamber. This device has a strain rate measurement range of 10–1000 s^{-1}. The sample still uses the specifications shown in Figure 1b. The above experimental conditions are listed in Table 1.

Table 1. Uniaxial tensile test conditions.

Strains (s^{-1})	Temperature (K)	Equipment Model	Loading Rate (mm/s)
0.01	238.15/273.15/293.15	MTS-810	0.15
1			15
10			150
100		ZwickRoell-5020	1500
1000			15,000

2.3. PC/ABS Composite Material Ballistic Impact Test

The PC/ABS composite material penetration test sample was designed as a 100 mm × 100 mm × 3 mm rectangular target plate. Four holes were punched into the sample's four corners to make it unmovable. The bullet used in the experiment is a cylindrical length of 24 mm with a hemispherical head measuring 12 mm in diameter. The bullet is made of Cr12MoV tool steel. Figure 2a provides the plate's and bullet's precise design.

The experiment of the ballistic impact test uses a high-speed air gun in combination with a high-speed camera system. Figure 2b displays the schematic of the experimental device arrangement. The caliber of the high-speed air gun launch tube is 12 mm, and the length of the gun tube is 4 m. A tachymeter was placed between the high-speed air gun barrel and the test simple to measure the average velocity of the bullet strikes. After research, it was found that the average speed of cars on the highway was approximately 30 m/s. To test the applicability of the PC/ABS composite material, the speed of the experiment should be greater than 30 m/s [22]. After many empty gun experiments, with the same bullet and the conditions of the pressure, the bullet's hitting speed is maintained at around 34 m/s.

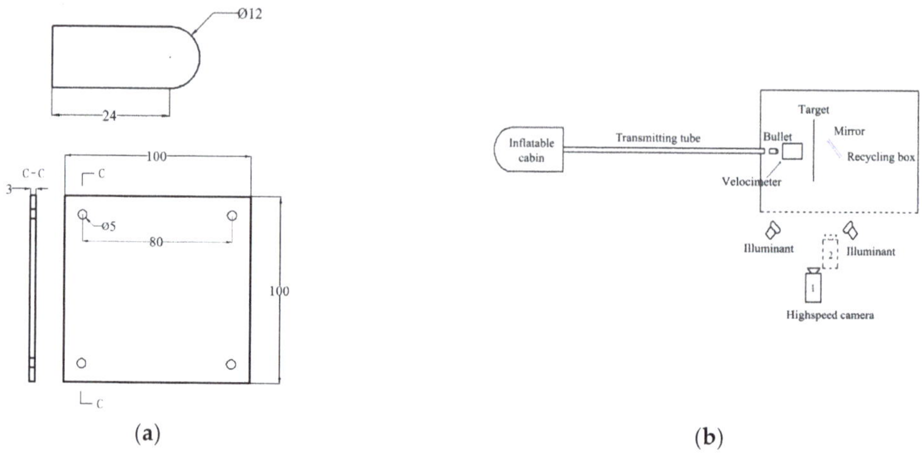

Figure 2. Ballistic impact test experimental design diagrams: (**a**) Design dimensions of target plate and bullet (unit: mm); (**b**) Layout of penetration test device.

The bullet penetration process is captured by a high-speed camera system. The camera is pointed toward the side of the target plate, and the camera's shooting direction is perpendicular to the path of the bullet's incidence. A mirror was placed on the target's rear to capture back-view pictures during the bullet impact, positioned at a 45° angle to both the direction of the bullet and the direction of high-speed photography. This allowed the high-speed camera to capture images of the deformation and fracture on the specimen's back. In all experiments, the sampling frequency of the high-speed camera was set at 4×10^4 Hz, the time interval between neighboring photos was 25 μs, and the shooting resolution was set at 384 × 288 pixels.

3. Results and Discussion

3.1. Uniaxial Tensile Test Results

Uniaxial tensile tests were conducted on the MTS-810 testing machine at varying temperatures and strain rates. Uneven material distribution can lead to unrepeatable experimental results, so this paper performs three replications under each experimental condition to assess the homogeneity of the composite [23]. The typical results are displayed in Figure 3, which illustrates the composite's good homogeneity. Figure 4a,b depict the results of averaging three sets of test results for different ambient temperatures at the same strain rate, as well as different strain rates at the same ambient temperature.

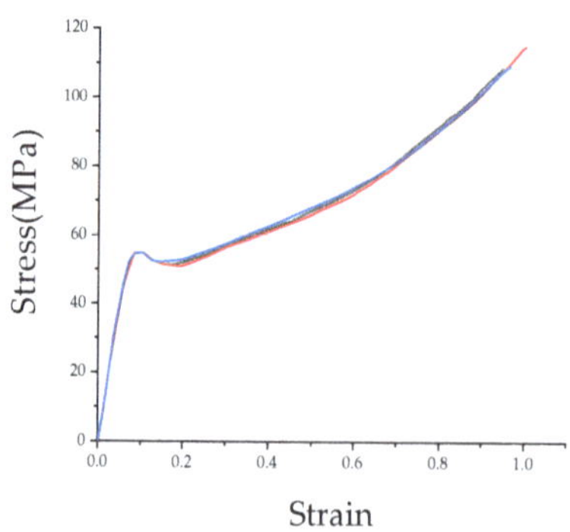

Figure 3. Three times quasi static test results of the PC/ABS composite under 273.15 K and a strain rate of 0.01 s^{-1}.

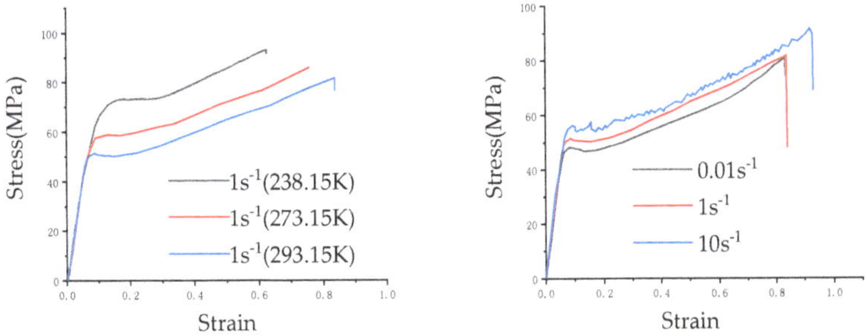

(**a**) (**b**)

Figure 4. Effect of a single variable for the failure behavior of the PC/ABS composite: (**a**) Force–displacement curve of the PC/ABS composite under the same strain rate (1 s^{-1}) and different temperatures; (**b**) Force–displacement curve of the PC/ABS composite under different strain rates at room temperature.

The figure shows that the mechanical properties of the PC/ABS composite are influenced by both the loading rate and ambient temperature. The temperature factor responds to these mechanical properties in a particularly noticeable way: as the ambient temperature rises, yield stress and fracture strain respectively decrease and increase. The mechanical properties' effects are relatively weak by the strain rate at the same temperature and the mechanical properties are similar to those of traditional ductile materials [24,25].

The yield stresses at different temperatures and strain rates are compared and analyzed as shown in Figure 5. The results show that the yield stress of the material is linear with the temperature and the logarithm of strain rate. The yield stress of the material increases with the increase in strain rate and decreases with the increase in temperature. When the temperature rises, the movement of polymer chains in the PC/ABS composite is activated, and under the action of external load, the molecular chains are rotated and displaced, the plastic flow is strengthened, and the yield stress is reduced. Under quasi-static tensile conditions, the material is in a steady state, allowing infinite plastic flow; under medium to high strain rates, the polymer chains in the PC/ABS composites cannot rotate and displace rapidly, resulting in the strengthening of the yield stress.

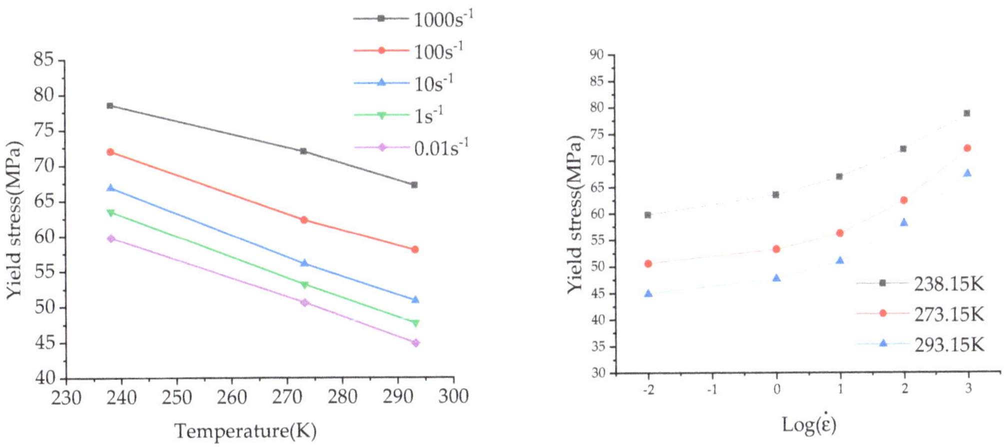

(**a**) (**b**)

Figure 5. Comparison of yield stress under different experimental conditions: (**a**) Yield stress temperature curve; (**b**) Yield stress strain rate (logarithmic) curve.

3.2. Experimental Results of Ballistic Impact Test

Target plate penetration experiments were conducted on the PC/ABS composite under the required experimental conditions. The high-speed impact process was recorded by a high-speed camera. For the convenience of observation, the instant of bullet impact (when the bullet was just in touch with the target plate) was defined as the start time. Figure 6 displays the experimental processes of the penetration procedure from the side and rear perspectives. It can be found that it has a large deformation and a long plastic stage during the experiment, rather than a brittle fracture. Therefore, it is found that the PC/ABS composite material is a typical ductile material by analyzing the deformation and failure mode of the target plate.

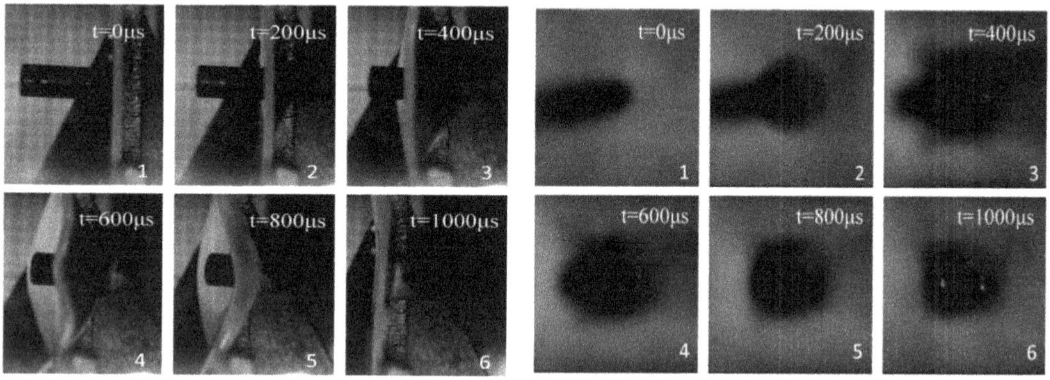

(**a**) (**b**)

Figure 6. Penetration process under different angles of view of the experiment: (**a**) Penetration process under the side angle of view of the experiment; (**b**) Penetration process from the experimental back view.

3.3. Determination of the Parameters of the J-C Constitutive

The number of viscoelastic thermodynamic coupling models under impact loads is small, and there are many problems such as too much parameter measurement [26–28]. Because PC/ABS is widely used in the preparation of vehicles, the study of its failure behavior should pay attention to the influence of strain rate and temperature, especially the plastic deformation and failure at a high strain rate, rather than its creep or relaxation behavior. Therefore, the J-C constitutive is selected in this paper. The J-C constitutive model is a phenomenological model that describes plastic hardening, strain rate effects, and thermal softening of materials [29]. These three phenomenological formulations are connected multiplicatively in the J-C constitutive. The J-C constitutive is mainly applied to materials with large deformations, high strain rates, and high temperatures, meaning it is suitable for numerical simulations of most materials. The form of Equation (1) is as follows [30–32]:

$$\sigma = (A + B\varepsilon_p^n)(1 + C\ln\frac{\dot{\varepsilon}}{\dot{\varepsilon}_0})[1 - (\frac{T - T_0}{T_m - T_0})^m] \quad (1)$$

where ε_p—equivalent plastic strain; $\dot{\varepsilon}$—equivalent plastic strain rate; $\dot{\varepsilon}_0$—reference strain rate; T_0—reference temperature; T_m—melting point temperature of the material; T—test temperature. A, B, C, n, m are the material parameters.

The following is the principle used in this research to determine the parameters A, B, C, n, m of the J-C constitutive: The stress–strain curve transformed by the force–displacement curve at the reference temperature of 238.15 K and strain rate of 0.01 s^{-1} gives the values of A, B and n, the stress–strain curve transformed by the various strain rates at 238.15 K gives the value of C; the stress–strain curve transformed by the various temperatures at 0.01 s^{-1} gives the value of m. Based on this principle, the initial parameters of the J-C constitutive are determined as A = 60 MPa, B = 115 MPa, C = 0.03, n = 1.75, m = 0.95. However, the parameters determined by this method have a large error and cannot accurately reproduce the failure behavior of the material [15]. The constitutive model is mainly used in the field of numerical simulation, so researchers will reproduce the experimental process by numerical simulation and modify the parameters by comparing with the results of uniaxial tensile experiment. The above parameters are put into the Abaqus 2020 to apply the same loading conditions as the test. Then, numerical simulation is compared with the tests, and the parameters are adjusted until they are in total agreement with the test. The specific process of determining the final parameters is shown in Figure 7. This method is hereafter referred to as the inversion method. The basic mechanical parameters of the PC/ABS composite are as follows: the density is 1120 kg/m^3, the modulus of elasticity is 1750 MPa, Poisson's ratio is 0.38, and the specific heat capacity is 1400 J/(kg·K). The Abaqus solver was used for numerical simulation, and the mesh type was C3D8R. After many attempts, the mesh size had no obvious influence on the simulation results.

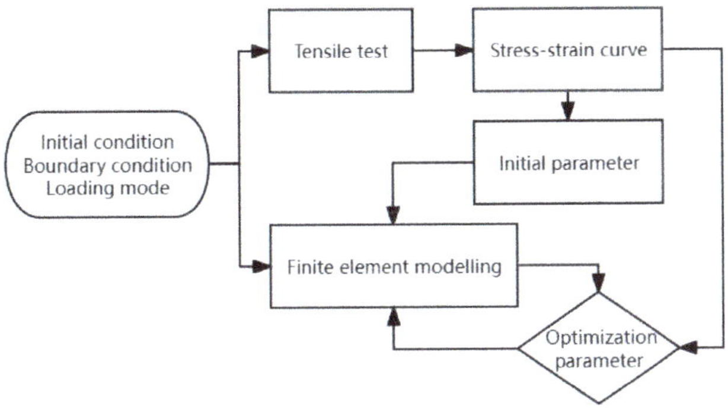

Figure 7. Process of determining parameters by the inversion method.

Through the above inversion method, J-C constitutive parameters are determined and shown in Table 2. A comparison between numerical and experimental results is shown in Figure 8, and it should be pointed out that there is no failure criterion, so there is no steep drop in the numerical simulation curve. At this time, the deformation process can be basically reproduced at low and medium strain rates.

Table 2. J-C constitutive parameters of the PC/ABS composite.

A	B	C	n	m
57.5 MPa	120 MPa	0.032	1.734	1.02

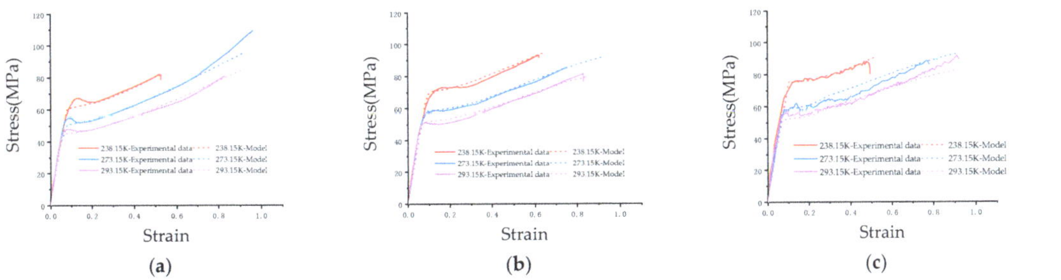

Figure 8. Comparison between numerical simulation and experiment after modification by inversion method: (**a**) Test results with strain rate of 0.01 s^{-1}; (**b**) Test results with strain rate of 1 s^{-1}; (**c**) Test results with strain rate of 10 s^{-1}.

3.4. Validation of J-C Constitutive Parameters

The results obtained using the inversion method are only applicable to low and medium strain rates. However, the cover material is often subjected to high-speed impact loading when the airbag is deployed. To test the validity of the J-C constitutive determined by the inversion method when applied to high strains, we calibrated it with high strain rate tensile tests. Using a ZwickRoell-5020 high-speed tensile tester, we performed uniaxial tensile tests at strain rates of 100 s^{-1} and 1000 s^{-1} at three ambient temperatures sustained by particular temperature chambers. The test results were compared with the J-C constitutive numerical simulation data obtained using the inversion approach. The numerical simulation results of the J-C constitutive were close to the test results. Figure 9 presents the comparative results. This shows that the J-C constitutive derived from the inversion approach is fit for large strain rates. It should especially be pointed out that the experimental process can be regarded as an adiabatic process under high-speed impact [33], and a large amount of heat will be generated in the experiment, and a large degree of temperature change will be generated in the pattern. Therefore, a mechanical thermal effect is added in the simulation to correct for the effects of adiabatic warming.

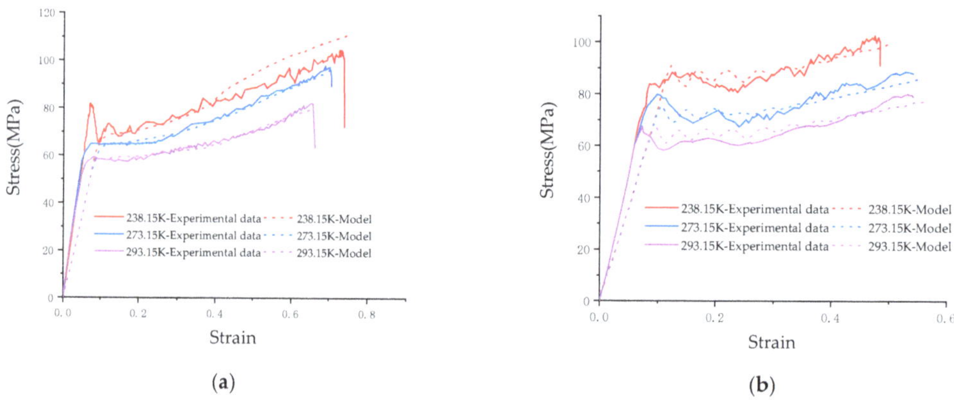

Figure 9. Experiments and simulations at high strain rates validate the accuracy of the constitutive model: (**a**) Comparison of results under different temperatures with strain rate of 100 s^{-1}; (**b**) Comparison of results under different temperatures with strain rate of 1000 s^{-1}.

3.5. Failure Behavior Analysis in One-Dimensional Tensile State

The J-C constitutive parameters constructed above do not include a failure criterion. Researchers are concerned with the failure behavior of materials in engineering. The failure behavior of this composite material is examined in this chapter. A high-speed video camera recorded the deformation and fracture process of the PC/ABS composite at room temperature with different strain rates. Note that other temperatures were achieved by an ambient temperature box. Therefore, the process for other temperatures could not be captured with a high-speed video camera. Figure 10 shows typical results. Digital image correlation (DIC) can be used to measure dimensional changes in drawing patterns in real-time using optical sensors. The deformation information of the specimen under tension at different strain rates and the local deformation of the specimen at the failure time was obtained with a DIC, and the fracture strain ε_{tr} is calculated (Equation (2) shows the calculation, where A_0 represents the original cross-sectional area of the material, and A represents the cross-sectional area of the material fracture) [34]. The local strains in the failure region are listed in Table 3.

$$\varepsilon_{tr} = \ln\left(\frac{A_0}{A}\right) \tag{2}$$

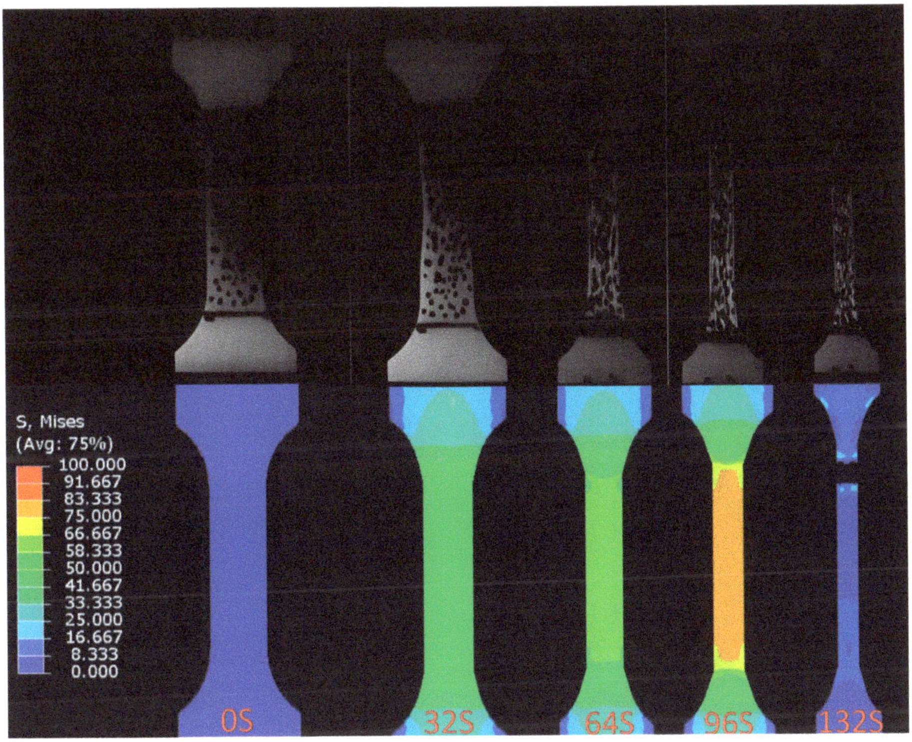

Figure 10. Comparison between the deformation process of numerical simulation specimen at different times and the record of the high-speed camera.

Table 3. Fracture strains and sample temperature.

strain rate (s^{-1})	0.01	1	10	100	1000
fracture strain	0.825	0.833	0.922	0.646	0.527
sample temperature by simulation (K)	293	313	328	311	309
sample temperature by theory (K)	293	320.6	326.5	313.9	312.1

In this paper, based on the J-C constitutive constructed by the inversion method combined with the fracture strain ε_{fr} calculated by the deformation information of the specimen. Numerical simulation was conducted to analyze the uniaxial tensile behavior of the PC/ABS composite at room temperature (293.15 K) under different strain rates. The simulation compares the process to the recording of a high-speed camera. Figure 10 displays the typical comparison results, while Figure 11 compares the force–displacement curves of the tests under different strain rates with the numerical simulations. It can be seen from the figures that the J-C constitutive determined by the inverse method and the failure parameters obtained by using DIC can reproduce the uniaxial tensile test process of the PC/ABS composite under different strain rates at room temperature.

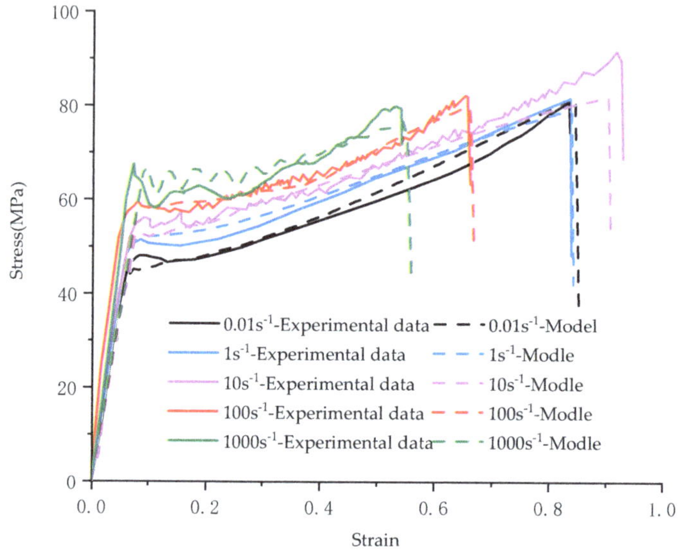

Figure 11. Comparison between uniaxial tensile test results and numerical simulation of the PC/ABS composite at room temperature and different strain rates.

The J-C failure model is shown in Equation (3) [35], which describes the effects of stress triaxiality, strain rate, and temperature in a decoupled form so that factors can be removed when they are not important to the study.

$$\varepsilon_{tr} = [D_1 + D_2 \exp D_3(\frac{\sigma_m}{\sigma_{eq}})][1 + D_4 \ln \frac{\dot{\varepsilon}}{\dot{\varepsilon}_0}][1 + D_5(\frac{T_s - T_r}{T_m - T_r})] \quad (3)$$

where σ_m—hydrostatic stress; σ_{eq}—equivalent strength; T_s—sample temperature; T_r—reference temperature. It should be noted that the temperature of the sample during the stretching process will change significantly due to the generation of a large amount of heat, so T_s refers to the internal temperature of the material before the sample fracture, rather than the ambient temperature.

Polymer materials are highly sensitive to temperature [36], and during the tensile process, a large amount of heat is generated in the sample, leading to a temperature rapid increase in material. To accurately analyze failure behavior, the failure temperature in the finite-element simulation is chosen as the specimen's fracture temperature. This paper focuses on the effect of strain rate and temperature on the failure behavior of the PC/ABS materials, for which the J-C failure model is degraded as shown in Equation (4).

$$\varepsilon_{tr} = [d_1 + d_2 \ln \frac{\dot{\varepsilon}}{\dot{\varepsilon}_0}][1 + d_3(\frac{T_s - T_r}{T_m - T_r})] \quad (4)$$

where $\dot{\varepsilon}_0 = 0.01\ \text{s}^{-1}$, $T_r = 293$ K, $T_m = 450$ K. After fitting, it can be determined that $d_1 = 0.825, d_2 = -0.043, d_3 = 2.6$. At medium strain rates, thermal softening takes a dominant role and the temperature rises rapidly, leading to a slight increase in the fracture strain. However, at high strain rates, high-velocity impacts dominate the fracture strain of the material, the fracture strain decreases, heat fails to accumulate in the material in large quantities, and the temperature rise before fracture is relatively few.

T_S is shown in Table 3, and its theoretical temperature can also be calculated from Equation (5) [37], where density $\rho = 1120\ \text{kg/m}^3$, specific heat capacity $c = 1400\ \text{J/(kg·K)}$ and mechanical thermal effect $\beta = 0.9$, and ΔT is the elevated temperature. At a strain rate of 0.01 s^{-1}, the sample is in thermal equilibrium with the outside world and the T_S can be regarded as the same as the ambient temperature.

$$\Delta T = \frac{\beta \int_0^{\varepsilon_{tr}} \sigma d\varepsilon_{tr}}{c\rho} \quad (5)$$

3.6. Failure Behavior Analysis under Ballistic Impacts

The ballistic impact experiment was numerically simulated. The inputs were the DIC-calculated failure fracture strain and the J-C constitutive parameters. A friction coefficient of 0.25 was used to describe the interaction between the bullet and the target plate.

Figure 12 shows the simulation results of the finite-element simulation of the PC/ABS composite for the penetration experiment, as well as a comparison to the experimental process of penetrating the target plate. Time (t_0 = 0 μs) was defined as the moment the bullet began contact with the target plate. Then, we compared the simulation with the target plate's damage shape and the bullet's location at each of the following times: t_0 = 0 μs, t_1 = 200 μs, t_2 = 400 μs, t_3 = 600 μs, t_4 = 800 μs, and t_5 = 1000 μs. The results of the finite-element simulation were found to be in good agreement with the experimental results. The finite-element simulation's damage shape and experimentally reclaimed target plate were compared, as shown in Figure 13. The finite-element simulation results can be accepted given the intricacy of the experimental procedure, the target plate fixation error, and several irresistible factors like air pressure instability, bullet ejection deviation, etc., [38]. In conclusion, the failure fracture strain computed by the deformation information of the specimen and the J-C constitutive parameters of the PC/ABS composite established using the inversion method are both accurate in this paper.

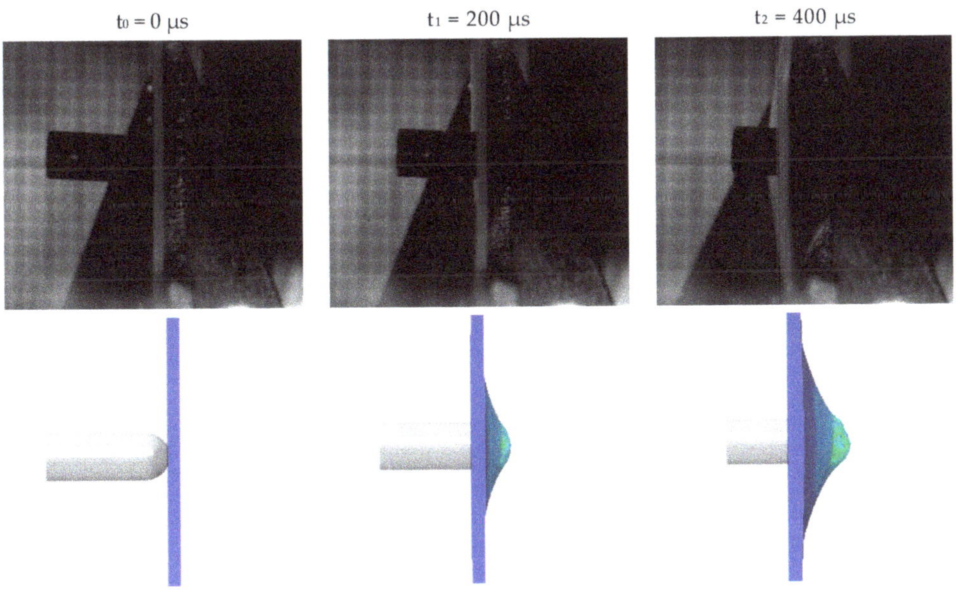

Figure 12. *Cont.*

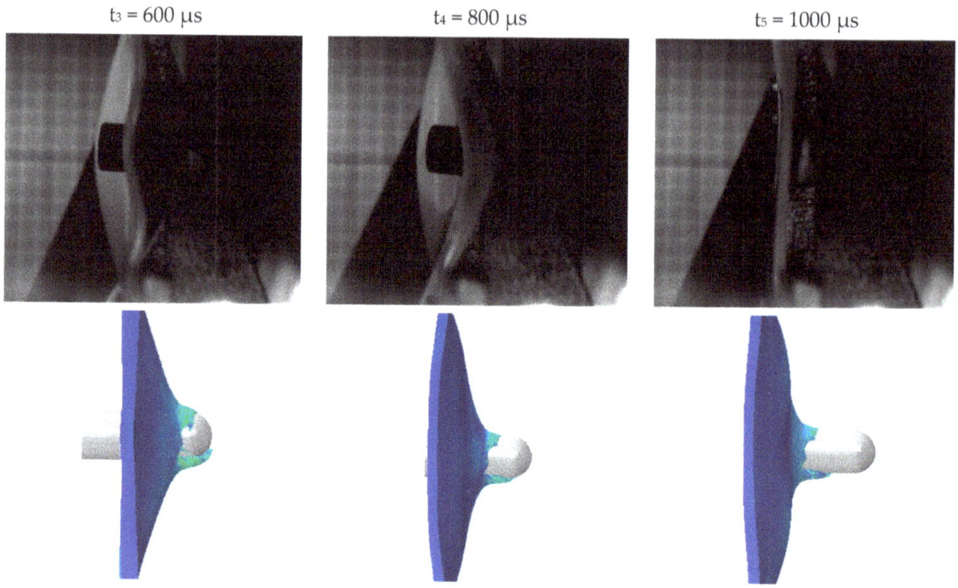

Figure 12. Comparison of the penetration failure process of the target plate in simulation and experiment.

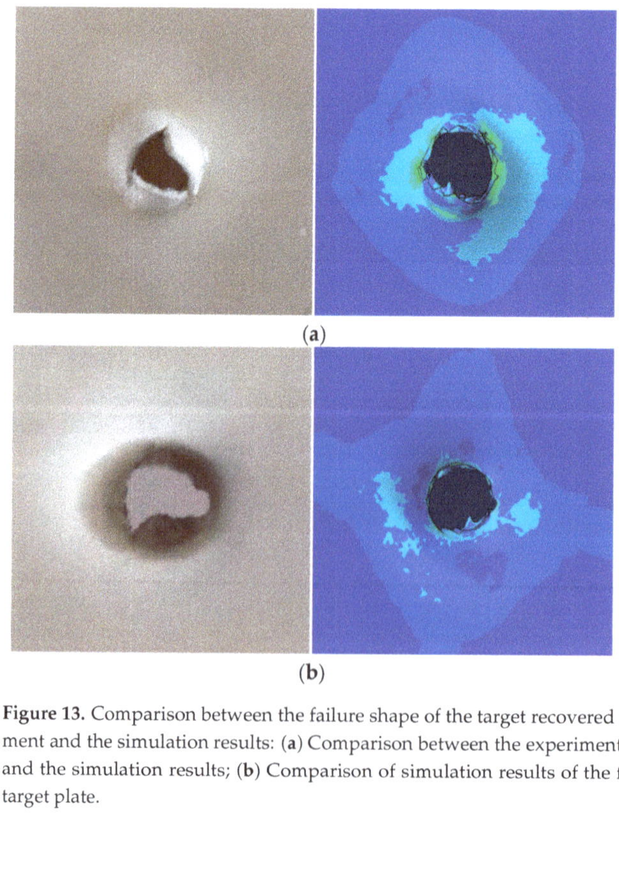

Figure 13. Comparison between the failure shape of the target recovered in the penetration experiment and the simulation results: (**a**) Comparison between the experiment on the back of the target and the simulation results; (**b**) Comparison of simulation results of the frontal experiment of the target plate.

4. Conclusions

PC/ABS composites were subjected to uniaxial tensile tests at different speeds and temperatures. A high-speed video camera filmed the samples' deformation process while they were at room temperature. Based on the experimental results, this paper used the inverse method to determine the J-C constitutive model that can be used to describe such a composite. The high-speed impact tensile tests of the PC/ABS composite were simulated numerically using Abaqus 2020 and the determined constitutive model and compared with the test results. In addition, the numerical simulation can reproduce the test process recorded by a high-speed camera in combination with the failure parameters obtained from the DIC analysis. We carried out the design of target penetration experiments simulation. The penetration simulation reproduces the experimental process and the failure model is correct. The main conclusions are as follows:

(1) The results of tensile tests at different temperatures and different strain rates show that such composite have more obvious temperature effect and strain rate effect, the yield stress decreases with the increase in temperature and increases with the increase in strain rate. The yield stress is linearly dependent on temperature and the logarithm of the strain rate;
(2) The J-C constitution established by the inversion method in this paper has high accuracy and is applicable to the PC/ABS composite. The failure behavior of the material at different temperatures and strain rates can be predicted;
(3) Based on the local deformation of the sample recorded by the DIC technique, the fracture strain of the PC/ABS composite can be deduced. This fracture strain can accurately reproduce the fracture behavior of uniaxial tensile materials.

Author Contributions: J.Z.: investigation, writing—original draft, visualization. Z.X.: investigation, writing—review and editing. D.M.: funding acquisition, writing—review and editing. H.W.: writing—review and editing, data curation. All authors have read and agreed to the published version of the manuscript.

Funding: This research was financially supported by "the Ningbo 2025 Science and Technology Innovation Major Project (2021Z100)".

Institutional Review Board Statement: Not applicable.

Informed Consent Statement: Not applicable.

Data Availability Statement: The raw/processed data required to reproduce these findings cannot be shared at this time as the data also form part of an ongoing study.

Conflicts of Interest: The authors declare that they have no known competing financial interests or personal relationships that could have appeared to influence the work reported in this paper.

References

1. Dundar, M.A.; Dhaliwal, G.S. Investigation for impact behavior of acrylonitrile-butadiene-styrene amorphous thermoplastic. *Polym. Test.* **2020**, *89*, 106624. [CrossRef]
2. Cao, K.; Ma, X.; Zhang, B.; Wang, Y.; Wang, Y. Tensile behavior of polycarbonate over a wide range of strain rates. *Mater. Sci. Eng. A* **2010**, *527*, 4056–4061. [CrossRef]
3. Graziano, A.; Titton Dias, O.A.; Petel, O. High-strain-rate mechanical performance of particle-and fiber-reinforced polymer composites measured with split Hopkinson bar: A review. *Polym. Compos.* **2021**, *42*, 4932–4948. [CrossRef]
4. Tambrallimath, V.; Keshavamurthy, R.; Saravanabavan, D.; Koppad, P.G.; Kumar, G.P. Thermal behavior of PC-ABS based graphene filled polymer nanocomposite synthesized by FDM process. *Compos. Commun.* **2019**, *15*, 129–134. [CrossRef]
5. Yin, Z.N.; Wang, T.J. Deformation response and constitutive modeling of PC, ABS and PC/ABS alloys under impact tensile loading. *Mater. Sci. Eng. A* **2010**, *527*, 1461–1468. [CrossRef]
6. Zheng, B.; Wang, H.; Huang, Z.; Zhang, Y.; Zhou, H.; Li, D. Experimental investigation and constitutive modeling of the deformation behavior of Poly-Ether-Ether-Ketone at elevated temperatures. *Polym. Test.* **2017**, *63*, 349–359. [CrossRef]
7. Wang, H.; Zhou, H.; Huang, Z.; Zhang, Y.; Qiao, H.; Yu, Z. Experimental investigation and modeling of the mechanical behavior of PC/ABS during monotonic and cyclic loading. *Polym. Test.* **2016**, *50*, 216–223. [CrossRef]

8. Louche, H.; Piette-Coudol, F.; Arrieux, R.; Issartel, J. An experimental and modeling study of the thermomechanical behavior of an ABS polymer structural component during an impact test. *Int. J. Impact Eng.* **2009**, *36*, 847–861. [CrossRef]
9. Hund, J.; Naumann, J.; Seelig, T. An experimental and constitutive modeling study on the large strain deformation and fracture behavior of PC/ABS blends. *Mech. Mater.* **2009**, *124*, 132–142. [CrossRef]
10. Duan, Y.; Saigal, A.; Greif, R. Analysis of multiaxial impact behavior of polymers. *Polym. Eng. Sci.* **2002**, *42*, 395–402. [CrossRef]
11. Duan, Y.; Saigal, A.; Greif, R. Impact behavior and modeling of engineering polymers. *Polym. Eng. Sci.* **2003**, *43*, 112–124. [CrossRef]
12. Rittel, D. On the conversion of plastic work to heat during high strain rate deformation of glassy polymers. *Mech. Mater.* **1999**, *31*, 131–139. [CrossRef]
13. Lam, Y.C.; Khoddam, S.; Thomson, P.F. Inverse com-putational method for constitutive parameters ob-tained from torsion,plane-strain and axisymmetric compression tests. *J. Mater. Process. Technol.* **1998**, *83*, 62–71. [CrossRef]
14. Zhou, J.M.; Qi, L.H.; Chen, G.D. New inverse method for identification of constitutive parameters. *Trans. Nonferrous Met. Soc. China* **2006**, *16*, 148–152. [CrossRef]
15. Samantaray, D.; Mandal, S.; Bhaduri, A.K. A comparative study on Johnson Cook, modified Zerilli–Armstrong and Arrhenius-type constitutive models to predict elevated temperature flow behaviour in modified 9Cr–1Mo steel. *Comput. Mater. Sci.* **2009**, *47*, 568–576. [CrossRef]
16. Milani, A.S.; Dabboussi, W.; Nemes, J.A.; Abeyaratne, R.C. An improved multi-objective identification of Johnson-Cook material parameters. *Int. J. Impact Eng.* **2009**, *36*, 294–302. [CrossRef]
17. Springmann, M.; Kuna, M. Identification of material parameters of the Gurson-Tvergaard-Needleman model by combined experimental and numerical techniques. *Comput. Mater. Sci.* **2005**, *32*, 544–552. [CrossRef]
18. Lauro, F.; Bennani, B.; Croix, P.; Oudin, J. Identification of the damage parameters for anisotropic materials by in-verse technique:Application to an aluminium. *J. Mater. Process. Technol.* **2001**, *118*, 472–477. [CrossRef]
19. Kim, H.; Park, E.; Kim, S.; Park, B.; Kim, N.; Lee, S. Experimental study on mechanical properties of single-and dual-material 3D printed products. *Procedia Manuf.* **2017**, *10*, 887–897. [CrossRef]
20. Huh, H.; Kang, W.J.; Han, S.S. A tension split Hopkinson bar for investigating the dynamic behavior of sheet metals. *Exp. Mech.* **2002**, *42*, 8–17. [CrossRef]
21. Rusinek, A.; Zaera, R.; Klepaczko, J.R.; Cheriguene, R. Analysis of inertia and scale effects on dynamic neck formation during tension of sheet steel. *Acta Mater.* **2005**, *53*, 5387–5400. [CrossRef]
22. Islam, M. An exploratory analysis of the effects of speed limits on pedestrian injury severities in vehicle-pedestrian crashes. *J. Transp. Health* **2023**, *28*, 101561. [CrossRef]
23. Tiwari, A.; Wiener, J.; Arbeiter, F.; Pinter, G.; Kolednik, O. Application of the material inhomogeneity effect for the improvement of fracture toughness of a brittle polymer. *Eng. Fract. Mech.* **2020**, *224*, 106776. [CrossRef]
24. Lim, S.J.; Huh, H. Ductile fracture behavior of BCC and FCC metals at a wide range of strain rates. *Int. J. Impact Eng.* **2022**, *159*, 104050. [CrossRef]
25. Lewis, J.; Pasco, J.; McCarthy, T.; Chadha, K.; Harding, M.; Aranas Jr, C. High strain rate and high temperature mechanical response of additively manufactured alloy 625. *J. Manuf. Process.* **2022**, *81*, 922–944. [CrossRef]
26. Xu, H.; Zhou, J.; Cao, X.; Miao, C. A viscoelastic-viscoplastic thermo-mechanical model for polymers under hypervelocity impact. *Int. J. Mech. Sci.* **2024**, *272*, 109205. [CrossRef]
27. Li, B.; Zhu, Z.; Ning, J.; Li, T.; Zhou, Z. Viscoelastic–plastic constitutive model with damage of frozen soil under impact loading and freeze–thaw loading. *Int. J. Mech. Sci.* **2022**, *214*, 106890. [CrossRef]
28. Nahar, C.; Sanariya, S.; Gurrala, P.K. Numerical simulation of polymers at low and moderate strain rates. *Mater. Today Proc.* **2021**, *44*, 696–700. [CrossRef]
29. Somarathna, H.M.C.C.; Raman, S.N.; Mohotti, D.; Mutalib, A.A.; Badri, K.H. Hyper-viscoelastic constitutive models for predicting the material behavior of polyurethane under varying strain rates and uniaxial tensile loading. *Constr. Build. Mater.* **2020**, *236*, 117417. [CrossRef]
30. Korkmaz, M.E.; Yaşar, N.; Günay, M. Numerical and experimental investigation of cutting forces in turning of Nimonic 80A superalloy. *Eng. Sci. Technol. Int. J.* **2020**, *23*, 664–673. [CrossRef]
31. Jia, Z.; Guan, B.; Zang, Y.; Wang, Y.; Mu, L. Modified Johnson-Cook model of aluminum alloy 6016-T6 sheets at low dynamic strain rates. *Mater. Sci. Eng. A* **2021**, *820*, 141565. [CrossRef]
32. Zhu, Q.; Sun, B.; Zhou, Y.; Sun, W.; Xiang, J.W. Sample Augmentation for Intelligent Milling Tool Wear Condition Monitoring Using Numerical Simulation And Generative Adversarial Network. *IEEE Trans. Instrum. Meas.* **2021**, *70*, 1–10. [CrossRef]
33. Pan, Z.; Sun, B.; Shim, V.P.; Gu, B. Transient heat generation and thermo-mechanical response of epoxy resin under adiabatic impact compressions. *Int. J. Heat Mass Transfer* **2016**, *95*, 874–889. [CrossRef]
34. Golewski, G.L. Comparative measurements of fracture toughgness combined with visual analysis of cracks propagation using the DIC technique of concretes based on cement matrix with a highly diversified composition. *Theor. Appl. Fract. Mech.* **2022**, *121*, 103553. [CrossRef]
35. Chakraborty, P.; Kumar, R.; Tiwari, V. Parametric optimization of constitutive and failure model to numerically simulate the dynamic fracture response of AA7475-T7351 under extreme thermomechanical loading. *Eng. Fract. Mech.* **2024**, *298*, 109930. [CrossRef]

36. Van Breemen, L.C.; Engels, T.A.; Klompen, E.T.; Senden, D.J.; Govaert, L.E. Rate-and temperature-dependent strain softening in solid polymers. *J. Polym. Sci. Part B Polym. Phys.* **2012**, *50*, 1757–1771. [CrossRef]
37. Kapoor, R.; Nemat-Nasser, S. Determination of temperature rise during high strain rate deformation. *Mech. Mater.* **1998**, *27*, 1–12. [CrossRef]
38. Wu, H.; Fang, Q.; Chen, X.W.; Gong, Z.M.; Liu, J.Z. Projectile penetration of ultra-high performance cement based composites at 510–1320 m/s. *Constr. Build. Mater.* **2015**, *74*, 188–200. [CrossRef]

materials

MDPI

Article

Identification of Apple Fruit-Skin Constitutive Laws by Full-Field Methods Using Uniaxial Tensile Loading

Teresa Campos [1,2], Rafael Araújo [3], José Xavier [4,5,*], Quyền Nguyễn [6], Nuno Dourado [1,2], José Morais [3] and Fábio Pereira [3]

1 CMEMS-UMINHO, Universidade do Minho, 4800-058 Guimarães, Portugal; nunodourado@dem.uminho.pt (N.D.)
2 LABBELS–Associate Laboratory, 4800-058 Guimarães, Portugal
3 CITAB/UTAD, Departamento de Engenharias, Quinta de Prados, 5001-801 Vila Real, Portugal; jmorais@utad.pt (J.M.); famp@utad.pt (F.P.)
4 UNIDEMI, Department of Mechanical and Industrial Engineering, NOVA School of Science and Technology, Universidade NOVA de Lisboa, 2829-516 Caparica, Portugal
5 LASI, Intelligent Systems Associate Laboratory, 4800-058 Guimarães, Portugal
6 2C2T-Centro de Ciência e Tecnologia Têxtil, Universidade do Minho, 4800-058 Guimarães, Portugal
* Correspondence: jmc.xavier@fct.unl.pt

Abstract: The protective and preservative role of apple skin in maintaining the integrity of the fruit is well-known, with its mechanical behaviour playing a pivotal role in determining fruit storage capacity. This study employs a combination of experimental and numerical methodologies, specifically utilising the digital image correlation (DIC) technique. A specially devised inverse strategy is applied to evaluate the mechanical behaviour of apple skin under uniaxial tensile loading. Three apple cultivars were tested in this work: Malus domestica Starking Delicious, Malus pumila Rennet, and Malus domestica Golden Delicious. Stress–strain curves were reconstructed, revealing distinct variations in the mechanical responses among these cultivars. Yeoh's hyperelastic model was fitted to the experimental data to identify the coefficients capable of reproducing the non-linear deformation. The results suggest that apple skin varies significantly in composition and structure among the tested cultivars, as evidenced by differences in elastic properties and non-linear behaviour. These differences can significantly affect how fruit is handled, stored, and transported. Thus, the insights resulting from this research enable the development of mathematical models based on the mechanical behaviour of apple tissue, constituting important data for improvements in the economics of the agri-food industry.

Keywords: apple skin; hyperelasticity; digital image correlation; uniaxial tensile loading; finite element analysis

Citation: Campos, T.; Araújo, R.; Xavier, J.; Nguyễn, Q.; Dourado, N.; Morais, J.; Pereira, F. Identification of Apple Fruit-Skin Constitutive Laws by Full-Field Methods Using Uniaxial Tensile Loading. *Materials* **2024**, *17*, 700. https://doi.org/10.3390/ma17030700

Academic Editors: Madhav Baral and Charles Lu

Received: 27 December 2023
Revised: 25 January 2024
Accepted: 26 January 2024
Published: 1 February 2024

1. Introduction

Skin appearance and cracking are significant causes of fruit value decrease and losses. It is well recognised that fruit skin (FS; exocarp) is subjected to a complex stress field during fruit growth, harvesting, storage and transportation. Additionally, the complex mechanical behaviour of FS makes the identification of its constitutive laws a challenging theme, presenting difficulties from both theoretical and experimental perspectives. Understanding the relationship between FS composition and morphology and its mechanical behaviour is an important research topic that has a clear economic impact on the agri-food industry.

The majority of existing works on this issue involve the evaluation of material parameters of fruit tissues (e.g., rupture force, rupture energy and firmness), aiming to establish relationships between mechanical properties and production and post-harvesting factors. Oey et al. [1] highlighted the influence of turgor on the structural and mechanical properties of apple tissue. Also, Juxia, W. and co-authors [2] studied the biomechanical characteristics of the peels of two apple cultivars, using tensile strength, tear strength and puncture resistance tests. The Allende team [3] focused on evaluating the relationship between the histology of tomato peel and its breaking strength. Grotte M. and his co-workers [4] measured the firmness of the skin and flesh of fruit using puncture tests.

In the above studies, the experimental and data-reduction methods used do not allow for the identification of intrinsic mechanical properties of FS, which are fundamental to numerical studies when estimating fruit mechanical behaviour.

The quality and risk of rupture of FS are usually evaluated by parameters such as rupture force, rupture energy, and firmness. These parameters are often associated with high scatter, primarily due to the mechanical tests as well as the coarse geometric and mechanical simplifications adopted in evaluating these parameters. The classical method used to determine the mechanical properties of FS is the uniaxial tensile test [5,6]. Different loading strategies have been used for strain partition into elastic, plastic, and viscoelastic components [7–9].

It is known that plant tissues have complex mechanical behaviour [10] and most of them are considered to be elastic or viscoelastic [7–9]. However, some reported values of ultimate strains [5,10] suggest that a hyperelastic approach is required, although the material behaviour can change from elastic (viscoelastic) to hyperelastic (hyperviscoelastic) through growth and ripening phases. Bargel and Neinhuis [5] confirm this behaviour, showing the decrease in extensibility of the tomato fruit cuticle at the final stages of ripening. Similarly, Bidhendi et al. [11] noted that stress–strain behaviour of the onion epidermis under tension was remarkably non-linear. These authors evaluated the capacity of a few hyperelastic models to reproduce the non-linear deformation of the onion epidermis using a fitting strategy. Over the years, an extensive amount of hyperelastic constitutive models have been proposed, such as the generalised Fung model [12], the Yeoh model [13,14], and the Holzapfel–Gasser–Ogden model [15,16]. In this context, it remains unclear which specific experimental tests should be conducted to precisely calibrate a hyperelastic model.

In this work, we investigated the hyperelastic behaviour of apple skin coupled with full-field displacement measurements based on the digital image correlation technique. Monotonic tensile tests were performed to obtain stress–strain curves on apple skin specimens in the longitudinal direction. A methodology based on inverse identification considering finite element model updating (FEMU) was adopted to minimise the difference between full-field numerical and experimental displacements. Through this process, the Yeoh model coefficients were determined to replicate the hyperelastic response of the skin accurately. This approach allows for the rational exploration of the connections between mechanical properties, material composition, and morphology.

2. Materials and Methods

2.1. Specimen Preparation

Three apple cultivars, Malus domestica Starking Delicious, Malus pumila Rennet, and Malus domestica Golden Delicious (Figure 1), were tested under the same conditions (optimal stage), one day after harvesting in the Armamar region (Portugal; 41°07′ N 7°41′ O). Five skin samples were extracted from each cultivar from the epicarp region, where lenticel zones were less concentrated. Skin samples were taken from trees with no shading area, without any control over the sunning position.

This was performed using a scalpel to form a rectangle shape (60 × 20 mm) in the longitudinal direction (Figure 2). Adequate razoring was then executed to remove apple pulp until a uniform thickness (0.2 mm) was attained along the entire length of the specimen. Subsequently, a speckle pattern suitable for DIC measurements was applied to the pill side (20 × 20 mm) by spraying black paint using an airbrush over the natural substrate of the apple skin (Figure 2).

(**a**)

(**b**)

(**c**)

Figure 1. Apple cultivars: (**a**), Malus pumila Rennet; (**b**) Malus domestica Golden Delicious; and (**c**) Malus domestica Starking Delicious.

Figure 2. Specimen orientation ($L = 60$ mm , $B = 20$ mm).

2.2. Uniaxial Tensile Test

Tensile tests were performed in the apple skin specimens (in the longitudinal direction: pedicel to calix) using a servo-electrical testing machine (Micro-Tester INSTRON 5848; 2 kN load-cell, INSTRON, MA, USA), with the crosshead displacement rate set to 3 mm/min (Figure 3a). To prevent water loss, the test was completed immediately after the specimen preparation, at room temperature (25 °C and 65 RH). Specimen slipping in the grips was avoided by cautious grip tightening during the specimen setting. Load–displacement curves were monitored by setting the acquisition rate to 5 Hz while capturing images for DIC (digital image correlation) measurements (Figure 3b) with a frequency of 1 Hz. These time steps allowed for the synchronisation of the load data with the acquired DIC images. Light distribution was adequately chosen to allow suitable contrast for DIC measurements (Figure 3c).

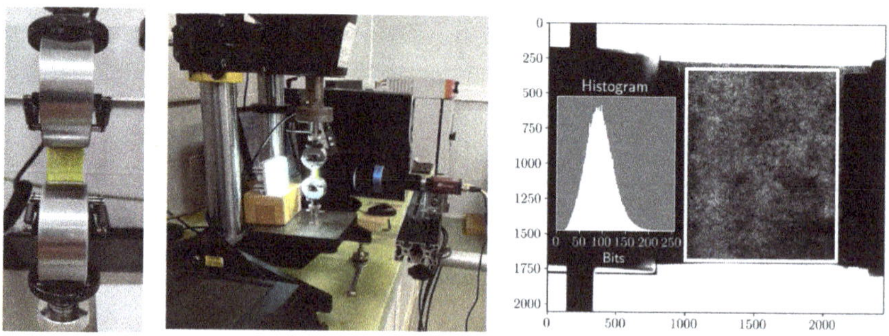

(**a**) (**b**) (**c**)

Figure 3. (**a**) Detail of the uniaxial tensile test; (**b**) Experimental setup showing DIC acquisition system; (**c**) DIC pattern with the corresponding histogram.

2.3. Digital Image Correlation

2.3.1. Optical System and Speckle Pattern

The integration of the 2D-DIC technique with mechanical tests was chosen due to its suitability for conducting contactless and full-field measurements across the fruit-skin substrate. This approach eliminates the need for gluing-based systems, such as strain gauges or Bragg gratings, traditionally used to assess strains [17–19]. An 8-bit camera (Baumer Optronic GmbH, Radeberg, Germany, model FWX20) with a telecentric lens (Opto Engineering SRL, Mantova, Italy, model TC 13 36) was used, as reported in Table 1. Images were recorded at an acquisition frequency of 1 Hz, and the working distance was set to 103.5 mm, with a fixed magnification factor of 4.4 μm/pixel. This camera sensor was positioned perpendicular to the flat surface of sample while ensuring there was appropriate lighting with a white-light LED (Table 1). The MatchID DIC software was used to process the DIC analysis.

Table 1. Optical devices and DIC data.

CCD Camera	Model	Baumer® Optronic FWX20 (8 bts)
	Pixel resolution	1624 × 1236 pixels, 4.4 μm/pixel
	Shutter time	0.7 ms
	Acquisition frequency	1 Hz
Lens	Model	Opto Engineering Telecentric lens TC 23 36
	Magnification	0.243 ± 3%
	Field of view (1/1.8″)	29.3 × 22.1 mm^2
	Working distance	103.5 ± 3 mm
	Working F-number	f/8
	Field depth	11 mm
	Conversation Factor	0.018 mm/pixel
Lighting	LED	Raylux 25
Speckle Pattern	Painting technique	Airbrush (nozzle set of 0.2 mm)
	Average speckle size	6 pixels ǀ 21.5 μm

2.3.2. DIC Setting: Parametric Analysis

When conducting DIC measurements, it is essential to perform a convergence analysis on the intrinsic parameters that govern the numerical imaging correlation method [17,20].

This is particularly crucial for biological materials that exhibit natural heterogeneities at the observation scale. The DIC parameters significantly impact spatial resolution and accuracy in both displacement and strain measurements [21]. A parametric study was conducted with the MatchID Performance Analysis Tool [22], allowing for multiple DIC analyses on the same image set and considering various parameter combinations. Each point on the DIC setting space corresponds to a specific spatial resolution. To quantify this metric, we used the Virtual Strain Gauge (VSG) measure [22,23]:

$$\mathrm{VSG} = [(\mathrm{SW} - 1) \times \mathrm{ST}] + \mathrm{SS}\ [\mathrm{pixels}] \quad (1)$$

where SW refers to the strain window, determining the number of data points in the fitting polynomial approach; ST stands for the subset step; and SS represents the subset size.

The physical units of millimetres for the VSG can be achieved by considering the image conversion factor of the optical system (see Table 1). Each parameter was systematically examined within a defined design space with specified minimum and maximum values (refer to Table 2). We assessed the reconstruction of the strain component ε_{xx} at the specimen centre, observing its variation in response to the preselected DIC setting parameters.

Table 2. Parameters used in the Performance Analysis Tool of MatchID 2D DIC.

Subset-based settings	
Subset size	SS ∈ {21 + 4n ∣ n =10} pixel
Subset step	ST = 10 pixel (fixed)
Shape function	{Affine, Quadratic}
Strain reconstruction-based settings	
Strain window	SW ∈ {3 + 2n ∣ n = 8} data points
Polynomial order *	Bilinear (Q4), Biquadratic (Q8)
Strain convention	Green–Lagrange

* Local least-squares fitting approach for the strain evaluation.

Figure 4 shows the ε_{xx} strain signal at the central reference point as a function of the VSG. The data points visibly converge to a plateau with increasing VSG values, suggesting an average strain at that specific point. This analysis revealed a compromise that balanced spatial resolution and accuracy. Finally, Table 3 summarises the DIC settings selected in this study to report the full-field data.

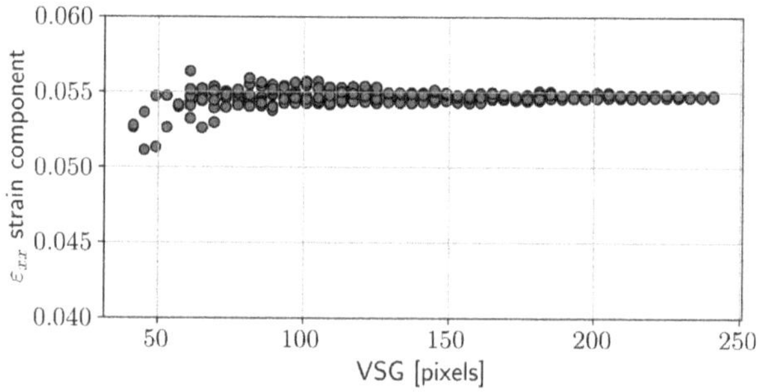

Figure 4. Evaluation of ε_{xx} as a function of the coordinate x for several Virtual Strain Gauge (VVSG) values for both tension and compression tests.

Table 3. DIC and strain settings selected in the full-field measurements.

DIC settings	
Correlation criterion Interpolant Subset shape function	ZNCC Bicubic spline Quadratic
Subset size	41 pixels
Step size	10 pixels
Image pre-filtering	Gaussian, 5-pixel kernel
Strain settings	
Strain window size Strain interpolation Strain convention	Nine data points Quadratic Q4 Green–Lagrange

2.4. *Procedure to Assess the Material Law*

Evolutionary algorithms (EAs) have proven to be effective when applied to the iterative optimisation of FEMU updating. This iterative process involves defining an objective function based on the differences between the nodal displacements in experimental and numerical datasets. Noteworthy applications encompass tasks such as determining viscoelastic properties for wood-based panels [24], identifying elastic parameters at fibre–matrix interfaces in composite materials [25], optimising parameters for hyperelastic cardiac materials [26], and establishing damage parameters for large-scale structural systems [27].

Furthermore, EAs offer the advantage of not being heavily reliant on the initial solution, thus mitigating the risk of becoming trapped in local minima during the optimisation process [24,28].

The Yoeh, Odgen, and Neo-Hookean hyperelastic models are the most recommended hyperelastic models in the literature for simulating the mechanical behaviour of various biological tissues. MCalibration 5.1.2 software (Veryst Engineering, Needham, MA, USA) was used to identify which of the three suggested models could most accurately replicate the experimental outcome (i.e., stress–strain curves). Experimental stress–strain ($\sigma = f(\varepsilon)$) data ensuing from the tensile tests was used as input to identify the material law. The objective function to be minimised was set as the mean square difference (MSD) between the numerical output and the experimental data ($\sigma = f(\varepsilon)$). An extensive search for optimal parameters minimising the fitness function was performed, which included an initial random search, followed by the application of an EA. This search strategy (i.e., EA) was offered by MCalibration as a suitable methodology to identify the hyperelastic coefficients, thus leading to the most suitable agreement with the experimental true $\sigma - \varepsilon$ curve. The accuracy of the obtained solutions was heavily contingent upon the selected evolutionary parameters, such as the population size, crossover rate, and mutation rate (Table 4). In this process, Yeoh's model demonstrated more suitability in replicating the experimental response.

Table 4. Parameters used in the evolutionary search.

Population Size	Crossover Rate	Mutation Rate	Maximum Number of Generations
100	0.05	0.95	1000

The output of this procedure is the identification of the set of coefficients defining the material law, allowing us to replicate the experimental (mechanical) behaviour under tensile loading. To this aim, in-plane stress analyses were conducted considering a finite element model formed by 3321 shell (6-node) quadrilateral finite elements, using boundary conditions to ensure admissible kinematic requirements for the tensile test, as illustrated in Figure 5a.

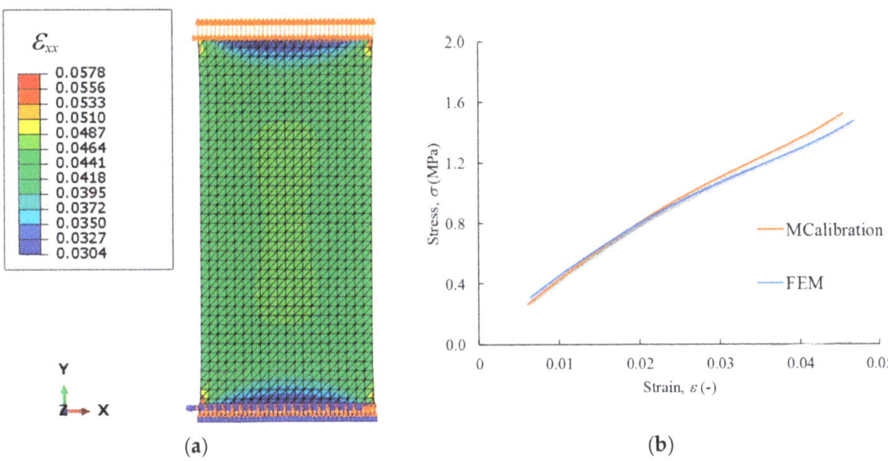

Figure 5. Numerical results: (**a**) FE mesh showing the strain field along the loading direction; and (**b**) the attained stress–strain agreement of FE and MCalibration.

3. Hyperelastic Law

The approach adopted in this work starts with the basic assumption that apple skin exhibits an isotropic behaviour. Concerning the hyperelastic models currently used in commercial FEM software, the deformation gradient F and the right Cauchy–Green tensor C are decomposed into dilatational and distortional components,

$$\mathbf{F} = J^{\,1/3}\,\mathbf{F}^* \tag{2}$$

$$\mathbf{C} = \mathbf{F}^{\mathrm{T}}\mathbf{F} = J^{\,2/3}\mathbf{C}^* \tag{3}$$

with F^* and C^* being the distortional components and $J = \det(F)$ being the dilatational component. Accordingly, the first and second invariants of C^* are related to the invariants of the right Cauchy–Green tensor by the equations:

$$I_1^* = J^{-\frac{2}{3}} I_1 \text{ and } I_2^* = J^{-\frac{4}{3}} I_2 \tag{4}$$

I_1, I_2 and J^2 are the invariants of the right Cauchy–Green tensor

$$\begin{gathered} I_1 = \lambda_1^2 + \lambda_2^2 + \lambda_3^2 \\ I_2 = \lambda_1^2\lambda_2^2 + \lambda_2^2\lambda_3^2 + \lambda_3^2\lambda_1^2 \\ I_3 = \lambda_1^2\lambda_2^2\lambda_3^2 = J^2 \end{gathered} \tag{5}$$

and λ_1, λ_2 and λ_3 are the principal extension ratios.

The general constitutive equation of hyperelastic materials provdes the Cauchy stress tensor σ as a function of the deformation gradient F by

$$\sigma(F) = \frac{1}{J}\frac{\partial\psi(F)}{\partial F}F^{\mathrm{T}} \tag{6}$$

$\psi(F)$ is the strain-energy density function. The restrictions imposed by the principle of material frame indifference imply that the strain-energy density function depends only on the stretch tensor (U) component of F or, equivalently, on $C = U^2$:

$$\psi(F) = \hat{\psi}(C) \tag{7}$$

The chain rule yields the following relationship between the partial derivatives of the two strain-energy density functions with respect to their tensor arguments:

$$\frac{\partial\psi(F)}{\partial F} = 2F\frac{\partial\hat{\psi}(C)}{\partial C} \tag{8}$$

Hence, combining Equations (6) and (8), the Cauchy stress can be expressed as

$$\sigma(C) = \frac{2}{J}F\frac{\partial\hat{\psi}(C)}{\partial C}F^T \tag{9}$$

For isotropic materials, the strain-energy density function is a function of the right Cauchy–Green tensor via its invariants (4 and 5), and can be written as $\widetilde{\psi}\left(I_1^*(I_1, J), I_2^*(I_2, J), J\right)$. Thus, the Cauchy stress tensor (9) becomes:

$$\sigma(I_1^*, I_2^*, J) = \frac{J}{2}\left(\frac{\partial\widetilde{\psi}}{\partial I_1^*}\frac{\partial I_1^*}{\partial I_1} + J^{2/3}I_1^*\frac{\partial\widetilde{\psi}}{\partial I_2^*}\frac{\partial I_2^*}{\partial I_2}\right)\boldsymbol{b} - \frac{2}{J}\frac{\partial\widetilde{\psi}}{\partial I_2^*}\frac{\partial I_2^*}{\partial I_2}\boldsymbol{b}^2 + \left(\frac{\partial\widetilde{\psi}}{\partial I_1^*}\frac{\partial I_1^*}{\partial J} + \frac{\partial\widetilde{\psi}}{\partial I_2^*}\frac{\partial I_2^*}{\partial J} + \frac{\partial\widetilde{\psi}}{\partial J}\right)\boldsymbol{I} \tag{10}$$

where $\boldsymbol{b} = \boldsymbol{F}\boldsymbol{F}^T$ is the left Cauchy–Green tensor and $\boldsymbol{I}$ is the identity tensor.

$$\hat{\psi}(C_{ij}; D_i) = \sum_{i+j=1}^{N} C_{ij}(I_1^* - 3)^i (I_2^* - 3)^j + \sum_{i=1}^{N}\frac{1}{D_i}(J-1)^{2i} \tag{11}$$

where C_{ij} and D_i represent material parameters. A particular version of the polynomial model is the Yeoh model:

$$\hat{\psi}(C_{10}, C_{20}, C_{30}, \kappa) = C_{10}(I_1^* - 3) + C_{20}(I_1^* - 3)^2 + C_{30}(I_1^* - 3)^3 + \frac{\kappa}{2}(J-1)^2 \quad (12)$$

with κ representing the bulk modulus.

4. Results and Discussion

4.1. Uniaxial Tensile Tests

Figure 6 shows the load–displacement (P-δ) curves obtained cyclically, with increasing load levels at a constant crosshead displacement rate (0.5 mm/min) until the final rupture (last cycle). These results show that the area enclosed by the loading–unloading curves, representing the energy per unit volume, gradually increases with the load value defined to invert the crosshead, thus configuring a hysteresis loop (a history-dependent property). This shows the increase in width in the hysteresis loop strain with load amplitude and the rise in compliance with the number of load cycles in the loading branch, with the latter effect being referred to as softening behaviour. Another characteristic identified in those plots regards the unrecoverable strain in the material following unloading, which is more visible as the load value defined to invert the crosshead increases. Figure 6 also illustrates that the material can no longer replicate the same non-linear behaviour, as the number of cycles increases, as observed from $N = 1$ to $N = 6$. As a result, it is obvious that the load–unload curves obtained from mechanical testing on apple skin exhibit non-linear elastic behaviour. This response suggests that following pure elasticity, these materials can tolerate a high intensity of load and even exceed 100% deformation without damage [29].

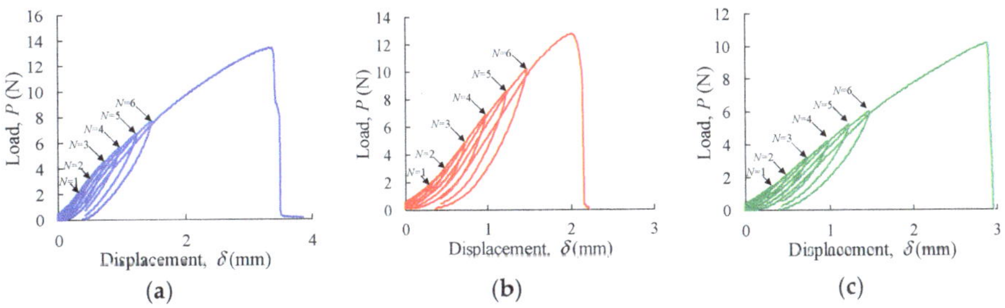

Figure 6. Load–unloading cycles in: (**a**) Malus domestica Starking Delicious; (**b**) Malus pumila Rennet; and (**c**) Malus domestica Golden Delicious.

Figure 7 shows the set of experimental P-δ curves obtained in monotonic tensile tests for the tested apple cultivars. These results evidence the existence of a linear trend in the load–displacement curve (up to point A), followed by non-linear behaviour until total failure.

DIC processing has revealed a set of strain cartographies along the loading axis, i.e., ε_{xx} (Figure 8a,d,g), where an extensive homogeneous region was formed in the specimen central area. Mean values of strain ε_{xx} were calculated considering the measurements performed in that region, which were correlated with the applied load to plot the stress–strain curves for each apple cultivar (Figure 9a–c). Those results allowed us to conclude the existence of a low scatter of the ensued experimental data. Figure 9a–c also presents the result of regressions derived from fitting procedures using power-law distributions for each cultivar, showing clear differences.

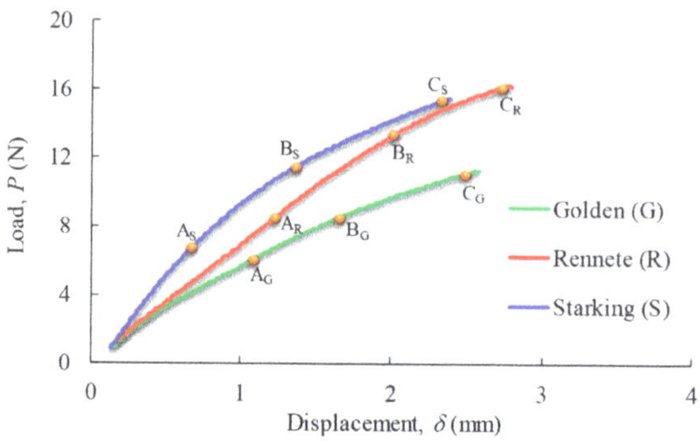

Figure 7. Typical load–displacement curves under tensile loading. Points A, B, and C identify the following stages: within the elastic response, the occurrence of relevant strain gradients (lenticels identification), and the crack onset of apple skin, respectively.

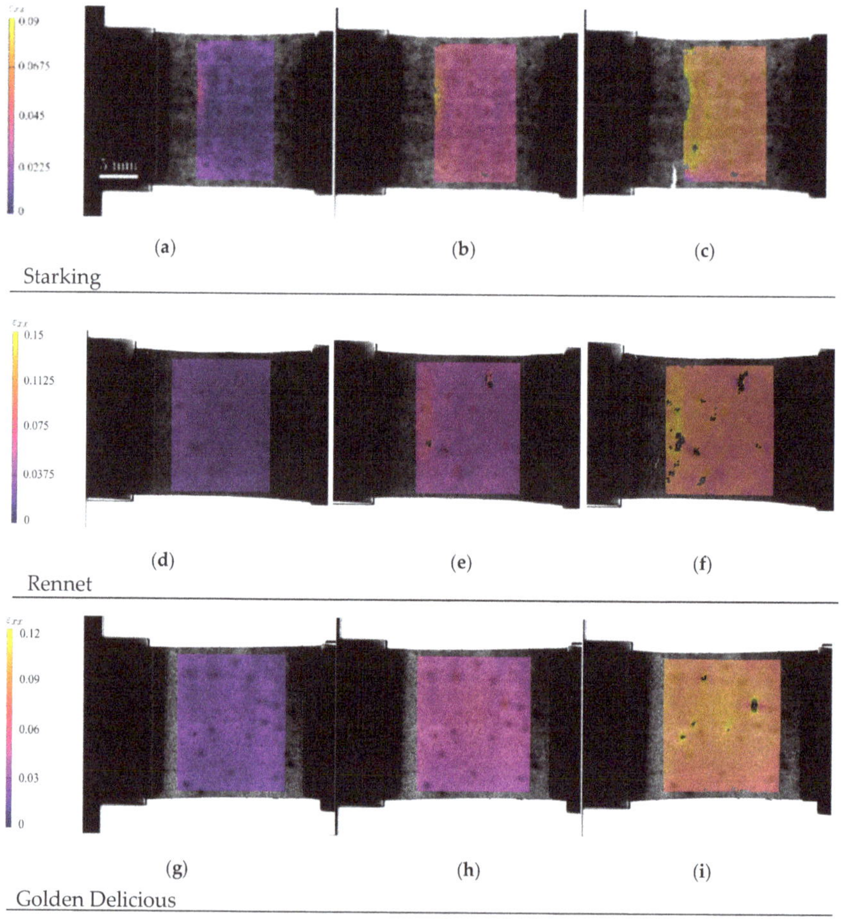

Figure 8. Strain cartographies of Starking Delicious, Rennet, and Golden Delicious in the x direction for longitudinal loading (ε_{xx}), obtained in three load phases according to Figure 7 in points: (**a**) A_S; (**b**) B_S; (**c**) C_S; (**d**) A_R; (**e**) B_R; (**f**) C_R; (**g**) A_G; (**h**) B_G; and (**i**) C_G.

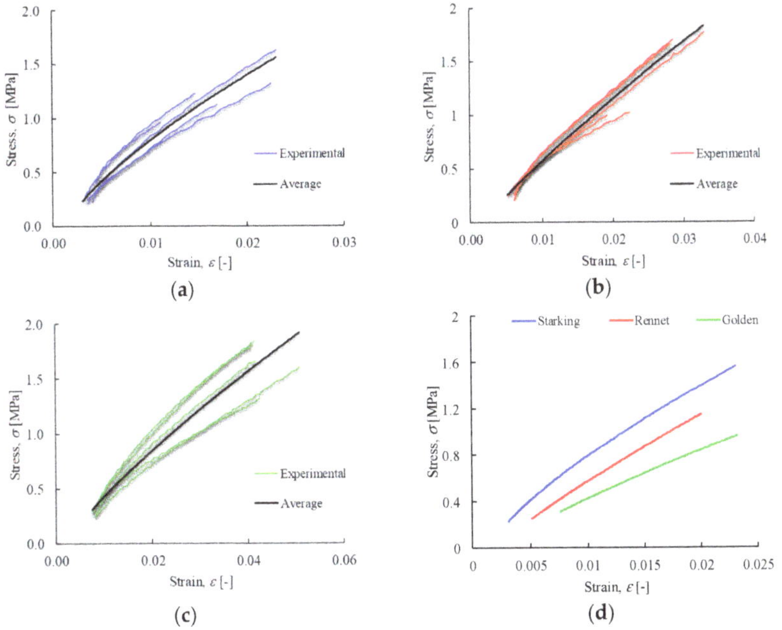

Figure 9. True stress–strain curves of apple cultivars: (**a**) Malus domestica Starking Delicious; (**b**) Malus pumila Rennet; (**c**) Malus domestica Golden Delicious; and (**d**) corresponding average curves.

Figure 8b,e,h allow us to identify strain gradients (points B in Figure 7) in particular regions of the skin apple samples, whose origin was later associated with existing anatomic structures in apple skin, namely lenticels (Figure 10). The existence of these structures in the apple skin allows for ensuring fruit oxygen and other gas exchanges, which are characteristic of the secondary plant body (rather than the stomata, or pores in the epidermis, in the primary plant). In fact, lenticels appear as a result of microcracks in the bark cuticle, similar to the formation of areas known as russet. While lenticels form small spots on the skin, russet covers large areas of the fruit. Both phenomena involve periderm formation in response to rupture in the cuticle, influencing skin mechanical behaviour [30].

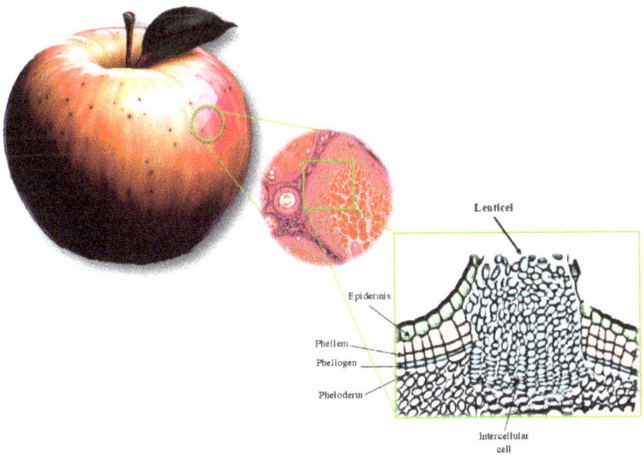

Figure 10. Lenticels schematic illustration.

In regards to Figure 8c,f,i, with correspondence to point C in Figure 7, crack initiation and propagation are visible in the set of regions previously identified as highly affected by strain concentrations (Figure 8b,e,h), where damage onset was under development.

4.2. Numerical Agreement

A procedure was developed to identify the most suitable constitutive model to mimic the hyperelastic behaviour of apple skin under uniaxial tensile loading. The methodology was based on the stress–strain response, exhibiting a pronounced non-linear S-shape (as noticed by [2]). Implementing this approach, the Yeoh model proved very effective in replicating the true stress–strain response, encompassing the non-linear domain (see Figure 11; rescaled to improve visibility). This observation suggests that the adopted methodology is suitable for extrapolating situations beyond experimentally tested conditions, including scenarios involving different strain rates or environmental conditions.

Table 5 presents the set of coefficients used to define the constitutive law characteristic of the Yeoh model (Equation (12)) for each cultivar identified by the inverse method.

Table 5. Material parameters of the Yeoh model identified by the inverse method for each apple cultivar (values C_{10}, C_{20}, C_{30} in MPa).

Starking Delicious							
Material parameters	Specimen 1	Specimen 2	Specimen 3	Specimen 4	Specimen 5	Average	CoV (%)
C_{10}	13.28	30.85	24.37	17.74	13.68	19.99	34%
C_{20}	−2289.75	−56,083.60	−17,502.80	−3771.46	−2056.59	−16,340.8	−127%
C_{30}	509,610	82,163,500	11,920,400	960,077	573,257	19,225,369	165%
Rennet Delicious							
Material parameters	Specimen 6	Specimen 7	Specimen 8	Specimen 9	Specimen10	Average	CoV (%)
C_{10}	11.69	13.61	13.53	13.56	13.68	13.22	6%
C_{20}	−908.19	−1672.44	−4534.94	−1476.87	−4779.09	−2674.31	−61%
C_{30}	123,028	311,363	137,444	252,213	198,770	204,563.6	34%
Golden Delicious							
Material parameters	Specimen 11	Specimen 12	Specimen 13	Specimen 14	Specimen 15	Average	CoV (%)
C_{10}	7.354	10.33	10.30	7.187	10.88	9.21	17%
C_{20}	−304.38	−503.81	−762.77	−384.85	−611.95	−513.55	−32%
C_{30}	16,769	33,889.7	65,864.2	23,801.7	39,268.8	35,918.68	47%

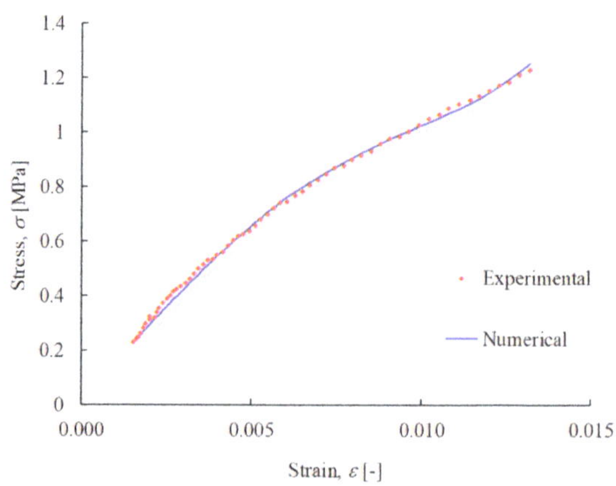

(**a**)

Figure 11. *Cont.*

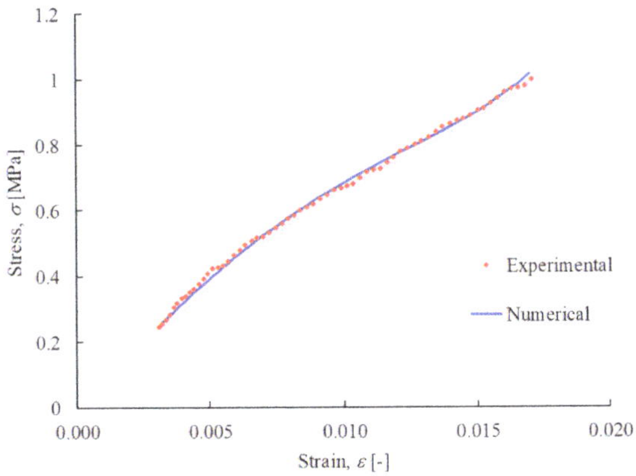

(**b**)

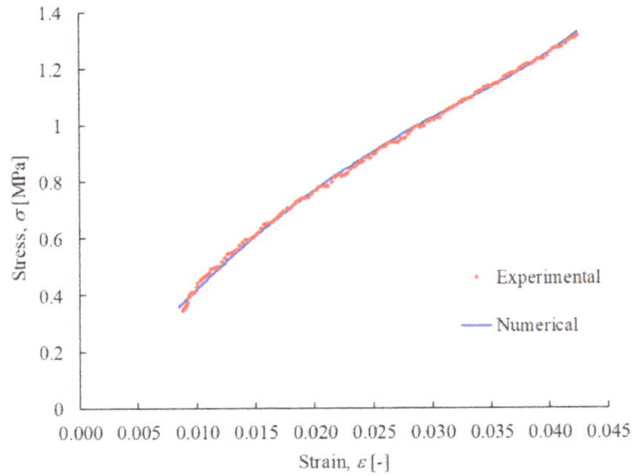

(**c**)

Figure 11. True stress–strain curve showing the obtained numerical agreement: (**a**) Malus domestica Starking Delicious; (**b**) Malus pumila Rennet; and (**c**) Malus domestica Golden Delicious.

The results presented in Table 6 reveal a significant scatter in the identified coefficients regardless of the apple cultivars. Despite this, coefficient C_{10} presents a smaller CoV in comparison to C_{20} and C_{30}, demonstrating a more consistent response to smaller strains (as noted by Yeoh (1990) [13]), both within the individual apple cultivar and among cultivars.

Also, mean values of coefficients C_{10}, C_{20} and C_{30} were determined to perceive eventual differences among the analysed cultivars (Figure 12). Analysing C_{20} and C_{30} for Starking, Rennet and Golden, the difference among these values is very significant, which evidences existing potential intrinsic differences in the apple skin microstructure or composition. Although a negative C_{20} may appear unconventional, one can affirm that there are no real physical issues with this observation [13,14]. The fact that coefficient C_{20} is negative, whereas C_{10} and C_{30} are positive, indicates that the secant shear modulus varies with the deformation in a characteristic way [13,14]. Fruit skin was considered isotropic and nearly incompressible. For the numerical analysis, compressibility values close to 0 were assigned to D_1 (in MPa).

The coefficients for Yeoh's hyperelastic model on the studied apple cultivars are compared in Table 6 with other research on various biological materials. A broad range of values in various biological materials was seen when compared with literature studies, underscoring the uniqueness of the coefficients for each kind of tissue. In comparison to apple skin, the epidermis of onions has much lower values, suggesting a more elastic response. These biological materials have specific mechanical properties that underscore the need to tailor constitutive models to capture their particularities. This underscores the requirement for a customised approach when examining various biological structures.

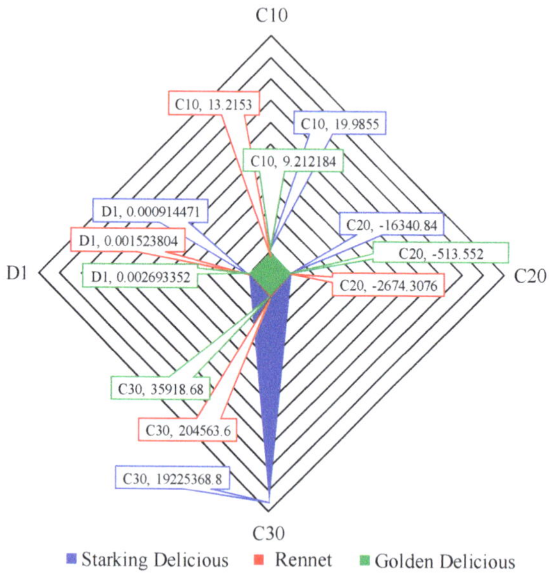

Figure 12. Average of three cultivars.

Table 6. Yeoh's coefficients of different biological tissues.

Material	Yeoh Coefficients		
	C10 (MPa)	**C20 (MPa)**	**C30 (MPa)**
Starking Delicious (This study)	19.99	−16,340.8	19,225,369
Rennet (This study)	13.22	−2674.31	204,563.6
Golden Delicious (This study)	9.21	−513.55	35,918.68
Onion epidermis [11]	0.406	6.68	34
Porcine skin [31]	0.26	15.5	1.75
Pig skin [32]	8.42	8.75	2.35
Arteries branches [33]	0.1067	5.1602	0.0

5. Conclusions

This work follows an innovative approach that harnesses a combination of experimental techniques based on full-field measurements (contactless) with appropriate numerical modelling to predict the mechanical response of apple skin under tensile loading.

The proposed methodology is used to identify the coefficients of a constitutive law of apple skin, to mimic the hyperelastic behaviour. In this process, the Yeoh's model appeared to be the most appropriate. The influence of the lenticel areas on the mechanical behaviour of the apple skin was emphasised by DIC measurements. According to this study, these areas are at risk of rupturing and are important in lowering the apple skin's resistance to external stress before achieving irreversible deformation. The identification of the influence of lenticel areas on the mechanical behaviour of apple skin adds an additional layer of practical relevance.

Moreover, modelling and predicting the mechanical behaviour of these materials was rendered attainable through the biomechanical interpretation of both Yeoh's model parameters and deformation cartographies. Understanding the coefficients in Yeoh's equation makes it possible to develop more accurate models to explain the mechanical response of apple skin, which enhances the comprehension of the non-linear and elastic skin characteristics. In the future, this mechanical characterisation may allow for establishing a relationship between constitutive parameters of fruit skin and qualitative attributes currently used to classify fruit (e.g., firmness and texture).

These findings constitute an important contribution to our knowledge of the mechanical behaviour of this complex biological structure (apple skin), outlining practical implications concerning the selection and application of apple cultivars in many technical and commercial contexts, from the food industry to biomaterials engineering. Additionally, these results not only allow for advances in the scientific understanding of apple skin behaviour but also have valuable practical implications in sectors that depend on the quality and durability of this fruit.

Author Contributions: Conceptualisation, J.X., J.M. and F.P.; methodology, J.X., J.M. and F.P.; software, J.X., Q.N. and N.D.; validation, J.M. and F.P.; formal analysis, T.C., J.X., and Q.N.; investigation, T.C., R.A. and N.D.; resources, N.D. and F.P.; data curation, T.C. and R.A.; writing—original draft preparation, T.C., N.D., J.X. and F.P.; writing—review and editing, T.C., J.X., N.D., J.M. and F.P.; supervision, N.D. and F.P.; project administration, F.P.; funding acquisition, F.P. All authors have read and agreed to the published version of the manuscript.

Funding: This research was funded by the Portuguese Foundation for Science and Technology (FCT), grant number EXPL/EME-APL/0587/2021, which corresponds to the following DOI: 10.54499/EXPL/EME-APL/0587/2021.

Institutional Review Board Statement: Not applicable.

Informed Consent Statement: Not applicable.

Data Availability Statement: Data are contained within the article.

Acknowledgments: The authors acknowledge the Portuguese Foundation for Science and Technology (FCT—MCTES) for the conceded financial support through the reference grant EXPL/EME-APL/0587/2021. The second author also acknowledges FCT for his grant ref. BI/UTAD/22/2022. The third author also acknowledges FCT for its financial support via the projects UIDB/00667/2020 and UIDP/00667/2020 (UNIDEMI).

Conflicts of Interest: The authors declare no conflicts of interest.

References

1. Oey, M.L.; Vanstreels, E.; Baerdemaeker, J.; Tijskens, E.; Ramon, H. Effect of turgor on micromechanical and structural properties of apple tissue: A quantitative analysis. *Postharvest Biol. Technol.* **2007**, *44*, 240–247. [CrossRef]
2. Wang, J.; Cui, Q.; Li, H.; Liu, Y. Experimental Research on Mechanical Properties of Apple Peels. *J. Eng. Technol. Sci.* **2015**, *47*, 688–705. [CrossRef]
3. Allende, A.; Desmet, M.; Vanstreels, E.; Verlinden, B.E.; Nicolaï, B.M. Micromechanical and geometrical properties of tomato skin related to differences in puncture injury susceptibility. *Postharvest Biol. Technol.* **2004**, *34*, 131–141. [CrossRef]
4. Grotte, M.; Duprat, F.; Loonis, D.; Piétri, E. Mechanical properties of the skin and the flesh of apples. *Int. J. Food Prop.* **2001**, *4*, 149–161. [CrossRef]
5. Bargel, H.; Neinhuis, C. Tomato (Lycopersicon esculentum Mill.) fruit growth and ripening as related to the biomechanical properties of fruit skin and isolated cuticle. *J. Exp. Bot.* **2005**, *56*, 1049–1060. [CrossRef] [PubMed]
6. Khanal, B.P.; Grimm, E.; Finger, S.; Blume, A.; Knoche, M. Intracuticular wax fixes and restricts strain in leaf and fruit cuticles. *New Phytol.* **2013**, *200*, 134–143. [CrossRef] [PubMed]
7. Khanal, B.P.; Grimm, E.; Knoche, M. Russeting in apple and pear: A plastic periderm replaces a stiff cuticle. *AoB Plants* **2013**, *5*, pls048. [CrossRef] [PubMed]
8. Petracek, P.D.; Bukovac, M.J. Rheological Properties of Enzymatically Isolated Tomato Fruit Cuticle. *Plant Physiol.* **1995**, *109*, 675–679. [CrossRef] [PubMed]

9. Lopez-Casado, G.; Salamanca, A.; Heredia, A. Viscoelastic nature of isolated tomato (Solanum lycopersicum) fruit cuticles: A mathematical model. *Physiol. Plant.* **2010**, *140*, 79–88. [CrossRef] [PubMed]
10. Domínguez, E.; Cuartero, J.; Heredia, A. An overview on plant cuticle biomechanics. *Plant Sci. Int. J. Exp. Plant Biol.* **2011**, *181*, 77–84. [CrossRef]
11. Bidhendi, A.J.; Li, H.; Geitmann, A. Modeling the nonlinear elastic behavior of plant epidermis. *Botany* **2020**, *98*, 49–64. [CrossRef]
12. Fung, Y.C. *Biomechanics: Mechanical Properties of Living Tissues*, 2nd ed.; Springer: New York, NY, USA, 1993; ISBN 978-1-4757-2257-4. [CrossRef]
13. Yeoh, O.H. Characterization of Elastic Properties of Carbon-Black-Filled Rubber Vulcanizates. *Rubber Chem. Technol.* **1990**, *63*, 792–805. [CrossRef]
14. Yeoh, O.H. Some Forms of the Strain Energy Function for Rubber. *Rubber Chem. Technol.* **1993**, *66*, 754–771. [CrossRef]
15. Gasser, T.C.; Ogden, R.W.; Holzapfel, G.A. Hyperelastic modelling of arterial layers with distributed collagen fibre orientations. *J. R. Soc. Interface* **2006**, *3*, 15–35. [CrossRef]
16. Holzapfel, G.A.; Gasser, T.C.; Ogden, R.W. A New Constitutive Framework for Arterial Wall Mechanics and a Comparative Study of Material Models. *J. Elast.* **2000**, *61*, 1–48. [CrossRef]
17. Pereira, J.L.; Xavier, J.; Ghiassi, B.; Lousada, J.; Morais, J. On the identification of earlywood and latewood radial elastic modulus of Pinus pinaster by digital image correlation: A parametric analysis. *J. Strain Anal. Eng. Des.* **2018**, *53*, 566–574. [CrossRef]
18. Pereira, J.; Xavier, J.; Morais, J.; Lousada, J. Assessing wood quality by spatial variation of elastic properties within the stem: Case study of Pinus pinaster in the transverse plane. *Can. J. For. Res.* **2014**, *44*, 107–117. [CrossRef]
19. Xavier, J.; de Jesus, A.M.P.; Morais, J.J.L.; Pinto, J.M.T. Stereovision measurements on evaluating the modulus of elasticity of wood by compression tests parallel to the grain. *Constr. Build. Mater.* **2012**, *26*, 207–215. [CrossRef]
20. Henriques, J.; Xavier, J.; Andrade-Campos, A. Identification of Orthotropic Elastic Properties of Wood by a Synthetic Image Approach Based on Digital Image Correlation. *Materials* **2022**, *15*, 625. [CrossRef]
21. Lava, P.; Jones, E.M.C.; Wittevrongel, L.; Pierron, F. Validation of finite-element models using full-field experimental data: Levelling finite-element analysis data through a digital image correlation engine. *Strain* **2020**, *56*, e12350. [CrossRef]
22. MatchID, MatchID Manual. MatchID: Metrology Beyond Colors. 2017. Available online: https://www.matchid.eu/ (accessed on 25 January 2024).
23. Jones, E.M.C.; Iadicola, M.A. (Eds.) *A Good Practices Guide for Digital Image Correlation*; International Digital Image Correlation Society, 2018; Available online: https://idics.org/guide/DICGoodPracticesGuide_PrintVersion-V5h-181024.pdf (accessed on 25 January 2024). [CrossRef]
24. Magoroua, L.; Le Bosa, F.; Rougerb, F. Identification of constitutive laws for wood-based panels by means of an inverse method. *Compos. Sci. Technol.* **2002**, *62*, 591–596. [CrossRef]
25. Kang, Y.L.; Lin, X.H.; Qin, Q.H. Inverse/genetic method and its application in identification of mechanical parameters of interface in composite. *Compos. Struct.* **2004**, *66*, 449–458. [CrossRef]
26. Nair, A.U.; Taggart, D.G.; Vetter, F.J. Optimizing cardiac material parameters with a genetic algorithm. *J. Biomech.* **2007**, *40*, 1646–1650. [CrossRef] [PubMed]
27. Perera, R.; Ruiz, A. A multistage FE updating procedure for damage identification in large-scale structures based on multiobjective evolutionary optimization. *Mech. Syst. Signal Process.* **2008**, *22*, 970–991. [CrossRef]
28. Rahmani, B.; Mortazavi, F.; Villemure, I.; Levesque, M. A new approach to inverse identification of mechanical properties of composite materials: Regularized model updating. *Compos. Struct.* **2013**, *105*, 116–125. [CrossRef]
29. Alami, A.H. (Ed.) *Mechanical Energy Storage for Renewable and Sustainable Energy Resources*, 1st ed.; Springer: Cham, Switzerland, 2020; p. 98. ISBN 978-3-030-33787-2. [CrossRef]
30. Khanal, B.P.; Si, Y.; Knoche, M. Lenticels and apple fruit transpiration. *Postharvest Biol. Technol.* **2020**, *167*, 111221. [CrossRef]
31. Remache, D.; Caliez, M.; Gratton, M.; Dos Santos, S. The effects of cyclic tensile and stress-relaxation tests on porcine skin. *J. Mech. Behav. Biomed. Mater.* **2018**, *77*, 242–249. [CrossRef]
32. Dwivedi, K.K.; Lakhani, P.; Kumar, S.; Kumar, N. The Effect of Strain Rate on the Stress Relaxation of the Pig Dermis: A Hyper-Viscoelastic Approach. *J. Biomech. Eng.* **2020**, *142*, 091006. [CrossRef]
33. Costalat, V.; Sanchez, M.; Ambard, D.; Thines, L.; Lonjon, N.; Nicoud, F.; Brunel, H.; Lejeune, J.P.; Dufour, H.; Bouillot, P.; et al. Biomechanical wall properties of human intracranial aneurysms resected following surgical clipping. *J. Biomech.* **2011**, *44*, 2685–2691. [CrossRef]

materials

MDPI

Article

Numerical and Experimental Study into Paper Compression Test

Leszek Czechowski [1,*], Paweł Pełczyński [2], Maria Bieńkowska [2] and Włodzimierz Szewczyk [2]

1 Department of Strength of Materials, Lodz University of Technology, Stefanowskiego Street 1/15, 90-537 Lodz, Poland
2 Centre of Papermaking and Printing, Lodz University of Technology, Wolczanska Street 221, 93-005 Lodz, Poland; pawel.pelczynski@p.lodz.pl (P.P.); maria.bienkowska@p.lodz.pl (M.B.); wlodzimierz.szewczyk@p.lodz.pl (W.S.)
* Correspondence: leszek.czechowski@p.lodz.pl

Abstract: The study aims to present the results of paper compression under an axial load. Different heights of samples subjected to compression were taken into account. The main goal of the analysis was to determine experimentally the maximum compression load. In addition, numerical models based on the finite element method (FEM) were validated to refer to empirical results. The performed numerical simulations were founded on Green–Lagrangian nonlinear equations for large displacements and strains. The progressive failure of the compressed orthotropic material after exceeding maximum stresses was based on Hill's anisotropy theory. Nonlinear calculations were conducted by using a typical Newton–Raphson algorithm for achieving a sequence convergence. The accuracy of the developed model was confirmed experimentally in compression tests. The technique of analysing the shape of the compressed paper sample on the basis of images recorded during the measurement was used. The obtained test results are directly applicable in practice, especially in the calculation of the mechanical properties of corrugated cardboard and in determining the load capacity of cardboard packaging. Knowing the maximum compressive stress that packaging paper can withstand allows packaging to be properly designed and its strength assessed in the context of the transport and storage of goods.

Keywords: paper strength; digital image analysis; finite element method; Hill's anisotropy potential theory

Citation: Czechowski, L.; Pełczyński, P.; Bieńkowska, M.; Szewczyk, W. Numerical and Experimental Study into Paper Compression Test. *Materials* **2023**, *16*, 7513. https://doi.org/10.3390/ma16247513

Academic Editors: Madhav Baral and Charles Lu

Received: 19 October 2023
Revised: 1 December 2023
Accepted: 3 December 2023
Published: 5 December 2023

1. Introduction

Paper is an orthotropic material widely used for the production of corrugated and cellular cardboard, from which a variety of packaging is created [1]. The strength properties of these packages depend to a large extent on the properties of the paper. Much effort is devoted to researching these properties of paper. Authors of [2] explore the material's response to compressive forces. They enhance our understanding of its structural integrity under pressure. Reference [3] sheds light on the complex phenomenon of creep in corrugated board, which is a key aspect when packaging fresh produce. This research provides valuable information on the challenges of maintaining product integrity during transportation and storage. The research presented in [4] integrates experimental and numerical approaches to comprehensively study paperboard tube failure. The synergy of methods enhances the robustness of findings, contributing to a deeper understanding of lateral stress impacts on paper structures. The work [5] introduces a novel method for assessing the radial crush strength of paper cores; this source marks a notable advancement in evaluating paper strength properties. The innovative approach presented holds promise for redefining industry standards in the assessment of paper core integrity. Study [6] presents a comprehensive approach, examining the crushing dynamics of double-wall corrugated cardboard and its impact on the load-bearing capacity of boxes. This allows us to look at packaging materials from an energy perspective. In [7], the response of a paper honeycomb to changing relative humidity was investigated. This helps us to understand how environmental conditions affect the stresses of these structures. Works [8–10] acknowledge the significance of environmental conditions, specifically humidity, in influencing the properties of honeycomb structures. However, paper [11] introduces a new method for measuring the edgewise compression properties of paper. It laid the groundwork for understanding the fundamental compression characteristics of paper, serving as a historical reference in the evolution of testing methodologies. On the other hand, work [12] presents a modelling framework for understanding both global and local buckling phenomena in corrugated board panels subjected to edge-to-edge compression. The research sheds light on the structural intricacies of corrugated materials under compression, contributing to the broader understanding of their mechanical behaviour. There is a need to minimize the weight of the packaging while maintaining the required strength. To achieve this, it is increasingly customary to predict the strength properties of packaging by calculating the properties of the papers used in its production [13–18]. One of the most important properties is the resistance to edge crushing, which is measured in a short-plug compression test known as an SCT [19,20]. However, the SCT test does not take into account the phenomenon of buckling of the sample, which occurs at longer fastening lengths and is commonly observed under the operating conditions of corrugated packaging. The phenomenon of sample buckling causes a significant reduction in its load capacity and it is necessary to take it into account in the prediction of this quantity [21–27]. Thanks to modern methods of processing and analysis of digital images [28–31], it is possible to observe changes in the shape of paper samples during the test of resistance to crushing by compressive forces [32,33]. This allows the change in the shape of the sample to be linked to the force acting on it during the test. To achieve this, images of the tested sample are recorded. As a result, it is possible to observe and evaluate the buckling parameters of the sample, in particular its shape and the size of the deflection arrow. The results of the image analysis of the paper sample compression process provide valuable data for numerical prediction of their load capacity. This allows for a more effective search for a numerical model describing the process of paper destruction in a unidirectional compression test.

The research presented in this article focused on the development and validation of a numerical model describing the process of crushing paper with compressive forces acting in its plane on the basis of experimental data from a testing machine and resulting from the analysis of a sequence of sample images recorded during the measurement [34]. Thanks to this approach, it was possible to verify the prediction of the shape of the paper sample subjected to the test, which is a strong confirmation of the validity of the developed model. Moreover, taking into account Green–Lagrangian nonlinear equations for large displacements and strains, progressive failure based on Hill's anisotropy potential theory was assumed to validate the numerical model and to achieve numerically maximum loads.

2. Materials and Methods

2.1. Measuring Stand

The SCT test of paper samples is usually made using a universal testing machine (UTM). This allows us to record the shortening of the sample and the accompanying force. To enable the assessment of the shape of the sample during the measurement, the sample was recorded using a vision system based on a single camera. In order to synchronize the shape registration of the paper sample subjected to the test and the measurement of the displacement of the movable UTM handle, a measuring station as shown in Figure 1a was developed. The stand is a modification of a measurement setup developed for registering the local deformation of paper in a one-directional tensile test [35]. It consists of a Zwick Roell Z 010 testing machine (89079 Ulm, Germany) with SCT test holders and a camera. The design of the holders allows for much longer connection lengths than in a standard SCT. Thanks to this, it was possible to perform the presented experiments and observe the phenomenon of buckling of paper samples. The camera used is a Canon EOS 6D Mark II digital SLR camera (Hong Kong, China) with a Canon MP-E 65 mm f/2.8 1-5x Macro lens. The camera is placed on a tripod that provides stable conditions for photo recording. The sample is illuminated by a white light source ensuring constant illumination. The camera is attached to the tripod via a setting shoe with the possibility of adjusting the position of the parallel and perpendicular axis to the optical axis of the camera. Chessboard patterns of a single square length of 1 mm (Figure 1b) are glued onto the handles of the testing machine. This allows us to track the movement of the handle and determine the spatial resolution of the recorded images. The measurement begins by placing the test paper sample in the holders of the testing machine and positioning the camera ensuring the visibility of the sample and fragments of chessboards in its field of view. The recording of the photo sequence is then triggered and, after recording several images, the measurement is initiated in the machine. Reaching the desired displacement of the movable handle completes the measurement. The number of recorded images depends on the duration of the compression test. The photo capture rate is 4 fps, but is not constant. In some cases, it dropped after the camera cache was full. Nevertheless, thanks to reading information from photo EXIF about the time of taking the photo with an accuracy of 1/100 of a second, it was possible to precisely synchronize the photos with the UTM measurement results. The measuring station and the measurement procedure are described in more detail in [34].

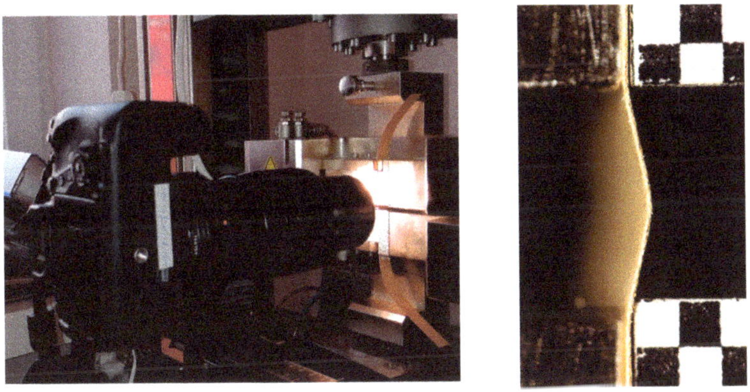

(a) (b)

Figure 1. The measurement setup composed of UTM and DSLR camera (**a**) and UTM handles with chessboard patterns glued to them and a paper sample under test (**b**).

2.2. Image Processing and Analysis Technique

The recorded sequence of images was subjected to preprocessing and analysis. Preprocessing consisted of converting images to grayscale, extracting images of the examined paper sample, a chessboard on a movable handle and a chessboard on a fixed handle, linear filtering of the sample image using a filter highlighting vertical edges in the image with the mask shown in Figure 2a and linear filtering of chessboard images using a filter highlighting the corners of the chessboard corners in the image with the mask shown in Figure 2b.

The image resulting from the filtering of the paper sample image is shown in Figure 3a. In Figure 3b, the original image of the sample with the left (green) and right (red), detected edges, is presented. The left edge is detected with low efficiency due to low contrast in the image. For further analysis, the right edge was selected.

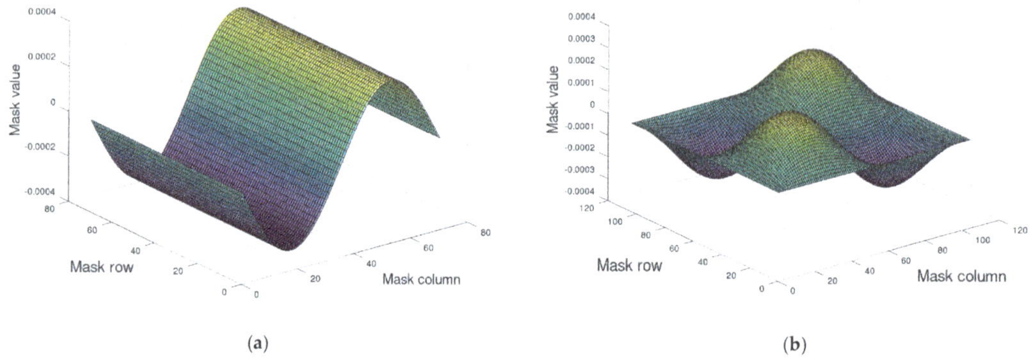

Figure 2. The shape (**a**) of the filter mask to emphasize the edges in the image; (**b**) the shape of the filter mask to emphasize the corners of the chessboard.

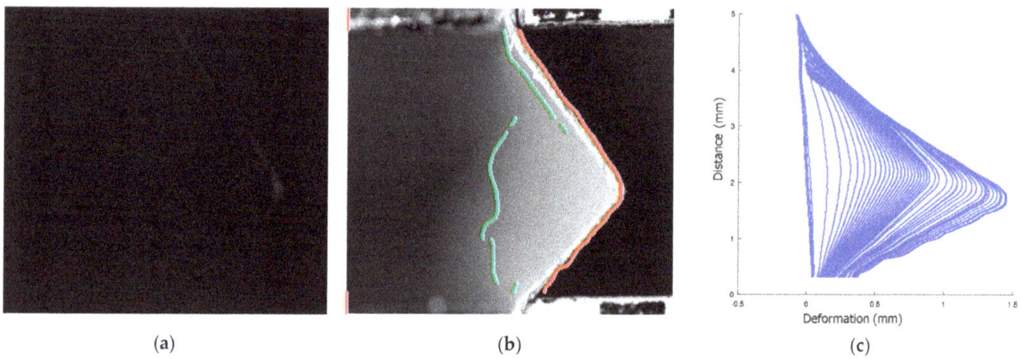

Figure 3. The result of the image filtering of the paper sample (**a**); the image of the sample with the edges detected (**b**); and the shape of the right edge of the tested sample obtained in subsequent images of the analysed sequence (**c**).

Sets of coordinates of points of the detected edges and corners were input information to the algorithm for image data analysis. Image coordinates were converted to metric coordinates (mm) based on information about the spatial resolution of the images and the arbitrary place of origin of the coordinate system (related to the place where the sample is attached to the stationary holder). The shape of the edges of the tested sample in subsequent images of the recorded sequence is shown in Figure 3c. Next, the average values of the coordinates of the corners of the moving and stationary chessboard were calculated. They were used to determine the movement of the handles as a function of the time of recording individual photos. On the basis of this information, the time of image recording as a function of the measurement data acquisition time in the testing machine was estimated. The result was the synchronization of data recording times from two different sources.

2.3. *Hill's Anisotropic Potential Theory*

Material properties of orthotropic paper were assumed: E_x = 5600 MPa, E_y = E_z = 2450 MPa, G_{xy} = G_{xz} = G_{yz} = 2120 MPa, Poisson's ratio 0.3. Index x denotes the direction of compression. This means that moduli E_x and E_y represent the properties in longitudinal and perpendicular directions with regard to the compression direction, respectively. In order to model the plasticity of the analysed paper, Hill's anisotropic criterion was applied [36–38] to describe the anisotropy material behaviour after exceeding the yield stress of paper. If this criterion was considered for isotropic hardening, the yield function could be determined by [36–38]:

$$f(\sigma) = \sqrt{\{\sigma\}^T[M]\{\sigma\}} - \sigma_0(\bar{\varepsilon}^p) \tag{1}$$

where σ_0 and $\bar{\varepsilon}^p$ represent a reference yield stress and equivalent plastic strain, respectively. For the material with the three orthogonal planes of symmetry (sym.), the plastic compliance matrix $[M]$ can be given as:

$$[M] = \begin{bmatrix} G+M & -H & -G & 0 & 0 & 0 \\ & F+H & -F & 0 & 0 & 0 \\ & & F+G & 0 & 0 & 0 \\ & & & 2N & 0 & 0 \\ & & & & 2L & 0 \\ & & & & & 2M \end{bmatrix} \tag{2}$$

The coefficients mentioned above, F, G, H, L, M and N, are the material constants and should be defined experimentally. These magnitudes are shown in the following form:

$$\begin{gathered} F = \tfrac{1}{2}\left(-\tfrac{1}{R_{xx}^2} + \tfrac{1}{R_{yy}^2} + \tfrac{1}{R_{zz}^2}\right),\ G = \tfrac{1}{2}\left(\tfrac{1}{R_{xx}^2} - \tfrac{1}{R_{yy}^2} + \tfrac{1}{R_{zz}^2}\right),\ H = \tfrac{1}{2}\left(\tfrac{1}{R_{xx}^2} + \tfrac{1}{R_{yy}^2} - \tfrac{1}{R_{zz}^2}\right) \\ L = \tfrac{3}{2}\left(\tfrac{1}{R_{yz}^2}\right),\ M = \tfrac{3}{2}\left(\tfrac{1}{R_{xz}^2}\right),\ N = \tfrac{3}{2}\left(\tfrac{1}{R_{xy}^2}\right) \end{gathered} \tag{3}$$

The yield stress ratios can be determined with the use of the following expressions as below:

$$R_{xx} = \frac{\sigma_{xx}^y}{\sigma_0},\ R_{yy} = \frac{\sigma_{yy}^y}{\sigma_0},\ R_{zz} = \frac{\sigma_{zz}^y}{\sigma_0},\ R_{xy} = \sqrt{3}\frac{\sigma_{xy}^y}{\sigma_0},\ R_{yz} = \sqrt{3}\frac{\sigma_{yz}^y}{\sigma_0},\ R_{xz} = \sqrt{3}\frac{\sigma_{xz}^y}{\sigma_0} \tag{4}$$

where $\overline{\sigma}_{ij}^y$ represents the yield stress for appropriate directions. In the case of the application of Hill's criterion, it was assumed that the tangential modulus E^t after coming to the limit strength was 56 MPa and subsequently stress ratios R_{ij} amounted to: R_{xx} = 1, R_{yy} = R_{zz} = 0.44 and R_{xy} = R_{xz} = R_{yz} = 0.3. On the basis of one-directional tensile tests of the paper sample, the yield stress σ_0 was determined to be 36 MPa.

2.4. Finite Element Model

FE (finite element) calculations were carried out by using Ansys 18.2 software [38]. The dimensions a (0.7 mm, 1.3 mm, 2 mm, 2.5 mm, 3 mm, 3.5 mm, 4 mm, 4.5 mm, 5 mm) and b (15 mm) represent the height and the width of the sample, respectively (Figure 4). To elaborate adequate discrete models of the compressed samples, the 4-node 181-shell element was used. The size of finite element was assumed $a/50$. The buckling analysis (BA) was performed based on the linear block Lanczos algorithm to determine the critical loads. The nonlinear analysis was conducted for large strains and deflections on the basis of Green–Lagrangian equations. The number of substeps for the single calculation was assumed to be from 1000 up to 50,000. The maximum number of iterations for each substep was set up to 5000. Figure 4 shows the FE model and the boundary conditions. The compression load applied to the sample was realised by using a master node associated with slave nodes lying on the outer edges of the sample. The preliminary imperfection w_0 of columns for the sake of considered model was assumed: 0.1 t, 0.5 t, 0.75 t, 1 t, 1.5 t, 2 t, 4 t, where t (0.14 mm) denotes the thickness of the sample. The initial deflection was related to the mode of the first buckling force of the analysed sample. Nonlinear estimations and convergence analysis were conducted by using the Newton–Raphson algorithm.

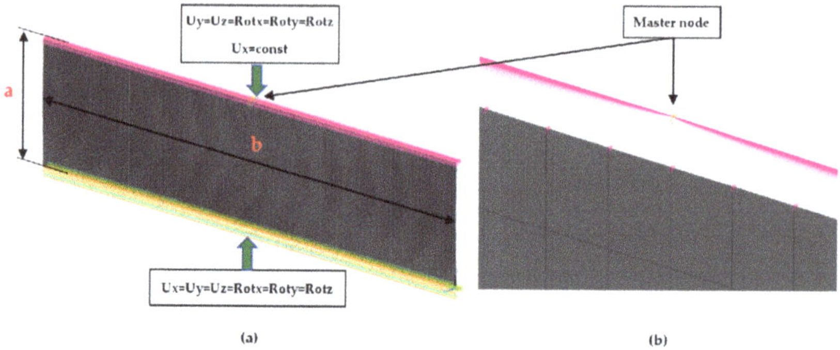

Figure 4. Model FE (**a**) and view of master node connected with outer nodes (**b**).

3. Results and Discussion

3.1. Shape of Paper Samples Subjected to Compressive Forces in Subsequent Measurement Phases

In order to be able to compare the results of paper modelling in compression tests with the results of real measurements in the UTM, the shapes of samples in measurement phases characterized by a change in sample behaviour were presented. The moment when the buckling started, the moment when the maximum compressive force was reached, and the moment when the sample broke and its shape of the buckled sample ceased to resemble a fragment of a sinusoid, were selected. Images of the samples are shown in Figures 5–7.

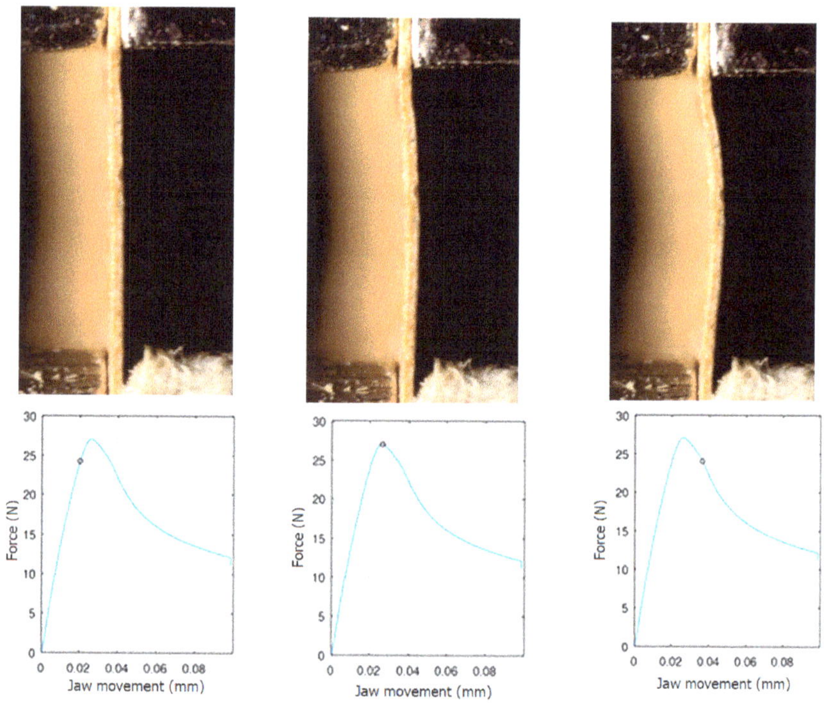

Figure 5. The shape of the paper sample and the corresponding point on the force-shortening curve for a clamping height of 3 mm at the beginning of buckling (**a**); at maximum force (**b**); at the end of the sine wave shape (**c**).

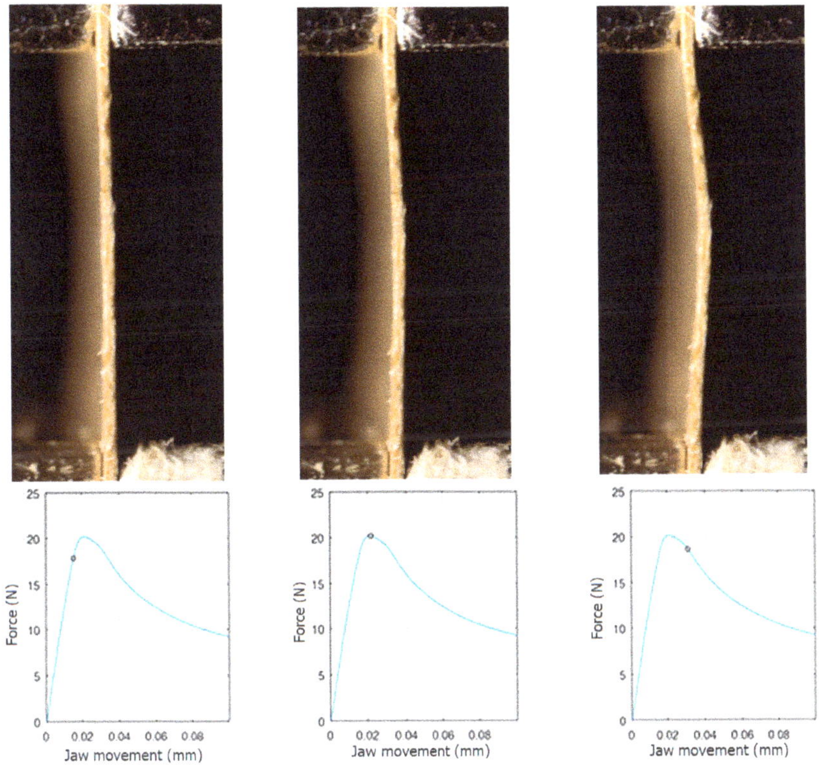

(**a**) (**b**) (**c**)

Figure 6. The shape of the paper sample and the corresponding point on the force-shortening curve for a clamping height of 4 mm at the beginning of buckling (**a**); at maximum force (**b**); at the end of the sine wave shape (**c**).

The analysis of the recorded images reveals the occurrence of four phases of destruction of the samples, characterized by the following aspects:

1. compression without visible deflection—linear dependence of force on shortening;
2. significant deflection (possible appearance of buckling force) and consequent nonlinear increase in force with further shortening;
3. visible deformation of sample and nonlinear drop of force vs. shortening;
4. appearance of a joint in the middle of the sample or destruction in the area of the joints.

Figures 5–7 show the shape of the samples and the corresponding points on the force-shortening curves at the boundaries between the mentioned destruction phases.

3.2. Critical Forces

This subsection shows the results of critical load calculations for a compressed paper sample with clamped supports (Figure 4). The estimations were performed for several heights a taking into account the first five buckling loads. Critical forces expressed in a Newton unit have been inserted in Table 1. Based on the results, it can be easily seen that for the shortest a, the critical forces are the greatest. It is interesting that the next critical loads are very close to each other. This means that the orthotropic material allows us to attain different modes with a small changing a force. In Table 2, total deformation maps are shown. Taking a look at the deformation modes, the number of half-waves for the appropriate buckling load are the same regardless of the considered height. This means that a small orthotropy (in the case of the analysed paper two to three times) does not matter significantly. Furthermore, these shown shapes at the application of the initial deflection were assumed.

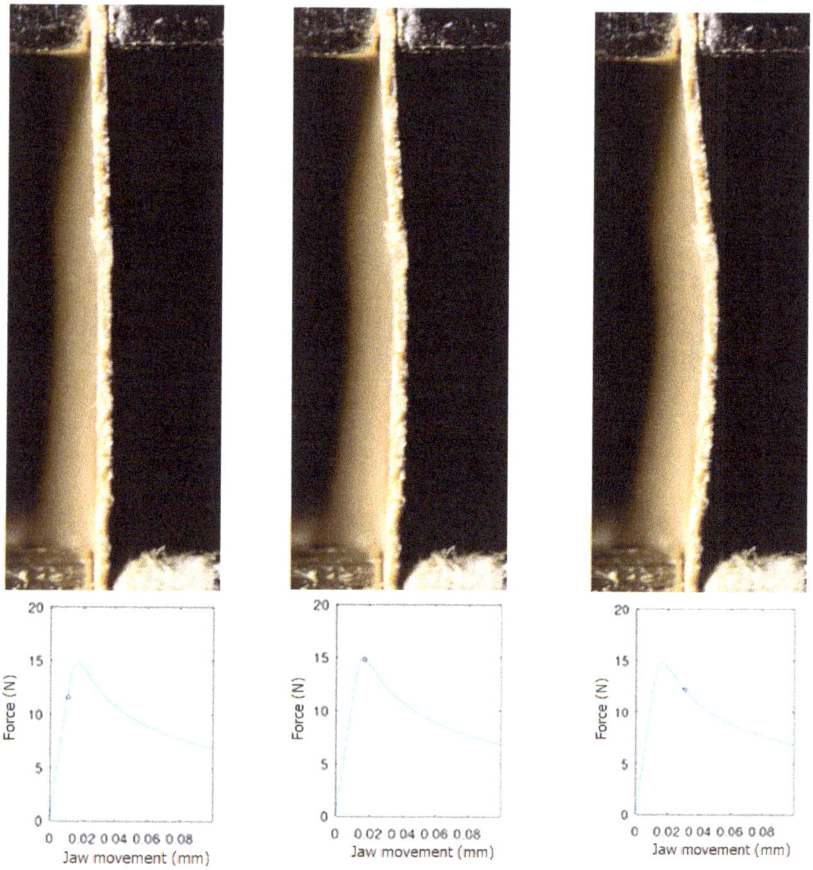

(a) (b) (c)

Figure 7. The shape of the paper sample and the corresponding point on the force-shortening curve for a clamping height of 5 mm at the beginning of buckling (**a**); at maximum force (**b**); at the end of the sine wave shape (**c**).

Table 1. The critical forces for different heights *a* between jaws.

	F_{cr} (N)								
Mode	a = 0.7 mm	a = 1.3 mm	a = 2 mm	a = 2.5 mm	a = 3 mm	a = 3.5 mm	a = 4 mm	a = 4.5 mm	a = 5 mm
1	1124.3	415.3	187.5	122.2	85.7	63.4	48.7	38.6	31.3
2	1125.8	416.9	189.1	123.8	87.3	64.9	50.2	40.1	32.8
3	1128.4	420.2	192.6	127.4	90.9	68.6	54.0	43.9	36.7
4	1132.2	425.3	198.3	133.4	97.1	75.0	60.6	50.8	43.8
5	1137.4	432.5	206.5	142.1	106.3	84.7	70.8	61.4	54.9

Table 2. The buckling modes of compressed paper with several heights *a*.

a (mm)	Number of Mode				
	1	**2**	**3**	**4**	**5**
0.7	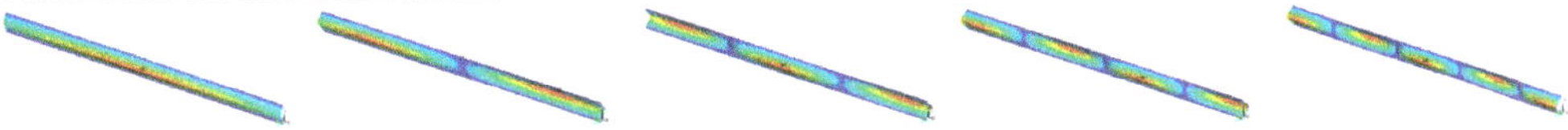				
1.3	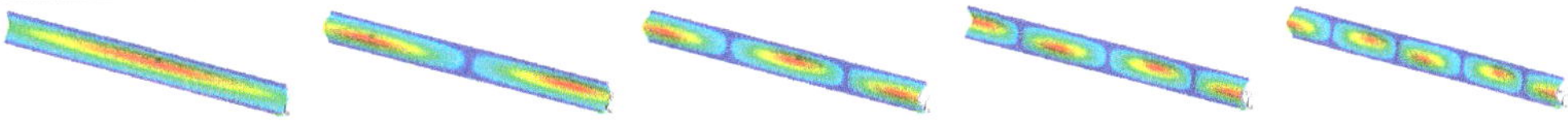				
2	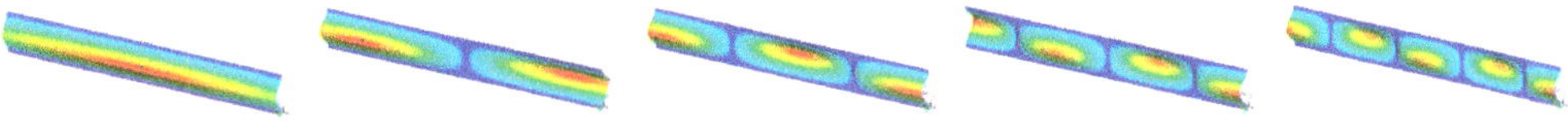				
2.5	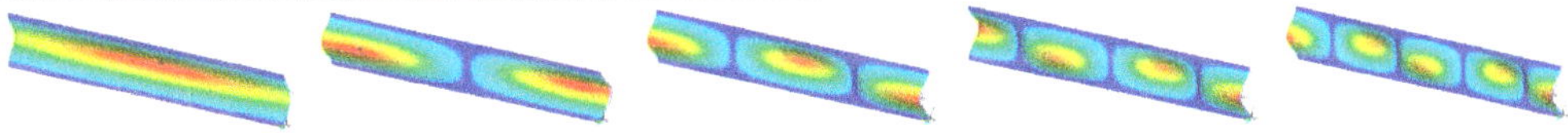				
3	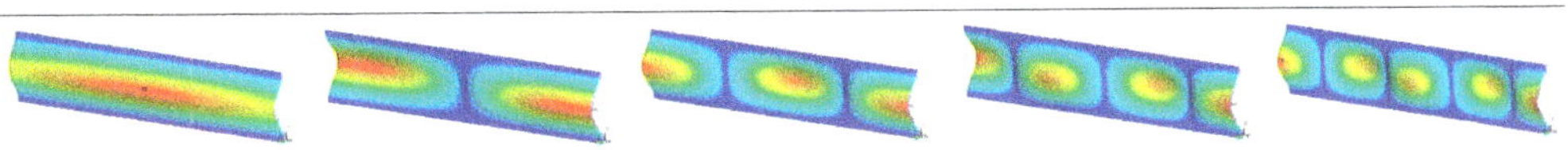				
3.5	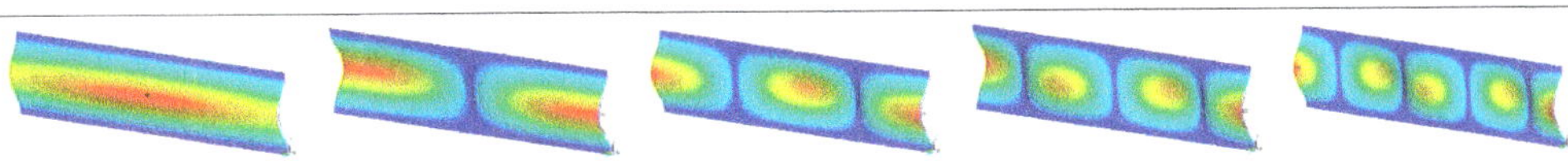				
4	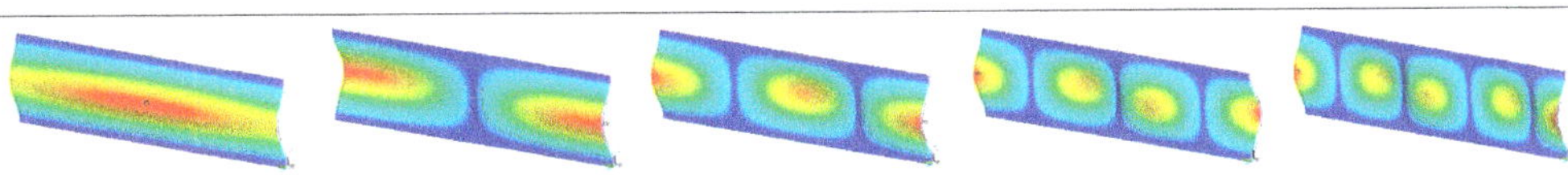				
4.5	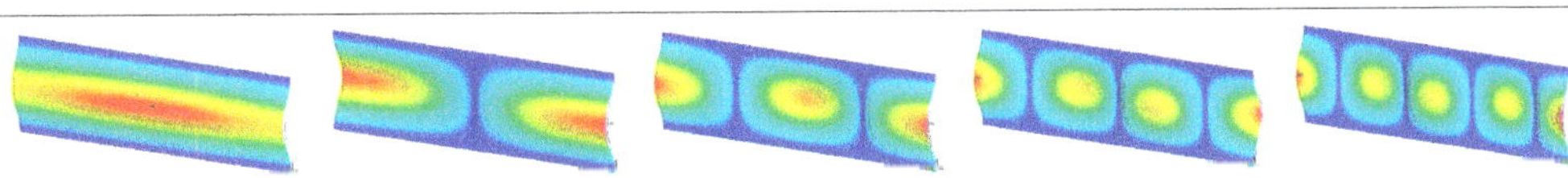				
5	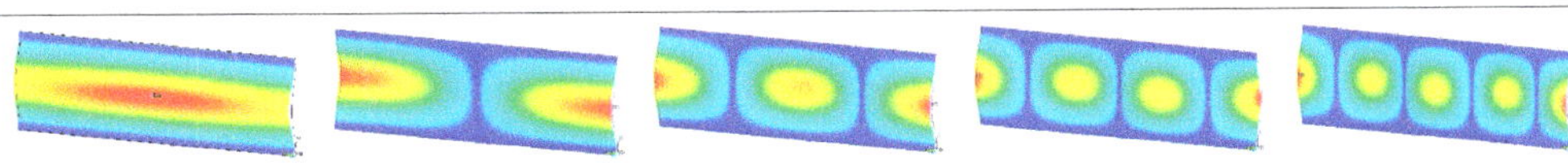				

3.3. Full Curves of Compression

This subsection shows the diagrams of paper compression as the shortening s vs. the compression force F_{comp} and gives maximum values of the forces. Based on the results, Figure 8a illustrates the curves of compression for the smallest considered height, i.e., $a = 0.7$ mm. The average peak of the force amounted to 38.2 N. It is 30 to 40 times smaller than the first critical force. If one can take into account the numerical curves, a significant divergence is seen in contrast with the experimental ones. Firstly, the stiffnesses (regardless of the magnitude of the initial imperfection) are meaningfully greater. Secondly, it was not possible to obtain maximum loads if even the initial deflection was equal to four thicknesses. However, it should be noted that at this variant the curves are the closest (between the experiment and the simulation). To justify this situation, one should note that the initial imperfection of the sample in the numerical model was based on modes achieved for linear buckling (See Table 2—for $a = 0.7$, first mode). Nevertheless, in the case of the experiment, the initial imperfection in this case could differ or be of another mode (through the whole width of the sample almost uniform curvature might have occurred, for example). This could mean that the resistance of paper under compression just for this variant caused a significant drop in stiffness and simultaneously a decrease in the maximum load that was observed in the experiment looking at registered curves. In the next graph (Figure 8b), if a = 1.3 mm was taken into consideration, in the numerical model some peaks were achieved. The mean maximum load in the experiment came to about 32 N. In the case of numerical estimations (for w_0 = 2 t), the maximum load is 36 N. It can be seen that the curves go similarly, but in the case of the FE curve, the stiffness is still greater. By increasing the height a, the curves based on the FEA are getting closer to experimental curves. It is just observed on the next chart (Figure 9a) for a = 2 mm. Indeed, this effect is visible just for a greater preliminary deflection (between w_0 = 1 t and w_0 = 1.5 t). In the case of numerical curves, the maximum load is noticed at a smaller shortening (circa 0.01 mm, but for the experiment it is at 0.03 mm). Better consistency at the maximum load (experiment vs. simulation) is observed for height a = 2.5 mm (Figure 9b), because peaks amounted to 28 N for both cases (for w_0 = 1 t). Looking through the next diagrams (Figure 10a,b, Figure 11a,b and Figure 12) where height a increases, it can be easily seen that there is better convergence between curves both in the maximum compression load and stiffness. Nevertheless, it should be admitted that initial imperfections used in numerical simulation are still great. In Figure 12, apart from the curves created based on Hill's theory, for a comparison, the curve without application of this theory is shown. In this case, a different trend in the curve is seen after exceeding a limit stress (yielding of the paper might have been possible). Hence, this means that achieving a maximum load was not possible. The comparison of maximum loads obtained by both methods is set out in Table 3. Based on charts for greater a, it can be noted that within an increase in the initial deflection, both maximum loads and stiffnesses decrease; therefore, numerical curves begin to cover or be close to experimental ones.

Table 3. The maximum compression forces for different length a.

	F_{max}								
Variant	a = 0.7 mm	a = 1.3 mm	a = 2 mm	a = 2.5 mm	a = 3 mm	a = 3.5 mm	a = 4 mm	a = 4.5 mm	a = 5 mm
EXP_{mean}	36.5 ± 3.1	33.4 ± 1.3	30.8 ± 1.0	28.7 ± 0.4	26.8 ± 0.7	24.0 ± 1.7	20.4 ± 1.4	17.5 ± 1.5	15.9 ± 0.9
FEM/ w_0 = 0.1 t	-	71.9	65.7	59.3	51.6	44.8	38.2	32.2	27.2
FEM/ w_0 = 0.5 t	-	58.3	46.7	38.8	32.4	28.6	25.3	22.5	19.9
FEM/ w_0 = 0.75 t	-	-	-	-	27.3	24.2	-	-	17.4
FEM/ w_0 = 1 t	-	48.1	35.7	28.8	23.6	21.1	19.0	17.1	15.5
FEM/ w_0 = 1.5 t	-	-	29.1	-	-	-	-	-	-
FEM/ w_0 = 2 t	-	35.6	-	-	-	-	-	-	-
FEM/ w_0 = 4 t	-	-	-	-	-	-	-	-	-

3.4. Deformation Edge Points

This subsection shows some diagrams as a deflection w (growth of deflection) vs. the distance between the jaws for several stages (before buckling, at maximum load, at greater shortenings of sample) of compression for a = 2.5 mm, 3 mm, 4 mm and 5 mm (Figures 13 and 14; for shorter samples, no buckling was observed). The deflections illustrated in the diagrams denote a displacement of the lateral edge of the sample. Based on these results, it can be seen that deflections obtained numerically are greater than in the experiment. For higher samples, these differences are not significant. Moreover, the modes in both methods seem to be similar. Of course, there are some discrepancies because peak points might have been lightly shifted with regard to the symmetry axis but in the experiment with regard to the imperfect structure such a behaviour of paper is possible and accessible. In the case of results for paper compression with a = 2.5 mm, the experimental displacements in sequential stages are not clearly visible as far as typical buckling modes are concerned. Therefore, the resistance of the paper is more noticeable because initial imperfections could be small and did not matter significantly. In general, the greater the lengths of samples that were considered, the better the convergences in deflections were achieved (e.g., for a = 4 mm or 5 mm).

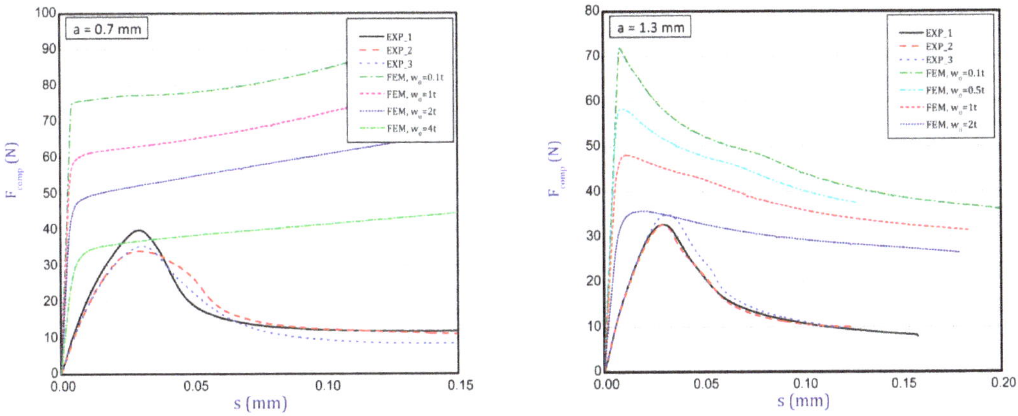

(a) (b)

Figure 8. Shortening (s) vs. compression force (F_{comp}) for height of 0.7 mm (**a**) and 1.3 mm (**b**).

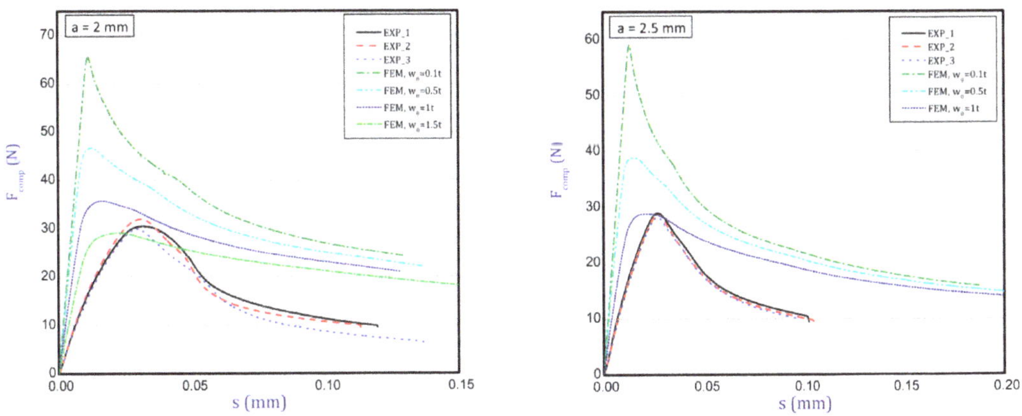

(a) (b)

Figure 9. Shortening (s) vs. compression force (F_{comp}) for height of 2 mm (**a**) and 2.5 mm (**b**).

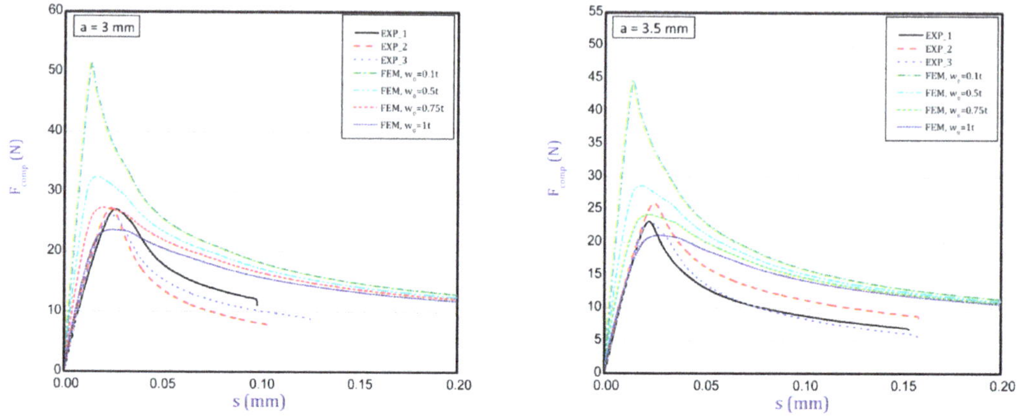

(a) (b)

Figure 10. Shortening (s) vs. compression force (F_{comp}) for height of 3 mm (**a**) and 3.5 mm (**b**).

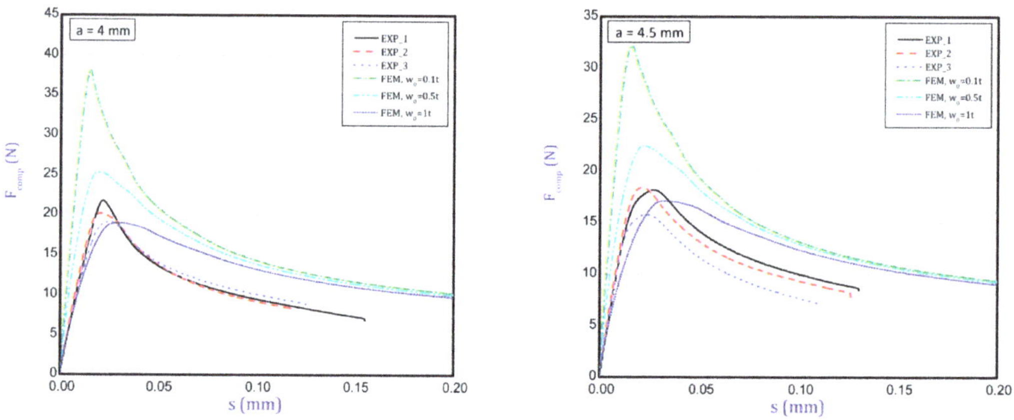

(a) (b)

Figure 11. Shortening (s) vs. compression force (F_{comp}) for height of 4 mm (**a**) and 4.5 mm (**b**).

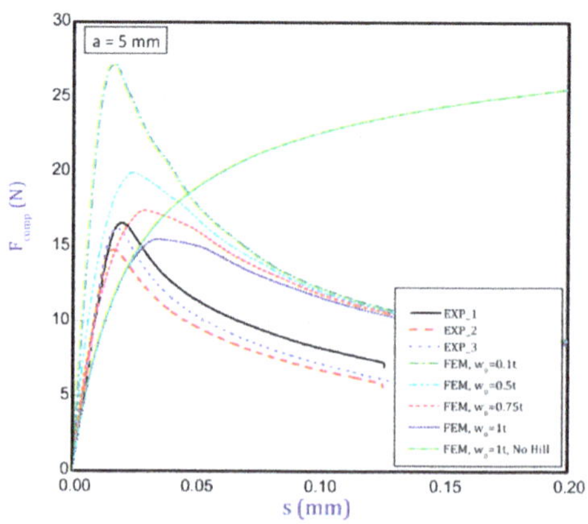

Figure 12. Shortening (s) vs. compression force (F_{comp}) for height of 5 mm.

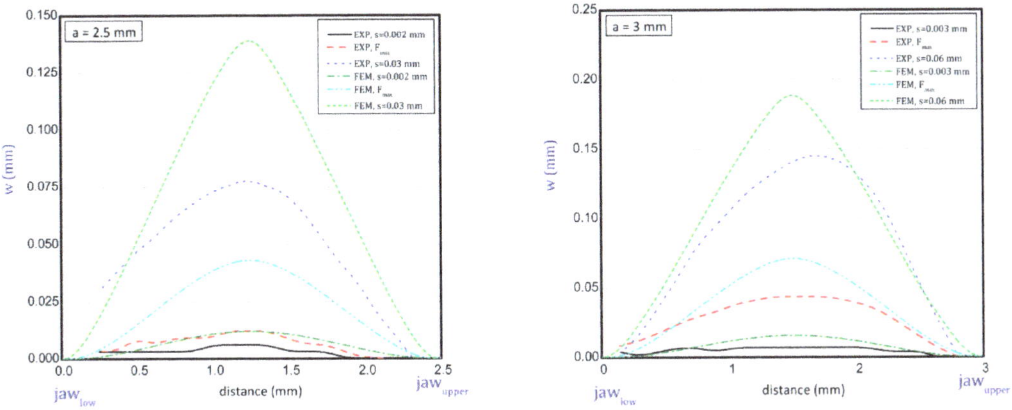

(a) (b)

Figure 13. Deflection (w) vs. distance between jaws for height of 2.5 mm (**a**) and 3 mm (**b**).

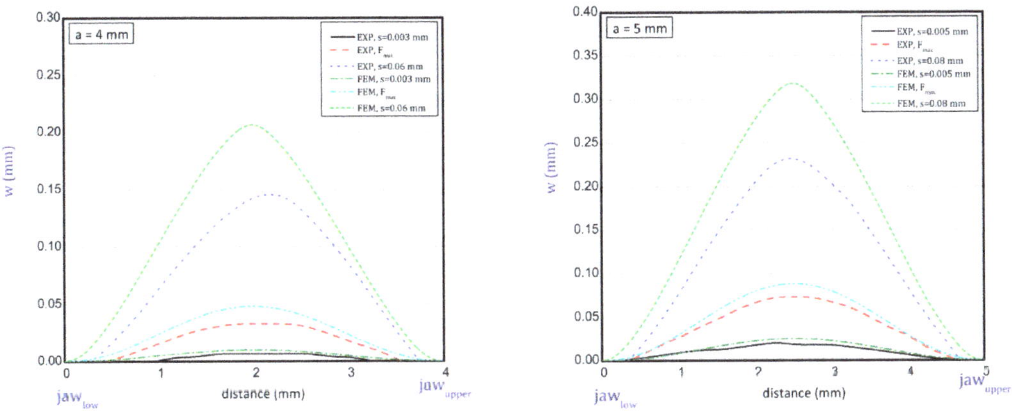

(a) (b)

Figure 14. Deflection (w) vs. distance between jaws for height of 4 mm (**a**) and 5 mm (**b**).

4. Conclusions

The presented study was conducted to explore the compression behaviour of paper samples under an axial load, considering a few heights of samples. The SCT test assumed a pure compression without taking into account material buckling. However, this work involved the analyses both of buckling and its impact on the load-bearing capacity of paper. Through an empirical study, we determined the maximum compression loads and curves of compression whose results were the base of a validation of numerical models by using a finite element method, among others. The results of simulations, where numerical models were based on Green–Lagrangian nonlinear equations and Hill's anisotropy theory, showed pretty good coincidence, especially for greater heights of compression; however, in the case of the smallest considered height, numerical determination of the maximum load was not possible even for a large initial imperfection. The reason this situation could result in an improper mode of initial deflection which finally might have been influenced by the entire stiffness. Nevertheless, the present study is the experimental confirmation of the developed model's accuracy in compression tests, coupled with the innovative technique of shape analysis based on recorded images, and enhanced the practical applicability of the authors' findings. These results directly benefit various industries, particularly in the evaluation of mechanical properties for corrugated cardboard and the determination of load capacity for cardboard packaging. Understanding the maximum compressive stress that packaging paper can withstand empowers the design of robust packaging solutions and aids in assessing their strength. Moreover, the conducted analysis of paper compression in the present paper allows us to explain and describe several phenomena occurring in paper goods. Firstly, one of the essential aspects can be obtaining different bending stiffnesses of corrugated cardboards for opposite signs of bending moments whose effect can be easily explained just by testing a deflection in the flat layer of cardboard taking into consideration the compression due to forces acting in their plane. Secondly, the performed tests of a paper compression determining the buckling effect and maximum loads may have defined the demanded lengths between the peaks in the wave (core) in multilayer corrugated cardboard.

Author Contributions: P.P. and W.S. developed this study. M.B. and P.P. conducted the tests. L.C. carried out the numerical simulations. L.C. and P.P. prepared the first version of this manuscript. All authors participated in the discussion and revised the manuscript. All authors have read and agreed to the published version of the manuscript.

Funding: This research received no external funding.

Institutional Review Board Statement: Not applicable.

Informed Consent Statement: Not applicable.

Data Availability Statement: Data are contained within the article.

Conflicts of Interest: The authors declare no conflict of interest.

References

1. Rzepa, S. Parametry papieru. In Proceedings of the Sympozjum Mondi Packaging "From Fibre to Corrugated Board", Świecie, Poland, 1–2 December 2004.
2. Bai, J.; Wang, J.; Pan, L.; Lu, L.; Lu, G. Quasi-static axial crushing of single wall corrugated paperboard. *Compos. Struct.* **2019**, *226*, 111237. [CrossRef]
3. Fadiji, T.; Coetzee, C.J.; Oparab, U.L. Analysis of the creep behaviour of ventilated corrugated paperboard packaging for handling fresh produce—An experimental study. *Food Bioprod. Process.* **2019**, *117*, 126–137. [CrossRef]
4. Czechowski, L.; Bieńkowska, M.; Szewczyk, W. Paperboard tubes failure due to lateral compression—Experimental and numerical study. *Compos. Struct.* **2018**, *203*, 132–141. [CrossRef]
5. Kołakowski, Z.; Szewczyk, W.; Bieńkowska, M.; Czechowski, L. New method for evaluation of radial crush strength of paper cores. *Mechanika* **2018**, *24*, 169–173. [CrossRef]
6. Gajewski, T.; Garbowski, T.; Staszak, N.; Kuca, M. Crushing of double-walled corrugated board and its influence on the load capacity of various boxes. *Energies* **2021**, *14*, 4321. [CrossRef]
7. Yu-Ping, E.; Wang, Z.-W. Plateau stress of paper honeycomb as response to various relative humidities. *Packag. Technol. Sci.* **2010**, *23*, 203–216. [CrossRef]
8. Abbès, B.; Guo, Y.Q. Analytic homogenization for torsion of orthotropic sandwich plates: Application to corrugated cardboard. *Compos. Struct.* **2010**, *92*, 699–706. [CrossRef]
9. Semple, K.E.; Sam-Brew, S.; Deng, J.; Cote, F.; Yan, N.; Chen, Z.; Smith, G.D. Properties of commercial Kraft paper honeycomb furniture stock panels conditioned under 65 and 95 percent relative humidity. *For. Prod. J.* **2015**, *65*, 106–122. [CrossRef]
10. Kmita-Fudalej, G.; Szewczyk, W.; Kołakowski, Z. Calculation of honeycomb paperboard resistance to edge crush test. *Materials* **2020**, *13*, 1706. [CrossRef]
11. Calvin, S.; Fellers, C. A new method for measuring the edgewise compression properties of paper. *Sven. Papperstidng* **1975**, *78*, 329–332.
12. Åslund, P.E.; Hägglund, R.; A Carlsson, L.; Isaksson, P. Modeling of global and local buckling of corrugated board panels loaded in edge-to-edge compression. *J. Sandw. Struct. Mater.* **2014**, *16*, 272–292. [CrossRef]
13. Fadiji, T.; Ambaw, A.; Coetzee, C.J.; Berry, T.M.; Opara, U.L. Application of finite element analysis to predict the mechanical strength of ventilated corrugated paperboard packaging for handling fresh produce. *Biosyst. Eng.* **2018**, *174*, 260–281. [CrossRef]
14. Li, X.; Wang, J.; Chuang, C.; Gao, D.; Lu, G.; Lu, L.; Wang, Z. Mathematical models for predicting the quasi-static stress characteristics of corrugated paperboard with sinusoidal core along the longitudinal compression. *Int. J. Mech. Sci.* **2018**, *149*, 136–149. [CrossRef]

15. Kubiak, T.; Kołakowski, Z.; Swiniarski, J.; Urbaniak, M.; Gliszczynski, A. Local buckling and post-buckling of composite channel-section beams—Numerical and experimental investigations. *Compos. Part B Eng.* **2016**, *91*, 176–188. [CrossRef]
16. Mou, X.N.; Lu, L.X.; Zhou, Y.L. Evaluation of in-plane compressive densification strain of honeycomb paperboard. *Adv. Mech. Eng.* **2020**, *12*, 1–11. [CrossRef]
17. Wang, D.M.; Wang, Z.-W. Experimental investigation into the cushioning properties of honeycomb paperboard. *Packag. Technol. Sci.* **2008**, *21*, 309–316. [CrossRef]
18. Smardzewski, J.; Prekrat, S. Modelling of thin paper honeycomb panels for furniture. In Proceedings of the International Conference Ambienta, Wood Is Good—With Knowledge and Technology to a Competitive Forestry and Wood Technology Sector, Zagreb, Croatia, 12 October 2012; pp. 179–186.
19. *PN-EN ISO 3035:2011*; Tektura Falista Oznaczanie Odporności na Zgniatanie Płaskie. PKN: Warszawa, Poland, 2011. (In Polish)
20. Hagman, A.; Huang, H.; Nygårds, M. Investigation of shear induced failure during SCT loading of paperboards. *Nord. Pulp Pap. Res. J.* **2013**, *28*, 415–429. [CrossRef]
21. Murray, N.W.; Khoo, P.S. Some basic plastic mechanism in the local buckling of thin-walled steel structures. *Int. J. Mech. Sci.* **1981**, *23*, 703–713. [CrossRef]
22. Królak, M. (Ed.) *Stany zakrytyczne i nośność graniczna cienkościennych dźwigarów o ścianach płaskich*; PWN: Warszawa-Łódź, Poland, 1990; p. 553. (In Polish)
23. Volmir, A.S. *Nonlinear Dynamics of Plates and Shells*; Science Publishing House: Moscow, Russia, 1972; p. 432. (In Russian)
24. Reddy, J.N. *Mechanics of Laminated Composite Plates and Shells: Theory and Analysis*, 2nd ed.; CRC Press: Boca Raton, FL, USA, 2004.
25. Kołakowski, Z.; Jankowski, J. Some inconsistencies in the nonlinear buckling plate theories—FSDT, S-FSDT, HSDT. *Materials* **2021**, *14*, 2154. [CrossRef]
26. Fellers, C. The Significance of Structure on the Compression Behavior of Paper. Ph.D. Thesis, Department of Paper Technology, Royal Institute of Technology, Stockholm, Sweden, 1980.
27. Brandberg, A.; Kulachenko, A. Compression failure in dense non-woven fiber networks. *Cellulose* **2020**, *27*, 6065–6082. [CrossRef]
28. Gonzalez, R.C.; Woods, R.E. *Digital Image Processing*; Prentice Hall: Hoboken, NJ, USA, 2002.
29. Laermann, K.H. (Ed.) *Optical Methods in Experimental Solid Mechanics*; Springer: Vienna, Austria, 2000. [CrossRef]
30. Zhu, C.; Wang, H.; Kaufmann, K.; Vecchio, K.S. A computer vision approach to study surface deformation of materials. *Meas. Sci. Technol.* **2020**, *31*, 055602. [CrossRef]
31. Sutton, M.A.; Orteu, J.J.; Schreier, H. *Image Correlation for Shape, Motion and Deformation Measurements: Basic Concepts, Theory and Applications*; Springer Science & Business Media: New York, NY, USA, 2009. [CrossRef]
32. Pełczyński, P.; Szewczyk, W.; Bieńkowska, M.; Kołakowski, Z. A New Technique for Determining the Shape of a Paper Sample in In-Plane Compression Test Using Image Sequence Analysis. *Appl. Sci.* **2023**, *13*, 1389. [CrossRef]
33. Borgqvist, E.; Wallin, M.; Tryding, J.; Ristinmaa, M.; Tudisco, E. Localized deformation in compression and folding of paperboard. *Packag. Technol. Sci.* **2016**, *29*, 397–414. [CrossRef]
34. Pełczyński, P.; Szewczyk, W.; Bieńkowska, M. Single-Camera Sysytem for Measuring Paper Deformations Based on Image Analysis. *Metrol. Meas. Syst.* **2021**, *28*, 509–522. [CrossRef]
35. Hill, R. Theoretical plasticity of textured aggregates. *Math. Proc. Camb Philos. Soc.* **1979**, *85*, 179. [CrossRef]
36. Hill, R. A theory of the yielding and plastic flow of anisotropic metals. *Proc. R. Soc. Lond. Ser. A* **1948**, *193*, 281.
37. Hill, R. *The Mathematical Theory of Plasticity*; Oxford University Press: New York, NY, USA, 1983.
38. *ANSYS 18.1 User's Guide*; Ansys Inc.: Houston, TX, USA, 2017.

materials

MDPI

Article

Thermo-Chemo-Mechanical Modeling of Residual Stress in Unidirectional Carbon Fiber-Reinforced Polymers during Manufacture

Rui Bao [1,2], Junpeng Liu [2], Zhongmin Xiao [1,*] and Sunil C. Joshi [1]

[1] School of Mechanical and Aerospace Engineering, Nanyang Technological University, Singapore 639798, Singapore; baor1001@163.com (R.B.); mscjoshi@ntu.edu.sg (S.C.J.)
[2] College of Safety and Ocean Engineering, China University of Petroleum, Beijing 102249, China; liujp@cup.edu.cn
* Correspondence: mzxiao@ntu.edu.sg

Abstract: The application of carbon fiber-reinforced composite materials in marine engineering is growing steadily. The mechanical properties of unbonded flexible risers using composite tensile armor wire are highly valued. However, the curing process generates a certain amount of internal residual stress. We present a detailed analysis of epoxy resin laminates to assess the impact of thermal, chemical, and mechanical effects on the curing stress and strain. An empirical model that correlates temperature and degree of cure was developed to precisely fit the elastic modulus data of the curing resin. The chemical kinetics of the epoxy resin system was characterized using differential scanning calorimetry (DSC), while the tensile relaxation modulus was determined through a dynamic mechanical analysis. The viscoelastic model was calibrated using the elastic modulus data of the cured resin combining temperature and degree of the curing (thermochemical kinetics) responses. Based on the principle of time–temperature superposition, the displacement factor and relaxation behavior of the material were also accurately captured by employing the same principle of time–temperature superposition. Utilizing the empirical model for degree of cure and modulus, we predicted micro-curing-induced strains in cured composite materials, which were then validated with experimental observations.

Keywords: composite processing; epoxy prepreg; viscoelastic modulus; curing strain

Citation: Bao, R.; Liu, J.; Xiao, Z.; Joshi, S.C. Thermo-Chemo-Mechanical Modeling of Residual Stress in Unidirectional Carbon Fiber-Reinforced Polymers during Manufacture. *Materials* **2024**, *17*, 3040. https://doi.org/10.3390/ma17123040

Academic Editors: Madhav Baral and Charles Lu

Received: 15 May 2024
Revised: 14 June 2024
Accepted: 18 June 2024
Published: 20 June 2024

1. Introduction

As the exploration of offshore oil and gas resources extends into deeper waters, there is an increasing demand for high-performance equipment and longer transmission pipelines. This necessitates pipelines with high-pressure resistance (over 10,000 psi), a wide operating temperature range (−20 °C to +150 °C), excellent corrosion resistance, and high durability (with a service life exceeding 25 years). Furthermore, the pipelines must be capable of withstanding high flow rates, minimizing the environmental impact, and operating effectively at substantial depths (greater than 1000 m). Traditional unbonded flexible risers, typically reinforced with carbon steel tensile armor layers, are increasingly proving insufficient for deepwater operations due to their high density and significant weight. These armor layers, crucial for the structural integrity of the risers, consist of several flat steel bars spirally wound at angles between 30° and 50° relative to the axial direction. To maintain torsional and load balance, the adjacent armor layers are configured in reverse spirals with large angles. The main function of the tensile armor layers is to resist the axial tension and reduce the partial torque, ensuring that the riser remains operational and safe under its own weight and additional tensile force [1,2]. The riser structure is composed of multiple independent layers, each made from different materials and possessing unique sectional shapes, as shown in Figure 1.

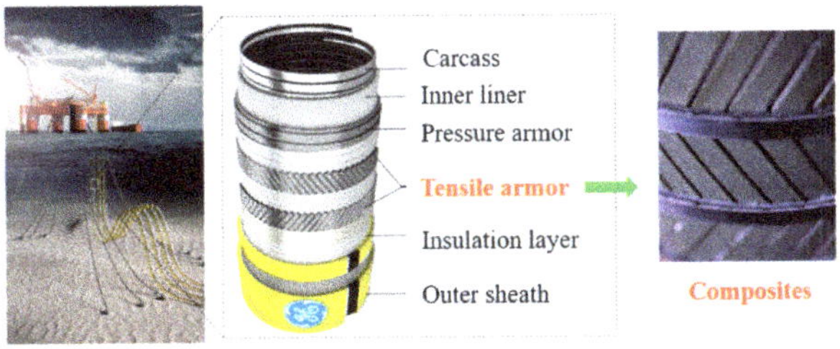

Figure 1. Structure of unbounded flexible risers in deep water.

The replacement of traditional steel with composite materials for manufacturing tensile armor layers presents a promising, yet under-researched, solution. Carbon fiber-reinforced polymers (CFRPs) are particularly suitable for this application based on their high strength, low weight, and corrosion resistance. CFRPs weigh approximately 77% less than traditional steel, which makes it an excellent choice for reducing the overall weight of flexible risers while maintaining the same load-bearing capacity. They possess an elevated strength-to-weight ratio, which enhances the structural efficiency of flexible risers by reducing the overall weight while maintaining the same load-bearing capacity. Additionally, CFRPs demonstrate superior fatigue and corrosion resistance. Extensive research on curing-induced thermal residual stress in composite materials has been conducted. However, few studies have specifically addressed the impact of residual stress on the strength of flexible riser composite tensile armor wires. A simple and effective computational model is needed to calculate the residual stress and strength failure in these composite materials. To do this, it is necessary to combine experimental research with numerical simulation through methods considering constitutive models, volume fraction, curing temperature, and service temperature of fibers and matrices. This will help establish a comprehensive computational theory and a calculation model for residual stress in composite tensile armor layers.

Compared to other materials, composite tensile armor wires present fewer inconveniences such as delamination, degumming, and voids, especially when manufactured using an autoclave. Pultrusion is a continuous manufacturing process for composite profiles, which significantly affects the quality and performance of the final product [3]. Due to the inherent anisotropy and heterogeneity of composite materials, combined with the mismatch in the thermal expansion coefficients between the internal fibers and the resin, residual stresses can develop. However, the residual stresses generated during the autoclave curing process can significantly impact the mechanical properties of the composites, potentially initiating matrix cracks, delamination, and warpage. These stresses may compromise the performance of the composite structure [4]. Dusi [5] analyzed the cure kinetics and viscosity of Fiberite 976 resin, determining the heat of reaction, rate of cure, and degree of cure at various temperatures using DSC and measuring viscosity over time with a viscometer. Mijović [6] analyzed the cure kinetics of three epoxy formulations composed of TGDDM and DDS using isothermal tests. They found that an auto-catalyzed mechanism with an overall reaction order of 2 adequately described the cure kinetics, noting increased reaction rates at higher temperatures and DDS concentrations. Fu [7], Che [8], and Singleton [9] analyzed cure deformations and residual stresses in 3D braided composites using a representative volume element (RVE) and thermochemical and thermodynamic models. They examined the effects of temperature rise rate and curing temperature on the degree of cure, deformation, and residual stresses. Therefore, it is crucial to consider these residual stresses in the design and engineering of composite material structures to ensure long-term reliability and performance.

Several factors affect the formation of residual stress and strain throughout the autoclave curing process. The accurate prediction of cure-induced residual stresses in composite laminates depends on a suitable constitutive model that incorporates factors such as thermal expansion, cure-induced chemical shrinkage, and material degradation or relaxation during curing. Traditionally, studies on residual stress have primarily been focused on elastic models during the cool-down stage only. These models overlook several key effects: chemical shrinkage, stress development before cooling, and stress relaxation during the cooling process. Studies have shown that the classic lamination theory (CLT) or its modified versions can effectively forecast the residual stress in thin laminates [10–12]. Stango and Wang [13] calculated thermal residual stresses by using CLT; they found that CLT overestimates residual stresses and discussed potential reasons for this discrepancy. However, these models do not account for residual stress development before cooling and fail to solve the complex coupling of temperature and curing degree in thick laminates. Consequently, an increasing number of scholars are paying attention to this critical issue.

It is important to recognize that most polymers, due to their viscoelastic properties like stress relaxation and strain creep, are not described by linear elastic models, particularly under conditions of high temperature and low degree of cure. Initially, Schapery [14] conducted series analyses on the viscoelastic stress of composite laminates. Following this, several researchers focused on studying residual stresses using viscoelastic models, predominantly during the cooling stage of the curing process [15]. Accurately simulating the viscoelastic behavior of materials throughout the entire curing process of laminated boards remains challenging. Subsequent studies incorporated factors such as the thermal expansion coefficient, cure-induced chemical shrinkage, and material stress relaxation, examining some fundamental characteristics of the matrix using the time–temperature superposition and equivalence principles [16–19]. Kim and White [20] investigated the stress relaxation behavior of 3501-6 epoxy resin during curing, using experimental data to model the relaxation process. Eom [21] applied time–temperature superposition to predict the instantaneous viscoelastic properties during cure, providing a method to anticipate material behavior throughout the curing process. Prasatya [22] developed a viscoelastic model to predict the isotropic residual stress in thermosetting materials, examining the effects of the processing parameters on stress development. As the thermal, physical, rheological, and mechanical properties of a resin change during curing, the analysis becomes more complex. To better represent these performance changes, advancements in modeling and further progress in experimental characterization are necessary.

In this paper, the reaction kinetics of CFRPs produced from unidirectional (UD) prepregs was characterized and modeled. The curing kinetics parameters of the material were obtained by non-isothermal differential scanning calorimetry (DSC). Additionally, a viscoelastic residual stress model for composite materials was developed, employing the time–temperature superposition principle to describe how the material's mechanical properties change with temperature and degree of cure. Finally, differential finite element (FE) codes were devised and incorporated into ABAQUS with a UMAT subroutine, which was then utilized to model the evolution of residual stress in composite laminates exposed to different curing cycles and validated through experimental verification.

2. Multiscale Modeling Process

2.1. Resin Cure Kinetics

The kinetics of the resin curing reaction was characterized using DSC. Resins undergo a series of phase transitions during the curing process, each accompanied by the release of exothermic heat due to polymerization reactions. From the DSC data, an empirical model that describes the cure rate of a resin can be developed by applying various kinetic models. In order to calculate the resin degree of cure α over time, it is necessary to measure the total heat ΔH_{whole} generated during the reaction process, as well as the instantaneous heat ΔH released:

$$\alpha = \frac{\Delta H}{\Delta H_{whole}}. \tag{1}$$

The autocatalytic model effectively captures the cross-linking dynamics of polymer chains by incorporating two reaction orders. The autocatalytic model of an epoxy resin system is as follows [6,23]:

$$\frac{d\alpha}{dt} = Ae^{\frac{-E_\alpha}{RT}}\alpha^m(1-\alpha)^n. \tag{2}$$

By applying the natural logarithm to both sides of Equation (2), we can obtain:

$$\ln\left(\frac{d\alpha}{dt}\right) = \ln A + m\ln\alpha + n\ln(1-\alpha) - \frac{E_\alpha}{RT}, \tag{3}$$

in which α represents the degree of cure, T is the temperature, A stands for the pre-exponential factor, E_α denotes the activation energy, R is the gas constant, and m, n characterize the reaction order of the cure kinetics. They are obtained from multivariable linear regression of the DSC data.

An Nth-order kinetic model suggests that a single reaction order prevails during the curing process, and the rate equation is expressed as follows [24]:

$$\frac{d\alpha}{dt} = Ae^{\frac{-E_\alpha}{RT}}(1-\alpha)^n. \tag{4}$$

The natural logarithm is calculated for both sides of Equation (4):

$$\ln\left(\frac{d\alpha}{dt}\right) = \ln A + n\ln(1-\alpha) - \frac{E_\alpha}{RT}. \tag{5}$$

2.2. Cure-Dependent Modulus

The elastic modulus of the resin experiences significant development in response to changes in the degree of cure during manufacture. The instantaneous isotropic resin modulus is described by Bogetti [23] as follows:

$$\begin{cases} E_m = E_m^0 & \alpha < \alpha_{gel} \\ E_m = (1-\alpha_{mod})E_m^0 + \alpha_{mod}E_m^\infty & \alpha \geq \alpha_{gel} \end{cases}, \tag{6}$$

in which $\alpha_{mod} = \frac{\alpha-\alpha_{gel}}{1-\alpha_{gel}}$, E_m^0 represents the resin modulus at the onset of the curing process, E_m^∞ denotes the modulus of the fully cured resin, and α_{gel} is the degree of cure at the gel point (theoretically 0.63 for epoxy resin [17]).

The shear modulus is given by:

$$G_m = \frac{E_m}{2(1+v_m)}, \tag{7}$$

where the Poisson ratio of the resin is specified as [25]:

$$v_m = \frac{1}{2}\left(1 - \frac{E_m(1-2v^\infty)}{E_m^\infty}\right). \tag{8}$$

Furthermore, the orthotropic elasticity of the laminate at specified degree of cure and temperature can be described through a micromechanical model [26]. The properties of T700 carbon fiber and the cured epoxy resin are listed in Table 1, which provides detailed material characteristics essential for accurate modeling and analysis.

Table 1. Mechanical characteristics of T700 and completely cured epoxy.

Laminate	Material Properties	
T700	Elastic modulus	115 GPa
	α_{1f}	-0.9×10^{-6}/K
	α_{2f}	7.2×10^{-6}/K
	Poisson's ratio	0.3
Epoxy	Thermal coefficient	5.27×10^{-5}/K
	Volumetric shrinkage	0.29

2.3. Viscoelastic Constitutive Model

During curing, the relaxation of residual stresses formed in composites can be characterized by a detailed viscoelastic constitutive equation that incorporates a hereditary integral. For situations involving variations in both temperature and degree of cure, the stress in a linearly viscoelastic material can be calculated with the equation below [16,27]:

$$\sigma(t) = \int_0^t E(\alpha, T, t-\tau)\frac{d}{d\tau}\left[\varepsilon^{total}(\tau) - \varepsilon^{tc}(\tau)\right]d\tau, \quad (9)$$

where ε^{tc} represents the free thermochemical strain resulting from alterations in both temperature and degree of cure. Composite materials exhibit thermal flow behavior, and the corresponding equation is written as [28]:

$$\sigma(t) = \int_0^t E(\xi(t) - \xi'(\tau))\frac{d}{d\tau}\left[\varepsilon^{total}(\tau) - \varepsilon^{tc}(\tau)\right], \quad (10)$$

in which ξ and ξ', termed reduced times, are a time representation combining temperature and degree of cure and are described as:

$$\xi(t) = \int_0^t \frac{1}{a_T(\alpha, T)}dt'; \;\; \xi'(\tau) = \int_0^\tau \frac{1}{a_T(\alpha, T)}dt', \quad (11)$$

where a_T is the shift factor and enables the time–temperature superposition. Thermal expansion ε^{th} and chemical shrinkage ε^{ch} are the major source of the primary cause of non-mechanical strains ε^{tc} which is expressed as:

$$\begin{cases} \varepsilon^{tc} = \varepsilon^{th} + \varepsilon^{ch} \\ \varepsilon^{th1} = \alpha_{T_1}(T - T_a) \\ \varepsilon^{th2} = \alpha_{T_2}(T - T_a). \end{cases} \quad (12)$$

The thermal expansion coefficients along the fiber direction α_{T_1} and the transverse direction α_{T_2} are defined as:

$$\begin{cases} \alpha_{T1} = \frac{\alpha_{1f}E_fV_f + \alpha_mE_mV_m}{E_fV_f + E_mV_m} \\ \alpha_{T2} = \left(\alpha_{2f} + v_{12f}\alpha_{1f}\right)V_f + (\alpha_m + v_m\alpha_m)V_m, \end{cases} \quad (13)$$

in which E_f and E_m are the elastic moduli, V_f and V_m are the volume fractions, and α_f and α_m are the thermal expansion coefficients.

During the curing process, the volume of the resin decreases because of the chemical cross-linking reaction. The chemical shrinkage strain of an isotropic homogeneous resin can be characterized as follows:

$$\varepsilon^r = \sqrt[3]{1 + \alpha v_{re}} - 1, \quad (14)$$

where ν_{re} is the volume resin reduction and equals to −0.03 [29]. This indicates that the effective thermal expansion and cure shrinkage of composites are impacted by the properties of their constituents and their volumetric proportions.

3. Model Verification and Numerical Implementation

3.1. Experimental Section

In this section, a UD prepreg suitable for high-temperature (180 °C) curing was used. The epoxy resin was tailored for high-temperature applications. The composite material was reinforced with T700 carbon fibers. The conventional DSC Q200 from TA Instruments (New Castle, DE, USA) was used to assess the exothermic flow of the epoxy prepreg during the manufacturing process and to determine the curing reaction parameters. In a nitrogen atmosphere, the uncured prepreg was placed in a DSC sample tray for dynamic scanning; the sample weight was approximately 18 mg. Subsequently, DSC was set at the different heating rates of 5 °C/min, 10 °C/min, 15 °C/min, and 20 °C/min within the temperature range of room temperature, up to 300 °C. Firstly, standard materials like indium were used for temperature calibration, recording their melting points and adjusting the instrument settings accordingly. Next, we conducted a heat flow calibration using a material with a known heat capacity, such as sapphire, and adjusted the heat flow settings. A stable baseline was established by running an empty crucible experiment. Finally, the calibration was verified with additional standard materials to ensure accuracy and precision.

The storage and loss moduli were evaluated by DMA-Q800 from TA Instruments (New Castle, DE, USA). A sample measuring 50 × 5 × 2 mm was tested in the 3-point bending mode. Before inserting the standard spline into the DMA, we accurately measured the spline using a vernier caliper and determined the average value. The calibrated device functioned across multiple cure cycles and heating rates in the 3-point bending mode with a testing frequency of 1 Hz and a heating rate of 5 °C/min, consistently maintaining an oscillation amplitude of 20 μm.

Directional strains in laminates during the curing reaction are measured using Fiber Bragg Grating (FBG) sensors. The conversion relationship among strain, temperature, and wavelength λ_B is as follows [30]:

$$\frac{\Delta\lambda_B}{\lambda_B} = \left(\alpha_f + \varsigma\right)\Delta T + (1 - p_e)\Delta\varepsilon \tag{15}$$

In which α_f denotes the thermal expansion, and ς and p_e are the thermo-optic and photo-elastic coefficients of the optical fibers. The coefficient p_e equals 0.22 for silica optical fibers oriented in both directions. At the same time, a thermocouple was placed to compensate for the temperature response of the FBG sensors on the laminate. Figure 2 shows the placement of the FBG sensors and the thermocouple. They were placed in the center of a 4 mm thick UD laminate, indicating the material directions.

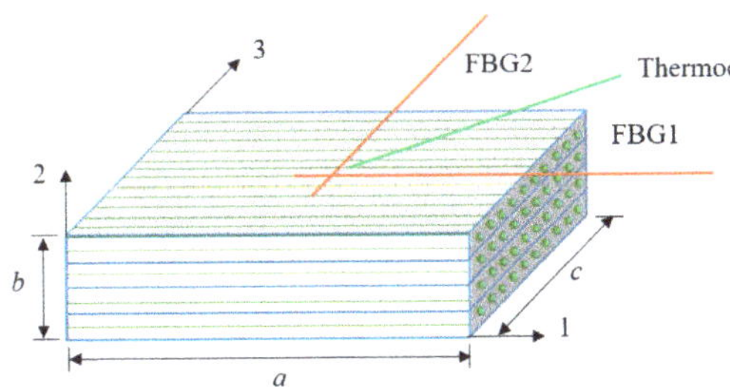

Figure 2. Location of FBG sensors and thermocouple for strain measurement.

3.2. Numerical Model

A multi-scale model was established to simulate the stress–strain state during the laminate curing process. At the macroscopic level, stress, strain, temperature, and degree of cure of the material were determined. The unidirectional laminate was assumed to be a uniform transversely isotropic material, and we used a viscoelastic constitutive model to determine the macroscopic mechanical response throughout the curing process. At the microscale, the resin's thermochemical properties were calculated. Temperature and degree of curing were calculated at the macroscale and transferred to the microscale. These calculations determined the corresponding thermochemical mechanical properties, which were fed back to the macroscopic level.

The mechanical model was executed and integrated into the commercial finite element software ABAQUS 2021, complete with the geometry and mesh configuration of the finite element model. The laminate plane was meshed using the C3D8T element. The user subroutines HETVAL and USFFLD were developed to investigate the curing kinetics of thermochemical models. Simultaneously, the user subroutines UMAT and UEXPAN were used to model the changes in stress and strain in relation to the degree of cure and temperature. A finite element model with 4 mm thick laminate boundary conditions was obtained following the guidelines provided in Figure 3.

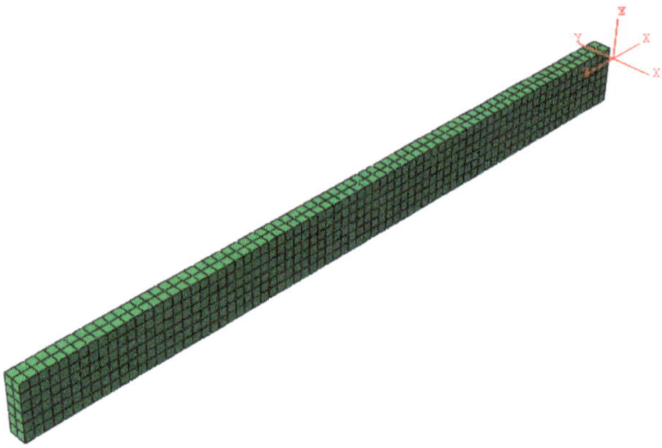

Figure 3. The geometry and mesh of the finite element model.

The dashed line in Figure 4 shows the manufacturer's recommended curing cycle. For thicknesses below 4 mm, curing cycle-1 is recommended, which involves heating from room temperature to 180 °C at a rate of 2.5 °C/min, holding this temperature for 1 h, and then cooling down to room temperature at a rate of 2 °C/min. For thicknesses greater than 4 mm, curing cycle-2 is recommended, initially heating up to 120 °C at a rate of 1.5 °C/min, maintaining this temperature for one hour, then increasing it to 180 °C at the same rate, holding it for 90 min, and finally returning to room temperature at a rate of 2.5 °C/min. The curves in the figure also show a comparison of the distribution of the degree of curing. With temperature changes, the curing rate in cycle-1 becomes more rapid, and the glass transition phase is less pronounced. However, for thicker laminates, this can lead to uneven curing. Using the *t*-test method, the t-statistic value was calculated to be 1.009, and the *p*-value was 0.320. Here, we compared the curing degrees of cycle-1 and cycle-2 at specific time points. At all time points, there was a small statistical difference in the overall solidification degree between the two cycles. The performance difference between the two cycles under real conditions could be evaluated through practical application testing.

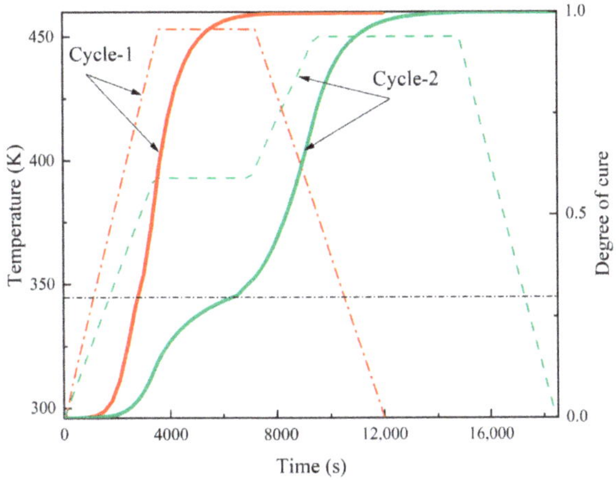

Figure 4. Development of the degree of cure in the central point of the laminate. During the curing process of thick laminates, the curing speed is different between the interior and the surface of the material. The temperature gradients cause inconsistent thermal expansion and contraction. This inconsistency can result in residual stress within the laminate, potentially causing warping, deformation, and even cracking of the laminate, thereby compromising its dimensional stability and structural integrity. Additionally, different degrees of cure between the inner and the outer layers can result in uneven crosslinking density, leading to disparities in the mechanical properties (such as strength and toughness) in different locations, thus reducing the overall performance and reliability of the material. Furthermore, uneven solidification during the curing process can introduce defects such as uncured areas or bubbles, increasing the risk of material failure during service. An optimized curing process is necessary to overcome these challenges, requiring real-time monitoring systems to ensure uniform curing and high-quality laminate performance.

4. Results and Discussion

4.1. Cure Kinetics

We examined the cure kinetics over a degree range from 0.05 to 0.9 to derive the related parameters. Both Nth-order and autocatalytic models were employed to analyze the cure kinetic parameters from the collected data. Initially, dynamic measurements were conducted using DSC at various heating rates (5 °C/min, 10 °C/min, 15 °C/min, 20 °C/min), and the heat flux curves consistently displayed an endothermic peak. Figure 5 shows this phenomenon at different heating rates. By employing various dynamic models, an empirical model depicting the curing rate can be derived from DSC data. The Nth-order kinetic model presumes a singular order reaction throughout the entire curing process. The cumulative data are graphed and linearly fitted to determine the reaction order. Figure 6 displays the curve obtained using an Nth-order dynamic model. It was observed that a linear relationship existed when the curing degree was below 0.7, but significant nonlinearity arose when the curing degree exceeded 0.7. This discrepancy suggests that when the curing degree surpasses 0.7, the Nth-order kinetic model fails to conform to Equation (4), thereby indicating that the Nth-order model cannot accurately represent the curing reaction process of the material. The Nth-order model began to fail at a degree of cure α of approximately 0.7, across all heating rates analyzed (5 K/min, 10 K/min, 15 K/min, and 20 K/min). At this threshold, the values of $\ln\left(\frac{d\alpha}{dt}\right) + \frac{E_a}{RT}$ ranged from 22.2 to 23.0, depending on the heating rate. Beyond these points, significant deviations indicated the Nth-order model's inadequacy for describing the curing process accurately.

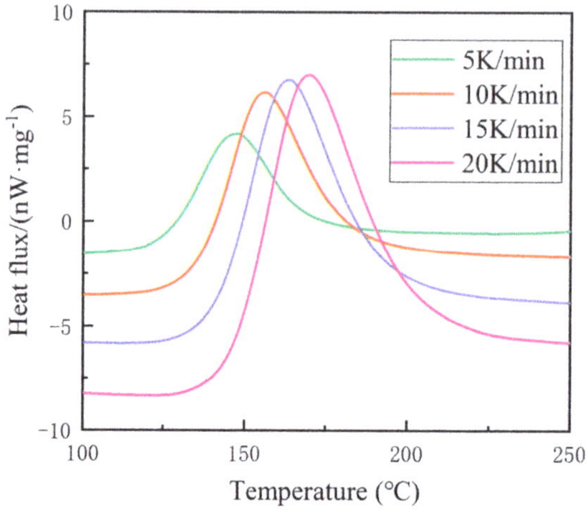

Figure 5. Heat flux based on the temperature at different heating rates.

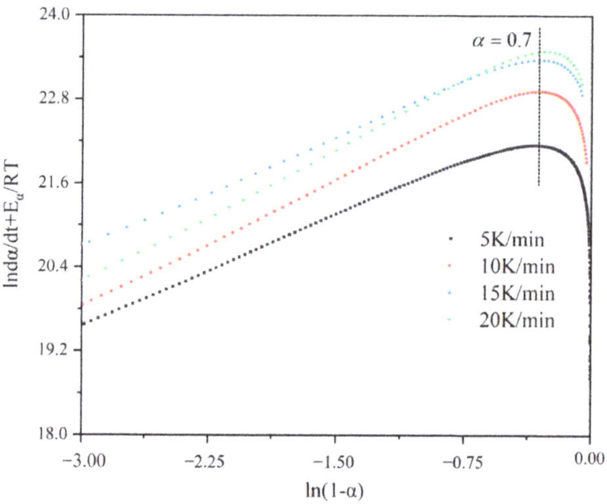

Figure 6. Fit relationship diagram for the Nth-order reaction kinetics model.

The activation energy E_α of the curing process showed a clear decreasing trend as the degree of cure increased from 0 to 1, as shown in Figure 7. Initially, at a degree of cure near 0, the activation energy was high, around 85 kJ/mol, indicating a significant energy requirement to initiate the curing reaction. As the degree of cure progressed to approximately 0.25, the activation energy decreased to about 80 kJ/mol and continued to drop gradually to around 70 kJ/mol at a degree of cure of 0.75. In the final phase, approaching full cure, the activation energy sharply declined to approximately 60 kJ/mol. This trend suggests that the curing process became progressively easier as the polymer network formed and crosslinking was facilitated, highlighting the importance of optimizing the curing conditions in different stages to ensure uniform and high-quality material properties.

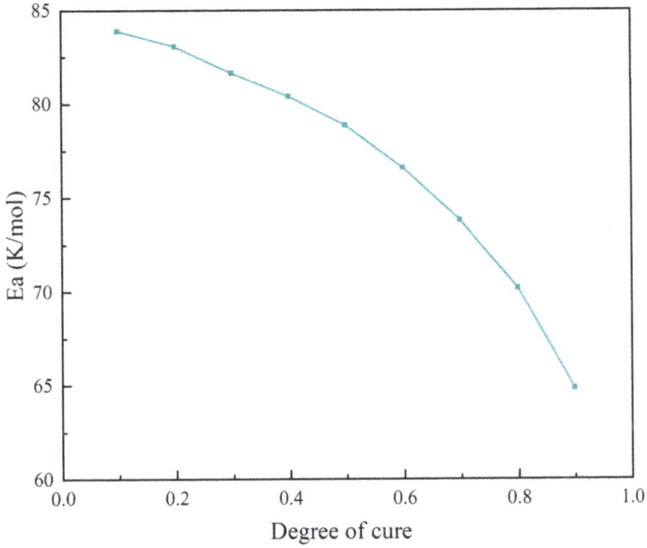

Figure 7. Relationship between activation energy and degree of cure for the curing process.

Nonlinear fitting was performed on other parameters in the model, and the curing kinetics parameters are shown in Table 2. Figure 8 shows the comparison between the experimental and the autocatalytic models. The high R^2 values across all heating rates indicated that the autocatalytic model fitted the experimental data well, maintaining a high predictive accuracy. The slight decrease in R^2 at higher heating rates suggests minor deviations, but overall, the model remained robust. It can be seen that the model and the experimental data were in good agreement in both early and late stages of the curing reaction. The autocatalytic model, which accounted for the concentration of resin and reaction products in calculating the reaction rate, could accurately predict the curing process of the prepreg system. Additionally, consistency was apparent, as shown in Figure 9: the temperature range where the curing degree curve increased coincided with the peak temperature on the reaction rate curve. As the temperature continued to rise, the degree of curing tended to saturate (close to 1), and the corresponding reaction rate curve started to decrease, reflecting the gradual completion of the reaction. The solidification progress was slow in the low-temperature zone, and the corresponding peak of the reaction rate curve appeared in a higher temperature range.

Table 2. Cure kinetic parameters of autocatalytic model and Nth-order model.

Heating Rate/(K·min^{-1})	Autocatalytic Model				Nth-Order Model		
	ln(A)/min^{-1}	m	n	R^2	ln(A)/min^{-1}	n	R^2
5	23.26	0.55	1.28	0.997	21.53	0.179	0.49
10	23.96	0.5	1.41	0.991			
15	24.23	0.4	1.21	0.987			
20	24.39	0.39	1.43	0.988			

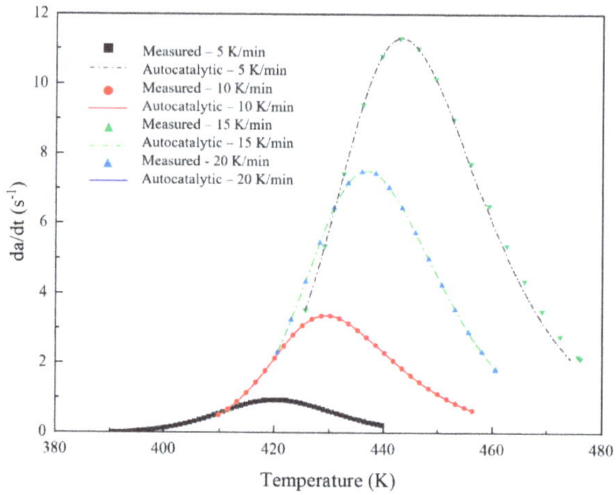

Figure 8. Comparison of the cure rate profiles at different heating rates.

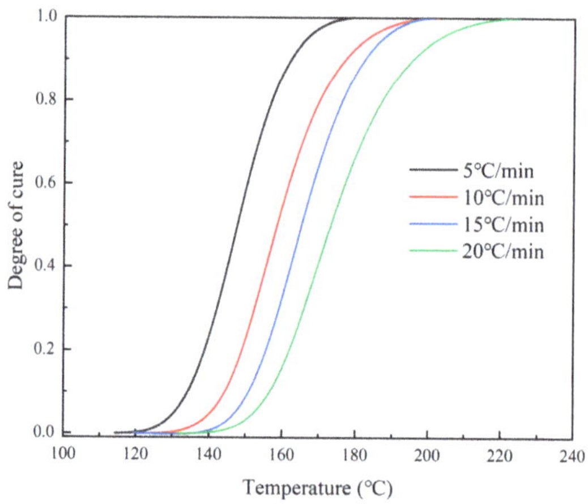

Figure 9. Degree of cure at different heating rates.

Using Table 3, we conducted a detailed statistical comparison. It involved examining the R^2 values of both models at different heating rates and calculating additional fit quality indices by using RMSE, AIC, and BIC. Taking the heating rate of 5 K/min as an example, the comparison is clearly shown in Table 3. The autocatalytic model consistently provided a better fit at different heating rates. This was evidenced by higher R^2 values and lower RMSE, AIC, and BIC values, as expected. The Nth-order model, with significantly lower R^2 values, indicated a much poorer fit to the experimental data. This detailed statistical analysis highlighted the superiority of the autocatalytic model for describing the curing process.

Table 3. Summary of the expanded comparison.

Model	R^2	RMSE	AIC	BIC
Autocatalytic	0.997	0.15	30.1	35.4
Nth-order	0.490	0.60	45.0	50.3

4.2. Material Modulus Characteristics

The mechanical properties of a resin change with polymerization and temperature. Understanding these changes is crucial for estimating residual stresses and the resins' final properties. The dynamic storage modulus $E'(w,T)$ is converted to the apparent tensile relaxation modulus $G(t_r,T)$ as follows [31,32]:

$$E'(w,T) \approx G(t_r,T) \tag{16}$$

in which t_r is the relaxation time. On the left side of Figure 10, we show the relaxation modulus against time in a logarithmic scale at different temperatures. Except for the curve at the reference temperature T_{ref} = 23 °C, the other curves are horizontally shifted, tending to overlap. As depicted on the right side of this figure, a single smooth master curve against the reduced time was obtained. The smoothness of the curve confirmed the applicability of the time–temperature superposition principle to the apparent tensile relaxation modulus.

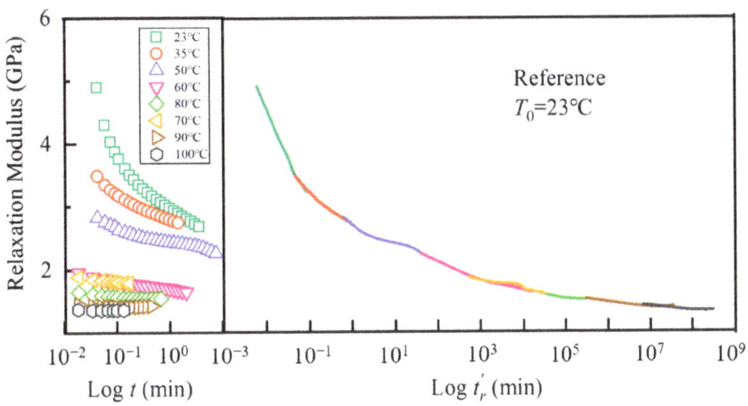

Figure 10. Relaxation modulus for the epoxy system and its master curve.

Here, we used the shift factor α_T of the Arrhenius equation at temperatures near or above the glass temperature:

$$\log \alpha_T = \frac{\Delta H}{2.303R}\left(\frac{1}{T} - \frac{1}{T_0}\right) \tag{17}$$

where R represents the gas constant, set at 8.314×10^{-3} [kJ/(K mol)]. The shift factors for the relaxation modulus G used in the master curve construction are shown in Figure 11. We set $x = r$ indicating the relaxation. The shift factors for the tensile relaxation modulus aligned well with the Arrhenius equation with $\Delta H_1 = 34$ kJ/mol and $\Delta H_2 = 400$ kJ/mol. Through their accurate fitting with the experimental data, reflecting the material physical behavior, it was proved that at low temperatures, a small activation energy corresponded to a middle thermal activation value, while a high activation energy at high temperatures indicated significant molecular mobility and rearrangement. These shift factors can construct an accurate master curve of the tensile relaxation modulus, which can then be used to analyze the relaxation behavior of materials over a wide temperature range.

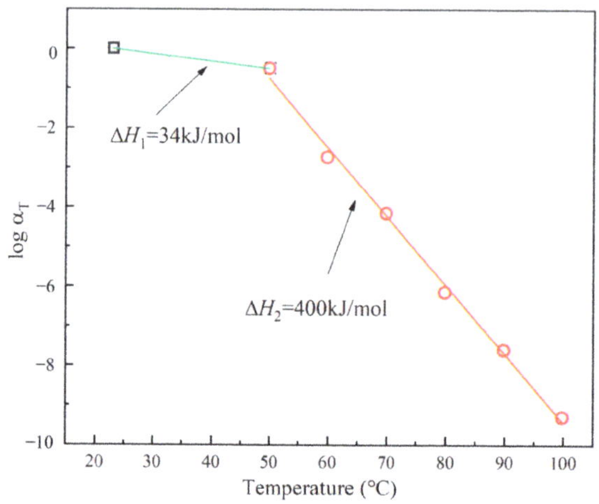

Figure 11. Shift factors for the relaxation modulus.

The fully cured samples were readied with a specific curing cycle for DMA analysis. Tension tests were conducted to evaluate how the elastic modulus of the specimens changed with temperature at varying frequencies. The temperature was incrementally raised, and isothermal pauses of 5 min were implemented to establish thermal equilibrium. Figure 12a shows the storage modulus and loss modulus of E of the cured resin. Figure 12b shows the loss factor tan δ in relation to temperature. The resin transformed from liquid to solid gel as the temperature increased. The storage modulus of the cured resin decreased at around 125 °C and tended to stabilize at around 170 °C. At temperatures above 180 °C, the resin was completely cured, and the modulus tended to stabilize. The loss factor remained low and constant at lower temperatures when the material was in its glassy state and rose at higher temperatures as the material became viscoelastic. The peak of the loss factor typically occurred near the glass transition temperature of the cured resin. Prior to vitrification, a significant variation was noted in the measured storage modulus as the frequency of the specimen increased. The experimental results indicated a difference in the storage modulus of 8 GPa, and the loss modulus was 0.6 GPa. After the vitrification of the resin, the difference between the storage modulus and the energy dissipation modulus at the two indicated frequencies was 5 GPa and 0.3 GPa.

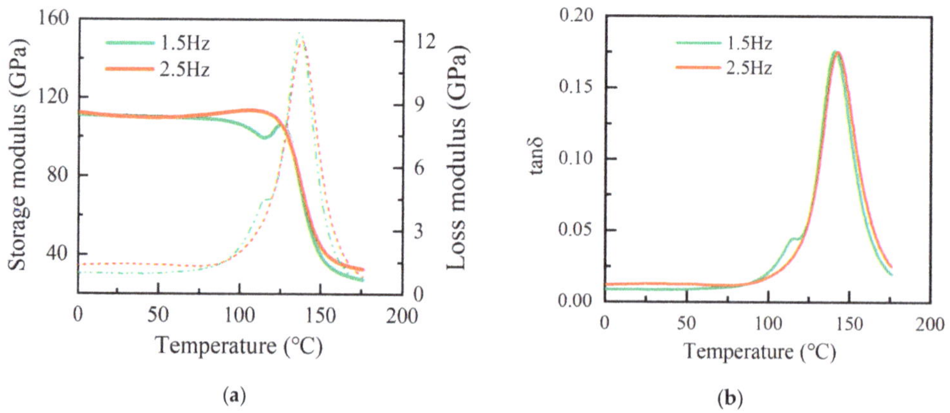

Figure 12. (**a**) Elastic modulus of the cured resin at different frequencies; (**b**) loss factor at different frequencies.

4.3. Residual Stress and Strain Simulation

In this section, we describe the finite element verification of the proposed viscoelastic model and the numerical simulation of composite laminates that we conducted to study the residual stress distribution during the curing of a CFRP. The composite material studied was T700/epoxy resin unidirectional prepreg with a thickness of 4 mm. To facilitate the comparison with the experiments, the finite element model was set to four layers. Under the recommended curing cycle, the mechanical state of the center point (a/2, b, c/2) of the laminate in Figure 2 was studied.

Figures 13 and 14 present a dual-axis plot tracking the temperature and stress of a carbon fiber-reinforced composite over time through two curing cycles. Cycle-1 is represented in orange, and cycle-2 in green. The orange dashed line represents the curing temperature curve for cycle-1, while the solid line represents the residual stress. At a certain time, the corresponding stress depended on the temperature. The curing time for cycle-1 was 3.33 h, and that for cycle-2 was 5 h. The key aspects to note are the points of vitrification and transition to a gel state, which are significant in the curing process. In both cycles, the stress response to the thermal treatment was apparent; however, the stress in cycle-2 suggests a more complex interaction, possibly due to the response to the dual-stage heating process. Notably, the stress did not completely revert to its initial state as the temperature returned to baseline in both cycles, suggesting that residual stress remained in the material after curing.

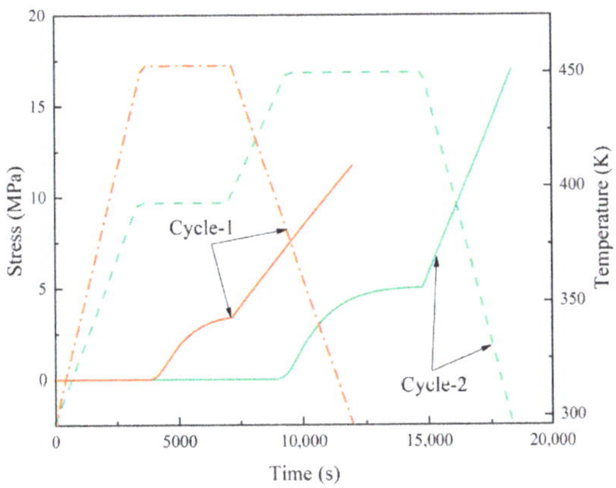

Figure 13. Development of longitudinal stress during the cure cycles in the examined laminates.

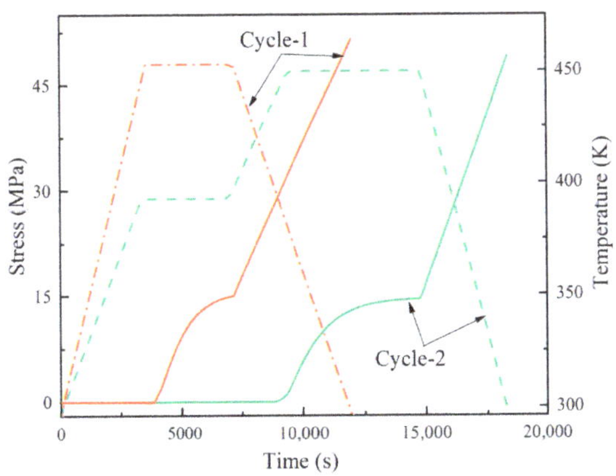

Figure 14. Development of transverse stress during the cure cycles in the examined laminates.

In the initial stage of cycle-2 curing, lasting less than 60 min, due to the fast stress relaxation and chemical shrinkage at a low degree of cure, the stress slightly decreased with the increasing temperature in the first heating stage (compressive stress). In the first stage, the resin began to cure and contract, and the shrinkage rate and thermal expansion rate of the material were similar; so, the stress increased but not significantly. In the second stage of heating and holding, due to the chemical hardening effect, the stress increased. In the later stage of holding in the second stage, the stress changes tended to stabilize, which can be attributed to the fact that a balance between the complete solidification of the material and stiffness relaxation was reached. During the cooling stage, the slow relaxation and thermal expansion of the material in the second stage resulted in a significant increase in stress. In the curing processes of cycle-1 and cycle-2, the total residual stresses were 11.7 and 17 MPa, respectively. Due to the fact that the composite material underwent two heating and insulation stages in the two cycles, many cross-linking reactions occurred, resulting in high stress levels in the material. At the same time, the second heating stage allowed more unreacted resins to participate in the reaction, increasing the curing degree and material rigidity.

Compared to Figure 14, the unidirectional direction of composite fibers resulted in a higher modulus and greater strength in the longitudinal direction (fiber direction) during the curing process. At the same time, the fibers could also withstand higher loads. In the longitudinal cycle-1 and cycle-2, the total residual stresses were 51.4 MPa and 49.1 MPa, respectively, with little difference between the two cycles. The comparison between the values measured for the FBG sensor and the numerical values of the curing shrinkage strain at the center of the laminate is shown in Figures 15 and 16. In the actual experimental testing, an oven was used for curing, with pressure applied to ensure the curing behavior of the material with cycle-1. The numerical simulation results revealed a dynamic interplay between temperature and strain during the curing of the prepreg. Initially, thermal expansion caused the strain to increase to a peak as the temperature rose. Concurrently, the semi-cured resin began its transition into a liquid state. As the curing process began, the material polymerization reaction caused resin shrinkage, resulting in a sharp decrease in the material strain. In this stage, the shrinkage in the laminate attributable to the curing process superseded the thermal expansion strain. Due to the slow chemical reaction of the polymer chain, a small delay in strain variation was observed after vitrification. After the resin curing was completed, the strain increased twice, proportionally to the temperature increase caused by thermal expansion. Finally, as the resin cooled, volume contraction due to the reduced temperature led to a minimization of the strain. At the end of curing, the difference between the transverse strain ($-4640\ \mu\varepsilon$) based on the proposed modulus and the experimentally measured transverse strain ($-4438\ \mu\varepsilon$) was very small. Therefore, the proposed viscoelastic modulus model can accurately predict the curing strain.

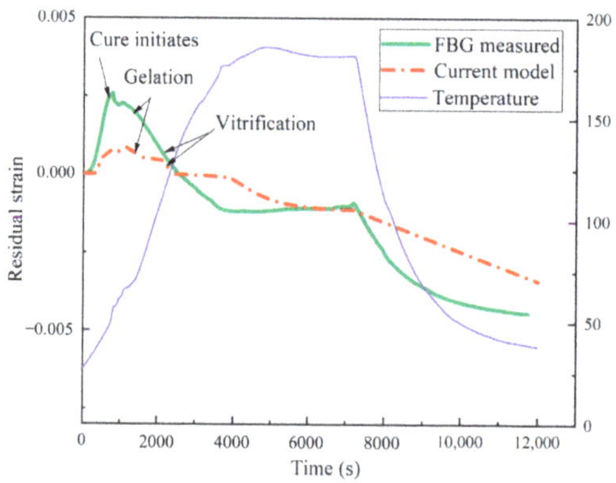

Figure 15. Longitudinal curing strain during the manufacturing process.

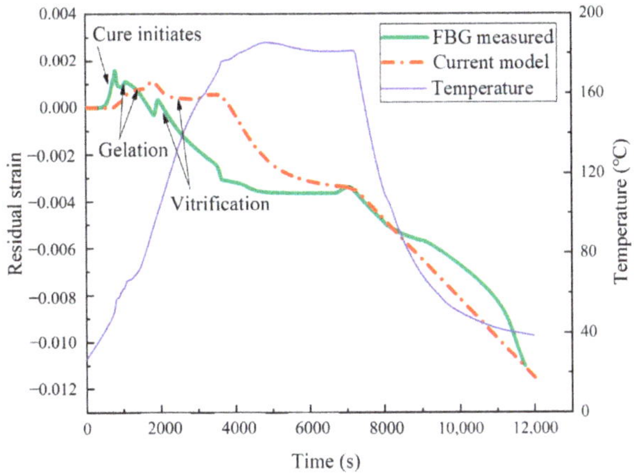

Figure 16. Transverse curing strain during the manufacturing process.

5. Conclusions

We comprehensively characterized the residual strain observed during the curing process of CFRPs. Subsequently, the kinetics of this process was meticulously analyzed using both *N*th-order and autocatalytic models. The autocatalytic model demonstrated a closer alignment with practical applications, and the temperature range increase in the curing degree curve was consistent with the peak temperature in the reaction rate curve. The alignment between the autocatalytic model and its applications was based on its accurate prediction of the kinetic parameters, the consistency of the obtained results with the experimental data, and its ability to describe the mechanical properties of laminate. An empirical model combining the curing degree and temperature was used to describe the thermochemical properties of the cured prepregs. This model effectively captured the viscoelastic behavior and shift factors by using the principle of time–temperature superposition. Following resin vitrification, the differential observed between the storage modulus and the loss modulus at varying frequencies was quantified as amounting to 5 GPa and 0.3 GPa. Viscoelastic models are utilized to investigate the material manufacturing process, predicting and analyzing the residual stress and strain during the curing process. For laminates measuring 4 mm in thickness, the difference between the two cycles was minimal. Comparing the curing cycles under the same conditions, the measured curing strain and the simulated curing strain showed overall consistency. This viscoelastic model assumes uniform material properties and curing, which may not be accurate for thicker laminates. This can cause inaccuracies in predictions. Thus, experimental validation and better modeling techniques are necessary to accurately predict complex laminate behaviors.

Author Contributions: Methodology, R.B., J.L. and S.C.J.; software, R.B.; validation, R.B.; investigation, R.B., J.L., Z.X. and S.C.J.; resources, Z.X.; writing—original draft, R.B.; writing—review and editing, Z.X. All authors have read and agreed to the published version of the manuscript.

Funding: This study is supported by the China Scholarship Council [CSC No.202206560023].

Institutional Review Board Statement: Not applicable.

Informed Consent Statement: Not applicable.

Data Availability Statement: The data are contained within the article.

Acknowledgments: The first author gratefully acknowledges Nanyang Technological University (NTU) for financial support and other types of support.

Conflicts of Interest: The authors declare no conflicts of interest.

References

1. Liu, J.; Ma, J.; Vaz, M.A.; Duan, M. Axisymmetric Structural Behaviours of Composite Tensile Armoured Flexible Pipes. *Mar. Struct.* **2020**, *74*, 102829. [CrossRef]
2. Li, D.; Jiang, W.; Xing, Q.; Liu, Q. Axial Tensile Ultimate Strength of an Unbonded Flexible Riser Based on a Numerical Method. *Materials* **2024**, *17*, 2286. [CrossRef] [PubMed]
3. Vedernikov, A.; Tucci, F.; Safonov, A.; Carlone, P.; Gusev, S.; Akhatov, I. Investigation on the Shape Distortions of Pultruded Profiles at Different Pulling Speed. *Procedia Manuf.* **2020**, *47*, 1–5. [CrossRef]
4. Lee, W.I.; Loos, A.C.; Springer, G.S. Heat of Reaction, Degree of Cure, and Viscosity of Hercules 3501-6 Resin. *J. Compos. Mater.* **1982**, *16*, 510–520. [CrossRef]
5. Dusi, M.R.; Lee, W.I.; Ciriscioli, P.R.; Springer, G.S. Cure Kinetics and Viscosity of Fiberite 976 Resin. *J. Compos. Mater.* **1987**, *21*, 243–261. [CrossRef]
6. Mijović, J.; Kim, J.; Slaby, J. Cure Kinetics of Epoxy Formulations of the Type Used in Advanced Composites. *J. Appl. Polym. Sci.* **1984**, *29*, 1449–1462. [CrossRef]
7. Fu, Y.; Gao, X.; Yao, X. Mesoscopic Simulation on Curing Deformation and Residual Stresses of 3D Braided Composites. *Compos. Struct.* **2020**, *246*, 112387. [CrossRef]
8. Che, L.; Fang, G.; Wu, Z.; Ma, Y.; Zhang, J.; Zhou, Z. Investigation of Curing Deformation Behavior of Curved Fiber Metal Laminates. *Compos. Struct.* **2020**, *232*, 111570. [CrossRef]
9. Singleton, T.; Saeed, A.; Strawbridge, L.; Khan, Z.A. Finite Element Analysis of Manufacturing Deformation in Polymer Matrix Composites. *Materials* **2024**, *17*, 2228. [CrossRef]
10. Stango, R.J.; Wang, S.S. Process-Induced Residual Thermal Stresses in Advanced Fiber-Reinforced Composite Laminates. *J. Eng. Ind.* **1984**, *106*, 48–54. [CrossRef]
11. Shokrieh, M.M.; Kamali, S.M. Theoretical and Experimental Studies on Residual Stresses in Laminated Polymer Composites. *J. Compos. Mater.* **2005**, *39*, 2213–2225. [CrossRef]
12. Schulz, W.A.; Myers, D.G.; Singer, T.N.; Ifju, P.G.; Haftka, R.T. Determination of Residual Stress and Thermal History for IM7/977-2 Composite Laminates. *Compos. Sci. Technol.* **2005**, *65*, 2014–2024. [CrossRef]
13. Shokrieh, M.M.; Akbari, S.; Daneshvar, A. A Comparison between the Slitting Method and the Classical Lamination Theory in Determination of Macro-Residual Stresses in Laminated Composites. *Compos. Struct.* **2013**, *96*, 708–715. [CrossRef]
14. Schapery, R.A. On the Characterization of Nonlinear Viscoelastic Materials. *Polym. Eng. Sci.* **1969**, *9*, 295–310. [CrossRef]
15. Wang, T.-M.; Daniel, I.M.; Gotro, J.T. Thermoviscoelastic Analysis of Residual Stresses and Warpage in Composite Laminates. *J. Compos. Mater.* **1992**, *26*, 883–899. [CrossRef]
16. White, S.R.; Hahn, H.T. Process Modeling of Composite Materials: Residual Stress Development during Cure. Part I. Model Formulation. *J. Compos. Mater.* **1992**, *26*, 2402–2422. [CrossRef]
17. Adolf, D.; Martin, J.E. Calculation of Stresses in Crosslinking Polymers. *J. Compos. Mater.* **1996**, *30*, 13–34. [CrossRef]
18. Kim, Y.K. Process-Induced Residual Stress Analysis by Resin Transfer Molding. *J. Compos. Mater.* **2004**, *38*, 959–972. [CrossRef]
19. Merzlyakov, M.; McKenna, G.B.; Simon, S.L. Cure-Induced and Thermal Stresses in a Constrained Epoxy Resin. *Compos. Part Appl. Sci. Manuf.* **2006**, *37*, 585–591. [CrossRef]
20. Kim, Y.K.; White, S.R. Stress Relaxation Behavior of 3501-6 Epoxy Resin during Cure. *Polym. Eng. Sci.* **1996**, *36*, 2852–2862. [CrossRef]
21. Eom, Y.; Boogh, L.; Michaud, V.; Sunderland, P.; Månson, J. Time-cure-temperature Superposition for the Prediction of Instantaneous Viscoelastic Properties during Cure. *Polym. Eng. Sci.* **2000**, *40*, 1281–1292. [CrossRef]
22. Prasatya, P.; McKenna, G.B.; Simon, S.L. A Viscoelastic Model for Predicting Isotropic Residual Stresses in Thermosetting Materials: Effects of Processing Parameters. *J. Compos. Mater.* **2001**, *35*, 826–848. [CrossRef]
23. Bogetti, T.A.; Gillespie, J.W. Process-Induced Stress and Deformation in Thick-Section Thermoset Composite Laminates. *J. Compos. Mater.* **1992**, *26*, 626–660. [CrossRef]
24. Shin, D.D.; Hahn, H.T. A Consistent Cure Kinetic Model for AS4/3502 Graphite/Epoxy. *Compos. Part Appl. Sci. Manuf.* **2000**, *31*, 991–999. [CrossRef]
25. Kravchenko, O.G.; Kravchenko, S.G.; Pipes, R.B. Cure History Dependence of Residual Deformation in a Thermosetting Laminate. *Compos. Part Appl. Sci. Manuf.* **2017**, *99*, 186–197. [CrossRef]
26. Kim, Y.K.; White, S.R. Process-Induced Stress Relaxation Analysis of AS4/3501-6 Laminate. *J. Reinf. Plast. Compos.* **1997**, *16*, 2–16. [CrossRef]
27. Behseresht, S.; Park, Y.H. Additive Manufacturing of Composite Polymers: Thermomechanical FEA and Experimental Study. *Materials* **2024**, *17*, 1912. [CrossRef]
28. Ganapathi, A.S.; Joshi, S.C.; Chen, Z. Influence of Cure Kinetic, Rheological and Thermo-Mechanical Behavior on Micro-Level Curing Strain of an Epoxy Prepreg. *J. Therm. Anal. Calorim.* **2016**, *124*, 305–316. [CrossRef]
29. Liu, Z.; Zheng, X.; Gao, L.; Yan, L.; Song, G.; Zhang, S. Comparative Study on the Effect of Cure Parameters on Residual Deformation for Thermoset Composite Laminates. *J. Compos. Mater.* **2021**, *55*, 2591–2604. [CrossRef]
30. Kim, S.S.; Murayama, H.; Kageyama, K.; Uzawa, K.; Kanai, M. Study on the Curing Process for Carbon/Epoxy Composites to Reduce Thermal Residual Stress. *Compos. Part Appl. Sci. Manuf.* **2012**, *43*, 1197–1202. [CrossRef]

31. Christensen, R.M. *Theory of Viscoelasticity*; Courier Corporation: New York, NY, USA, 2003.
32. Stankiewicz, A. Sampling Points-Independent Identification of the Fractional Maxwell Model of Viscoelastic Materials Based on Stress Relaxation Experiment Data. *Materials* **2024**, *17*, 1527. [CrossRef] [PubMed]

materials

MDPI

Article

A Multiphysics Thermoelastoviscoplastic Damage Internal State Variable Constitutive Model including Magnetism

M. Malki [1], M. F. Horstemeyer [2,*], H. E. Cho [2,*], L. A. Peterson [3], D. Dickel [4], L. Capolungo [5] and M. I. Baskes [6]

1 Aerospace and Automotive Department, International University of Rabat, Rabat 11103, Morocco; mounia.malki@uir.ac.ma
2 School of Engineering, Liberty University, Lynchburg, VA 24515, USA
3 Center for Advanced Vehicular Systems, Mississippi State University, Starkville, MS 39759, USA; lukep@cavs.msstate.edu
4 Department of Mechanical Engineering, Mississippi State University, Starkville, MS 39762, USA; doyl@me.msstate.edu
5 Materials Science and Technology Division, Los Alamos National Laboratory, Los Alamos, NM 87545, USA; laurent@lanl.gov
6 Department of Materials Science & Engineering, University of North Texas, Denton, TX 76203, USA; michael.baskes@unt.edu
* Correspondence: mhorstemeyer@liberty.edu (M.F.H.); hecho@liberty.edu (H.E.C.)

Abstract: We present a macroscale constitutive model that couples magnetism with thermal, elastic, plastic, and damage effects in an Internal State Variable (ISV) theory. Previous constitutive models did not include an interdependence between the internal magnetic (magnetostriction and magnetic flux) and mechanical fields. Although constitutive models explaining the mechanisms behind mechanical deformations caused by magnetization changes have been presented in the literature, they mainly focus on nanoscale structure–property relations. A fully coupled multiphysics macroscale ISV model presented herein admits lower length scale information from the nanoscale and microscale descriptions of the multiphysics behavior, thus capturing the effects of magnetic field forces with isotropic and anisotropic magnetization terms and moments under thermomechanical deformations. For the first time, this ISV modeling framework internally coheres to the kinematic, thermodynamic, and kinetic relationships of deformation using the evolving ISV histories. For the kinematics, a multiplicative decomposition of deformation gradient is employed including a magnetization term; hence, the Jacobian represents the conservation of mass and conservation of momentum including magnetism. The first and second laws of thermodynamics are used to constrain the appropriate constitutive relations through the Clausius–Duhem inequality. The kinetic framework employs a stress–strain relationship with a flow rule that couples the thermal, mechanical, and magnetic terms. Experimental data from the literature for three different materials (iron, nickel, and cobalt) are used to compare with the model's results showing good correlations.

Keywords: magnetism; magnetostrictive strain; magneto-mechanical effect; ferromagnets

Citation: Malki, M.; Horstemeyer, M.F.; Cho, H.E.; Peterson, L.A.; Dickel, D.; Capolungo, L.; Baskes, M.I. A Multiphysics Thermoelastoviscoplastic Damage Internal State Variable Constitutive Model including Magnetism. *Materials* **2024**, *17*, 2412. https://doi.org/10.3390/ma17102412

Academic Editor: Jordi Sort

Received: 9 February 2024
Revised: 30 April 2024
Accepted: 2 May 2024
Published: 17 May 2024

1. Introduction

In recent years, automotive electrification has served as an efficient technology to reduce fuel consumption, greenhouse gas emissions (GHG), and dependency on volatile resources, as well as maintain the high-power density and efficiency of a motor [1,2]. A new generation of electric propulsion motors is based on magnetic materials (iron (Fe), nickel (Ni), and cobalt (Co) for example), including soft magnetic laminations and Rare Earth (RE) elements (e.g., Neodymium or Cerium). Magnets exhibit a rich variety of material behavior originating from their type (diamagnet, paramagnet, ferromagnet, antiferromagnet, and ferrimagnet) and microstructural behavior (magnetic domains interaction and electron spin) that are strongly dependent on temperature, mechanical stress, external magnetic field, damage within the material, and time.

To meet the high automotive market requirements, the electric cars industry used Permanent Magnets (PM) and Rare-Earth (RE) magnets to increase the engine power at a low price. Permanent magnets are critical components for electric motors and power generators. However, RE elements are earth sources that decrease on a daily basis, thus requiring a high market value and becoming more expensive. According to the US Department of Energy (DOE) and other international institutes [3], RE elements are critical elements that are subjected to high supply risks, since the main location of RE elements is in China. To date, RE elements are non-recyclable elements; therefore, finding a cheap and more available alternative is an emerging issue to address. In an effort to fill this engineering gap, scientists dedicated considerable efforts to find an alternative to permanent magnets by understanding the physical behavior of magnets and modeling it in a mathematical framework that is used for various industrial applications.

Previous models describe the mechanical response of a magnetic sample when subjected to an external magnetic field [4,5]. The mechanical response comes, in most cases, as a mechanical dimensional elongation [6]. This effect is called magnetostriction [6,7] and is mostly found in ferromagnetic materials and giant magnets [8]. Bozorth (1945) [9] and Brown (1949) [10] first presented a theoretical analysis of the magneto-mechanical effect in ferromagnetic materials by replacing the applied stress with an equivalent field. Afterward, Cullity (1972) [11] researched such problems using Le Chatelier's principle. Sablik et al. (1988) [12] considered the changes in the hysteresis of ferromagnetic materials under constant stress. More systematic research on the magneto-mechanical effects was presented by Jiles (1995) [13] and Jiles and Atherton (1986) [14] based on the concepts of an "effective field theory" and "law of approach".

Although various models have been previously introduced to solve such intricate engineering problems [15–17], most studies individually considered mechanical, magnetic effects, and thermal effects. None of the previously cited models coupled the mechanical, thermal, and damage effects with magnetic effects in a consistent model.

Only a handful of studies on magnetoplasticity have been completed, and the history is fairly recent. Zagoruiko (1965) [18] was one of the first researchers to address the magneto-plasticity effect when he demonstrated that a pulsed magnetic field affected the plasticity of NaCl crystals. Later, Kravchenko (1970) [19] showed that the presence of a magnetic field inversely affects the metal's plasticity. Later, Al'shits (1987) [20] showed that a static magnetic field can affect the plastic behavior of NaCl. Molotskii (2000) [21] showed that when magnetic field transitions between singlet and triplet states occur, the depinning of dislocations from obstacles is facilitated, thus increasing plasticity. Mullner et al. (2003) [22] showed how a magnetic field changed the stress–strain behavior of a single crystal Ni-Mn-Ga. An experimental study also showed that the yield strength can decrease by ~5%, but the ductility can increase by ~15% [23]. In other studies, however, the hardness of steel increased under the magnetic field [24,25], and some studies even showed that both ductility and tensile strength increased [26,27]. Many laboratory experiments have shown that the mechanical properties of materials can be affected by precipitation and phase transformation kinetics altered by the applied magnetic field [28,29]. For instance, AISI 8620 steel showed some changes in the amount of cementite and its distributions with exposure to a magnetic field [28]. In their study, the cementite increased with the magnetic field while temperature was maintained low, which implies the altered precipitation is due to the magnetic field instead of temperature. Murase et al. (1993) [30] also showed that the fracture toughness of austenitic steels decreased by approximately 20% at a magnetic strength of 8 T. This embrittlement was due to the martensitic transformation induced by the magnetic field as it decreases austenite stability. Other than the effects on phase transformation and precipitation, several studies reported that magnetic fields influence fracture toughness [31–33], ductility [24,28,32,34,35], fatigue life [23,25,28,36–41], and creep [42] through various and complex mechanisms. Interestingly, Mullner et al. (2003) [22] reported that the crystallographic orientations (texture) and twins under a magnetic field change the magnetization across the twins and thus affect the dislocations in polycrystalline Ni-Mn-Ga. A useful review summarizing recent research progress regarding microstructure and property of metals influenced by magnetic field can be also found in Hu et al. (2022) [43].

Several studies have been conducted to develop constitutive models to account for the magnetostriction and magnetoplasticity behavior of magnetic materials. In recent studies, Zheng and Liu (2005) [44] proposed a constitutive relation for largely (non-linear) magnetostrictive straining, particularly for Terfenol-D rods. Li and Xu (2011) [45] modified the classical model by Jiles and Atherton (1984) [46] and Sablik (1997) [47] to account for asymmetry in magneto-mechanical behavior arising from different loading directions (tensile and compressive). They incorporated a stress demagnetization effect, a variable domain pinning coefficient, stress-dependent saturation magnetostriction, and a domain coefficient dependent on applied stress. Wang et al. (2011) [48] integrated the plastic deformation effect into a magneto-mechanical constitutive model. More recently, Daniel (2018) [49] presented a useful analytical model with only three parameters that captures stress-dependent magnetostriction deformation based on the energy consideration. Shi et al. (2020) [50] proposed a magnetoelastoplastic model from the standpoint of magnetic memory, and Shi (2020) [51] also presented a magnetoelastoplastic constitutive relation by using magnetoelastic energy and magnetoplastic energy due to domain pinning. In this historical context, a theoretical study on the thermomechanical effects coupled with damage and magnetic effects is of great interest to provide a physical basis for various magnetic testing techniques and gives a better understanding of the test results. Therefore, a comprehensive model framework for predicting magnetically influenced deformation behavior is needed. In the present study, we report a magnetism-dependent elastoviscosplastic and damage model in the framework of Internal State Variables constitutive theory, which incorporates aspects of kinematics, thermodynamics, and kinetics of deforming continuum body under magnetic, thermal, and mechanical loads.

The Internal State Variable (ISV) theory has been viable over the past years starting from the significant contribution of Onsager (1931) [52] related to thermodynamics. Then, Eckart (1948) [53] used ISV theory in continuum mechanics. Kroner (1960) [54] postulated that the ISV continuum theory can use physically observed behaviors although the complete microstructure arrangement is unnecessary as long as the macroscale ISV representation is complete. Later, Coleman and Gurtin (1967) [55] proposed the use of history-dependent variables to quantify dissipative mechanisms of internal deformation within a thermodynamically consistent framework. Hence, an ISV model employs a set of constitutive equations that capture the history effects of a material to predict its mechanical properties and its future behavior based on the already existing mathematical state description [56]. Many models are based on Coleman and Gurtin's (1967) [55] framework, and they are summarized by Horstemeyer and Bammann's (2010) [57] historical review of ISV theory.

The deformation gradient maps the deformation from the reference configuration to the current configuration. The multiplicative decomposition of the deformation gradient describes the deformation of elastic–thermal–magnetic behavior. Previous researchers used kinematic decomposition [58–73]; to establish a constitutive model for deformations of different materials (such as thermal effects [71]). Later, Dimitrov et al. (2020) [74] extended the thermomechanical description to electrothermomechanical constitutive equations to relate the electric effects on the thermomechanical hardening of the metals. Recently, Cho et al. (2022) [75] incorporated the nuclear irradiation effect on the elastoviscoplastic behavior of crystalline metals but not for magnetic effects heretofore.

The contribution of our work includes the development of an Internal State Variable (ISV) constitutive model that accounts for magnetism-dependent elasto-viscoplasticity and damage for magnetic materials that brings in three novel ideas: (1) introduction of kinematics for the deforming continuum body under an external magnetic field to account for elastic/inelastic deformation and vorticity affected by the magnetic field; (2) introduction of a new magnetic ISV constrained by the first and second laws of thermodynamics

(Clausius–Duhem inequality); and (3) this ISV-based constitutive model is a novel approach to address the kinetics of mechanical, thermal, and magnetic boundary conditions. The kinematics, thermodynamics, and kinetics of an ISV model should be independently developed but internally consistent, and we generate the theory for magnetic-influenced deformation herein. The contribution of this work is twofold. First, the physical establishment of how magnetic effects, resulting from a material's proper magnetization or a material subjected to an externally applied field, can change the behavior of a structure on the macroscale level. Second, the development of a consistent thermodynamic model following the Coleman and Gurtin (1967) [55] thermodynamic framework satisfies the first and second laws of thermodynamics. The first section of this document presents an introduction to the problem statement. Section 2 presents the macroscale deformation behavior exhibiting a response to an external magnetic field admitting subscale information from the mesoscale and the nanoscale. Section 3 provides a full description of the model's kinematics relations of the thermal–elastic–damage–magnetic ISV model. The Coleman and Gurtin (1967) [55] thermodynamic framework of the ISV model is presented in Section 4. To describe the macroscale behavior of the material, the kinetics of the ISV model are presented in Section 5. Finally, we present model comparison with lab experimental data for magnetostriction and magnetization behavior.

A standard notation is followed in this mathematical formulation. In this text, tensors are denoted by **boldface** font while scalar values have the standard weight. All tensor components are written with respect to a fixed Cartesian coordinate system. Special care is given to specify configurations throughout the derivation by using accent marks where the tilde ($\check{R}$), circumflex ($\hat{R}$), macron ($\overline{R}$), double macron ($\overline{\overline{R}}$), and overbrace ($\overbrace{R}$) represent different intermediate configurations. The following definitions are used in the text: $\boldsymbol{AB} => (\boldsymbol{A}.\boldsymbol{B})_{ij} = A_{ik}B_{kj}, \boldsymbol{A};\boldsymbol{B} = A_{ij}B_{ij}$, $tr(\boldsymbol{A}) = A_{ii}$, $\left(\boldsymbol{A}^T\right)_{ij} = A_{ji}$. The overdot denotes the material time derivative. The apostrophe denotes the co-rotational derivative.

2. Phenomenological Behavior

In nature, several types of magnetic materials exist. Differences in magnet types depend on material microstructure properties and response to external magnetic fields. The different types of magnets are diamagnets, paramagnets, ferromagnets, antiferromagnets, and ferrimagnets, and these are summarized in Table 1. Brugmans (1778) [76] characterized diamagnetism as the tendency of the material to oppose an applied magnetic field (H). Diamagnetism creates a repulsive force, and paramagnetism creates an attractive force when subjected to an applied magnetic field (H), making the total magnetic field stronger [77]. The third type of magnetism is ferromagnetism. Ferromagnetism is characterized by a spontaneous and strong magnetic field without the presence of an externally applied field. The three main ferromagnetic elements that exist in nature are the following: iron (Fe), nickel (Ni), and cobalt (Co), which are used to demonstrate our theory. Antiferromagnetism tends to have electrons with intrinsic magnetic moments that do not align parallel with each other but align in antiparallel orientations [78]. The fifth type of magnetism is ferrimagnetism. A ferrimagnet has the same properties as a ferromagnet, such that it retains a magnetic field even when no external magnetic field is applied but has a net magnetization less than that of ferromagnets alone.

The magnet types exhibit behavior that extends to multiscale levels. The effects of magnetism are described at the macroscale, mesoscale, and nanoscale as presented in the following section.

Table 1. Five different types of magnetic materials.

Type	Spin Alignment	Spin Illustrated in Simplified Plot	Examples
Ferromagnets	Electron spins align parallel to one another, resulting in a spontaneous magnetization.	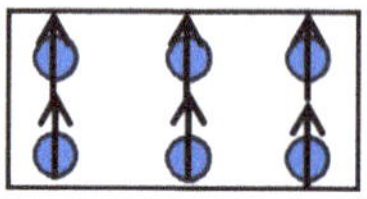	Fe, Co, Ni
Ferrimagnets	Majority of electron's spins parallel to one another, some spins are antiparallel, resulting in spontaneous magnetization.	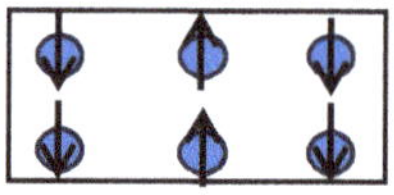	Magnetite (Fe_3O_4), yttrium iron garnet (YIG)
Antiferromagnets	Electron spins align antiparallel to each other, resulting in a null net magnetization.	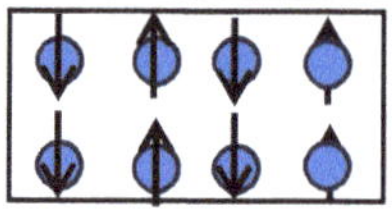	Cr
Paramagnets	Electron spins tend to align parallel when an external magnetic field is applied.	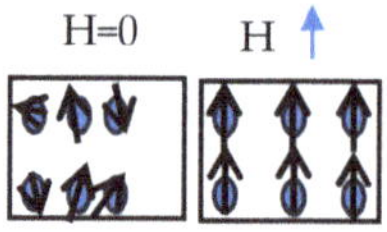	Oxygen, sodium, aluminum, calcium, uranium
Diamagnets	Electron spins tend to align antiparallel to an external magnetic field.	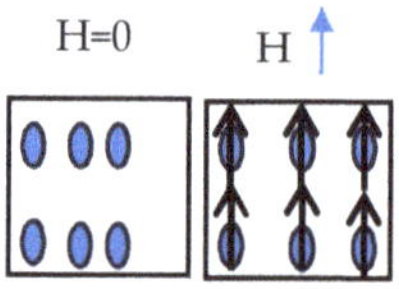	Copper, silver, gold, nitrogen

2.1. *Macroscale Level: The Magnetostriction Phenomenon*

Magnetostriction is a phenomenon found in ferromagnetic materials [7,15,79,80]. The magnetostriction phenomenon arises from the misalignment of magnetic domains such that, when subjected to an externally applied magnetic field, the domains align parallel to the magnetic field direction, resulting in a shape change at the macroscale level. Joule (1842) was the first to identify magnetostriction by observing a sample of nickel expanding when subjected to an external magnetic field. The concept of magnetostriction is a key feature employed in the magneto-mechanical coupled constitutive model described herein.

Magnetostriction phenomena involve elastic, magnetic, and thermal effects. Magnetostriction is of great industrial interest for use in sensors, actuators, adaptive structures, robotics, and transducers [81]. They are widely used in the field of nondestructive evaluation [82]. The essence of magnetostriction is the dependence of mechanical strain on magnetization. MagnetoStrictive Materials (MSMs) are a class of smart materials that transfer energy from one form to another form; for example, they can convert magnetic energy into mechanical energy (Joule effect [83,84]) and vice versa (Villari effect, c.f. [84]).

MSMs can exhibit large mechanical deformations in different directions when subjected to a strong external magnetic field. This behavior is due to the rotations of small magnetic domains (that exist inside of the grain) within the material, which are arbitrarily oriented when the material is not subjected to an external magnetic field. The orientations of these small domains change by the imposition of the magnetic field. The domain moments align themselves parallel to the externally applied field direction, thus creating a strain field, resulting in a noticeable mechanical elongation. As the intensity of the magnetic field increases, the magnetic domains tend to orient themselves in order to co-align their principal axes with the magnetic field in each region until saturation is reached. This effect is described in a small crystalline sample of a ferromagnetic material as illustrated in Figure 1.

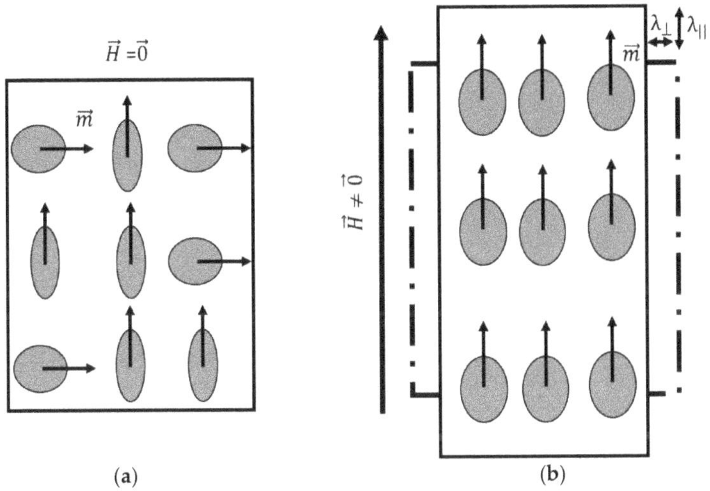

Figure 1. A nine-atom lattice showing the effect of an external magnetic field ($\vec{H}$) on the atom for which $\vec{m}$ is the magnetic moment. (**a**) Magnetic moments when no external magnetic field is applied, and (**b**) magnetic moments when subjected to a vertical external magnetic field. Two strain components appear: a parallel strain ($\lambda_{||}$) and a perpendicular one ($\lambda_{\perp}$).

Bieńkowski and Kulikowski (1980) and Jiles (1995) demonstrated the existence of a mechanism reciprocal to magnetostriction [13,85]. The mechanism, called the Villari effect, involves a change in magnetization induced by mechanical stress. The Joule and Villari effects are observed in ferromagnets, antiferromagnets, and ferrimagnetic objects. Figure 2 illustrates the Villari effect for a crystal lattice structure. When the lattice is subjected to a mechanical stress parallel to the original magnetic moments, the magnetization of the sample rotates. Note that up to this point, the previously described effects were limited to temperatures lower than the Curie temperature, which is the temperature above which the material loses its magnetic properties [86,87]. Once the temperature of these materials exceeds the Curie temperature, the magnetic properties of the material are lost.

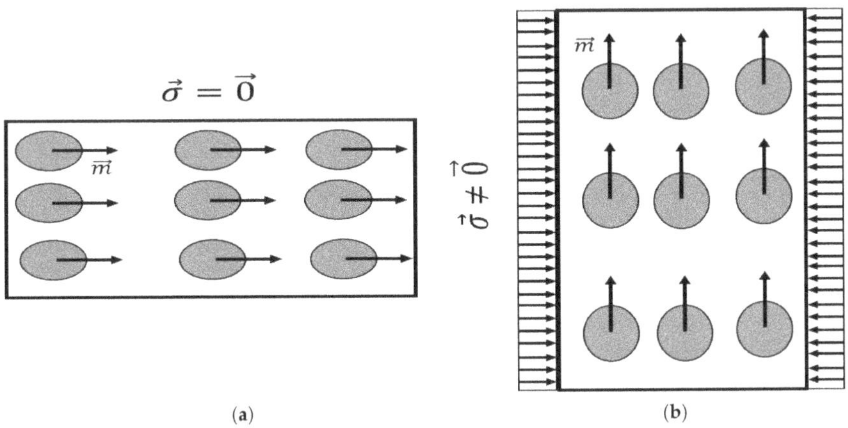

Figure 2. A nine-atom lattice showing the effect of a compressive uniformly distributed stress ($\vec{\sigma}$) on the magnetic properties of the atoms presented as the magnetic moment ($\vec{m}$). (**a**) Non-presence of stress illustrates a horizontal orientation of the magnetic moment. (**b**) The presence of compressive stress ($\vec{\sigma}$) results in a direction change in the magnetic moment ($\vec{m}$) (magnetic moments pointing up).

2.2. Mesoscale Level: Domain Wall Motion

In most cases, magnetostrictive strains exhibit a nonlinear behavior with respect to the external magnetic field. The nonlinear behavior is due to the domain structure within the grains and the grain orientation of the microstructure under study. Magnetic domains are the heart of magnetic material deformation. Figure 3 shows the domain structure of a magnetostrictive alloy. In Figure 3, the neighboring domains tend to have different magnetic moment orientations. The different alignment minimizes the magnetic energy within the specimen. Each domain (d) has a magnetization that can be expressed by [11]

$$\vec{M}_d = M_s \vec{\gamma}, \tag{1}$$

such that $\vec{M}_d$ is the magnetization of the domain, $\vec{\gamma}$ represents the vector orientation of the axis on which most of the material's magnetization is fixed, and M_s represents the saturation magnetization value of each domain.

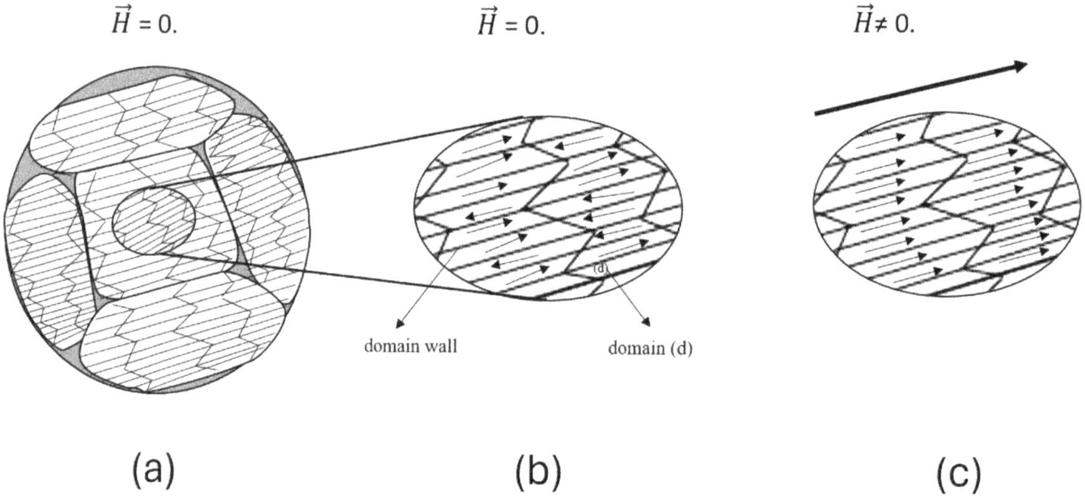

Figure 3. (**a**) Polycrystalline structure showing the magnetic domains and their appropriate magnetization ($\vec{m}$) direction when no external magnetic field ($\vec{H}$) is applied. (**b**) Magnified region of the polycrystalline structure, with no external magnetic field applied, and (**c**) magnified region when an external magnetic field ($\vec{H}$) is applied. The magnetic domain's direction aligns with the external magnetic field direction.

The moments of magnetic domains tend to rotate when exposed to a magnetic field until the magnetic domains' direction is aligned with the magnetic field direction [88]. Thus, the domain walls, which are considered the transition region between the domains, start to move and elongate due to domain wall motion [14]. Domains whose orientation is closer to the magnetic field direction tend to elongate through the process while the others tend to shrink. Domain elongation and shrinkage result in a dimensionless change on the macroscale level (the Joule effect). Domain growth stops once saturation magnetization is reached.

2.3. Nanoscale Level: Ising Model

At the nanoscale level, electron spins play an important role in moving domain walls. This physical behavior is explained through the Ising model [89–91]. The Ising model is a statistical model used to describe ferromagnetic behavior in terms of phase transitions and the magnetic domain motion. The Ising model was initially developed to solve a one-dimensional problem under the assumption of no phase transitions. The model is based on defining two spinning variables that represent the magnetic dipole moments of the atomic spins. For a two-dimensional (2D) lattice, each lattice site has a local magnetic moment and is represented by an arrow pointing up (for a positive magnetic moment) and an arrow pointing down (for a negative magnetic moment). The moment is assumed to be equal to +1 when the spins are pointing up or to −1 when the spins are pointing down. The Ising model is used to compute the magnetization order (O) using

$$O = \left\langle \frac{K_+ - K_-}{K} \right\rangle, \tag{2}$$

such that K represents the total number of spins in the lattice, K_+ is the number of positive spins, and K_- is the number of negative spins. The magnetization order in Equation (2) represents the expectation value of the magnetic moment ($\mu(K_+ - K_-)$) relative to the largest possible magnetic moment (μK) such that μ is the magnetic moment.

In terms of energy, the Ising model [89] includes two contributions: the first characterizes how neighboring spins affect the spin, and the second contribution characterizes how an applied magnetic field affects each spin within the lattice. This statement is written in the following way:

$$E = -J\sum\nolimits_{\langle i,j\rangle} g_i g_j - H\sum\nolimits_i g_i, \tag{3}$$

such that E is the total energy, J is the positive coefficient giving the interaction strength, and g_i is the spin variable corresponding to direction values (=+1 or −1). The first term of Equation (3) represents the neighboring spin's interaction, while the second term represents the effect of the applied field on each spin.

3. Kinematics

In continuum mechanics, a three-dimensional material subjected to a magneto-thermo-mechanical deformation can be described using the deformation gradient concept to map a deformation from the reference (initial) configuration (R_0) to the current configuration (R) with possible intermediate configurations in between. The deformation gradient mapping a particle from its initial position $\boldsymbol{X}$ to the current position $\boldsymbol{x}$ is given as follows [58,92]:

$$\boldsymbol{F} = \frac{\partial \boldsymbol{x}}{\partial \boldsymbol{X}}, \tag{4}$$

such that $\boldsymbol{X}$ is the displacement in the reference configuration (R_0) and $\boldsymbol{x}$ is the displacement in the current configuration (R). The deformation gradient assumes continuity, where the local deformation at $\boldsymbol{X}$ is characterized as the gradient of the motion, which is a second-order two-point tensor.

For the continuum model herein, we need to define the Eulerian and the Lagrangian strains in a classical manner [93]. The Lagrangian finite strain tensor with respect to the reference configuration is defined as follows:

$$\boldsymbol{E} = \frac{1}{2}\left(\boldsymbol{F}^T\boldsymbol{F} - \boldsymbol{I}\right), \tag{5}$$

with $\boldsymbol{I}$ as the identity matrix.

For large strains, a multiplicative decomposition of the deformation gradient [93] into plastic, damage, magnetic, thermal, and elastic parts is performed as schematically illustrated in Figure 4. The total deformation gradient is therefore written as

$$\boldsymbol{F} = \boldsymbol{F}_e\boldsymbol{F}_\theta\boldsymbol{F}_\varphi\boldsymbol{F}_H\boldsymbol{F}_p, \tag{6}$$

where the total deformation gradient can be multiplicatively decomposed into elastic ($\boldsymbol{F}_e$), thermal ($\boldsymbol{F}_\theta$), damage ($\boldsymbol{F}_\varphi$), magnetostrictive ($\boldsymbol{F}_H$), and plastic ($\boldsymbol{F}_p$) deformation gradients. Note that the thermomechanics in our constitutive model represents the thermal contribution to the deformation. For instance, our model takes the thermal contribution into consideration of the kinematics to track the elastic and inelastic deformation of the material as shown in Equation (6) and Figure 4.

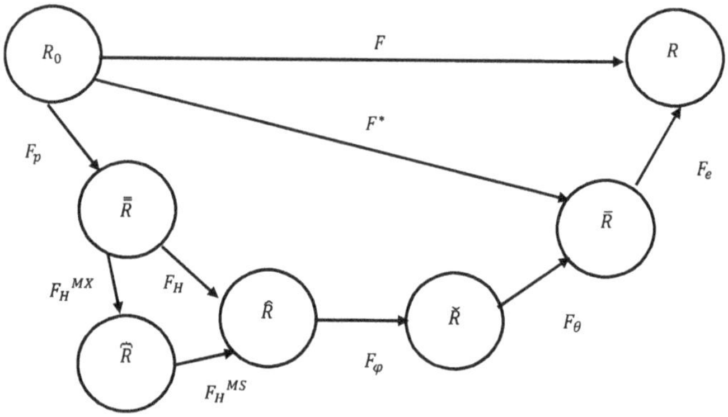

Figure 4. Multiplicative decomposition of the deformation gradient into the plastic (p), magnetic (H), damage (φ), thermal (θ), and elastic parts (e).

The magnetic deformation gradient ($\boldsymbol{F}_H$) is multiplicatively decomposed into two sub-deformation gradients in this model:

$$\boldsymbol{F}_H = \boldsymbol{F}_H^{MS} \boldsymbol{F}_H^{MX}, \tag{7}$$

where the first sub-deformation gradient ($\boldsymbol{F}_H^{MS}$) is related to the magnetostriction elongation effect and the second sub-deformation gradient ($\boldsymbol{F}_H^{MX}$) is related to the Maxwell magnetic field effects created by the externally applied field ($\boldsymbol{H}$). Generally, the Maxwell field effects on the deformation of the material are so small that they are not taken into consideration in previously developed models. However, the purpose of this model is to provide a full description of the magnetic material's behavior; therefore, all the effects are included. The total deformation gradient (in Equation (6)) can be simplified to a product of inelastic ($\boldsymbol{F}_*$) and elastic ($\boldsymbol{F}_e$) deformation gradient components,

$$\boldsymbol{F} = \boldsymbol{F}_e \boldsymbol{F}_*, \tag{8}$$

such that $\boldsymbol{F}_*$ represents all the inelastic deformations $\boldsymbol{F}_* = \boldsymbol{F}_\theta \boldsymbol{F}_\varphi \boldsymbol{F}_H \boldsymbol{F}_p$.

The first intermediate configuration ($\overline{\overline{R}}$) is defined by the plastic deformation gradient ($\boldsymbol{F}_p$). The second intermediate configuration ($\hat{R}$) is defined by the multiplication of the magnetic deformation gradient ($\boldsymbol{F}_H$) and the plastic deformation gradient ($\boldsymbol{F}_p$): $\boldsymbol{F}_H \boldsymbol{F}_p$.

The third intermediate configuration ($\check{R}$) is defined by the multiplication of the damage deformation gradient ($\boldsymbol{F}_\varphi$), the magnetic deformation gradient ($\boldsymbol{F}_H$), and the plastic deformation gradient ($\boldsymbol{F}_p$): $\boldsymbol{F}_\varphi\ \boldsymbol{F}_H \boldsymbol{F}_p$. The magnetic deformation gradient also depends on the damage since the presence of voids/cracks may modify the motion of the domain walls, known as the domain wall pinning effect [94]. Domain wall pinning can arrest material elongation caused by an external magnetic field. Moreover, a high number of heterogeneities (particles, voids) leads to a decrease in permeability (μ) and an increase in coercivity (H_c) [95].

The fourth intermediate configuration ($\overline{R}$) is defined by the multiplication of the thermal deformation gradient ($\boldsymbol{F}_\theta$), the magnetic deformation gradient ($\boldsymbol{F}_H$), the damage deformation gradient ($\boldsymbol{F}_\varphi$), and the plastic deformation gradient ($\boldsymbol{F}_p$):$\boldsymbol{F}^*$, such that $\boldsymbol{F}^* = \boldsymbol{F}_\theta \boldsymbol{F}_\varphi \boldsymbol{F}_H \boldsymbol{F}_p$. Both magnetic and damage behavior characteristics of a material are sensitive to temperature. A permanent magnet can lose its properties once a critical temperature (Curie temperature) is reached. Damage mechanisms and evolutionary rates vary with temperature. The elastic deformation gradient serves to describe unloading elastically through F_e^{-1}. The thermal deformation gradient and damage deformation follow Francis et al. (2014) [71]. Finally, the plastic deformation gradient is the last one since the inelastic flow rule is a function of thermal and damage effects.

For our interest, the constitutive equations are developed in intermediate configuration $\hat{R}$, where all magnetic deformations happen. The deformation gradient tensors in their corresponding intermediate configurations are mathematically defined as follows:

$$\boldsymbol{F}^{*}=\frac{\partial\overline{x}}{\partial\boldsymbol{X}},\boldsymbol{F}_{p}=\frac{\partial\overline{\overline{x}}}{\partial\boldsymbol{X}},\boldsymbol{F}_{H}=\frac{\partial\hat{x}}{\partial\overline{\overline{\overline{X}}}},\boldsymbol{F}_{\varphi}=\frac{\partial\widetilde{x}}{\partial\hat{X}},\ \boldsymbol{F}_{\theta}=\frac{\partial\overline{x}}{\partial\hat{X}},\boldsymbol{F}_{e}=\frac{\partial x}{\partial\overline{\overline{X}}}. \tag{9}$$

The Jacobian of the total deformation gradient, which is the change in volume between the reference (R_0) and current (R) configurations, is given as

$$J=det(\boldsymbol{F})=det(\boldsymbol{F}_{p})det(\boldsymbol{F}_{H})det(\boldsymbol{F}_{\varphi})det(\boldsymbol{F}_{\theta})det(\boldsymbol{F}_{e}), \tag{10}$$

such that the Jacobian of each deformation gradient represents the conservation of the mass of the system, given as follows:

$$\begin{gathered}det(\boldsymbol{F}_{p})=J_{p}=\tfrac{\overline{\overline{V}}}{V_{0}},det(\boldsymbol{F}_{H})=J_{H}=\tfrac{\hat{V}}{\overline{\overline{V}}},det({\boldsymbol{F}_{H}}^{MX})={J_{H}}^{MX}=\tfrac{\overbrace{V}}{\overline{\overline{V}}},det({\boldsymbol{F}_{H}}^{MS})={J_{H}}^{MS}=\tfrac{\hat{V}}{\overbrace{V}},det(\boldsymbol{F}_{\varphi})=J_{\varphi}=\tfrac{\widetilde{V}}{\hat{V}},\\ det(\boldsymbol{F}_{\theta})=J_{\theta}=\tfrac{\overline{V}}{\check{V}},\ \text{and}\ det(\boldsymbol{F}_{e})=J_{e}=\tfrac{V}{\overline{V}}.\end{gathered} \tag{11}$$

Based on previous work by Bammann and Aifantis (1989) [61], the damage deformation gradient is expressed as follows:

$$\boldsymbol{F}_{\varphi}=\frac{1}{(1-\varphi)^{\frac{1}{3}}}\boldsymbol{I}. \tag{12}$$

The Jacobian of the damage deformation gradient is the following [96]:

$$det(\boldsymbol{F}_{\varphi})=\frac{1}{(1-\varphi)}. \tag{13}$$

Similarly, Bammann and Solanki (2010) [97] defined the Jacobian of the thermal deformation gradient as follows:

$$det(\boldsymbol{F}_{\theta})={F_{\theta}}^{3}. \tag{14}$$

The developed model assumes a linear thermal expansion that can be assumed for the thermal deformation gradient tensor ($\boldsymbol{F}_\theta$) and is given as

$$\boldsymbol{F}_{\theta}=(1+\alpha_{th}\Delta\theta)\boldsymbol{I}, \tag{15}$$

where α_{th} is the thermal expansion coefficient and θ is the temperature.

Assuming deviatoric plastic deformation, the Jacobian of the plastic deformation gradient is unity,

$$det(\boldsymbol{F}_{p})=1. \tag{16}$$

From the total deformation gradient, the total Lagrangian strain tensor is obtained using additive decomposition in the reference configuration

$$\boldsymbol{E}=\boldsymbol{E}_{e}+\boldsymbol{E}_{\theta}+\boldsymbol{E}_{\varphi}+\boldsymbol{E}_{H}+\boldsymbol{E}_{p}, \tag{17}$$

where

$$\begin{gathered}\boldsymbol{E}=\tfrac{1}{2}(\boldsymbol{C}-\boldsymbol{I}),\overline{\boldsymbol{E}}_{e}=\tfrac{1}{2}(\overline{\boldsymbol{C}}_{e}-\boldsymbol{I}),\check{\boldsymbol{E}}_{\theta}=\tfrac{1}{2}(\check{\boldsymbol{C}}_{\theta}-\boldsymbol{I}),\hat{\boldsymbol{E}}_{\varphi}=\tfrac{1}{2}(\hat{\boldsymbol{C}}_{\varphi}-\boldsymbol{I}),\overline{\overline{\boldsymbol{E}}}_{H}=\tfrac{1}{2}(\overline{\overline{\boldsymbol{C}}}_{H}-\boldsymbol{I}),\\ \overbrace{\boldsymbol{E}}_{H}^{MS}=\tfrac{1}{2}(\overbrace{\boldsymbol{C}}_{H}^{MS}-\boldsymbol{I}),\ \overline{\overline{\boldsymbol{E}}}_{H}^{MX}=\tfrac{1}{2}(\overline{\overline{\boldsymbol{C}}}_{H}^{MX}-\boldsymbol{I}),\ \text{and}\ \boldsymbol{E}_{p}=\tfrac{1}{2}(\boldsymbol{C}_{p}-\boldsymbol{I}),\end{gathered} \tag{18}$$

and C is the Cauchy–Green deformation tensor. Pulling back all the intermediate Lagrangian tensors to the reference configuration, we obtain the following:

$$\begin{aligned} \boldsymbol{E}_e &= \boldsymbol{F}_p^T\boldsymbol{F}_H^T\boldsymbol{F}_\varphi^T\boldsymbol{F}_\theta^T\overline{\boldsymbol{E}}_e\boldsymbol{F}_\theta\boldsymbol{F}_\varphi\boldsymbol{F}_H\boldsymbol{F}_p, \\ \boldsymbol{E}_\theta &= \boldsymbol{F}_p^T\boldsymbol{F}_H^T\boldsymbol{F}_\varphi^T\check{\boldsymbol{E}}_\theta\boldsymbol{F}_\varphi\boldsymbol{F}_H\boldsymbol{F}_p, \\ \boldsymbol{E}_\varphi &= \boldsymbol{F}_p^T\boldsymbol{F}_H^T\hat{\boldsymbol{E}}_\varphi\boldsymbol{F}_H\boldsymbol{F}_p, \\ \boldsymbol{E}_H &= \boldsymbol{F}_p^T\overline{\overline{\boldsymbol{E}}}_H\boldsymbol{F}_p. \end{aligned} \tag{19}$$

The stretch tensors of each Lagrangian tensor are a strain measure in terms of material coordinates and can be obtained when the deformation gradients are determined as follows:

$$\boldsymbol{C} = \boldsymbol{F}^T\boldsymbol{F}, \overline{\boldsymbol{C}}_e = \boldsymbol{F}_e^T\boldsymbol{F}_e, \check{\boldsymbol{C}}_\theta = \boldsymbol{F}_\theta^T\boldsymbol{F}_\theta, \hat{\boldsymbol{C}}_\varphi = \boldsymbol{F}_\varphi^T\boldsymbol{F}_\varphi, \overline{\overline{\boldsymbol{C}}}_H = \boldsymbol{F}_H^T\boldsymbol{F}_H, \overline{\overline{\boldsymbol{C}}}_H^{MX} = \left(\boldsymbol{F}_H^{MX}\right)^T\boldsymbol{F}_H^{MX}, \overline{\overline{\boldsymbol{C}}}_H^{MS} = \left(\boldsymbol{F}_H^{MS}\right)^T\boldsymbol{F}_H^{MS}, \text{ and } \boldsymbol{C}_p = \boldsymbol{F}_p^T\boldsymbol{F}_p. \tag{20}$$

Each Cauchy–Green deformation tensor ($\boldsymbol{C}$) may be subjected to spectral decomposition of the form

$$\boldsymbol{C} = \sum_{i=1}^{3} \lambda_i^2 \boldsymbol{n}_i \otimes \boldsymbol{n}_i, \tag{21}$$

where the stretch ratio, λ_i, is the square root of each positive eigenvalue that corresponds to each orthonormal vector, $\boldsymbol{n}_i$. Each deformation gradient tensor has a polar decomposition of the form

$$\boldsymbol{F}_\bullet = \boldsymbol{R}_\bullet\boldsymbol{U}_\bullet, \tag{22}$$

where ($\bullet$) can be any of the terms resulting from the deformation gradient decomposition (p, H, φ, θ, e). The relationship between $\boldsymbol{C}$ and $\boldsymbol{U}$ is

$$\boldsymbol{U} = \sqrt{\boldsymbol{C}} = \sum_{i=1}^{3} \lambda_i \boldsymbol{n}_i \otimes \boldsymbol{n}_i, \tag{23}$$

where the directions (eigenvectors) ($\boldsymbol{n}_i$) remain unchanged, and the principal stretch ratios (λ_i) are used.

The scalar form of the damage right stretch tensor that affects the damage internal state variables, defined by Bammann and Solanki (2010) [97], is defined in the damage-associated configuration ($\hat{R}$) as follows:

$$\hat{t}_\varphi = \frac{1}{3}tr\left(\hat{C}_\varphi\right) = \frac{1}{(1-\varphi)^{\frac{1}{3}}}I, \tag{24}$$

for which the corresponding time derivative is given as follows:

$$\dot{\hat{t}}_\varphi = \frac{\dot{\varphi}}{3*(1-\varphi)^{\frac{4}{3}}}I = \frac{\dot{\varphi}}{3*(1-\varphi)}\hat{t}_\varphi I = \frac{1}{3}I : \dot{\check{C}}_\varphi. \tag{25}$$

The velocity gradient associated with the deformation of the current configuration is decomposed into elastic, thermal, magnetic, damage, and plastic components:

$$\boldsymbol{l} = \dot{\boldsymbol{F}}\boldsymbol{F}^{-1} = \boldsymbol{l}_e + \boldsymbol{l}_* = \boldsymbol{l}_e + \boldsymbol{l}_\theta + \boldsymbol{l}_\varphi + \boldsymbol{l}_H + \boldsymbol{l}_p = \boldsymbol{l}_e + \boldsymbol{l}_\theta + \boldsymbol{l}_\varphi + \boldsymbol{l}_H{}^{MS} + \boldsymbol{l}_H{}^{MX} + \boldsymbol{l}_p, \tag{26}$$

where ($\boldsymbol{l}_e$) is the elastic velocity gradient, ($\boldsymbol{l}_\theta$) is the thermal velocity gradient, ($\boldsymbol{l}_H$) is the magnetic velocity gradient, ($\boldsymbol{l}_\varphi$) is the damage velocity gradient, and ($\boldsymbol{l}_p$) is the plastic velocity gradient. Each velocity gradient can be written in terms of the deformation gradients as follows:

$$\begin{gathered} \boldsymbol{l}_e = \dot{\boldsymbol{F}}_e\boldsymbol{F}_e^{-1}, \boldsymbol{l}_\theta = \boldsymbol{F}_e\dot{\boldsymbol{F}}_\theta\boldsymbol{F}_\theta^{-1}\boldsymbol{F}_e^{-1}, \boldsymbol{l}_\varphi = \boldsymbol{F}_e\boldsymbol{F}_\theta\dot{\boldsymbol{F}}_\varphi\boldsymbol{F}_\varphi^{-1}\boldsymbol{F}_\theta^{-1}\boldsymbol{F}_e^{-1}, \\ \boldsymbol{l}_H = \boldsymbol{F}_e\boldsymbol{F}_\theta\boldsymbol{F}_\varphi\dot{\boldsymbol{F}}_H\boldsymbol{F}_H^{-1}\boldsymbol{F}_\varphi^{-1}\boldsymbol{F}_\theta^{-1}\boldsymbol{F}_e^{-1}, \text{ and } \boldsymbol{l}_p = \boldsymbol{F}_e\boldsymbol{F}_\theta\boldsymbol{F}_\varphi\boldsymbol{F}_H\dot{\boldsymbol{F}}_p\boldsymbol{F}_p^{-1}\boldsymbol{F}_H^{-1}\boldsymbol{F}_\varphi^{-1}\boldsymbol{F}_\theta^{-1}\boldsymbol{F}_e^{-1}. \end{gathered} \tag{27}$$

The velocity gradients in the intermediate $\hat{R}$ configuration is obtained by pulling back the elastic, thermal, and damage velocity gradients (F_e), (F_θ), and (F_φ) and pushing forward the plastic velocity gradient (F_p). This results in the following velocity gradients:

$$\begin{aligned}
\hat{l}_e &= F_\varphi^{-1} F_\theta^{-1} F_e^{-1} \dot{F}_e F_e^{-1} F_e F_\theta F_\varphi, \\
\hat{l}_\theta &= F_\varphi^{-1} F_\theta^{-1} \dot{F}_\theta F_\theta^{-1} F_\theta F_\varphi = F_\varphi^{-1} F_\theta^{-1} \dot{F}_\theta F_\varphi, \\
\hat{l}_\varphi &= F_H^{-1} \dot{F}_\varphi F_\varphi^{-1} F_H, \\
\hat{l}_H &= \dot{F}_H F_H^{-1}, \\
\hat{l}_p &= F_H \dot{F}_p F_p^{-1} F_H^{-1}.
\end{aligned} \tag{28}$$

Velocity gradient l can be decomposed into two parts, the skew and symmetric parts:

$$l = D + W, \tag{29}$$

where D is the symmetric rate of deformation tensor and W is the asymmetric spin tensor:

$$D = sym(l) = \frac{1}{2}\left(l + l^T\right), \text{and } W = skew(l) = \frac{1}{2}\left(l - l^T\right). \tag{30}$$

The total rate of deformation is additively decomposed into elastic, plastic, damage, magnetic, and thermal deformation rates by additive decomposition as follows:

$$D = D_e + D_\theta + D_\varphi + D_H + D_p, \tag{31}$$

where D_e, D_θ, D_φ, D_H, and D_p are the elastic, thermal, damage, magnetic, and plastic components of the rate of deformation. Likewise, the spin tensor is additively decomposed as follows:

$$W = W_e + W_\theta + W_\varphi + W_H + W_p, \tag{32}$$

where the thermal spin and the damage spin are assumed to be equal to zero because the nondiagonal components of the velocity gradient are zero. Therefore, the total spin is written as follows:

$$W = W_e + W_H + W_P. \tag{33}$$

The magnetic moment spin in this case refers to the spin moment resulting from the electron's intrinsic motion. The spin moment resulting from other subatomic elementary particles (such as quarks in the protons and neutrons of the atomic nuclei) is assumed to be neglected because of its small magnetic moment. The magnetic spin influences the ordering of the electrons, nuclei in atoms, and molecules. A change in the ordering of the molecules induces a change in the magnetic domain orientation, resulting in a dimensional change appearing on the macroscale level of the material. The spin of a complete body is the sum of the spins of the elementary particles (electrons, neutrons, and protons),

$$W_H = 1/\mathrm{g}[\chi D_H - D_H \chi], \tag{34}$$

where g is the orientation spin variable arising from the Ising model and χ and D_H are the kinematic magnetization term and the magnetic deformation rate tensor that are described in detail in the kinetics part of the model. This form is similar to the plastic spin [98]. Dafalias (1989) [98] showed that the plastic spin represents the rotation rate of the material with respect to its substructure during inelastic deformations. This physical behavior is expressed in terms of an equation relating the plastic spin to the plastic deformation rate tensor.

$$W_p = -1/\varsigma\left[\beta D_p - D_p \beta\right], \tag{35}$$

where ζ is the orientation coefficient and β is the kinematic hardening variable [98,99].

The strain rate is therefore given as follows:

$$\dot{\varepsilon} = \dot{\varepsilon}_e + \dot{\varepsilon}_\theta + \dot{\varepsilon}_\varphi + \dot{\varepsilon}_H + \dot{\varepsilon}_p. \tag{36}$$

The Cauchy stress (σ) is expressed as follows:

$$\sigma = J_e^{-1}\boldsymbol{\tau} = J_e^{-1}\boldsymbol{F}_e\hat{\boldsymbol{S}}\boldsymbol{F}_e^T, \tag{37}$$

where the Cauchy stress tensor (σ) and the first Piola Kirchhoff stress tensor ($\boldsymbol{\tau}$) are found in the current configuration, R, and the second Piola–Kirchhoff stress ($\hat{\boldsymbol{S}}$) invoked the intermediate configuration, $\hat{R}$.

4. Thermodynamics

In this section, a thermodynamic model with internal state variables is developed to capture the path-dependent inelastic deformation processes in the intermediate configuration ($\hat{R}$) (where all magnetic deformations occur) and then mapped to the current configuration (R) [100].

The law of conservation of energy dictates that the rate of change in the internal energy of any Representative Volume Element (RVE) is equal to the rate of mechanical work of the net external force acting on that volume plus all other energies (magnetic energy in this model) that enter or leave the RVE. In local form, the first law of thermodynamics is given as follows:

$$\rho\dot{u} = \boldsymbol{S} : \dot{\boldsymbol{E}} + \left(\boldsymbol{B}.\dot{\boldsymbol{H}} + \dot{\boldsymbol{B}}.\boldsymbol{H}\right) + \rho r - \nabla.\boldsymbol{q}, \tag{38}$$

such that u is the specific internal energy, $\boldsymbol{S}$ is the Piola–Kirchhoff stress tensor, $\boldsymbol{H} = \boldsymbol{H}(\boldsymbol{B}, \boldsymbol{M})$ is the external magnetic field, $\boldsymbol{B}$ is the magnetic flux density, $\dot{\boldsymbol{H}}$ is the external magnetic field rate, $\dot{\boldsymbol{B}}$ is the magnetic flux density rate, r is the specific heat generation rate, $\boldsymbol{q}$ is the heat flux vector, and ρ is the density. Term $\left(\boldsymbol{B}.\dot{\boldsymbol{H}} + \dot{\boldsymbol{B}}.\boldsymbol{H}\right)$ includes the magnetoelastic and the Zeeman energies [101]. The magnetoelastic energy results from magnetostriction, while the Zeeman energy represents the interaction of the magnetic material and the externally applied magnetic field.

In the intermediate configuration ($\hat{R}$), the first law of thermodynamics is written as follows [71,102]:

$$\hat{\rho}\dot{\hat{u}} = \hat{\boldsymbol{S}} : \dot{\hat{\boldsymbol{E}}} + \left(\hat{\boldsymbol{B}}.\dot{\hat{\boldsymbol{H}}} + \dot{\hat{\boldsymbol{B}}}.\hat{\boldsymbol{H}}\right) + \hat{\rho}\hat{r} - \hat{\nabla}.\hat{\boldsymbol{q}}. \tag{39}$$

The Clausius–Duhem (CD) inequality is given in the local form as follows [71,103]:

$$\rho\dot{s} - \frac{1}{\theta}\rho r + \frac{1}{\theta}\nabla.\boldsymbol{q} - \frac{1}{\theta^2}q.\nabla\theta \geq 0, \tag{40}$$

where s is the entropy of the material.

In the intermediate configuration ($\hat{R}$), the CD inequality is given as follows:

$$\hat{\rho}\dot{\hat{s}} - \frac{1}{\theta}\hat{\rho}\hat{r} + \frac{1}{\theta}\hat{\nabla}.\hat{\boldsymbol{q}} - \frac{1}{\theta^2}\hat{q}.\hat{\nabla}\hat{\theta} \geq 0. \tag{41}$$

The Helmholtz free energy in the intermediate configuration $\left(\hat{R}\right)$ is defined using the formulation of Coleman and Gurtin (1967) [55] as follow:

$$\hat{\psi} = \hat{u} - \theta\hat{s}, \tag{42}$$

and its time derivative is defined as follow:

$$\dot{\hat{\psi}} = \dot{\hat{u}} - \dot{\theta}\hat{s} - \theta\dot{\hat{s}}. \tag{43}$$

Substituting Equation (43) into the energy balance relation in Equation (39) yields

$$\hat{\rho}\left(\dot{\hat{\psi}} + \dot{\theta}\hat{s} + \theta\dot{\hat{s}}\right) = \hat{\mathbf{S}} : \dot{\hat{\mathbf{E}}} + \left(\hat{\mathbf{B}}.\dot{\hat{H}} + \dot{\hat{\mathbf{B}}}.\hat{H}\right) + \hat{\rho}\hat{r} - \hat{\nabla}.\hat{q}. \tag{44}$$

Substituting Equation (44) into the Clausius–Duhem inequality (Equation (41)) produces inequality

$$-\hat{\rho}\dot{\hat{\psi}} - \hat{\rho}\dot{\theta}\hat{s} + \hat{\mathbf{S}} : \dot{\hat{\mathbf{E}}} + \left(\hat{\mathbf{B}}.\dot{\hat{H}} + \dot{\hat{\mathbf{B}}}.\hat{H}\right) - \frac{1}{\hat{\theta}}\hat{q}.\hat{\nabla}\theta \geq 0. \tag{45}$$

The Helmholtz free energy is assumed as a locally defined function and can be characterized by observable variables such as temperature and strain and other non-observable variables that characterize internal rearrangements of a material's microstructure such as isotropic hardening and kinematic hardening (*cf.* [97,104]). In this model, the Helmholtz free energy is assumed to be a function of the following independent state variables: the product of elastic strain and damage stretch $\boldsymbol{E}_e\boldsymbol{C}_\varphi$, temperature θ, magnetic field flux density B, and a set of i number of strain-like internal variables ISVs $\hat{\Pi}_i$ that are given as follows:

$$\psi = \hat{\psi}(\boldsymbol{E}_e\boldsymbol{C}_\varphi, B, \theta, \hat{\Pi}_i). \tag{46}$$

The ISVs ($\hat{\Pi}_i$), are chosen to represent irreversible mechanisms related to the internal rearrangement of the material microstructure caused by externally applied magnetic, thermal, and mechanical fields. The evolution of ISVs induces strain fields within the domains and changes the electron spin motion on an electronic scale. The ISVs of this model are given as follows:

$$\hat{\Pi}_i = \boldsymbol{\beta C_\varphi}, \varepsilon_s t_\varphi, \hat{\boldsymbol{M}} \tag{47}$$

where $\boldsymbol{\beta}$ is the strain-like quantity due to the kinematic hardening describing the effects of geometrically necessary dislocation density (GND) evolution, ε_s is the strain-like quantity due to the isotropic hardening describing the statistically stored dislocation density (SSD) effects, and $\hat{\boldsymbol{M}}$ is the total magnetization of the material. Magnetization nonlinearity occurs due to the rotation and the growth of the magnetic domains. Magnetization refers to which the material can be magnetized when subjected to an external magnetic field. Therefore, the Helmholtz free energy function in Equation (47) may be expressed as

$$\psi = \hat{\psi}\left(\boldsymbol{E}_e\boldsymbol{C}_\varphi, B, \theta\,, \hat{\boldsymbol{\beta}}\hat{\boldsymbol{C}}_{\boldsymbol{\varphi}}, \hat{\varepsilon}_s\hat{t}_\varphi, \hat{\boldsymbol{M}}\right). \tag{48}$$

Assuming that the Helmholtz free energy can be additively decomposed,

$$\begin{aligned} \hat{\psi} &= \hat{\psi}(\boldsymbol{E}_e\boldsymbol{C}_\varphi, B, \theta, \hat{\boldsymbol{\beta}}\hat{\boldsymbol{C}}_{\boldsymbol{\varphi}}, \hat{\varepsilon}_s\hat{t}_\varphi, \hat{\boldsymbol{M}}) \\ &= \hat{\psi}_{E_eC_\varphi}\left(\hat{\boldsymbol{E}}_e\hat{\boldsymbol{C}}_\varphi, \theta\right) + \hat{\psi}_{\hat{B}}\left(\hat{B}, \theta\right) + \hat{\psi}_{\beta C_{\boldsymbol{\varphi}}}\left(\hat{\boldsymbol{\beta}}\hat{\boldsymbol{C}}_{\boldsymbol{\varphi}}, \theta\right) + \hat{\psi}_{\varepsilon_s t_\varphi}\left(\hat{\varepsilon}_s\hat{t}_\varphi, \theta\right) + \hat{\psi}_{\hat{M}}\left(\hat{\boldsymbol{M}}, \theta\right). \end{aligned} \tag{49}$$

Given these ISVs, the time rate of change in Helmholtz free energy is derived as

$$\dot{\hat{\psi}} = \frac{\partial\hat{\psi}}{\partial(\hat{\mathbf{E}}_e\hat{\mathbf{C}}_\varphi)} : \dot{\hat{\mathbf{E}}}_e\hat{\mathbf{C}}_\varphi + \frac{\partial\hat{\psi}}{\partial(\hat{\mathbf{E}}_e\hat{\mathbf{C}}_\varphi)} : \hat{\mathbf{E}}_e\dot{\hat{\mathbf{C}}}_\varphi + \frac{\partial\hat{\psi}}{\partial(\hat{B})}\dot{\hat{B}} + \frac{\partial\hat{\psi}}{\partial(\hat{\varepsilon}_s\hat{t}_\varphi)}\dot{\hat{\varepsilon}}_s\hat{t}_\varphi + \frac{\partial\hat{\psi}}{\partial(\hat{\varepsilon}_s\hat{t}_\varphi)}\hat{\varepsilon}_s\dot{\hat{t}}_\varphi + \frac{\partial\hat{\psi}}{\partial\left(\hat{\boldsymbol{\beta}}\hat{\boldsymbol{C}}_{\boldsymbol{\varphi}}\right)}\dot{\hat{\boldsymbol{\beta}}}\hat{\boldsymbol{C}}_{\boldsymbol{\varphi}} + \frac{\partial\hat{\psi}}{\partial\left(\hat{\boldsymbol{\beta}}\hat{\boldsymbol{C}}_{\boldsymbol{\varphi}}\right)}\hat{\boldsymbol{\beta}}\dot{\hat{\boldsymbol{C}}}_{\boldsymbol{\varphi}} + \frac{\partial\hat{\psi}}{\partial(\theta)}\dot{\theta} + \frac{\partial\hat{\psi}}{\partial(\hat{M})}\dot{\hat{\boldsymbol{M}}}. \tag{50}$$

The setting of the thermodynamic conjugates corresponding to the aforementioned magnetism internal state variables is as follows:

$$\hat{y} = \frac{\partial\hat{\psi}}{\partial(\hat{\boldsymbol{M}})}. \tag{51}$$

Substituting the free energy rate (Equation (50)) and (Equation (51)) into the C-D Inequality (Equation (42)) yields

$$-\hat{\rho}(\frac{\partial\hat{\psi}}{\partial(\hat{\boldsymbol{E}}_e\hat{\boldsymbol{C}}_\varphi)}:\dot{\hat{\boldsymbol{E}}}_e\hat{\boldsymbol{C}}_\varphi+\frac{\partial\hat{\psi}}{\partial(\hat{B})}\dot{\hat{B}}+\frac{\partial\hat{\psi}}{\partial(\hat{\boldsymbol{E}}_e\hat{\boldsymbol{C}}_\varphi)}:\hat{\boldsymbol{E}}_e\dot{\hat{\boldsymbol{C}}}_\varphi+\frac{\partial\hat{\psi}}{\partial(\hat{\varepsilon}_s\hat{t}_\varphi)}\dot{\hat{\varepsilon}}_s\hat{t}_\varphi+\frac{\partial\hat{\psi}}{\partial(\hat{\varepsilon}_s\hat{t}_\varphi)}\hat{\varepsilon}_s\dot{\hat{t}}_\varphi+\frac{\partial\hat{\psi}}{\partial(\hat{\boldsymbol{\beta}}\hat{\boldsymbol{C}}_{\boldsymbol{\varphi}})}\dot{\hat{\boldsymbol{\beta}}}\hat{\boldsymbol{C}}_{\boldsymbol{\varphi}}+\frac{\partial\hat{\psi}}{\partial(\hat{\boldsymbol{\beta}}\hat{\boldsymbol{C}}_{\boldsymbol{\varphi}})}\hat{\boldsymbol{\beta}}\dot{\hat{\boldsymbol{C}}}_{\boldsymbol{\varphi}}+\frac{\partial\hat{\psi}}{\partial(\hat{\theta})}\dot{\theta}+\hat{\boldsymbol{y}}.\dot{\hat{M}})- \\ \hat{\rho}\dot{\hat{\theta}}\hat{s}+\hat{\boldsymbol{S}}:\dot{\hat{\boldsymbol{E}}}+(\dot{\hat{\boldsymbol{B}}}.\hat{\boldsymbol{H}}+\hat{\boldsymbol{B}}.\dot{\hat{\boldsymbol{H}}})-\frac{1}{\hat{\theta}}\hat{\boldsymbol{q}}.\hat{\nabla}\hat{\theta}\geq 0. \tag{52}$$

Based on the model developed by Bammann and Solanki (2010) [97], an increasingly strong interaction between some individual dislocation strain fields and their neighboring dislocations induces more dislocation motion that causes material hardening. Therefore, the thermodynamic conjugates, which are stress-like quantities, of the ISVs associated with the stored dislocation and geometrically necessary densities are $\hat{\kappa}$ and $\hat{\boldsymbol{\alpha}}$, and are given as follows:

$$\hat{\kappa}=\hat{\rho}\frac{\partial\hat{\psi}}{\partial(\hat{\varepsilon}_s\hat{t}_\varphi)}\hat{t}_\varphi,\ \hat{\boldsymbol{\alpha}}=\hat{\boldsymbol{\rho}}\frac{\partial\hat{\psi}}{\partial(\hat{\boldsymbol{\beta}}\hat{\boldsymbol{C}}_{\boldsymbol{\varphi}})}\hat{\boldsymbol{C}}_\varphi^T. \tag{53}$$

Substituting Equation (50) into Equation (51) yields

$$(-\hat{\rho}\frac{\partial\hat{\psi}}{\partial(\hat{\boldsymbol{E}}_e\hat{\boldsymbol{C}}_{\boldsymbol{\varphi}})}:\hat{\boldsymbol{U}}_\varphi+\hat{\boldsymbol{S}}):\dot{\hat{\boldsymbol{E}}}_e+(-\hat{\rho}\frac{\partial\hat{\psi}}{\partial(\hat{\theta})}-\hat{\rho}\hat{s}+\zeta(\theta)\underline{I})\dot{\theta}+(\frac{1}{2}\hat{S}-\rho\frac{\partial\hat{\psi}}{\partial(\hat{\boldsymbol{\beta}}\hat{\boldsymbol{C}}_{\boldsymbol{\varphi}})}\hat{\boldsymbol{\beta}}- \\ \frac{1}{3}I\rho\frac{\partial\hat{\psi}}{\partial(\hat{\varepsilon}_s\hat{t}_\varphi)}\hat{\varepsilon}_s-\hat{\rho}\frac{\partial\hat{\psi}}{\partial(\hat{\boldsymbol{E}}_e\hat{\boldsymbol{C}}_\varphi)}\hat{\boldsymbol{E}}_e)\dot{\hat{\boldsymbol{C}}}_\varphi+(\hat{\boldsymbol{H}}-\hat{\rho}\frac{\partial\hat{\psi}}{\partial(\hat{B})})\dot{\hat{\boldsymbol{B}}}+(\hat{\boldsymbol{B}}.\dot{\hat{\boldsymbol{H}}})+\hat{\boldsymbol{S}}:\dot{\hat{\boldsymbol{E}}}_H-\hat{\rho}\hat{\boldsymbol{y}}.\dot{\hat{M}}-\hat{\kappa}\dot{\hat{\varepsilon}}_s- \\ \hat{\boldsymbol{\alpha}}\dot{\hat{\boldsymbol{\beta}}}+\hat{\boldsymbol{S}}:\dot{\hat{\boldsymbol{E}}}_p-\frac{1}{\hat{\theta}}\hat{\boldsymbol{q}}.\hat{\nabla}\hat{\theta}\geq 0, \tag{54}$$

where

$$\dot{\hat{\boldsymbol{E}}}=\dot{\hat{\boldsymbol{E}}}_e+\dot{\hat{\boldsymbol{E}}}_\theta+\dot{\hat{\boldsymbol{E}}}_\varphi+\dot{\hat{\boldsymbol{E}}}_H+\dot{\hat{\boldsymbol{E}}}_p. \tag{55}$$

In Equation (54), the damage and the thermal strain rates are given by Dimitrov et al. (2019) as

$$\dot{\hat{\boldsymbol{E}}}_\varphi=\frac{1}{2}\dot{\hat{\boldsymbol{C}}}_\varphi,\ and\ \dot{\hat{\boldsymbol{E}}}_\theta=\varrho(\theta)\underline{I}\dot{\theta}. \tag{56}$$

Unlike the other listed strains, the thermal expansion strain is considered a nonlocal variable in this study. We assume the thermal expansion is adequately represented by the linear coefficient of thermal expansion (ϱ) and the temperature increment ($\Delta\theta$), as previously presented by Dimitrov et al. (2019) [105]:

$$\hat{\boldsymbol{E}}_\theta=\hat{\boldsymbol{E}}_\theta(\theta)=\frac{1}{2}(\hat{\boldsymbol{C}}_\theta-\boldsymbol{I})=\frac{1}{2}\left[2\varrho\Delta\theta\boldsymbol{I}+(\varrho\Delta\theta)^2\boldsymbol{I}\right]. \tag{57}$$

For most practical applications, the coefficient of thermal expansion exhibits minimal temperature dependence and is considered constant within a small temperature range, below the Curie temperature for magnetic materials [106]. The material time derivative of the thermal expansion strain ($\dot{\hat{\boldsymbol{E}}}_\theta$) in the local form is then approximated as follows:

$$\dot{\hat{\boldsymbol{E}}}_\theta=\frac{\partial}{\partial\theta}\hat{\boldsymbol{E}}_\theta\dot{\theta}=\varrho(\theta)\boldsymbol{I}\dot{\theta}+\frac{\partial\varrho}{\partial\theta}\boldsymbol{I}\dot{\theta}\approx\varrho(\theta)\boldsymbol{I}\dot{\theta}. \tag{58}$$

Based on the scheme used by Coleman and Gurtin (1967) [55] and Kratochvil and Dillon (1969) [107], the constitutive equations for stress, entropy, and magnetism for this continuum model are given as follows:

$$\begin{aligned}&\hat{\boldsymbol{S}}=\hat{\rho}\frac{\partial\hat{\psi}}{\partial(\hat{\boldsymbol{E}}_e\hat{\boldsymbol{C}}_\varphi)}:\hat{\boldsymbol{C}}_\varphi,\\&\hat{s}=-\frac{\partial\hat{\psi}}{\partial(\theta)}+\frac{1}{\hat{\rho}}\tilde{\varrho}tr(\underline{\tilde{S}}),\\&\left(\frac{1}{2}\hat{\boldsymbol{S}}-\rho\frac{\partial\hat{\psi}}{\partial(\hat{\boldsymbol{\beta}}\hat{\boldsymbol{C}}_\varphi)}\hat{\boldsymbol{\beta}}-\frac{1}{3}\boldsymbol{I}\rho\frac{\partial\hat{\psi}}{\partial(\hat{\varepsilon}_s\hat{t}_\varphi)}\hat{\varepsilon}_s-\hat{\rho}\frac{\partial\hat{\psi}}{\partial(\hat{\boldsymbol{E}}_e\hat{\boldsymbol{C}}_\varphi)}\hat{\boldsymbol{E}}_e\right)\dot{\hat{\boldsymbol{C}}}_\varphi=0,\\&-\hat{\rho}\frac{\partial\hat{\psi}}{\partial(\hat{B})}+\hat{\boldsymbol{H}}=\boldsymbol{0},\end{aligned} \tag{59}$$

where $\boldsymbol{H}$, $\boldsymbol{S}$ and $\boldsymbol{s}$ are considered as thermodynamic forces associated with $\boldsymbol{B}$, $\boldsymbol{E_e}$, and $\boldsymbol{\theta}$, respectively.

Using Equation (59), the dissipation energy inequality (Equation (54)) can be reduced to

$$(\hat{\boldsymbol{B}}.\dot{\hat{\boldsymbol{H}}}) + \hat{\mathbb{S}} : \dot{\hat{\boldsymbol{E}}}_H - \hat{\rho}\hat{\boldsymbol{y}}.\dot{\hat{M}} - \hat{\kappa}\dot{\varepsilon}_s - \hat{\boldsymbol{\alpha}} : \dot{\boldsymbol{\beta}} + \hat{\mathbb{S}} : \dot{\hat{\boldsymbol{E}}}_p - \frac{1}{\theta}\hat{\boldsymbol{q}}.\hat{\nabla}\theta \geq 0. \tag{60}$$

Following the classical definition of entropy and neglecting second-order effects, Equation (40) can be approximated as part of the internal energy that dissipates as specific heat. Equation (41) is assumed to equal the portion of the internal energy that is stored as reversible processes or converted to irreversible damage and dislocation structure evolution. Consequently, from the definition of specific heat per unit mass ($c_M = du/d\theta$), we write the temperature evolution equation:

$$\dot{\theta} = \frac{1}{\widetilde{\rho}\widetilde{c}_M}\left\{\widetilde{\mathbb{S}} : \dot{\widetilde{\boldsymbol{E}}}_p - \widetilde{\kappa}_s\dot{\widetilde{\varepsilon}}_s - \widetilde{\boldsymbol{\alpha}} : \dot{\widetilde{\boldsymbol{\beta}}} - \widetilde{\nabla}.\widetilde{q} + \widetilde{\rho}\widetilde{r} + \hat{\rho}\frac{\partial\hat{\psi}}{\partial(\hat{\boldsymbol{B}})}\dot{\hat{\boldsymbol{B}}} + (\dot{\hat{\boldsymbol{H}}}.\hat{\boldsymbol{B}} + \hat{\boldsymbol{H}}.\dot{\hat{\boldsymbol{B}}}) + \hat{\mathbb{S}} : \dot{\hat{\boldsymbol{E}}}_H - \hat{\rho}\hat{a}.\dot{\hat{\boldsymbol{M}}}\right\}. \tag{61}$$

5. Kinetics

5.1. Experimental Magnetostriction Test

In order to examine the effects of mechanical stress and magnetic field, experiments were conducted on the three rod specimens (length of 185 mm and diameter of 6 mm) of iron, nickel, and cobalt. The specimen dimension is appropriate to place them inside of the magnetic coils. The experiments quantified the magnetostriction of the rod specimens. The apparatus used in this experiment is the Michelson Interferometer. The Michelson Interferometer is an optical method used to measure the magnetostriction. The Michelson Interferometer emits a laser wave that is then divided into two parts. Each of the new light beams travels a different path that recombines together. The magnetostriction strain is equal to the small mirror movement once the sample is subjected to a magnetic field and starts to elongate. Figure 5 shows the Michelson Interferometer used in this study. We applied various intensities of external magnetic fields (various electrical currents), and the obtained relationship among external magnetic field, magnetostriction, and magnetization is presented in Figures 6 and 7, respectively.

Figure 5. The Michelson Interferometer setup used to measure the magnetostriction in this study. The iron (Fe), nickel (Ni), and cobalt (Co) specimens were placed in the magnetic coil.

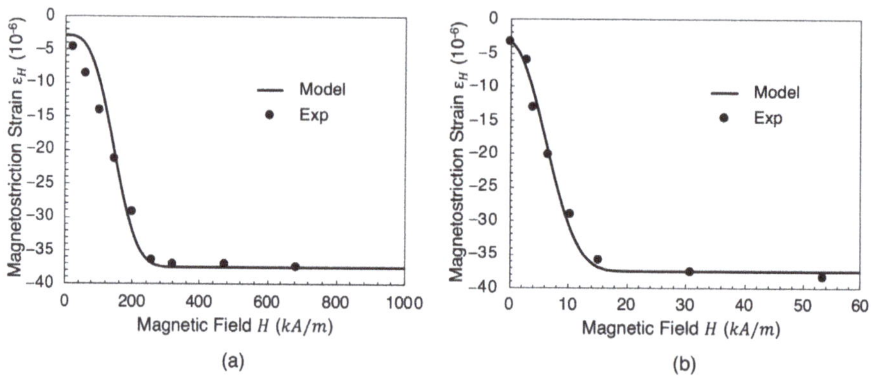

Figure 6. Magnetostriction variation (ε^H) with respect to the external magnetic field H (kA/m) for soft magnets: (**a**) cobalt (Co) and (**b**) nickel (Ni). Symbols are experimental data obtained in the part of the present study and lines are for the model.

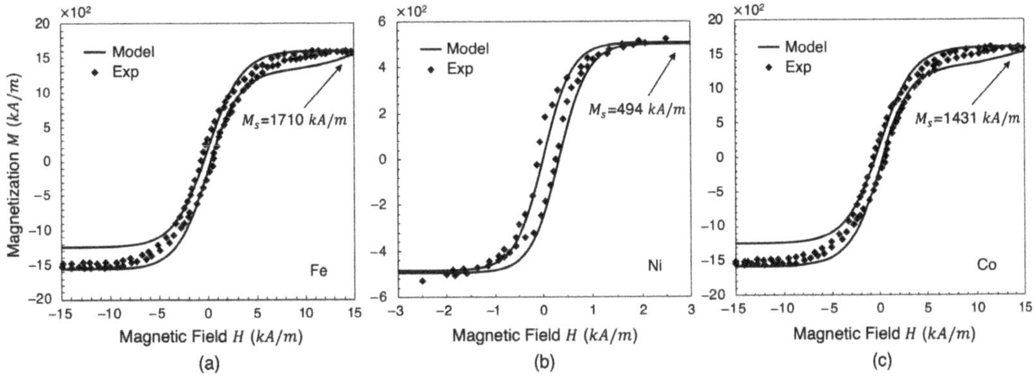

Figure 7. Magnetization variation M (kA/m) with respect to the external magnetic field H (kA/m), for soft magnets: (**a**) iron (Fe), (**b**) nickel (Ni), and (**c**) cobalt (Co). Symbols are experimental data [108] and lines are for the model.

5.2. Cauchy Stress Tensor

The frame indifferent Jaumann rate of the Cauchy stress was initially developed by Bammann (1990) [99] as a function of kinematics and elastic properties. It was then extended by Horstemeyer and Gokhale (1999) [109] to capture the degradation of a material's effective stiffness by damage. Therefore, the frame indifferent elastic stress rate in the current configuration (R) is given as follows:

$$\overset{\circ}{\sigma} = \dot{\sigma}\text{-}\boldsymbol{W}_e\sigma + \sigma\boldsymbol{W}_e{}^T = \dot{\sigma}\text{-}\boldsymbol{W}_e\sigma + \sigma\boldsymbol{W}_e = \lambda(\mathbf{1}-\varphi)tr(\boldsymbol{D}_e)\boldsymbol{I} + 2\mu(\mathbf{1}-\varphi)\boldsymbol{D}_e - \frac{\dot{\varphi}}{(\mathbf{1}-\boldsymbol{\varphi})}\sigma, \tag{62}$$

where λ, μ are the Lamé constants, φ is the total damage, $\boldsymbol{D}_e$ is the elastic rate deformation, and $\boldsymbol{W}_e$ is the elastic spin which is given as follows:

$$\boldsymbol{W}_e = \boldsymbol{W} - \boldsymbol{W}_H - \boldsymbol{W}_p, \tag{63}$$

where $\boldsymbol{W}_p$ is the plastic spin and $\boldsymbol{W}_H$ is the magnetic spin. The magnetic spin term is nonzero because of the electron spin motion distribution of the electrons within the atoms.

The elastic rate of deformation $\boldsymbol{D}_e$, is given as the difference between the total rate of deformation and the plastic, magnetic, damage, and thermal rates of deformation ($\boldsymbol{D}_p, \boldsymbol{D}_H, \boldsymbol{D}_\varphi$, and $\boldsymbol{D}_\theta$):

$$\boldsymbol{D}_e = \boldsymbol{D} - \boldsymbol{D}_p - \boldsymbol{D}_H - \boldsymbol{D}_\varphi - \boldsymbol{D}_\theta. \tag{64}$$

The plastic deformation rate is given using the strain flow rule, which was initially developed by Bammann (1990) [99] in order to relate the deviatoric rate of deformation to the applied stress and ISVs, then extended by Horstemeyer and Gokhale (1999) [109]. The plastic strain flow rule is the tensor rate at which the distances between a point (P) and its neighboring particles deform plastically, and it is given as follows:

$$\boldsymbol{D}_p = \sqrt{\frac{3}{2}}f(\theta) \times \sin h\left[\frac{\sqrt{\frac{3}{2}}\|\sigma' - \sqrt{\frac{2}{3}}\boldsymbol{\alpha}\| - \{R + Y(\theta)\}\{1-\varphi\}}{V(\theta)\{1-\varphi\}}\right] \times \frac{\sigma' - \sqrt{\frac{2}{3}}\boldsymbol{\alpha}}{\|\sigma' - \sqrt{\frac{2}{3}}\boldsymbol{\alpha}\|}. \tag{65}$$

The thermal and damage deformation rate were developed in a similar way by Bammann (1990) [99] and Horstemeyer et al. (2000) [110], respectively, and given as follows:

$$\boldsymbol{D}_\theta = \alpha_{th}\dot{\theta}\boldsymbol{I}, \tag{66}$$

$$\boldsymbol{D}_{\varphi} = \frac{\varphi}{3(1-\varphi)}\boldsymbol{I}. \tag{67}$$

The magnetic deformation rate is derived in this work as follows:

$$\boldsymbol{D}_H = \pm c\left(-bp\exp\left(\frac{(H\otimes H^T+\chi)^p}{\|H\otimes H^T+\chi\|^q}\right)*\frac{(H\otimes H^T+\chi)^{p-1}}{\|H\otimes H^T+\chi\|^q}*\dot{H}\right)*\frac{H\otimes H^T+\chi}{\|H\otimes H^T+\chi\|}+\boldsymbol{D}_H^{MX}, \tag{68}$$

For the sake of simplicity, we set the magnetic field vector and its transpose dyadic product as follows:

$$\boldsymbol{\Gamma} = H\otimes H^T. \tag{69}$$

Therefore, the magnetic deformation rate is written as

$$\boldsymbol{D}_H = \pm c\left(-bp\exp\left(\frac{(\boldsymbol{\Gamma}+\chi)^p}{\|\boldsymbol{\Gamma}+\chi\|^q}\right)*\frac{(\boldsymbol{\Gamma}+\chi)^{p-1}}{\|\boldsymbol{\Gamma}+\chi\|^q}*\dot{H}\right)*\frac{\boldsymbol{\Gamma}+\chi}{\|\boldsymbol{\Gamma}+\chi\|}+\boldsymbol{D}_H^{MX}. \tag{70}$$

In this case, the Maxwell-associated deformation rate ($\boldsymbol{D}_H^{MX}$) is assumed to be zero, since the deformation caused by the Maxwell stress is zero; thus, the magnetic deformation rate is written as follows:

$$\boldsymbol{D}_H = \pm c\left(-bp\exp\left(\frac{(\boldsymbol{\Gamma}+\chi)^p}{\|\boldsymbol{\Gamma}+\chi\|^q}\right)*\frac{(\boldsymbol{\Gamma}+\chi)^{p-1}}{\|\boldsymbol{\Gamma}+\chi\|^q}*\dot{H}\right)*\frac{\boldsymbol{\Gamma}+\chi}{\|\boldsymbol{\Gamma}+\chi\|}. \tag{71}$$

Functions $f(\theta)$, $Y(\theta)$, and $V(\theta)$ are functions that have an Arrhenius-type temperature dependence. They were developed by Bammann (1990) [99] and are given as follows:

$$\begin{aligned} f(\theta) &= C_5\exp(-C_6/\theta), \\ Y(\theta) &= C_3\exp(C_4/\theta), \\ V(\theta) &= C_1\exp(-C_2/\theta), \end{aligned} \tag{72}$$

where *Y(θ)* is the rate-independent yield stress. Function $f(\theta)$ determines when the rate dependences affect initial yielding. Function $V(\theta)$ determines the magnitude of the rate dependence on yielding. These functions are easily determined from simple isothermal compression tests with different strain rates and temperatures. C_1, C_2, C_3, C_4, C_5, and C_6 are Arrhenius-type temperature-dependent calibration constants.

Kinematic hardening internal state variable $\boldsymbol{\alpha}$ represents the anisotropic effect of the dislocation density while isotropic hardening internal state variable R mimics the global dislocation density effect. The kinematic hardening rate equation was developed by Bammann (1990) [99] and then extended by Tucker et al. (2010) [111] to account for the grain size effect,

$$\overset{\circ}{\boldsymbol{\alpha}} = \dot{\boldsymbol{\alpha}} - \boldsymbol{W}_e\boldsymbol{\alpha} + \boldsymbol{\alpha}\boldsymbol{W}_e = h(\theta)\boldsymbol{D}_p - \left[\sqrt{\frac{2}{3}}r_d(\theta)\|\boldsymbol{D}_p\| + r_s(\theta)\right]\sqrt{\frac{2}{3}}\|\boldsymbol{\alpha}\|\boldsymbol{\alpha})\left(\frac{DCS0}{DCS}\right)^Z, \tag{73}$$

where

$$\begin{aligned} r_d(\theta) &= C_7\left(1 - C_{19}\left[\frac{4}{27} - \frac{J_3^2}{J_2^3}\right] - C_{20}\frac{J_3}{J_2^{1.5}}\right)\exp\left(\frac{-C_8}{\theta}\right), \\ h(\theta) &= C_9\left(1 + C_{19}\left[\frac{4}{27} - \frac{J_3^2}{J_2^3}\right] + C_{20}\frac{J_3}{J_2^{1.5}}\right)\exp\left(\frac{-C_{10}}{\theta}\right), \\ r_s(\theta) &= C_{11}\exp(-C_{12}/\theta). \end{aligned} \tag{74}$$

The isotropic hardening rate equation is prescribed in a hardening minus recovery format which accounts for the grain size effect and is presented by Tucker et al. (2010) [111] as follows:

$$\dot{\kappa} = \sqrt{\frac{2}{3}}H(\theta)\boldsymbol{D}_p - \left[\sqrt{\frac{2}{3}}R_d(\theta)\|\boldsymbol{D}_p\| + R_s(\theta)\right]\kappa^2\left(\frac{DCS0}{DCS}\right)^Z, \tag{75}$$

where κ is the isotropic hardening, H is the work hardening modulus, $R_d(\theta)$ is the dynamic recovery that captures the dislocation glide effect, $R_s(\theta)$ is the static recovery that captures the dislocation climb or the diffusion effect, and $D_{in} = \sqrt{\frac{2}{3}}\dot{\varepsilon}^p N$ is the deviatoric inelastic strain rate. *DCS0* and *DCS* represent the initial average grain size and the average grain size that directly influence the dislocation density and thereby interact with the hardening parameters, respectively. Z is a constant exponent for the grain size effect on hardening.

The parameters of these mechanisms are given [72] as follows:

$$\begin{aligned} R_d(\theta) &= C_{13}\left(1 - C_{19}\left[\frac{4}{27} - \frac{J_3^2}{J_2^3}\right] - C_{20}\frac{J_3}{J_2^{1.5}}\right)\exp\left(\frac{-C_{14}}{\theta}\right), \\ H(\theta) &= C_{15}\left(1 + C_{19}\left[\frac{4}{27} - \frac{J_3^2}{J_2^3}\right] + C_{20}\frac{J_3}{J_2^{1.5}}\right)\exp\left(\frac{-C_{16}}{\theta}\right), \\ R_s(\theta) &= C_{17}\exp(-C_{18}/\theta), \end{aligned} \tag{76}$$

where J_2 and J_3 are the second and third invariants of deviatoric stress, respectively.

The equations describing the material's degradation (or total damage) were developed by Horstemeyer et al. (2000) [72] (void volume fraction) based on the consideration of the microphysical damage mechanism. They are given as follows:

$$\phi = \eta v c, \tag{77}$$

where η, v, and c represent void nucleation, growth, and coalescence, respectively. The total damage rate of the void volume fraction within a ductile metal is given as follows:

$$\dot{\phi} = \dot{\eta} v c + \eta \dot{v} c + \eta v \dot{c}. \tag{78}$$

The rate evolution of the void nucleation/growth and coalescence were described independently by Horstemeyer et al. (2000) [72]. The void nucleation rate is given as follows:

$$\dot{\eta} = \frac{d^{\frac{1}{2}}}{K_{ic}f^{\frac{1}{3}}}\eta\left(a\left[\frac{4}{27} - \frac{J_3^2}{J_2^3}\right] + b\frac{J_3}{J_2^{\frac{3}{2}}} + c\left\|\frac{I_1}{\sqrt{J_2}}\right\|\right)\|\boldsymbol{D}_d\|\exp\left(\frac{C_{\eta T}}{T}\right), \tag{79}$$

where d and f are material property constants of the initial secondary phase particle size and volume fraction, respectively. I_1, J_2, and J_3 are the first, the second, and third stress invariants representing the stress dependence of the void nucleation rate. Calibration constants a, b, and c represent the material's torsional for void nucleation, the difference between the tension and compression, and the stress triaxiality sensitivity for void nucleation, respectively, and they are all determined experimentally (based on tension, compression, and torsion tests at different strain levels). $C_{\eta T}$ is the calibration constant used to control thermal sensitivity during the void nucleation phase.

Void nucleation for a bar subjected to uniaxial stress for which the deformation is isothermal and happens at a constant strain rate can be obtained by an integration of Equation (79). This is given, as follows, by Bammann (1990) [99]:

$$\eta = \eta_0 \exp\left[\|\boldsymbol{E}\|\frac{d^{\frac{1}{2}}}{K_{ic}f^{\frac{1}{3}}}\eta\left(a\left[\frac{4}{27} - \frac{J_3^2}{J_2^3}\right] + b\frac{J_3}{J_2^{\frac{3}{2}}} + c\left\|\frac{I_1}{\sqrt{J_2}}\right\|\right)\exp\left(\frac{C_{\eta T}}{T}\right)\right], \tag{80}$$

where $\|\boldsymbol{E}\|$ is the norm of the total Lagrangian strain tensor.

The rate evolution of the second phase particle growth was developed by McClintock (1964) [112]. Among all damage models, McClintock's model is the most used one since it can be used at different strain/hardening rates, different temperatures, and different stress triaxialities. The void growth rate is therefore given as follows:

$$\dot{v} = \frac{4\pi}{3}\left(\frac{\sqrt{3}d_{v0}}{4(1-n)}\sin h\left[\sqrt{3}(1-n)\frac{\sqrt{2}I_1}{3\sqrt{J_2}}\right]\|\boldsymbol{D}_d\|\right)^3 \tag{81}$$

such that d_{v0} is the initial void diameter and n is the McClintock growth rate constant originally motivated by the material hardening rate.

The void growth equation for an increasing strain and/or stress triaxiality is given by McClintock (1964) [112] as follows:

$$v = \frac{4\pi}{3}\left(d_{v0}\exp\left[\|\boldsymbol{E}\|\frac{\sqrt{3}}{2(1-n)}\sin h\left(\sqrt{3}(1-n)\frac{\sqrt{2}I_1}{3\sqrt{J_2}}\right)\right]\right)^3. \tag{82}$$

As the applied stress increases, and as the voids nucleate within the material, voids tend to coalesce, resulting in a void sheet or a natural void. The coalescence rate evolution is described by Tucker et al. (2010) [111] and is given as follows:

$$\dot{c} = \left[c_{d1} + c_{d2}(\eta\dot{v} + \dot{\eta}v)\right]\exp^{C_{CT}T}\left(\frac{DCS0}{DCS}\right)^z, \tag{83}$$

where c_{d1} and c_{d2} are calibration constants, and DCS0 and DCS represent the initial average grain size and the average grain size that directly influence the dislocation density and as such interact with the hardening parameters, respectively. Z is a constant exponent for the grain size effect on hardening. C_{CT} is a thermal sensitivity calibration constant for void coalescence.

The co-rotational Jaumann rate is given as follows:

$$\overset{\circ}{\sigma} = Y\dot{\varepsilon}^e(1-\varphi) + Y\varepsilon^e(1-\dot{\varphi}) + \dot{Y}\varepsilon^e(1-\varphi) - \boldsymbol{W}_e\sigma + \sigma\boldsymbol{W}_e, \tag{84}$$

$$\overset{\circ}{\sigma} = Y\dot{\varepsilon}^e(1-\varphi) + Y\varepsilon^e(1-\dot{\varphi}) + \dot{Y}\varepsilon^e(1-\varphi) - \boldsymbol{W}_e[(Y^M\varepsilon^e + Y^M\varepsilon^t + Y^M\varepsilon^\varphi + Y^M\varepsilon^{MS} + Y^M\varepsilon^{MX})(1-\varphi)] + \sigma\boldsymbol{W}_e \tag{85}$$

where the elastic strain rate is given as

$$\dot{\varepsilon}^e = \dot{\varepsilon} - \dot{\varepsilon}^p - \dot{\varepsilon}^\varphi - \dot{\varepsilon}^\theta - \dot{\varepsilon}^H. \tag{86}$$

Assuming isotropic damage-induced deformation, the damage-induced strain (the volumetric strain related to the nucleation, growth, and coalescence of voids) is given by Horstemeyer et al. (1999) [109] as follows:

$$\dot{\varepsilon}^\varphi = \frac{1}{3}(1-\varphi)^{-1}\dot{\varphi}I, \tag{87}$$

which illustrates the damage-related strain rate change with respect to the damage parameter, in this case related to the nucleation, growth, and coalescence of the voids within the material.

The strain arising from thermal expansion and contraction is given by Francis et al. (2014) [71] as follows:

$$\dot{\varepsilon}^\theta = \alpha_{th}\Delta\theta. \tag{88}$$

Equation (72) shows the main relationship between the magnetostriction strain and the magnetic field for ferromagnetic material.

$$\boldsymbol{\varepsilon}^{MS} = \pm\left(1 - \exp\left(\frac{(\boldsymbol{\Gamma} + \boldsymbol{a})^{p}}{\|\boldsymbol{\Gamma} + \boldsymbol{a}\|^{q}}\right) * b\right) * c \pm d, \tag{89}$$

where a is the normal component χ, $c, d, p,$ and q are calibration constants, b is the magnetostriction constant at the saturation level, and H is the external magnetic field known to vary with respect to time. The magnetostriction strain rate is given as follows:

$$\dot{\boldsymbol{\varepsilon}}^{MS} = \pm c\left(-bp\exp\left(\frac{(\boldsymbol{\Gamma} + a)^{p}}{\|\boldsymbol{\Gamma} + a\|^{q}}\right) * \frac{(\boldsymbol{\Gamma} + a)^{p-1}}{\|\boldsymbol{\Gamma} + a\|^{q}} * \dot{H}\right). \tag{90}$$

The magnetostriction strain equation is compared with experimental data obtained from the Michelson interferometer for the two ferromagnetic materials: nickel (Ni) and cobalt (Co), as shown in Figure 6. Since the magnetic flux density strain is so small, it is assumed that $\boldsymbol{\varepsilon}^{MX}$ takes a constant value that depends on the magnetic field applied in the material. Therefore, the magnetic flux density strain rate ($\dot{\boldsymbol{\varepsilon}}^{MX}$) is assumed to be equal to zero.

$$\dot{\boldsymbol{\varepsilon}}^{MX} = 0. \tag{91}$$

5.3. An Internal State Variable for Magnetization

To capture the dissipative and the hysteretic response of magnetostrictive materials, the use of internal state variables is necessary. Magnetization is defined as the material's response to an external magnetic field (H). It is the average of the magnetic domains' moments. Paramagnetic and diamagnetic materials have no magnetization (or if they do it is a negligible one), unless it is subjected to a magnetic field. Once the magnetic field is removed, the material loses its magnetization. Ferromagnetic, ferrimagnetic, and antiferromagnetic materials all have magnetization even when no magnetic field (H) is applied. When an external magnetic field (H) is applied, the ferromagnetic, ferrimagnetic, and antiferromagnetic materials exhibit a nonlinear magnetization with respect to the magnetic field (H), as shown in Figure 7.

Previous models were developed to describe the hysteresis behavior of magnetic material. The most known model is the Jiles–Atherton Model [14,46,113] which describes the magnetization (M) behavior with respect to the magnetic field (H) through an ordinary differential equation. Differential equations require significant computational resources.

In this work, magnetization is assumed to be one of the internal state variables describing the magnetic domain behavior when subjected to magnetic field (H). Based on the hysteresis behavior, the magnetization rate evolution is written in a simpler form than previous models, which allows for simple numerical implementation and is given as follows:

$$\dot{\boldsymbol{M}}(H) = \dot{\xi}(H) + \dot{\chi}(H), \tag{92}$$

such that $\dot{\xi}(H)$ is the isotropic magnetization rate (needs identity matrix) and $\dot{\chi}(H)$ is the anisotropic magnetization rate given as follows:

$$\begin{aligned} \dot{\xi}(H) &= \dot{H}*\left(\tfrac{M_s}{A}*\sec h^2\left(\tfrac{H}{A}\right) - \tfrac{1}{B}\exp\left(-\tfrac{H}{B}\right)\right) \\ \dot{\chi}(H) &= \dot{\Gamma}*\left(\tfrac{M_s}{Q}*\sec h^2\left(\tfrac{\Gamma}{P}\right) + \tfrac{R}{P}\left(\sin h\left(\tfrac{\Gamma}{Q}\right)\right)\right), \end{aligned} \tag{93}$$

where A, B, P, Q, and R are constants of the material, M_s is the saturation magnetization.

These equations are compared with available experimental data [108] for magnetization of iron, nickel, and cobalt at various magnetic field strengths as shown in Figure 7. The results show an acceptable approximation to the experimental results.

6. Discussion

In this paper, we developed an Internal State Variable (ISV) constitutive model to account for magnetism-dependent elasto-viscoplastic and damage model for magnetic materials that brings in three novel ideas: (1) it is new to analyze the kinematics of the deforming continuum body under the external magnetic field to account for elastic/inelastic deformation and vorticity affected by the magnetic field; (2) this is the first paper (to the best of our knowledge) that introduces a new magnetic internal state variable constrained by the first and second laws of thermodynamics (Clausius–Duhem inequality); and (3) this ISV-based constitutive modeling is a novel approach, by itself, to this particular problem. The ISV model integrates the effect of a new magnetic observable variable: magnetic flux. Magnetic flux is the main macroscale constraint to which ferromagnetic materials are subjected during operation in most applications. To predict the behavior of ferromagnetic materials, the model included magnetization variations with respect to the magnetic field as an ISV that was then experimentally validated. Even though the strain response due to the magnetic field was small compared to the mechanical and thermal responses, the latter was included as the magnetostriction strain. The magnetostriction strain was experimentally validated for cobalt and nickel. More experiments related to different materials, boundary conditions, and non-monotonic sequences can be explored for future evaluation of the theoretical model.

7. Conclusions

In this study, a novel macroscopic constitutive theory is presented to describe the thermal–elastic–plastic damage behavior of magnetic materials. A multiscale and fully coupled multiphysics Internal State Variable (ISV) model is created to describe the effects of magnetic field forces and moments under thermomechanical deformations based on a kinematics, thermodynamics, and kinetics independently developed and subsequently coupled to provide an internally consistent theory for magnetic influenced deformation.

The major contributions are related to developing a model that captures the magnetic effects on deformation using a thermodynamically consistent framework developed by Coleman and Gurtin (1967) [55]. The ISV model features a kinematics description of the deformation using a multiplicative decomposition of the total deformation gradient into elastic, thermal, magnetostrictive, damage, and plastic components. Thermodynamic restrictions are employed using the Clausius–Duhem Inequality which combines the first and second laws of thermodynamics. The kinetic framework enables the prediction of magnetically influenced stresses and strains in materials exposed to magnetic fields. The novel ISV model framework couples elastic, thermal, damage, and plastic effects to magnetic effects.

To describe the mechanical deformation resulting from the magnetic field, an equation describing the magnetostriction variation with respect to the magnetic field is introduced. The magnetostriction strain is a simple equation, with one variable (magnetic field) and other calibration constants, that can predict the nonlinear behavior of soft and hard magnets. To describe the magnetic behavior of the magnet, magnetization is introduced as an internal state variable for which an equation is developed.

The developed magnetism-dependent ISV constitutive model is compared with experimental data of nickel, cobalt, and iron. From the experiments, we measure the mechanical deformation (magnetostriction strain) of nickel and cobalt and magnetization of iron, nickel, and cobalt when they are subjected to magnetic fields. The magnetostriction strain and the magnetization equations are developed in the framework of ISV theory and compared with the obtained experimental data, and both show good agreement.

For future considerations, we encourage researchers to conduct experiments in which the applied magnetization levels, temperatures, and applied strains at different strain rates and paths are varied. This ISV model framework should be admissible to address the aforementioned topics for use in design. In addition, we are planning to further the model development with a multiscale analysis of large strain deformation with different levels of magnetization. Finally, we recommend that different alloyed systems be examined in the context of this ISV model as only pure metals are used in this study.

Author Contributions: Conceptualization, M.F.H. and M.I.B.; Methodology, M.M., H.E.C. and L.A.P.; Investigation, M.M.; Resources, L.A.P. and L.C.; Data curation, M.M. and H.E.C.; Writing—original draft, M.M.; Writing—review & editing, M.F.H., H.E.C., D.D. and L.C.; Supervision, M.F.H., D.D., L.C. and M.I.B.; Project administration, M.F.H.; Funding acquisition, M.F.H. All authors have read and agreed to the published version of the manuscript.

Funding: This research received no external funding.

Institutional Review Board Statement: Not Applicable.

Informed Consent Statement: Not Applicable.

Data Availability Statement: The raw data supporting the conclusions of this article will be made available by the authors on request.

Acknowledgments: M. Malki, M.F. Horstemeyer, H.E. Cho, L.A. Peterson, and D. Dickel would like to thank the Center for Advanced Vehicular Systems (CAVS) at Mississippi State University. Most of the work herein was done at CAVS. H.E. Cho and M.F. Horstemeyer are also grateful to Liberty University for supporting this work.

Conflicts of Interest: The authors declare no conflicts of interest.

References

1. Zakaria, H.A.J.I.; Hamid, M.O.U.N.I.R.; Abdellatif, E.M.; Imane, A.M.A.R.I.R. Recent advancements and developments for electric vehicle technology. In Proceedings of the 2019 International Conference of Computer Science and Renewable Energies (ICCSRE), Agadir, Morocco, 22–24 July 2019; pp. 1–6.
2. Sacchi, R.; Bauer, C.; Cox, B.; Mutel, C. When, where and how can the electrification of passenger cars reduce greenhouse gas emissions? *Renew. Sustain. Energy Rev.* **2022**, *162*, 112475. [CrossRef]
3. Cui, J.; Kramer, M.; Zhou, L.; Liu, F.; Gabay, A.; Hadjipanayis, G.; Balasubramanian, B.; Sellmyer, D. Current progress and future challenges in rare-earth-free permanent magnets. *Acta Mater.* **2018**, *158*, 118–137. [CrossRef]
4. Avakian, A.; Ricoeur, A. Constitutive modelling of nonlinear reversible and irreversible ferromagnetic behaviors and applictions of multiferroic composites. *J. Intell. Mater. Syst. Struct.* **2016**, *27*, 2536–2554. [CrossRef]
5. Marvalova, B. *Modelling of Magnetosensitive Elastomers*; Technical University of Liberec: Liberec, Czech Republic, 2008.
6. Olabi, A.G. A Grunwald, Design and application of magnetostrictive materials. *Mater. Design.* **2008**, *29*, 469–483. [CrossRef]
7. Joule, J. On a new class of magnetic forces. *Ann. Electr. Magn. Chem.* **1842**, *8*, 219–224.
8. Gao, Z.; Zhou, Y. *A Magneto-Mechanical Fully Coupled Model for GIANT magnetostriction in High Temperature Superconductor*; Key Laboratory of Mechanics on Disaster and Environment in Western China attached to the Ministry of Education of China, Department of Mechanics and Engineering Sciences, College of Civil Engineering and Mechanics, Lanzhou University: Lanzhou, China, 2015.
9. Bozorth, R.M.; Williams, H.J. Effect of small stresses on magnetic properties. *Rev. Mod. Phys.* **1945**, *17*, 72. [CrossRef]
10. Brown, W.F., Jr. Irreversible magnetic effects of stress. *Phys. Rev.* **1949**, *75*, 147. [CrossRef]
11. Cullity, B.D. Introduction to Magnetic Materials. In *Addison Wesley*; Mass: Reading, MA, USA, 1972; p. 268.
12. Sablik, M.J.; Burkhardt, G.L.; Kwun, H.; Jiles, D.C. A model for the effect of stress on the low frequency harmonic content of the magnetic induction in ferromagnetic materials. *J. Appl. Phys.* **1988**, *63*, 3930. [CrossRef]
13. Jiles, D.C. Theory of Magneto-mechanical effect. *J. Phys. D Appl. Phys.* **1995**, *28*, 1537. [CrossRef]

14. Jiles, D.C.; Atherton, D.L. Theory of Ferromagnetic Hysteresis. *J. Magn. Magn. Mater.* **1986**, *61*, 48–60. [CrossRef]
15. Joule, J. On the Effects of Magnetism upon the Dimensions of Iron and Steel Bars. *Lond. Edinb. Dublin Philos. Mag. J. Sci.* **1847**, *76–87*, 225–241.
16. Penpeintner, M.; Holländer, R.B.; Seitner, M.J.; Weig, E.M.; Gross, R.; Gönnenwein, S.T.; Huebl, H. A universal platform for magnetostriction measurements in thin films. *arXiv* **2015**, arXiv:1509.00647.
17. Pislaru-Danescu, L.; Morega, A.M.; Morega, M. A novel magnetostrictive injection actuator based on new giant magnetostrictive materials. In Proceedings of the 2011 7th International Symposium on Advanced Topics in Electrical Engineering (ATEE), Bucharest, Romania, 12–14 May 2011.
18. Zagoruiko, N.V. Effect of an Electrostatic Field and a Pulsed Magnetic Field on the Movements of Dislocations in Sodium Chloride. *Sov. Phys. Crystallogr.* **1965**, *10*, 63–67.
19. Kravchenko, V.Y. Influence of the Magnetic Field on Electronic Deceleration of Dislocations. *JEPT Lett.* **1970**, *12*, 551–554.
20. Al'Shits, V.I.; Liubimov, V.N.; Sarychev, A.V.; Shuvalov, A.L. The topological characteristics of the singular points of an electrical field accompanying the propagation of sound in piezoelectrics. *Zhurnal Eksperimentalnoi I Teor. Fiz.* **1987**, *93*, 723–732.
21. Molotskii, M.I. Theoretical basis for electro-and magnetoplasticity. *Mater. Sci. Eng. A* **2000**, *287*, 248–258. [CrossRef]
22. Müllner, P.; Chernenko, V.A.; Kostorz, G. A microscopic approach to the magnetic-field-induced deformation of martensite (magnetoplasticity). *J. Magn. Magn. Mater.* **2003**, *267*, 325–334. [CrossRef]
23. Klypin, A.A. Effect of magnetic and electric fields on creep. *Met. Sci. Heat Treat.* **1974**, *15*, 639–642. [CrossRef]
24. Gavrilov, G.M. Change in the properties of quenched steel in magnetic field. *Met. Sci. Heat Treat.* **1977**, *19*, 442–446. [CrossRef]
25. Akram, S.; Babutskyi, A.; Chrysanthou, A.; Montalvão, D.; Pizurova, N. Effect of Alternating Magnetic Field on the Fatigue Behaviour of EN8 Steel and 2014-T6 Aluminium Alloy. *Metals* **2019**, *9*, 984. [CrossRef]
26. Li, G.R.; Wang, F.F.; Wang, H.M.; Zheng, R.; Xue, F.; Cheng, J.F. Influence of high pulsed magnetic field on tensile properties of TC4 alloy. *Chin. Phys. B* **2017**, *26*, 046201. [CrossRef]
27. Wang, H.-M.; Zhu, Y.; Li, G.-R.; Zheng, R. Plasticity and microstructure of AZ31 magnesium alloy under coupling action of high pulsed magnetic field and external stress. *Acta Phys. Sin.* **2016**, *65*, 146101. [CrossRef]
28. Choi, K.J.; Yoo, S.C.; Ham, J.; Kim, J.H.; Jeong, S.Y.; Choi, Y.S. Fatigue behavior of AISI 8620 steel exposed to magnetic field. *J. Alloys Compd.* **2018**, *764*, 73–79. [CrossRef]
29. Nawaz, B.; Long, X.; Yang, Z.; Zhao, J.; Zhang, F.; Li, J. Effect of magnetic field on microstructure and mechanical properties of austempered $70Si_3MnCr$ steel. *Mater. Sci. Eng. A* **2019**, *759*, 11–18. [CrossRef]
30. Murase, S.; Kobatake, S.; Tanaka, M.; Tashiro, I.; Horigami, O.; Ogiwara, H.; Shibata, K.; Nagai, K.; Ishikawa, K. Effects of a high magnetic field on fracture toughness at 4.2 K for austenitic stainless steels. *Fusion Eng. Des.* **1993**, *20*, 451–454. [CrossRef]
31. Zhang, Y.; Gey, N.; He, C.; Zhao, X.; Zuo, L.; Esling, C. High temperature tempering behaviors in a structural steel under high magnetic field. *Acta Mater.* **2004**, *52*, 3467–3474. [CrossRef]
32. Kovaleva, T.G.; Shevchuk, A.D. Effect of a magnetic field on the elasticity characteristics for certain steels and alloys. *Strength Mater.* **1983**, *15*, 701–703. [CrossRef]
33. Guo, Y.; Lee, Y.J.; Zhang, Y.; Sorkin, A.; Manzhos, S.; Wang, H. Effect of a weak magnetic field on ductile–brittle transition in micro-cutting of single-crystal calcium fluoride. *J. Mater. Sci. Technol.* **2022**, *112*, 96–113. [CrossRef]
34. Sidhom, A.A.; Sayed, S.A.; Naga, S.A. The influence of magnetic field on the mechanical properties & microstructure of plain carbon steel. *Mater. Sci. Eng. A* **2017**, *682*, 636–639. [CrossRef]
35. Hou, M.; Li, K.; Li, X.; Zhang, X.; Rui, S.; Cai, Z.; Wu, Y. Effects of Pulsed Magnetic Fields of Different Intensities on Dislocation Density, Residual Stress, and Hardness of Cr4Mo4V Steel. *Crystals* **2020**, *10*, 115. [CrossRef]
36. Bose, M.S.C. Effect of saturated magnetic field on fatigue life of carbon steel. *Phys. Status Solidi (A)* **1984**, *86*, 649–654. [CrossRef]
37. Fahmy, Y.; Hare, T.; Tooke, R.; Conrad, H. Effects of a Pulsed Magnetic Treatment on the Fatigue of Low Carbon Steel. *Scr. Mater.* **1998**, *38*, 1355–1358. [CrossRef]
38. Bhat, I.K.; Muju, M.K.; Mazumdar, P.K. Possible effects of magnetic fields in fatigue. *Int. J. Fatigue* **1993**, *15*, 193–197. [CrossRef]
39. Mohin, M.A.; Toofanny, H.; Babutskyi, A.; Lewis, A.; Xu, Y.G. Effect of Electromagnetic Treatment on Fatigue Resistance of 2011 Aluminum Alloy. *J. Multiscale Model.* **2016**, *7*, 1650004. [CrossRef]
40. Gu, Q.; Huang, X.; Xi, J.; Gao, Z. The Influence of Magnetic Field on Fatigue and Mechanical Properties of a 35CrMo Steel. *Metals* **2021**, *11*, 542. [CrossRef]
41. Aksenova, K.; Zaguliaev, D.; Konovalov, S.; Shlyarov, V.; Ivanov, Y. Influence of Constant Magnetic Field upon Fatigue Life of Commercially Pure Titanium. *Materials* **2022**, *15*, 6926. [CrossRef]
42. Skvortsov, A.A.; Pshonkin, D.E.; Luk'Yanov, M.N.; Rybakova, M.R. Influence of permanent magnetic fields on creep and microhardness of iron-containing aluminum alloy. *J. Mater. Res. Technol.* **2019**, *8*, 2481–2485. [CrossRef]
43. Hu, Y.; Zhao, H.; Yu, X.; Li, J.; Zhang, B.; Li, T. Research Progress of Magnetic Field Regulated Mechanical Property of Solid Metal Materials. *Metals* **2022**, *12*, 1988. [CrossRef]
44. Zheng, X.J.; Liu, X.E. A nonlinear constitutive model for Terfenol-D rods. *J. Appl. Phys.* **2005**, *97*, 053901. [CrossRef]
45. Li, J.; Xu, M. Modified Jiles-Atherton-Sablik model for asymmetry in magnetomechanical effect under tensile and compressive stress. *J. Appl. Phys.* **2011**, *110*, 063918. [CrossRef]

46. Jiles, D.C.; Atherton, D.L. Theory of ferromagnetic hysteresis (invited). *J. Appl. Phys.* **1984**, *55*, 2115–2120. [CrossRef]
47. Sablik, M.J. A Model for Asymmetry in Magnetic Property Behavior under Tensile and Compressive Stress in Steel. *IEEE Trans. Magn.* **1997**, *33*, 3958–3960. [CrossRef]
48. Wang, Z.D.; Deng, B.; Yao, K. Physical model of plastic deformation on magnetization in ferromagnetic materials. *J. Appl. Phys.* **2011**, *109*, 083928. [CrossRef]
49. Daniel, L. An analytical model for the magnetostriction strain of ferromagnetic materials subjected to multiaxial stress. *Eur. Phys. J. Appl. Phys.* **2018**, *83*, 30904. [CrossRef]
50. Shi, P.; Bai, P.; Chen, H.-E.; Su, S.; Chen, Z. The magneto-elastoplastic coupling effect on the magnetic flux leakage signal. *J. Magn. Magn. Mater.* **2020**, *504*, 166669. [CrossRef]
51. Shi, P. Magneto-elastoplastic coupling model of ferromagnetic material with plastic deformation under applied stress and magnetic fields. *J. Magn. Magn. Mater.* **2020**, *512*, 166980. [CrossRef]
52. Onsager, L. Reciprocal relations in irreversible processes. I and II. *Phys. Rev.* **1931**, *37*, 405. [CrossRef]
53. Eckart, C. The thermodynamics of irreversible processes. III and IV Relativistic Theory of the Simple Fluid. *Phys. Rev.* **1948**, *58*, 919. [CrossRef]
54. Kroner, E. General continuum theory of dislocations and proper stresses. *Arch. Rat. Mech. Anal.* **1960**, *4*, 273–334.
55. Coleman, B.D.; Gurtin, M.E. Thermodynamics with internal state variables. *J. Chem. Phys.* **1967**, *47*, 597–613. [CrossRef]
56. Cho, H.E.; Hammi, Y.; Francis, D.K.; Stone, T.; Mao, Y.; Sullivan, K.; Wilbanks, J.; Zelinka, R.; Horstemeyer, M.F. Microstructure-Sensitive, History-Dependent Internal State Variable Plasticity-Damage Model for a Sequential Tubing Process. In *Integrated Computational Materials Engineering (ICME) for Metals*; Wiley: Hoboken, NJ, USA, 2018; pp. 199–234.
57. Horstemeyer, M.F.; Bammann, D.J. Historical review of internal state variable theory for inelasticity. *Int. J. Plast.* **2010**, *26*, 1310–1334. [CrossRef]
58. Lee, E.H.; Liu, D.T. Finite Strain Elastic-plastic Theory with Application to Plane-Wave, Analysis. *J. Appl. Phys.* **1967**, *38*, 391–408. [CrossRef]
59. Murakami, S. Mechanical modeling of material damage. *J. Appl. Mech.* **1988**, *55*, 280–286. [CrossRef]
60. Rice, J.R. Inelastic constitutive relations for solids: An internal-variable theory and its application to metal plasticity. *J. Mech. Phys. Solids* **1971**, *19*, 433–455. [CrossRef]
61. Bammann, D.J.; Aifantis, E.C. A damage model for ductile metals. *Nucl. Eng. Des.* **1989**, *116*, 355–362. [CrossRef]
62. Marin, E.B.; McDowell, D.L. Associative versus non-associative porous viscoplasticity based on internal state variable concepts. *Int. J. Plast.* **1996**, *12*, 629–669. [CrossRef]
63. Steinmann, P.; Carol, I. A framework for geometrically nonlinear continuum damage mechanics. *Int. J. Eng. Sci.* **1998**, *36*, 1793–1814. [CrossRef]
64. Voyiadjis, G.Z.; Park, T. The kinematics of damage for finite-strain elasto-plastic solids. *Int. J. Eng. Sci.* **1999**, *37*, 803–830. [CrossRef]
65. Brünig, M. Numerical analysis and elastic–plastic deformation behavior of anisotropically damaged solids. *Int. J. Plast.* **2002**, *18*, 1237–1270. [CrossRef]
66. Regueiro, R.A.; Bammann, D.J.; Marin, E.B.; Garikipati, K. A nonlocal phenomenological anisotropic finite deformation plasticity model accounting for dislocation defects. *J. Eng. Mater. Technol.* **2002**, *124*, 380–387. [CrossRef]
67. Solanki, K.N. *Physically Motivated Internal State Variable Form of a Higher Order Damage Model for Engineering Materials with Uncertainty*; Mississippi State University: Starkville, MS, USA, 2008.
68. Bammann, D.J.; Aifantis, E.C. An anisotropic hardening model of plasticity. *Comput. Mech.* **1988**, *88*, 459–462.
69. Bammann, D.J.; Chiesa, M.; Johnson, G. Modeling large deformation and failure in manufacturing processes. In *Theoretical and Applied Mechanics*; Tatsumi, T., Watanabe, E., Kambe, T., Eds.; Mathematical Institute SANU: Belgrade, Serbia, 1996; pp. 359–376.
70. Bammann, D.J.; Chiesa, M.L.; Horstemeyer, M.F.; Weingarten, L.I. Failure in ductile materials using finite element methods. In *Structural Crashworthiness and Failure*; CRC Press: Boca Raton, FL, USA, 1993; pp. 1–54.
71. Francis, D.K.; Bouvard, J.L.; Hammi, Y.; Horstemeyer, M.F. Formulation of a damage internal state variable model for amorphous glassy polymers. *Int. J. Solids Struct.* **2014**, *51*, 2765–2776. [CrossRef]
72. Horstemeyer, M.F.; Lathrop, J.; Gokhale, A.M.; Dighe, M. Modelling stress state dependent damage evolution in a cast Al-Si-Mg aluminium alloy. *Theor. Appl. Fract. Mech.* **2000**, *3*, 31–47. [CrossRef]
73. Cho, H.E.; Hammi, Y.; Bowman, A.L.; Karato, S.; Baumgardner, J.R.; Horstemeyer, M.F. A unified static and dynamic recrystallization Internal State Variable (ISV) constitutive model coupled with grain size evolution for metals and mineral aggregates. *Int. J. Plast.* **2019**, *112*, 123–157. [CrossRef]
74. Dimitrov, N.; Liu, Y.; Horstemeyer, M.F. An electroplastic internal state variable (ISV) model for nonferromagnetic ductile metals. *Mech. Adv. Mater. Struct.* **2020**, *29*, 761–772. [CrossRef]
75. Cho, H.E.; Zbib, H.M.; Horstemeyer, M.F. An Internal State Variable Elastoviscoplasticity-Damage Model for Irradiated Metals. *J. Eng. Mater. Technol.* **2022**, *144*, 1–44. [CrossRef]
76. Brugmans, A. Magnetismus, seu de affinitatibus magneticis observationes academicae. In *Apud Luzac & Van Damme*; Lugduni Batavorum: Leiden, The Netherlands, 1778.

77. Stoner, E.C. Collective Electron Specific Heat and Spin Paramagnetism in Metals. *R. Soc. Publ.* **1935**, *154*, 656–678.
78. Nunez, A.S.; Duine, R.A.; Haney, P.; MacDonald, A.H. Theory of spin torques and giant magnetoresistance in antiferromagnetic metals. *Phys. Rev.* **2006**, *73*, 214426. [CrossRef]
79. Linnemann, K.; Klinkel, S.; Wagner, W. A constitutive model for magnetostrictive and piezoelectric materials. *Int. J. Solids Struct.* **2009**, *46*, 1149–1166. [CrossRef]
80. Dapino, M.J.; Smith, R.C.; Flatau, A.B. A structural magnetic strain model for magnetostrictive transducers. *IEEE Trans. Magn.* **2000**, *36*, 545–556. [CrossRef]
81. Burgoyne, J.W.; Daniels, D.; Timms, K.W.; Vale, S.H. Advances in superconducting magnets for commercial and industrial applications. *IEEE Trans. Appl. Supercond.* **2000**, *10*, 703–709. [CrossRef]
82. Kim, Y.Y.; Kwon, Y.E. Review of magnetostrictive patch transducers and applications in ultrasonic nondestructive testing of waveguides. *Ultrasonics* **2015**, *62*, 3–19. [CrossRef] [PubMed]
83. Calkins, F.T.; Flatau, A.B.; Dapino, M.J. Overview of magnetostrictive sensor technology. *J. Intell. Mater. Syst. Struct.* **2007**, *18*, 1057–1066. [CrossRef]
84. Ekreem, N.B.; Olabi, A.G.; Prescott, T.; Rafferty, A.; Hashmi, M.S.J. An overview of magnetostriction, its use and methods to measure these properties. *J. Mater. Process Technol.* **2007**, *191*, 96–101. [CrossRef]
85. Bieńkowski, A.; Kulikowski, J. The magneto-elastic Villari effect in ferrites. *J. Magn. Magn. Mater.* **1980**, *19*, 120–122. [CrossRef]
86. Skomski, R.; Sellmyer, D.J. Curie temperature of multiphase nanostructures. *J. Appl. Phys.* **2000**, *87*, 4756–4758. [CrossRef]
87. Arrott, A. Criterion for ferromagnetism from observations of magnetic isotherms. *Phys. Rev.* **1957**, *108*, 1394. [CrossRef]
88. Dey, P.; Roy, J.N. Magnetic Domain Wall Motion. *Spintronics* **2021**, 145–161. [CrossRef] [PubMed]
89. Ising, E. Beitrag zur Theorie des Ferromagnetismus. *Z. Phys.* **1925**, *31*, 253–258. [CrossRef]
90. Yang, C.N. The spontaneous magnetization of a two-dimensional Ising model. *Phys. Rev.* **1952**, *85*, 808. [CrossRef]
91. McCoy, B.M.; Wu, T.T. *The Two-Dimensional Ising Model*; Courier Corporation: North Chelmsford, MA, USA, 2014.
92. Bammann, D.J.; Johnson, G.C. On the kinematics of finite-deformation plasticity. *Acta Mech.* **1987**, *70*, 1–13. [CrossRef]
93. Hashiguchi, K. *Nonlinear Continuum Mechanics for Finite Elasticity-Plasticity: Multiplicative Decomposition with Subloading Surface Model*; Elsevier: Amsterdam, The Netherlands, 2020.
94. Sun, Q.C.; Song, T.; Anderson, E.; Brunner, A.; Förster, J.; Shalomayeva, T.; Taniguchi, T.; Watanabe, K.; Gräfe, J.; Stöhr, R.; et al. Magnetic domains and domain wall pinning in atomically thin CrBr3 revealed by nanoscale imaging. *Nat. Commun.* **2021**, *12*, 1–7. [CrossRef] [PubMed]
95. Fiorillo, F.; Küpferling, M.; Appino, C. Magnetic Hysteresis and Barkausen Noise in Plastically Deformed Steel Sheets. *Metals* **2018**, *8*, 15. [CrossRef]
96. Hammi, Y.; Horstemeyer, M.F.; Bammann, D.J. An anisotropic damage model for ductile metals. *Int. J. Damage Mech.* **2003**, *12*, 245–262. [CrossRef]
97. Bammann, D.J.; Solanki, K. On kinematic, thermodynamic, and kinetic coupling of a damage theory of polycrystalline material. *Int. J. Solids Struct.* **2010**, *26*, 775–924. [CrossRef]
98. Dafalias, F. The plastic spin in viscoplasticity. *Int. J. Solids Struct.* **1989**, *26*, 149–163. [CrossRef]
99. Bammann, D.J. Modeling temperature and strain rate dependent large deformations of metals. *Appl. Mech. Rev.* **1990**, *43*, S312–S319. [CrossRef]
100. Horstemeyer, M. *Integrated Computational Materials Engineering (ICME) for Metals: Using Multiscale Modeling to Invigorate Engineering Design with Science*; Wiley: Hoboken, NJ, USA, 2012.
101. Hanson, R.; Witkamp, B.; Vandersypen, L.M.K.; van Beveren, L.W.; Elzerman, J.M.; Kouwenhoven, L. Zeeman energy and spin relaxation in a one-electron quantum dot. *Phys. Rev. Lett.* **2003**, *91*, 196802. [CrossRef]
102. Gaskell, D.R.; Laughlin, D.E. *Introduction to the Thermodynamics of Materials*; CRC Press: Boca Raton, FL, USA, 2017.
103. Feinberg, M.; Lavine, R. Thermodynamics based on the Hahn-Banach theorem: The Clausius inequality. *Arch. Ration. Mech. Anal.* **1983**, *82*, 203–293. [CrossRef]
104. Salehghaffari, S.; Rais-Rohani, M.; Marin, E.; Bammann, D. A new approach for determination of material constants of internal state variable based plasticity models and their uncertainty quantification. *Comput. Mater. Sci.* **2011**, *55*, 237–244. [CrossRef]
105. Dimitrov, N.; Liu, Y.; Horstemeyer, M.F. On the thermo-mechanical coupling of the Bammann plasticity-damage internal state variable model. *Acta Mech.* **2019**, *230*, 1855–1868. [CrossRef]
106. Clark, A.E.; DeSavage, B.F.; Bozorth, R. Anomalous thermal expansion and magnetostriction of single-crystal dysprosium. *Phys. Rev.* **1965**, *138*, A216. [CrossRef]
107. Kratochvil, J.; Dillon, O.W., Jr. Thermodynamics of elastic-plastic materials as a theory with internal state variables. *J. Appl. Phys.* **1969**, *40*, 3207–3218. [CrossRef]
108. Paulo, V.I.M.; Neves-Araujo, J.; Revoredo, F.A.; Padrón-Hernández, E. Magnetization curves of electrodeposited Ni, Fe and Co nanotubes. *Mater. Lett.* **2018**, *223*, 78–81. [CrossRef]
109. Horstemeyer, M.F.; Gokhale, A.M. A void-crack nucleation model for ductile metals. *Int. J. Solids Struct.* **1999**, *36*, 5029–5055. [CrossRef]

110. Horstemeyer, M.; Matalanis, M.; Sieber, A.; Botos, M. Micromechanical finite element calculations of temperature and void configuration effects on void growth and coalescence. *Int. J. Plast.* **2000**, *16*, 979–1015. [CrossRef]
111. Tucker, M.T.; Horstemeyer, M.F.; Whittington, W.R.; Solanki, K.N. *Structure/property Relations of Aluminum under Varying Rates and Stress States*; No. LA-UR-10-07799; LA-UR-10-7799; Los Alamos National Lab. (LANL): Los Alamos, NM, USA, 2010.
112. McClintock, F.A. A criterion for ductile fracture by the growth of holes. *J. Appl. Mech.* **1964**, *35*, 363–371. [CrossRef]
113. Wang, X.; Thomas, D.W.P.; Sumner, M.; Paul, J.; Cabral, S.H.L. Characteristics of Jiles-Atherton Model Parameters and Their Application to Transformer Inrush Current Simulation. *IEEE Trans. Magn.* **2008**, *44*, 340–345. [CrossRef]

materials

MDPI

Article

Crashworthiness of Foam-Filled Cylindrical Sandwich Shells with Corrugated Cores

Pengbo Su [1], Bin Han [2,*], Yiming Wang [1], Hui Wang [1], Bo Gao [1] and Tian Jian Lu [3,4]

1 Xi'an Institute of Space Radio Technology, Xi'an 710100, China; su_pengbo@126.com (P.S.); wangyiming0920@126.com (Y.W.); 13519122235@139.com (H.W.); gaob_2004@163.com (B.G.)
2 School of Mechanical Engineering, Xi'an Jiaotong University, Xi'an 710049, China
3 State Key Laboratory of Mechanics and Control of Mechanical Structures, Nanjing University of Aeronautics and Astronautics, Nanjing 210016, China; tjlu@nuaa.edu.cn
4 Nanjing Center for Multifunctional Lightweight Materials and Structures (MLMS), Nanjing University of Aeronautics and Astronautics, Nanjing 210016, China
* Correspondence: hanbinghost@xjtu.edu.cn

Abstract: Inspired by material hybrid design, novel hybrid sandwich shells were developed by filling a corrugated cylindrical structure with aluminum foam to achieve higher energy absorption performance. The crushing behavior of the foam-filled corrugated sandwich cylindrical shells (FFCSCSs) was investigated using theoretical and numerical methods. Numerical results revealed a significant enhancement in the energy absorption of FFCSCSs under axial compression, showcasing a maximum specific energy absorption of 60 kJ/kg. The coupling strengthening effect is highly pronounced, with a maximum value of $\overline{F}_c/\overline{F}$ reaching up to 40%. The mechanism underlying this phenomenon can be approached from two perspectives. Firstly, the intrusion of folds into the foam insertions allows for more effective foam compression, maximizing its energy absorption capacity. Secondly, foam causes the folds to bend upwards, intensifying the mutual compression between the folds. This coupling mechanism was further investigated with a focus on analyzing the influence of parameters such as the relative density of the foam, the wall thickness of the sandwich shell, and the material properties. Moreover, a theoretical model was developed to accurately predict the mean crushing force of the FFCSCSs. Based on this model, the influence of various variables on the crushing behavior of the structure was thoroughly investigated through parametric studies.

Keywords: foam-filled corrugated sandwich cylindrical shells; coupling strengthening effect; energy absorption; theoretical model

Citation: Su, P.; Han, B.; Wang, Y.; Wang, H.; Gao, B.; Lu, T.J. Crashworthiness of Foam-Filled Cylindrical Sandwich Shells with Corrugated Cores. *Materials* **2023**, *16*, 6605. https://doi.org/10.3390/ma16196605

Academic Editors: Madhav Baral and Charles Lu

Received: 7 September 2023
Revised: 5 October 2023
Accepted: 7 October 2023
Published: 9 October 2023

1. Introduction

Researchers have consistently aimed to design high-performance protective equipment and enhance the crashworthiness of various modes of transportation to reduce injuries and property damage resulting from collisions while achieving lightweight designs. Thin-walled shells are widely employed as collision-resistant structures due to their high energy absorption efficiency, reliability, and low manufacturing cost [1]. During collisions, thin-walled shells absorb kinetic energy through significant plastic deformation, safeguarding public safety and protecting property; examples of such shells include energy-absorbing boxes in cars, bumpers in high-speed trains, and crash-resistant landing gears in helicopters. Comprehensive research has been conducted on the energy absorption capacity of single-cell, multicell, and foam-filled shells. Single-cell tubular structures, such as circular, square, and polygonal tubes, have been extensively studied. Corresponding theoretical models for estimating the energy absorption capacity of these structures have been gradually established. When subjected to axial compression, single-cell tubular structures typically exhibit three collapse modes: progressive mode, global mode, and transition mode [2,3]. The progressive collapse mode is the primary focus of most studies because it exhibits stable deformation processes and possesses higher energy absorption efficiency. In the case of circular tubes, Guillow et al. divided the progressive collapse mode into axisymmetric mode, nonsymmetric mode, and mixed mode [4]. Alexander [5] derived an approximate theory to estimate the collapse load specifically for the axisymmetric mode. For square tubes, Weirzbickihe and Abramowicz [6,7] proposed a fundamental folding mode known as the "super folding element" based on their observation of the folding deformation process. This mode was utilized to predict the mean crushing force. Subsequently, the "super folding element" was extended to polygonal tubes with arbitrary interior angles [8]. Triangular tubes [9–11], hexagonal tubes [12–15], octagonal tubes [16], nonconvex multicorner tubes [17,18], and star-shaped tubes [15,19] have been extensively investigated based on this work. It is demonstrated that energy dissipation primarily occurs at horizontal plastic hinges. Consequently, increasing the number of folds would significantly improve the energy absorption capacity of tubular structures.

To further enhance energy absorption, researchers proposed tubular structures filled with foam or honeycomb cores. Corresponding studies indicated that filled tubular structures exhibited significantly more folds than unfilled ones, resulting in higher energy absorption efficiency [20]. Reid et al. [21,22] conducted a series of static and dynamic experiments on circular and square tubes filled with polyurethane foam. Their findings revealed that the specific energy absorption of foam-filled structures was twice as high as that of nonfilled structures. Compared to polyurethane foam, metal foam, specifically aluminum foam, exhibits higher platform stress levels. With the commercial preparation process for aluminum foam gradually maturing, subsequent researchers conducted extensive studies on aluminum-foam-filled tubes with various cross sections, including square tubes [20,23–25], circular tubes [26–28], and polygonal tubes [23,29]. Furthermore, the potential of honeycomb-filled tubular structures in energy absorption was thoroughly examined by Hussein et al. [30] and Yin [31]. These studies effectively demonstrated the advantages of honeycomb filling to enhance energy absorption capabilities. Recently, novel fillings made of cellular materials were proposed based on the continuous development of configuration and preparation processes. These materials include functionally graded foam [32–35], auxetic foam [36], composite foam [37,38], ex situ aluminum foam [39], and liquid nanofoam [40].

To further improve the energy absorption capacity of tubular structures, researchers designed sandwich tubular structures filled with cellular materials, such as foams and honeycombs. The advantageous energy absorption capabilities of these sandwich structures were demonstrated by Seltzberger et al. [23], Li et al. [41], Zhang et al. [42], Zheng et al. [43], Gao et al. [44], Djamaluddin et al. [45], and Goel [46]. These investigations revealed that the sandwich structures exhibited a higher mean crushing force due to the coupling effect between the face sheets and the filling materials. Meanwhile, research findings revealed that sandwich structures featuring two-dimensional (2D) corrugated or honeycomb cores offered superior weight reduction and design benefits, as evidenced by studies [47–52]. In our previous work, the energy absorption capacity of corrugated sandwich shells was investigated through a combined experimental, theoretical, and numerical approach [53,54].

Moreover, due to the interconnected nature of the corrugated channels, some researchers filled these channels with cellular materials such as foam and aluminum honeycomb to enhance structural performance. Foam-filled corrugated sandwiches were designed and fabricated by Yan et al. [55] and Han et al. [56]. Their work revealed that the energy absorption performance of these structures under out-of-plane compression surpassed the combined energy absorption of the hollow corrugation and the foam individually, attributed to the coupling effect between the foam and the corrugated core. Similar coupling effects were also observed in honeycomb–corrugate hybrid structures [57], ceramic–corrugate hybrid structures [58–60], and other hybrid sandwich structures [61–64].

To further enhance the energy-absorbing properties of the tubular structure, a novel hybrid sandwich shell was proposed by incorporating aluminum foam into a corrugated sandwich cylindrical shell. To characterize the crushing behavior, the finite element (FE) method was employed, and its accuracy was verified using experimental data in reference [53]. The energy absorption of the FFCSCS under axial compression was investigated through numerical simulations. The collapse behavior and folding modes were analyzed, and the coupling strengthening mechanism between foam and shell wall was explored. Based on FE simulations, a theoretical model was developed to predict the mean crushing force of FFCSCSs. Parametric analysis was conducted using the theoretical model to explore the influence of different parameters on the coupling strengthening effect. This paper is organized as follows: Section 2 introduces the definition of terminology. Section 3 presents and validates the finite element model. Section 4 provides an analysis of the coupling strengthening effect and its mechanism. Section 5 discusses the influence of wall thickness, material, and foam density on the energy absorption performance of FFCSCSs based on finite element analysis. Section 6 presents a theoretical model for predicting the mean crushing force and conducts a parametric study based on the mechanistic model.

2. Terminology Definition in the Crushing Process

This section presents the relevant physical quantities and their definitions used to describe the crushing process and assess energy absorption characteristics of cylindrical shells under axial compression. As illustrated in Figure 1, H_s represents the initial height of the cylindrical shell, d is employed to signify the compressive displacement of the structure, and F characterizes the corresponding crushing force within the structure.

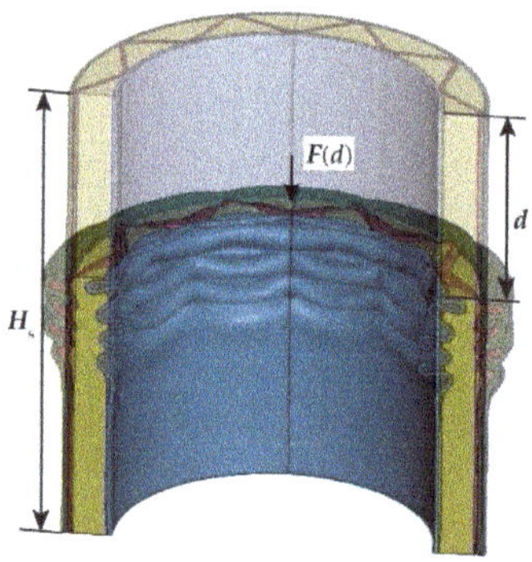

Figure 1. The FFCSCS is under crushing process.

The maximal value of F within the interval from 0 to d is denoted as $F_{max}(d)$. Simultaneously, the energy absorption within this stage is defined as $E(d)$, which can be expressed as

$$E(d) = \int_0^d F(d)\mathrm{d}x \tag{1}$$

Building upon this, the mean crushing force within the interval from 0 to d is denoted as $\overline{F}(d)$, which can be expressed as follows:

$$\overline{F}(d) = \frac{E(d)}{d} \tag{2}$$

Meanwhile, T_E (d) represents the energy absorption efficiency of the structure, and its expression is given as follows:

$$T_E(d) = \frac{E(d)}{F_{max}(d)H_s} \tag{3}$$

When under compression, T_E (d) exhibits a trend where it initially increases with the increasing value of d and subsequently decreases [20]. During this progression, T_E (d) reaches a maximum value, corresponding to the peak energy absorption efficiency of the structure [20]. The compression displacement at which T_E (d) reaches its maximum value is defined as d_{max} [20]. The values of $\overline{F}(d)$ and $F_{max}(d)$ at this specific compression displacement d_{max} are subsequently designated as the ultimate mean crushing force $\overline{F}$ and maximum crushing force F_{max} of the structure. Expanding on this concept, the crushing force efficiency, labeled as A_E, is defined as $\overline{F}/F_{max}$. In this study, SEA (specific energy absorption) represents the energy absorbed per unit mass by the structure during the compression failure process, and its expression is as follows:

$$\text{SEA} = \frac{E(d_{\max})}{M_s} \tag{4}$$

where $E(d_{max})$ signifies the energy absorption of the structure at d_{max} and M_s represents the mass of the structure.

For an ideally energy-absorbing structure, the objective is to maximize energy absorption within a specified compression displacement while keeping the crushing force during compression within acceptable limits. Simultaneously, the structure should possess lightweight characteristics. Translating these prerequisites into energy absorption parameters, the goal is to attain higher values for $\overline{F}$, SAE, and A_E while striving for a lower value of F_{max}.

3. Finite Element Modeling

3.1. Descriptions of the Geometric Model

Critical geometric parameters of the FFCSCS are presented in Figure 2, including inner radius (R_i), outer radius (R_o), number of corrugations (N), thickness of the inner and outer face sheets (t_f), thickness of the corrugated core (t_c), width of the corrugated core (w), and height of the shell along z direction (H_s). In the present study, t_c = t_f = t is specifically emphasized. The mass of the FFCSCS can be determined as follows:

$$M_s = 2tH\rho_s[\pi(R_i + R_o) + Nw] + \rho_f H\left[\pi(R_o^2 - R_i^2) - Nwt\right] \tag{5}$$

where the mass of FFCSCSs is divided into two components: the first component represents the corrugated sandwich cylindrical shell's mass, and the second component represents the mass of the filled foam.

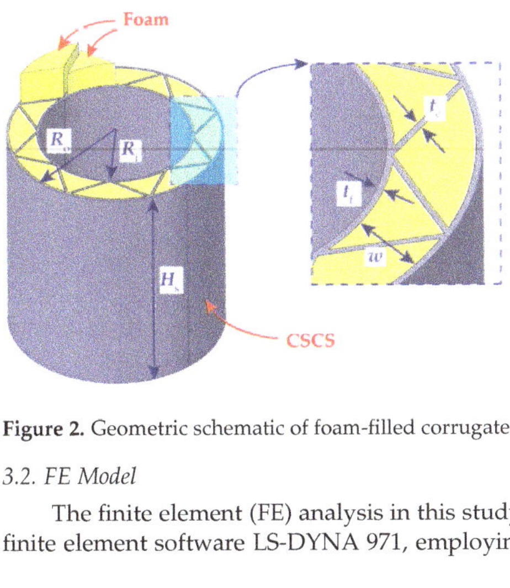

Figure 2. Geometric schematic of foam-filled corrugated sandwich cylindrical shells.

3.2. FE Model

The finite element (FE) analysis in this study was performed using the commercial finite element software LS-DYNA 971, employing its explicit algorithm. The FE model is given in Figure 3. Both the upper and lower plates were modeled as rigid bodies and simulated in LS-DYNA utilizing the *MAT_RIGID card. To impose appropriate boundary conditions, the lower plate was in its position, while the upper plate was constrained to have all degrees of freedom except for translational motion in the z direction. A displacement load was applied to the upper plate in the negative z-axis direction. The loading rate of 1 m/s was employed. At this rate, the kinetic energy within the structure represented less than 1% of the total energy, leading to an approximately quasi-static process.

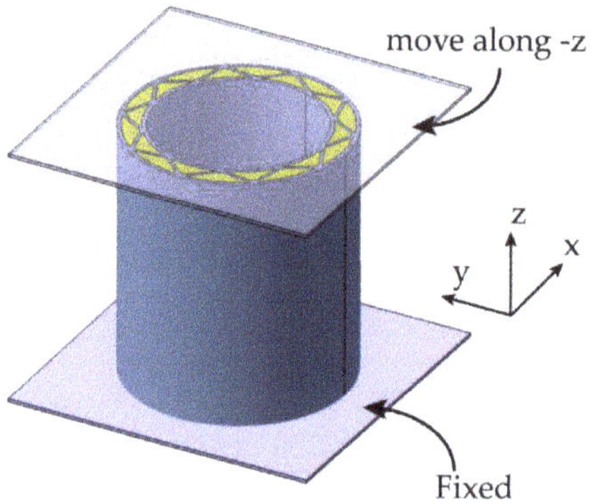

Figure 3. The finite element model of the FFCSCS under axial compression condition.

In the finite element model, the corrugated sandwich shell, aluminum foam, and rigid plates were meshed with SOLID164 solid elements. The corrugated sandwich shell was meshed with a minimum element size of 0.2 mm × 0.2 mm × 0.2 mm. This mesh size ensured that at least three layers of elements were present along the wall thickness direction of the shell, allowing for an appropriate resolution. For the aluminum foam, a mesh size of 0.25 mm × 0.25 mm × 0.25 mm was used. The upper and lower pressure plates, treated as rigid bodies, were meshed with a size of 0.3 mm × 0.3 mm × 0.3 mm. Mesh sensitivity study showed that further mesh refining did not yield improvements in the accuracy of the simulation results. Therefore, the current mesh size achieves a balance between computational accuracy and efficiency.

During the simulation, the contact interactions between the upper or lower plates and FFCSCS were defined by the *CONTACT_AUTOMATIC_SURFACE_TO_SURFACE card. To model the internal self-contact within the composite cylindrical sandwich shell structure, the *CONTACT_AUTOMATIC_SINGLE_SURFACE card was employed. To simulate the bonding relationship between foam and corrugated sandwich shell walls, the *CONTACT_AUTOMATIC_SURFACE_TO_SURFACE_TIEBREAK card was utilized. When significant normal and tangential stress occurred at the interface between the adhesive interfaces, this bonding relationship automatically degraded to *CONTACT_AUTOMATIC_SURFACE_TO_SURFACE according to the following degradation criterion [65]:

$$\left(\frac{|\sigma_n|}{\text{NFLS}}\right)^2 + \left(\frac{|\sigma_s|}{\text{SFLS}}\right)^2 \geq 1 \quad (6)$$

where σ_n and σ_s denote the normal and tangential stresses between the adhesive interfaces. NFLS and SFLS are the tensile and shear strengths of the bonding material. These strengths are determined based on the Loctite Hysol E-120HP two-component epoxy adhesive from Henkel, with measured values of 41 MPa for NFLS and 33 MPa for SFLS [66].

3.3. Material Properties

The materials considered for the face sheets and core in this study are 6063Al, 6061Al, and 304L stainless steel, respectively. The real stress–strain curves measured in experiments are sourced from references [54,67], depicted in Figure 4. The solid black lines in the figure represent the experimentally obtained data, with the corresponding material parameters denoted as follows: density (ρ_s), elastic modulus (E_s), yield strength ($\sigma_{0.2}$ or σ_y), ultimate strength (σ_u), and ultimate strain (ε_u). The hardening behavior of the materials is described by a power-hardening model represented by the equation $\sigma = \sigma_u(\varepsilon/\varepsilon_u)^n$, which is depicted by the red dashed line in Figure 4. The parameter n represents the power-law-hardening exponent, which is determined through fitting to the experimental data. σ_o represents the flow stress, considering the strain-hardening effects of the metal material. In the case of the power-law-hardening model, σ_o can be expressed as [24]

$$\sigma_o = \sqrt{\frac{\sigma_u \sigma_y}{1+n}} \tag{7}$$

The materials mentioned above were considered isotropic elastic–plastic solids with isotropic hardening in the finite element analysis. The Mises yield rule and J2 flow law were employed. The *MAT_PIECEWISE_LINEAR_PLASTICITY intrinsic model was utilized for the three materials in LS-DYNA, while the material's dynamic strengthening effect was not considered.

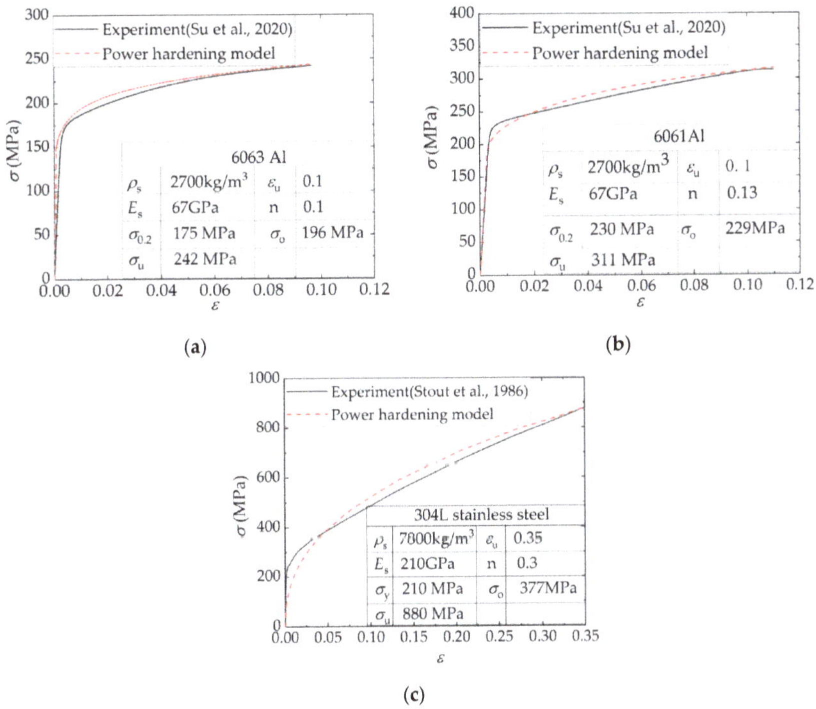

Figure 4. Material properties and corresponding power-hardening model of parent materials for corrugated cores and face sheets: (**a**) 6063 Al [54]; (**b**) 6061Al [54]; (c) 304L stainless steel [67].

This research investigated the effects of varying relative densities of aluminum foam on the coupled strengthening effect of FFCSCSs. To mitigate the errors arising from uncertainties in the aluminum foam processing process and substrate material, a theoretical model developed by Hanssen et al. [68] was used to derive the material parameters of foam material as follows:

$$\sigma_{\rm p} = 720\overline{\rho}_{\rm f}^{\,2.33} \tag{8}$$

$$E_{\rm p} = 330\overline{\rho}_{\rm f}^{\,2.45} \tag{9}$$

where $\overline{\rho}_{\rm f}$ represents the relative density of the foam, $\sigma_{\rm p}$ denotes the yield strength of the foam in MPa, and $E_{\rm p}$ refers to the modulus of elasticity of the foam in GPa.

The stress–strain curve of the foam after yielding was also obtained using the equation given by Hanssen et al. [68]:

$$\sigma = \sigma_{\rm p} + 42\overline{\rho}_{\rm f}^{\,1.42}\frac{e}{e_{\rm D}} + 251\overline{\rho}_{\rm f}\ln\left[\frac{1}{1-(e/e_{\rm D})^{\beta}}\right] \tag{10}$$

where e is the engineering strain of the foam, σ is the compressing stress of the foam, $\beta = 1/\left(0.1 + 15.7\overline{\rho}_{\rm f}^{\,3}\right)$, and $e_D = 1 - \overline{\rho}_{\rm f}$. This formula applies to the relative density of the foam ranging from 0.05 to 0.2.

Figure 5 displays the compressive stress–strain curves of seven relative density foams obtained using the equations mentioned above. The influence of these foams on the coupling enhancement effect of the FFCSCSs will be examined and discussed. In LS-DYNA, the foam material was defined using *MAT_CRUSHABLE_FOAM and identified as Material Type 63 in LS-DYNA. This material model requires the specification of mass density, Young's modulus, Poisson's ratio, and a load curve. The load curve encompasses both the plateau and densification stages that occur after the foam material reaches its yield point. It should be noted that in the finite element analysis conducted in this paper, the failure of the foam material, specifically in terms of fracture, was not taken into consideration.

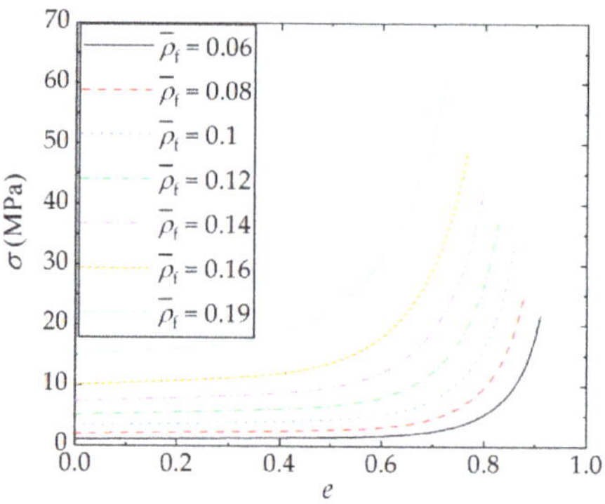

Figure 5. Material properties of the filled foams with different relative densities ($\overline{\rho}_{\rm f}$).

3.4. Validation against Experiments

In reference [53], preliminary axial compression tests were conducted on the FFCSCS. The experimental findings demonstrated a significant coupling enhancement effect of the FFCSCS when subjected to axial compression, in contrast to the independent foam and shell components. Figure 6 presented the comparison between the experimental and numerical results. In the experiment [53], the mean crushing force of the structure was 28.76 N, and the mean crushing force obtained through simulation in this work was 31.28 N. The excellent agreement between the two values indicates that the simulation approach employed in this study effectively characterizes the energy absorption characteristics of the structure. Furthermore, the final collapse mode captured from the FE results closely resembles the experimental photo, except for localized debonding observed in the experiments.

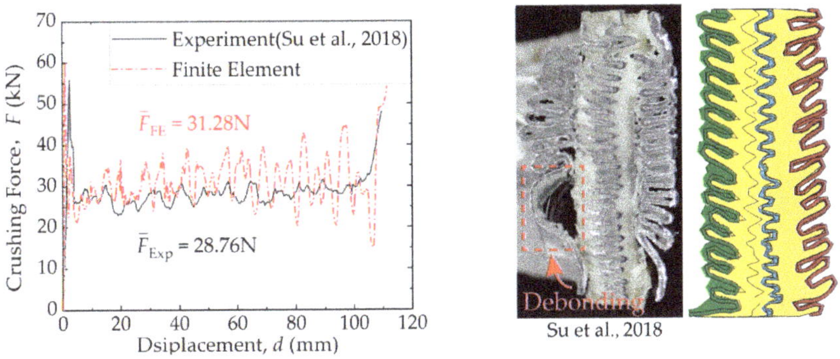

(a) (b)

Figure 6. Comparison between experiments [53] and FE results of PMI foam-filled 1060 Al sandwich cylindrical shell: (**a**) force–displacement curves; (**b**) final collapse mode.

4. Numerical Results

Based on FE simulations, a comprehensive study of the crushing behavior of the FFCSCS, encompassing the crushing process, the coupling enhancement effect, and its underlying mechanisms, is presented in this section.

4.1. Crushing Process

A representative structure is selected for a detailed analysis of the crushing process of FFCSCSs under axial compression. The shell material of this structure is 6063 aluminum, the wall thickness of the shell component is 0.8 mm, and the foam has a relative density of 0.1. This representative structure is labeled as 6063-08-01, signifying the shell material (6063), the shell-wall thickness (08), and the foam relative density (01).

Figure 7a depicts the crushing force–compression displacement (*F-d*) curve for the 6063-08-01 structure. Figure 7b presents the energy absorption–compression displacement curve (*E-d* curve). Figure 7c shows the deformation configurations corresponding to the peak and trough values of the *F-d* curve (Only 1/4 of the structure is shown to facilitate the observation of internal deformation).

Initially, the *F-d* curve exhibits a linear increase, indicating the structure is primarily in the elastic stage. Subsequently, a bifurcation point emerges on the *F-d* curve, signifying the transition from the linear–elastic stage to the nonlinear zone, where the crushing force *F* continues to increase. At point 2 on the *F-d* curve, the first fold in the structure begins to form and reaches its peak value before sharply declining. As compression progresses from point 2 to point 3, the folding area expands, resulting in a gradual decrease in *F*. From point 3 to point 4, mutual compression occurs within the first fold, causing the *F-d* curve to rise. As the new fold starts to form between point 4 and point 5, the *F-d* curve once again declines. Throughout the compression process, the *F-d* curve alternates between peak and valley values as folds form, expand, and extrude layer by layer. Upon reaching point 8, the curve enters the densification stage, exhibiting a rapid increase.

A clear trend can be observed that the energy absorption (E_{total}) increases linearly with the increase in compression displacement (*d*). The corrugated core exhibits the highest energy absorption capacity (E_{cc}), surpassing that of the outer face sheet (E_{OF}). At smaller *d* values, the inner face sheet absorbs slightly more energy (E_{IF}) compared to the foam (E_{Foam}). However, as *d* increases, E_{IF} gradually becomes equivalent to E_{Foam}.

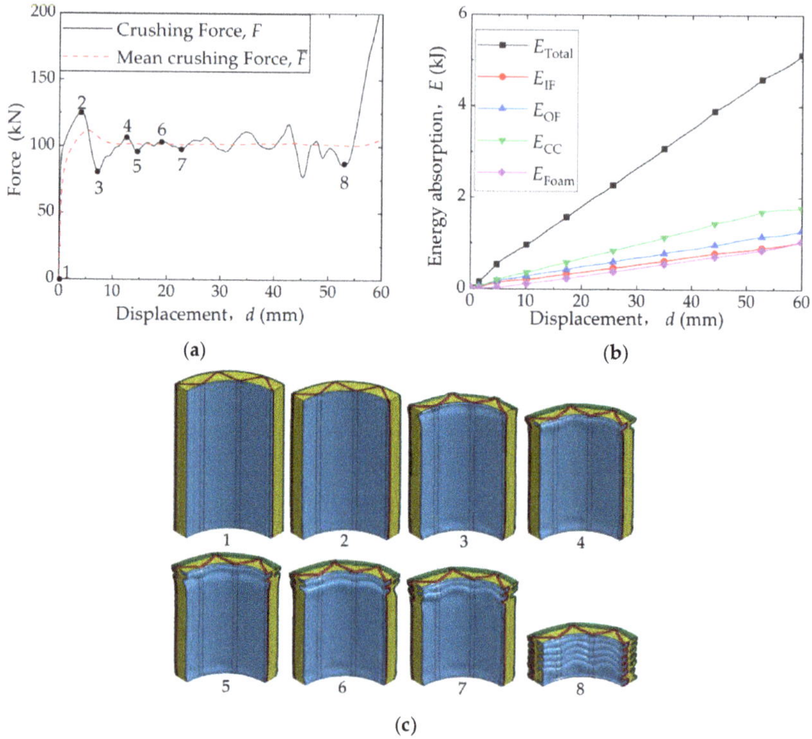

Figure 7. Crushing process and energy absorption of FFCSCS 6063-08-01: (**a**) force–displacement curves; (**b**) energy absorption–displacement curves; (**c**) collapse configurations with labels corresponding to those marked in the force–displacement curves.

4.2. Coupling Enhancement Effect

The coupling strengthening behavior and the strengthening mechanism of the FFCSCS 6063-08-01 are explored here. Additionally, separate simulation analyses are conducted for the two constituents of the FFCSCS: the corrugated core shell (CSCS) and the foam column (foam). This enables the analysis of the energy absorption contributions arising from the coupling strengthening effect. In this study, "foam + CSCS" represents the algebraic sum of the energy absorption characteristics of the two components when analyzed independently and does not represent an actual physical structure.

Figure 8a presents the crushing force–displacement (*F*-*d*) curves for 6063-08-01 and its individual constituents when subjected to independent compression. The *F*-*d* curve for the FFCSCS is represented by a solid black line, while those for the independently compressed corrugated sandwich cylindrical shell (CSCS) and foam column are depicted by a dashed blue line and a dotted green line, respectively. The curve for the "Foam + CSCS" combination is shown as a dashed red line. The shaded region between the solid black line and the red dashed line depicts the pronounced coupling strengthening effect observed between the components of the FFCSCS.

Based on the *F*-*d* curves depicted in Figure 8a, the mean crushing force $\overline{F}(d)$ for the FFCSCS and its individual components as a function of the compression displacement *d* is calculated and shown in Figure 8b. The shaded region in Figure 8b represents the coupling strengthening effect characterized by $\overline{F}(d)$, which exhibits a progressive increase with the increasing *d* until it reaches a stable state.

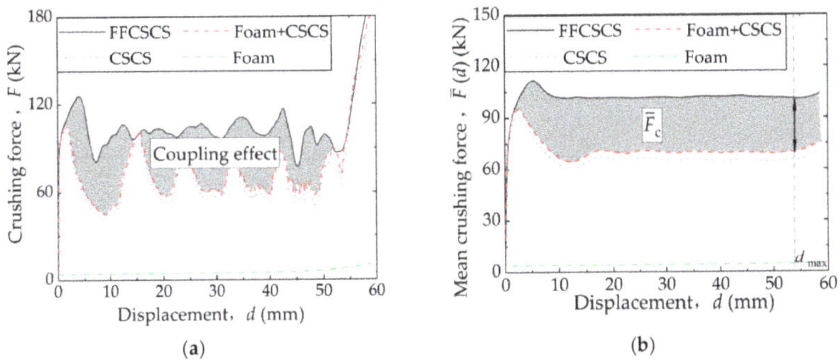

Figure 8. Coupling enhancement effect in FFCSCS 6063-08-01: (**a**) force–displacement; (**b**) mean crushing force–displacement curves.

The final mean crushing forces for the FFCSCS, foam, and CSCS are denoted as $\overline{F}$, $\overline{F}_f$, and $\overline{F}_s$, respectively. Therefore, the coupling strengthening effect is characterized by the coupling mean crushing force, expressed as $\overline{F}_c$, and is determined as follows:

$$\overline{F}_c = \overline{F} - \overline{F}_s - \overline{F}_f \tag{11}$$

Based on the definitions and compression curves mentioned above, several parameters are calculated for the CSCS, foam, and FFCSCS, including the mean crushing force $\overline{F}$, coupling mean crushing force $\overline{F}_c$, specific energy absorption SEA, and the crushing force efficiency A_E. The values for these parameters are presented in Table 1. It is evident that in the case of FFCSCS, $\overline{F}_c$ accounts for 31% of $\overline{F}$, indicating a significant coupling strengthening effect between the components. Furthermore, a comparison between the different configurations reveals that the SEA and A_E of the FFCSCS are considerably enhanced compared to CSCS and the foam.

Table 1. Comparison of energy absorption of FFCSCS 6063-08-01 and its individual components.

Structures	$\overline{F}$		$\overline{F}_c$		SEA (kJ/kg)	A_E
	Value (kN)	Percentage	Value (kN)	Percentage		
CSCS	65.7	65%	/	/	33.38	0.64
Foam	4.29	4%	/	/	24.79	/
FFCSCS	101.19	/	31.2	31%	42.79	0.8

4.3. Mechanism of Coupling Enhancement

In the preceding section, it is observed that a pronounced coupling strengthening effect exists among the components of FFCSCS 6063-08-01. To further reveal the coupling strengthening mechanism, the energy absorption properties and folding mode of FFCSCS, individual CSCS, and individual foam components are analyzed separately. It is worth mentioning that the individual foam component utilized in the analysis is in the form of a solid cylindrical structure, aligning its height with that of the FFCSCS. Its cross-sectional area encompasses the total area occupied by all foam sections within the corrugated channels.

Table 2 presents the energy absorbed by each component in the FFCSCS, the CSCS, and the foam. It is evident that in the FFCSCS, the corrugated core absorbs the highest amount of energy, followed by the outer face sheet, the inner face sheet, and the foam. Similarly, in the CSCS, the corrugated core absorbs the most energy, followed by the outer and inner face sheets. When comparing the energy absorption of the corresponding components in the FFCSCS and CSCS, it is notable that the corrugated core, outer face sheet, and inner face sheet of the FFCSCS exhibit increased energy absorption. Specifically, the inner face sheet shows a 26% increase, the outer face sheet shows a 14% increase, and the corrugated core shows a 5% increase. Moreover, when comparing the energy absorption of the foam in the FFCSCS with that of an equally massed individual foam column, it is evident that the filled foam in the FFCSCS exhibits a remarkable improvement. The energy absorption of the filled foam is enhanced by 283% compared to that of the individual foam column.

Table 2. Energy absorption of each component in FFCSCS, CSCS and the foam.

Component	FFCSCS		CSCS		Foam (kJ)	*E* Enhancement
	Value (kJ)	Percentage	Value (kJ)	Percentage		
Outer face	1.2	25%	1.056	30%	/	14%
Inner face	0.96	20%	0.76	22%	/	26%
Corrugation	1.76	36%	1.68	48%	/	5%
Foam	0.92	19%	/	/	0.24	283%

In Figure 9, the collapse mode of FFCSCS 6063-08-01 after compaction is depicted, with a quarter of the structure intercepted to facilitate observation of internal deformation. The collapse mode of FFCSCS 6063-08-01 exhibits an axisymmetric pattern, while independent local folds form along the corrugated core. Upon closer examination, it becomes apparent that shell folds intrude into the foam region during compression. This deformation mode allows for a more thorough foam compression, resulting in increased energy absorption compared to an individual foam column. Moreover, the foam alters the deformation modes of the folds in the corrugated core and face sheets. Figure 9 clearly illustrates the upward bending of the folds in the inner face sheet across all folding layers.

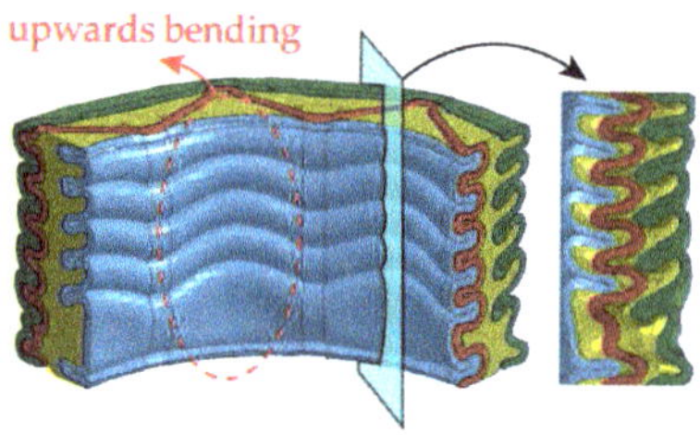

Figure 9. Collapse configuration of the FFCSCS 6063-08-01.

Table 3 provides an overview of the collapse configurations observed in the inner face sheet (IF), outer face sheet (OF), and corrugated core within both the FFCSCS and CSCS structures. In the initial row of the table, the collapse configurations of the IF within both structures are depicted, with the folds in each layer highlighted by a red line. It is evident that in the CSCS, the folds of the IF exhibit minimal deformation in the compression direction. In contrast, in the FFCSCS, all the folds are observed to bend upwards along the compression direction. Moving to the second row of Table 3, a comparative analysis of the collapse mode of the OF is presented.

Similarly, in the FFCSCS, the folds in the OF exhibit bending along the compression direction, whereas the folds in the CSCS display minimal deformation in this direction. The bending deformation of the folds in the FFCSCS results in increased plastic deformation of the material and enhanced interfolding compression. These two factors synergistically contribute to the improved energy absorption properties of the FFCSCS. Furthermore, the collapse mode of the corrugated core, as depicted in the third row of Table 3, exhibits a nearly identical behavior in both the CSCS and FFCSCS. In summary, the coupling effect is more pronounced for the inner and outer face sheets of the FFCSCS, whereas it is comparatively weaker for the corrugated core.

To elucidate the observed folding phenomenon in the inner and outer face sheets of the FFCSCS, Figure 10 provides a visual representation of the formation of the second layer of folds within the inner face sheet.

Table 3. Comparison of collapse configuration of each component in the FFCSCS and CSCS.

	CSCS	FFCSCS
IF		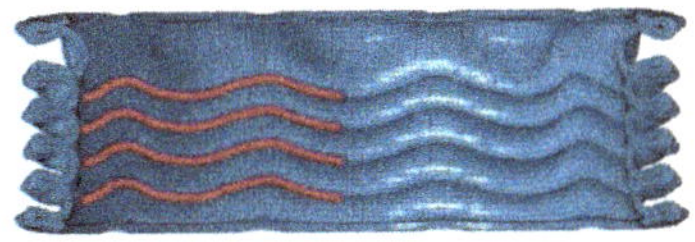
OF		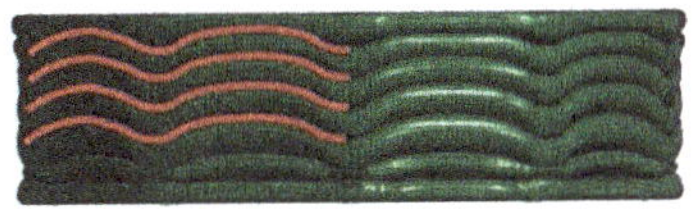
Core		

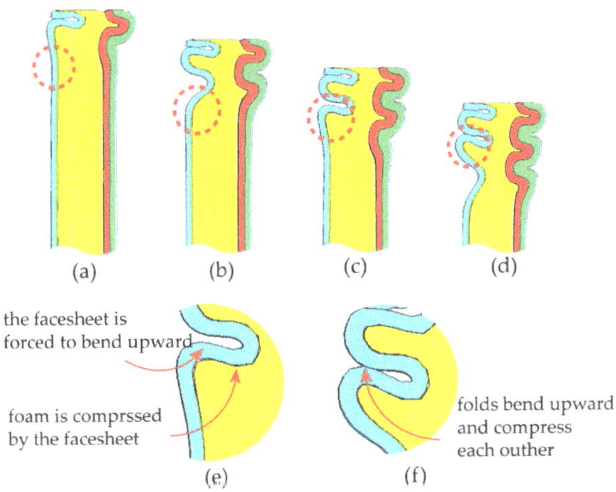

Figure 10. The formation process of folds (enclosed by the red dash circle) in FFCSCS (6063-08-01): (**a**) initial stage; (**b**) beginning of formation; (**c**) bending upwards; (**d**) compressing each other; (**e**) partially enlarged view of (**c**); (**f**) partially enlarged view of (**d**).

In Figure 10a, the initial stage of the crushing process is depicted, where the formation of the second layer of folds has not yet commenced. As depicted in Figure 10b, the initiation of the second layer of folds begins as the compression displacement increases. With further compression, as shown in Figure 10c, the second layer of folds becomes progressively compressed, resulting in their upward bending along the direction of compression. Moving to Figure 10d, it is evident that the second layer of folds is fully developed, and the folds bend upwards along the compression direction. Additionally, it is observed that the second layer of folds comes into contact with the first layer of folds, giving rise to mutual compression due to the bending deformation.

For a more detailed examination of this process, Figure 10e partially magnifies Figure 10c. It becomes apparent that the foam adjacent to the fold undergoes compression due to the pressure exerted by the face sheet, causing the foam material to flow upwards. This upwards flow of foam material subsequently forces the adjacent region of the face sheet to bend upwards. Similarly, Figure 10f partially enlarges Figure 10d, illustrating how the folds are forced to contact and compress each other. This collapse mode further enhances the energy absorption properties of the structure.

5. Discussion

The preceding analysis reveals the coupling strengthening effect and its underlying mechanisms. This section discusses the influence of the foam's relative density, the wall thickness, and the materials of the corrugated sandwich on the coupling strengthening effect.

5.1. Influence of Foam Density on the Coupling Effect

The influence of the relative density of the foam ($\overline{\rho}_f$) on the coupling enhancement effect of the FFCSCSs is investigated in this section. The relative density of the foam $\overline{\rho}_f$ varies within the range of 0.06 to 0.19 while maintaining a constant shell material (6063 Al) and a shell-wall thickness of 0.8 mm.

Figure 11 illustrates the crushing force–compression displacement curves (*F-d* curves) and compression mean crushing force–displacement curves ($\overline{F}$-*d* curves) for FFCSCSs with varying $\overline{\rho}_f$ ranging from 0.06 to 0.19. In Figure 11a, it is evident that the *F-d* curves for $\overline{\rho}_f$ ranging from 0.06 to 0.14 exhibit similar patterns. Initially, each curve reaches its peak value, followed by fluctuations around a stable value, and it finally undergoes a rapid rise due to compaction. Notably, both the peak and stable values in the *F-d* curves increase with an increasing value of $\overline{\rho}_f$. However, in Figure 11b, the shape of the *F-d* curves changes as $\overline{\rho}_f$ increases to 0.16 and 0.19. The curves display overall fluctuations without any distinct peak or stable values. Moving to Figure 11c, the $\overline{F}$-*d* curves for FFCSCSs with $\overline{\rho}_f$ values of 0.06 to 0.14 are presented. It is observed that as displacement (*d*) increases, the mean crushing force ($\overline{F}(d)$) also increases and gradually converges to the constant value ($\overline{F}$). Furthermore, it is evident that $\overline{F}$ increases with increasing $\overline{\rho}_f$. However, in Figure 11d, the $\overline{F}$-*d* curves exhibit a rising and falling pattern with an increase in *d* for structures with $\overline{\rho}_f$ values of 0.16 and 0.19, without converging to a constant value.

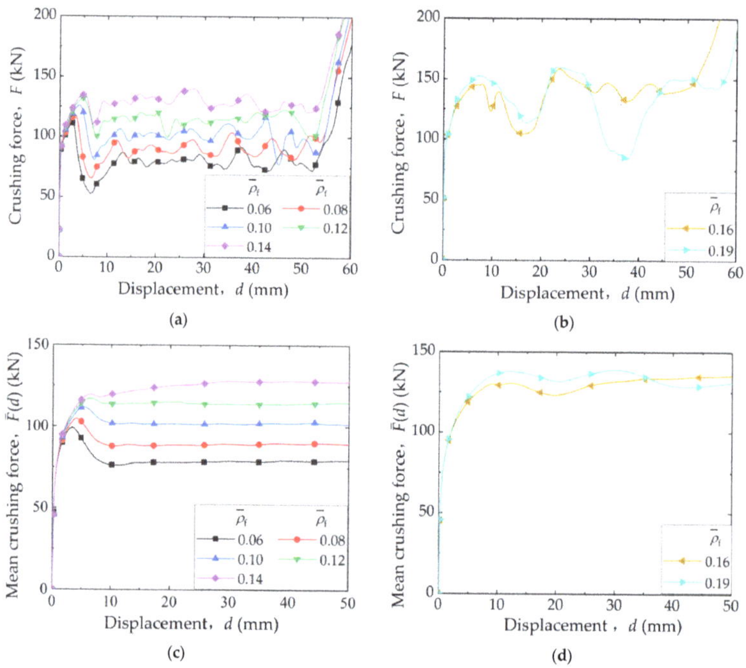

Figure 11. Crushing response of FFSCSCs with different relative foam density $\overline{\rho}_f$: (**a**) force–displacement curves for $\overline{\rho}_f$ = 0.06~0.14; (**b**) force–displacement curves for $\overline{\rho}_f$ = 0.16~0.19; (**c**) mean crushing force–displacement curves for $\overline{\rho}_f$ = 0.06~0.14; (**d**) mean crushing force–displacement curves for $\overline{\rho}_f$ = 0.16~0.19.

Figure 12 illustrates the collapse modes of FFCSCSs with $\overline{\rho}_f$ values ranging from 0.08 to 0.19. For $\overline{\rho}_f$ between 0.08 and 0.14, the structures exhibit the progressive folding mode, where the folds occur layer by layer along the compression direction. This phenomenon is depicted in Figure 11a, where the *F*-*d* curves show fluctuations, indicating the layer-by-layer formation of folds. However, as shown in Figure 12d, when $\overline{\rho}_f$ reaches 0.14, some of the folds in the FFCSCSs are not fully developed, and a tendency toward global deformation begins to emerge. As $\overline{\rho}_f$ increases to 0.16 and 0.19, the deformation mode of the structure transitions to a global folding mode, as demonstrated in Figure 12e,f. In this global folding mode, the *F*-*d* curves no longer exhibit fluctuations around a stable value, as observed in Figure 11b.

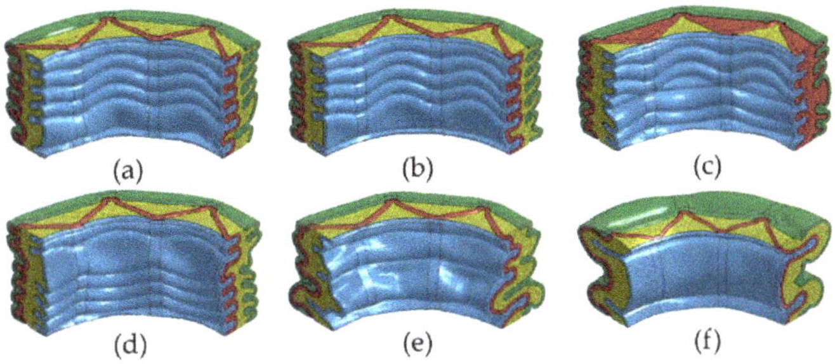

Figure 12. Collapse configuration of FFSCSCs with $\overline{\rho}_f$ values ranging from 0.08 to 0.19: (**a**) 0.08; (**b**) 0.10; (**c**) 0.12; (**d**) 0.14; (**e**) 0.16; (**f**) 0.19.

Table 4 presents the crushing performance of the FFCSCSs for $\overline{\rho}_f$ values ranging from 0 to 0.19, where $\overline{\rho}_f = 0$ represents the CSCS. It is evident that the FFCSCSs exhibit significantly higher mean crushing force ($\overline{F}$) and specific energy absorption (SEA) compared to the CSCS. For $\overline{\rho}_f \leq 0.16$, $\overline{F}$, $\overline{F}_c$, SEA and A_E of the FFCSCSs increase with an increasing value of $\overline{\rho}_f$. However, as $\overline{\rho}_f$ further increases to 0.19, $\overline{F}$, $\overline{F}_c$, SEA, and A_E decrease. This decline can be attributed to the global deformation, as depicted in Figure 12f, when $\overline{\rho}_f$ exceeds a certain threshold.

Table 4. Crushing properties of FFSCSCs with $\overline{\rho}_f = 0 \sim 0.019$.

$\overline{\rho}_f$	$\overline{F}$ (kN)	F_c (kN)	SEA (kJ/kg)	A_E
0	65.70	/	33.30	0.64
0.06	80.11	13.03	36.85	0.72
0.08	90.59	22.28	39.34	0.81
0.10	101.19	31.20	42.79	0.80
0.12	114.14	41.97	45.32	0.86
0.14	127.25	52.34	48.80	0.99
0.16	135.91	57.62	49.00	0.94
0.19	132.99	48.15	45.26	0.87

As indicated in Equation (2), the mean crushing force of the FFCSCS ($\overline{F}$) is composed of the mean crushing force of the CSCS ($\overline{F}_s$), the mean crushing force of the foam ($\overline{F}_f$), and the coupling mean crushing force ($\overline{F}_c$).

Figure 13a presents the absolute values of $\overline{F}_s$, $\overline{F}_f$, and $\overline{F}_c$, while their respective proportions in $\overline{F}$ are illustrated in Figure 13b. Throughout this section, the shell-wall thickness and material of the FFCSCSs remain constant, ensuring that $\overline{F}_s$ remains consistent for each FFCSCS. In Figure 13a, as $\overline{\rho}_f$ increases, both $\overline{F}_f$ and $\overline{F}_c$ initially increase, followed by a subsequent decrease for each structure. In Figure 13b, it can be observed that when $\overline{\rho}_f$ is equal to 0.6, $\overline{F}_s$ accounts for the highest proportion (82%), followed by $\overline{F}_c$ (16%), and $\overline{F}_f$ represents the lowest proportion (2%). As $\overline{\rho}_f$ increases, the proportion of $\overline{F}_s$ decreases, while the proportions of $\overline{F}_f$ and $\overline{F}_c$ increase. This observation indicates that the contribution of the foam itself and the coupling effects in energy absorption grow with increasing foam density. However, when $\overline{\rho}_f$ reaches 0.19, the structure undergoes global deformation, resulting in a reduction in the coupling effect and subsequently a decrease in the proportion of $\overline{F}_c$ in $\overline{F}$.

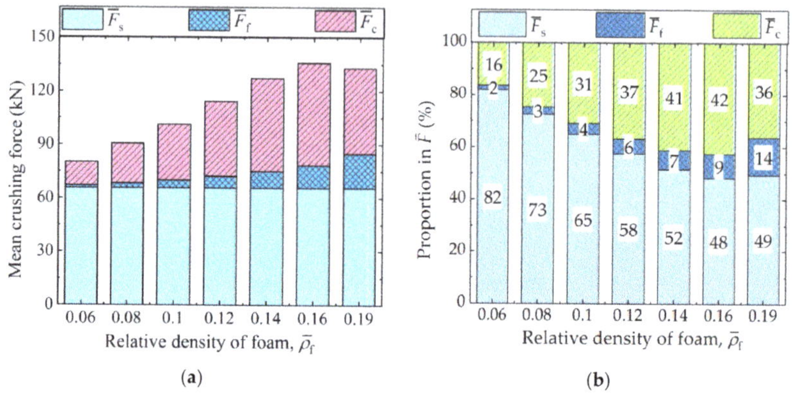

Figure 13. Composition of the mean crushing force ($\overline{F}$) for FFCSCSs with $\overline{\rho}_f$ = 0.06~0.19: (**a**) absolute value; (**b**) proportion in $\overline{F}$.

5.2. *Influence of Shell-Wall Thickness on the Coupling Effect*

This section examines the influence of the wall thickness (t) on the coupling strengthening effect in FFCSCSs. Three distinct wall thicknesses are considered: 0.6 mm, 0.8 mm, and 1.0 mm, respectively. For the structures discussed in this section, 6063 Al is employed for both the corrugated core and face sheets while maintaining the relative foam density within the range of 0.06 to 0.19.

Figure 14 illustrates the influence of shell-wall thickness on the crushing performance and coupling strengthening effect of FFCSCSs, considering a range of $\overline{\rho}_f$ values from 0 to 0.19. It is important to note that $\overline{\rho}_f = 0$ represents the CSCS structure. The results clearly demonstrate that the FFCSCSs exhibit higher values of $\overline{F}$, $\overline{F}_c$, A_E, and SEA compared to the CSCSs. For a given value of t, as $\overline{\rho}_f$ increases before global deformation occurs, there is a corresponding increase in $\overline{F}$, $\overline{F}_c$, A_E, and SEA. However, when the value of $\overline{\rho}_f$ exceeds a certain threshold, global deformation occurs, leading to a decrease in $\overline{F}$, $\overline{F}_c$, A_E, and SEA. Furthermore, for a given value of $\overline{\rho}_f$, prior to the occurrence of global deformation, higher values of t are associated with increased values of $\overline{F}$, $\overline{F}_c$, A_E, and SEA.

Moreover, as illustrated in Figure 14, the critical threshold of $\overline{\rho}_f$ at which FFCSCSs undergo global deformation varies depending on the values of t. A higher t value is associated with a lower critical threshold of $\overline{\rho}_f$. This finding indicates that structures with thicker walls are more prone to global deformation. Furthermore, once the structure undergoes global deformation, a higher value of t results in a more substantial decline in $\overline{F}$, $\overline{F}_c$, A_E, and SEA. To illustrate this, let us consider $\overline{F}_c$ as an example. When t values are set at 0.6 mm, 0.8 mm, and 1.0 mm, the corresponding reductions in $\overline{F}_c$ during global deformation are 8%, 15%, and 28%, respectively. Overall, FFCSCSs exhibit superior energy absorption capabilities compared to CSCSs. Notably, FFCSCSs with greater wall thicknesses demonstrate a pronounced coupling strengthening effect, resulting in higher energy absorption capacities.

Figure 15 illustrates the absolute values of $\overline{F}_s$, $\overline{F}_f$, and $\overline{F}_c$, as well as their respective proportions in $\overline{F}$, for various combinations of t and $\overline{\rho}_f$. In Figure 15a, when a specific value of $\overline{\rho}_f$ is considered, the bar charts represent $\overline{F}_s$, $\overline{F}_f$ and, $\overline{F}_c$ for FFCSCSs with different t values (0.6 mm, 0.8 mm, and 1.0 mm), arranged from left to right. It is evident that when a specific value of $\overline{\rho}_f$ is provided, $\overline{F}_s$ exhibits an increasing trend as t increases. In the case of $\overline{F}_f$, as t increases, the foam-filled area within the corrugated channel decreases, resulting in a reduction in $\overline{F}_f$. However, within the discussed range, the differences in t values are relatively small, resulting in less noticeable variations in $\overline{F}_f$ for different t values. For $\overline{F}_c$, before the global deformation occurs, a larger t corresponds to a larger $\overline{F}_c$ for the same $\overline{\rho}_f$. Additionally, when t is held constant, an increase in $\overline{\rho}_f$ results in no change in $\overline{F}_s$, while $\overline{F}_f$ and $\overline{F}_c$ increase. In Figure 15b, for a specific value of $\overline{\rho}_f$, the bar charts, from left to right, represent the proportions of $\overline{F}_s$, $\overline{F}_f$, and $\overline{F}_c$ in $\overline{F}$ for t values of 0.6 mm, 0.8 mm, and 1.0 mm, respectively. It is evident that when a specific value of $\overline{\rho}_f$ is given, an increase in t results in a higher proportion of $\overline{F}_s$, while the proportions of $\overline{F}_f$ and $\overline{F}_c$ decrease. Conversely, when t is held constant, before global deformation occurs, the proportion of $\overline{F}_s$ decreases, and the proportion of $\overline{F}_f$ and $\overline{F}_c$ increases as $\overline{\rho}_f$ increases.

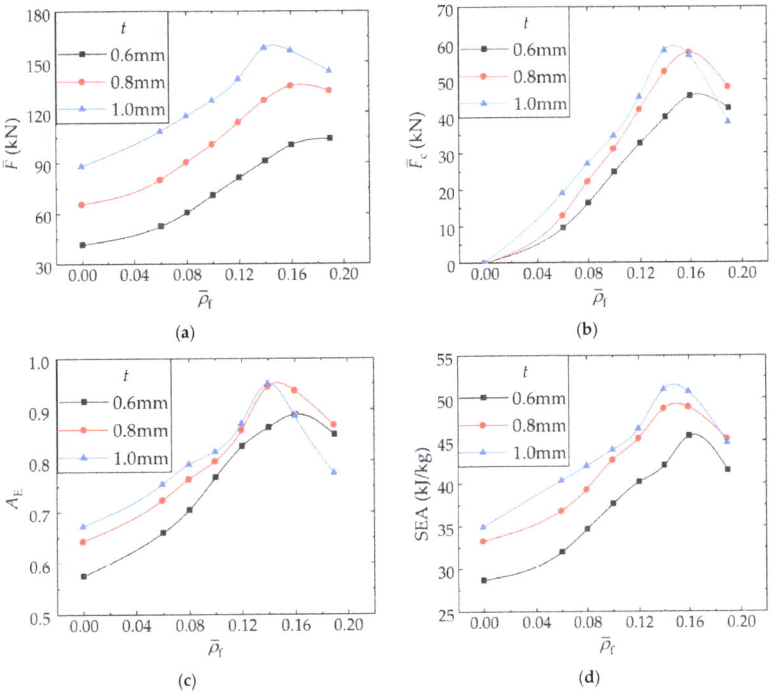

Figure 14. Influence of wall thickness and relative density of foam on the crushing properties of FFCSCSs: (**a**) mean crushing force, $\overline{F}$ (**b**) coupling mean crushing force, $\overline{F}_c$; (**c**) crushing force efficiency, A_E; (**d**) specific energy absorption, SEA.

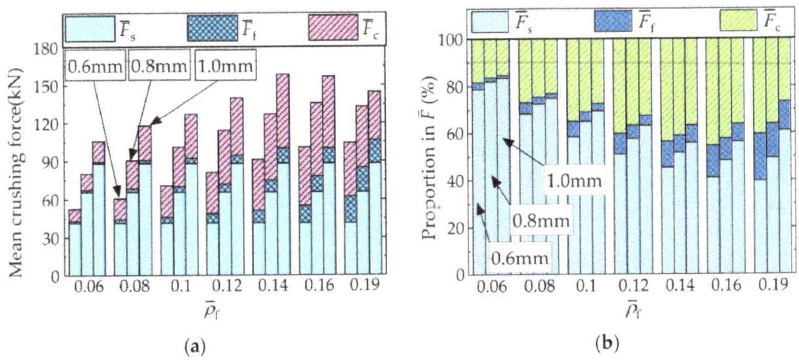

Figure 15. Composition of the mean crushing force for FFSCSCs with different wall thicknesses and relative densities of foam: (**a**) absolute value; (**b**) proportion in $\overline{F}$.

It is observed that, prior to the occurrence of global deformation, higher relative foam density and greater shell-wall thickness contribute to a strengthened coupling effect among the structural components. Conversely, greater relative foam density and smaller shell-wall thickness result in an increased proportion of the coupling strengthening effect in structural energy dissipation.

5.3. Influence of Shell Material on the Coupling Effect

This section investigates the influence of shell materials in FFCSCSs on the coupling strengthening effect. Three different materials are considered: 6063 Al, 6061 Al, and 304L stainless steel. Among these materials, there is a gradual increase in both yield stress and flow stress, progressing from 6063 Al to 6061 Al and finally to 304L stainless steel. In the considered structure, the corrugated core and face sheet wall thickness t is fixed at 0.8 mm, while the relative density of the foam $\overline{\rho}_f$ ranges from 0.06 to 0.19.

Figure 16 illustrates the impact of shell material on the crushing performance and coupling strengthening effect of FFCSCSs. It is observed that all FFCSCSs exhibit higher values for $\overline{F}$, $\overline{F}_c$, A_E, and SEA compared to the CSCSs. For each material, as the relative density $\overline{\rho}_f$ increases up to 0.16, $\overline{F}$, $\overline{F}_c$, A_E, and SEA increase accordingly. However, when $\overline{\rho}_f$ exceeds 0.16, the structure experiences global deformation, resulting in a decrease in energy absorption performance and a subsequent decline in $\overline{F}$, $\overline{F}_c$, A_E, and SEA.

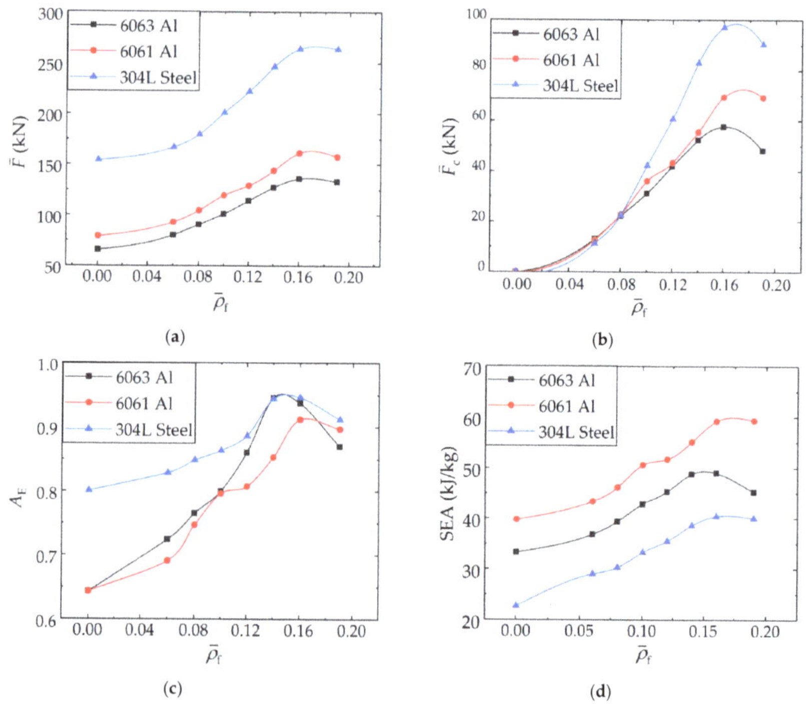

Figure 16. Influence of wall material and relative density of foam on the crushing properties of FFCSCs: (**a**) mean crushing force, $\overline{F}$; (**b**) coupling mean crushing force, $\overline{F}$; (**c**) crushing force efficiency, A_E; (**d**) specific energy absorption, SEA.

Furthermore, for a given $\overline{\rho}_f$, the performance of the structure is significantly influenced by the strength of the shell material, with higher strength materials exhibiting greater values for $\overline{F}$, $\overline{F}_c$, and A_E. However, in Figure 16d, it is observed that for the same $\overline{\rho}_f$, the SEA of the 6061 Al structure is the highest, followed by the 6063 Al, while the 304L stainless steel exhibited the lowest SEA. The reason for this phenomenon can be analyzed as follows.

Referring to Figure 4, although the flow stress of 304L stainless steel is 1.9 times that of the 6063 Al, its density is 2.9 times that of the 6063 Al as well, resulting in a lower SEA. On the other hand, both 6063 Al and 6061 Al have the same density, but the 6061 Al has higher yield stress and flow stress compared to the 6063 Al, leading to a higher SEA for the 6061 Al structures. Overall, the FFCSCSs consistently demonstrate superior energy absorption performance compared to the CSCSs for all materials considered. The coupling strengthening effect and mean crushing force of the FFCSCSs are strengthened with higher flow stress in the shell material. The specific energy absorption of the structure is influenced by both the flow stress and the density of the base material.

Figure 17 provides an analysis of the absolute values of $\overline{F}_s$, $\overline{F}_f$, and $\overline{F}_c$ with different shell materials and $\overline{\rho}_f$, along with their respective proportions in $\overline{F}$. In Figure 17a, for a specific value of $\overline{\rho}_f$, the bar charts depict $\overline{F}_s$, $\overline{F}_f$, and $\overline{F}_c$ for the FFCSCSs with different shell materials (6063 Al, 6061 Al, and 304L stainless steel), arranged in ascending order of material flow stress from left to right. It is observed that when a specific $\overline{\rho}_f$ value is assigned, both $\overline{F}_s$ and $\overline{F}_c$ increase with an increase in material flow stress, while $\overline{F}_f$ remains constant. Similarly, for a given material, as $\overline{\rho}_f$ increases before global deformation occurs, $\overline{F}_s$ remains constant, while both $\overline{F}_f$ and $\overline{F}_c$ increase. In Figure 17b, when $\overline{\rho}_f$ is assigned, the bar charts from left to right represent the proportion of $\overline{F}_s$, $\overline{F}_f$, and $\overline{F}_c$ in $\overline{F}$ for 6063 Al, 6061 Al, and 304L stainless steel, respectively. With a constant $\overline{\rho}_f$ value, an increase in material flow stress results in a higher proportion of $\overline{F}_s$, accompanied by a lower proportion of $\overline{F}_c$ and $\overline{F}_f$. Likewise, when a specific material is given, before the structure undergoes global deformation, the proportion of $\overline{F}_s$ decreases with an increasing value of $\overline{\rho}_f$, while the proportions of $\overline{F}_f$ and $\overline{F}_c$ increase.

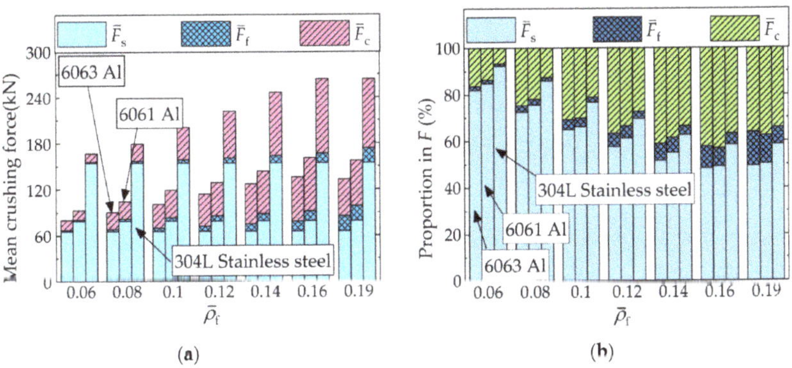

Figure 17. Composition of the mean crushing force for FFSCSCs with different wall materials and relative densities of foam: (**a**) absolute value; (**b**) proportion in $\overline{F}$.

Consequently, an increase in foam relative density and shell material strength results in a stronger coupling strengthening effect among the components of FFCSCSs. Conversely, higher foam relative density and weaker shell material lead to a greater proportion of the coupling strengthening effect in energy absorption.

6. Theoretical Analysis

Based on the findings above, it is evident that the coupling strengthening effect increases with higher foam density, greater wall thickness of the shells, and higher flow stress of the shell material. In this section, a theoretical model is derived for predicting the mean crushing force of the FFCSCSs. The development of this model builds upon our previous work [54] for predicting the mean crushing force of CSCSs and incorporates insights regarding the coupling effect of foam-filled square tube structures [69].

6.1. Theoretical Model

According to Equation (2), the mean crushing force ($\overline{F}$) of the FFCSCS is determined as the sum of the mean crushing forces of the shell ($\overline{F}_s$) and foam core ($\overline{F}_f$) and the coupling contribution ($\overline{F}_c$). The calculation of $\overline{F}_s$ is based on our former theoretical model, the detailed solution process of which can be referred to in Ref. [54]. During the solution process, the energy absorption of each folded cell within a folding cycle of $2H$ is cumulatively calculated. Based on the principle of energy balance (which ensures that the work performed by external forces is equal to the internal energy dissipation), the expression for $\overline{F}_s$ is derived as follows:

$$\overline{F}_s = W_{total}/2H\xi \tag{12}$$

where W_{total} is determined as the function in terms of H and b. Here, b refers to the radius of the toroidal surface in the super folding elements (not shown for brevity), while H represents the half-length of the fold. ξ denotes the effective crush distance coefficient.

The actual crushing mode of FFCSCSs should minimize the mean crushing force [5]. Therefore, it is crucial to ensure that

$$\frac{\partial \overline{F}_s}{\partial H} = 0 \quad \frac{\partial \overline{F}_s}{\partial b} = 0 \tag{13}$$

After solving the aforementioned equation, the resulting values of H and b are then used in Equation (8) to calculate $\overline{F}_s$.

The foam mean crushing force $\overline{F}_f$ can be calculated from [69]:

$$\overline{F}_f = \sigma_f S_{foam} \tag{14}$$

where σ_f is defined as the plateau stress of foam when compressed to 50%, and S_{foam} represents the cross-sectional area of the foam perpendicular to the compression direction.

To compute σ_f for foams with a range of $\overline{\rho}_f$ values spanning from 0.05 to 0.2, Equation (5) can be applied as follows:

$$\sigma_f = 2\int_0^{0.5}\left\{\sigma_p + 42\overline{\rho}_f{}^{1.42}\frac{e}{e_D} + 251\overline{\rho}_f \ln\left[\frac{1}{1-(e/e_D)^{\beta}}\right]\right\}de \tag{15}$$

According to reference [69], the general expression of the coupling mean crushing force $\overline{F}_c$ is provided as follows:

$$\overline{F}_c = NC_{avg}\sigma_f^{\alpha}\sigma_o^{(1-\alpha)}w^{\beta}t^{(2-\beta)} \tag{16}$$

where N represents the number of corrugated cells. C_{avg}, α, and β are dimensionless parameters that describe the coupling strengthening effect. The equation involves the plateau stress of foam σ_f, the flow stress of the shell material σ_o, the width of the corrugated core w, and the thickness of shell walls t.

Thus, the mean crushing force $\overline{F}$ can be expressed as follows:

$$\overline{F} = \overline{F}_s + \sigma_f S_{foam} + NC_{avg}\sigma_f^{\alpha}\sigma_o^{(1-\alpha)}w^{\beta}t^{(2-\beta)} \tag{17}$$

The first two terms in Equation (13) can be computed directly from the geometric and material parameters of FFCSCSs. However, the third term, representing the coupling mean crushing force, depends on three dimensionless parameters: C_{avg}, α, and β. In reference [69], these dimensionless parameters were obtained through fitting the experimental data. In this study, a similar fitting approach is employed to ascertain the values of C_{avg}, α, and β. This is achieved using MATLAB's built-in multiple nonlinear regression function, "nlinfit", which is based on numerical simulations. The goodness of fit is assessed using the coefficient of determination R^2 between the theoretical predictions and simulated results. A higher R^2

value, closer to 1, indicates a more reliable and higher-quality fit. The expression for R^2 is provided as follows:

$$R^2 = 1 - \frac{\sum_{i=1}^{M} (y_i - \hat{y}_i)^2}{\sum_{i=1}^{M} (y_i - \overline{y}_i)^2} \tag{18}$$

where y_i represents the simulated results, $\overline{y}_i$ denotes the mean value of simulations, $\hat{y}_i$ represents the theoretical predictions, and M corresponds to the number of fitted samples.

6.2. Comparison with Simulated Results

The finite element analysis presented in Section 4 shows that the structure undergoes global deformation when $\overline{\rho}_f$ exceeds a certain threshold value. Consequently, the theoretical model mentioned above is found to be inapplicable in such cases. Therefore, in developing the theoretical model, only FFCSCSs with $\overline{\rho}_f$ values ranging from 0.06 to 0.14 were considered. Based on the simulated results, through multivariate nonlinear regression analysis, the values of C_{avg}, α, and β were determined as 6.6051, 0.6796, and 1.3236, respectively. The predicted values were calculated by substituting these values into Equations (13) and (14).

Figure 18 displays both the simulated and theoretical results for the mean crushing force $\overline{F}$ and the coupling mean crushing force $\overline{F}_c$. The x axis represents the simulated results, and the y axis represents the theoretical predictions. The solid black line (45° diagonal line) represents perfect agreement between the theoretical and simulated results, while the gray dashed line represents an error margin of ±20% between the simulated and theoretical results. In Figure 18a, the theoretical results and corresponding simulated results for $\overline{F}_c$ are shown. All data points are distributed on both sides of the 45° diagonal line, indicating strong agreement between the theoretical and simulated results. The coefficient of determination R^2, obtained using Equation (15), is 0.956, affirming the reliability of the obtained values of C_{avg}, α, and β. Figure 18b displays the theoretical and corresponding simulated results for $\overline{F}$. The agreement between the simulated and theoretical results of $\overline{F}$ is higher compared to $\overline{F}_c$, with data points more closely aligned to the 45° diagonal line. Additionally, the R^2 value for the predicted value of $\overline{F}$ is 0.965. Consequently, the proposed theoretical model for predicting $\overline{F}$ and $\overline{F}_c$ within the discussed range of $\overline{\rho}_f$ values is deemed reliable.

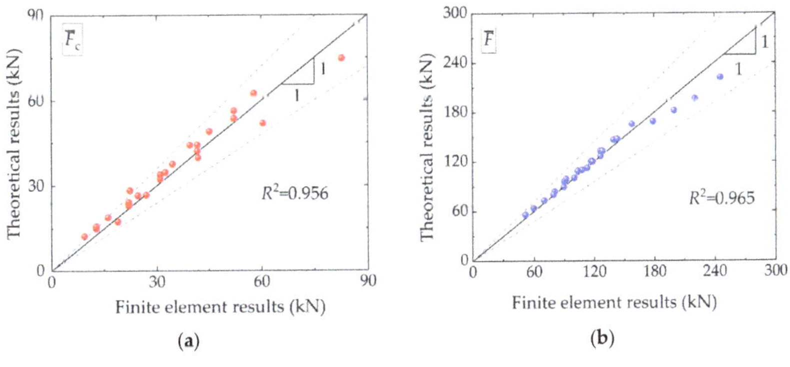

Figure 18. Comparison of theoretical predicated and finite element results: (**a**) coupling mean crushing force, $\overline{F}_c$; (**b**) mean crushing force, $\overline{F}$.

6.3. Parametric Studies

In Section 4, the influence of foam relative density $\overline{\rho}_f$ on the coupling strengthening effect was discussed using the finite element method. However, due to computational limitations and the complexity of numerical models, the compared structures did not adhere to the principle of equal mass, and the interval of $\overline{\rho}_f$ is relatively large (0.02). In this section, the theoretical model is employed to investigate the influence of $\overline{\rho}_f$ on $\overline{F}$, $\overline{F}_s$, $\overline{F}_f$, and $\overline{F}_c$ while adhering to the principle of equal mass. To achieve this, the mass of FFCSCSs is equated with that of the CSCS with a wall thickness of t = 1 mm. As $\overline{\rho}_f$ increases, the mass of FFCSCSs remains constant by reducing the value of t. In this section, $\overline{\rho}_f$ varies within the range of 0.6 to 0.14, with a finer interval of 0.002, allowing for a more precise analysis of the influence of $\overline{\rho}_f$ on the mentioned parameters.

Figure 19 presents the variation characteristics of $\overline{F}$, $\overline{F}_s$, $\overline{F}_f$, and $\overline{F}_c$ in different FFCSCSs with equal mass, as a function of $\overline{\rho}_f$. The base materials considered here are 1060 Al, 6063 Al, 6061 Al, and 304L stainless steel. The flow stress for 1060 Al is set to 140 MPa, while the values for the other three materials can be found in Section 3.3. Figure 19 shows that for a given shell material, increasing $\overline{\rho}_f$ necessitates a reduction in t to maintain the same structural mass. Throughout this process, $\overline{F}$, $\overline{F}_f$, and $\overline{F}_c$ show an increase, while $\overline{F}_s$ experiences a decrease. Among the FFCSCSs made of 6063 Al, 6061 Al, and 304L stainless steel, the mean crushing force can be arranged in descending order as $\overline{F}_s$, $\overline{F}_c$, and $\overline{F}_f$. Additionally, the difference between $\overline{F}_s$ and $\overline{F}_s$ decreases with an increasing $\overline{\rho}_f$. In the case of the FFCSCSs made of 1060 Al, when $\overline{\rho}_f$ is less than 0.13, $\overline{F}_s$ is greater than both $\overline{F}_c$ and $\overline{F}_f$. However, when $\overline{\rho}_f$ exceeds 0.13, $\overline{F}_s$ becomes smaller than $\overline{F}_c$. In general, based on equal mass, $\overline{F}$ increases with $\overline{\rho}_f$, corresponding to the increased energy absorption. Simultaneously, $\overline{F}_s$ decreases, while $\overline{F}_f$ and $\overline{F}_c$ increase.

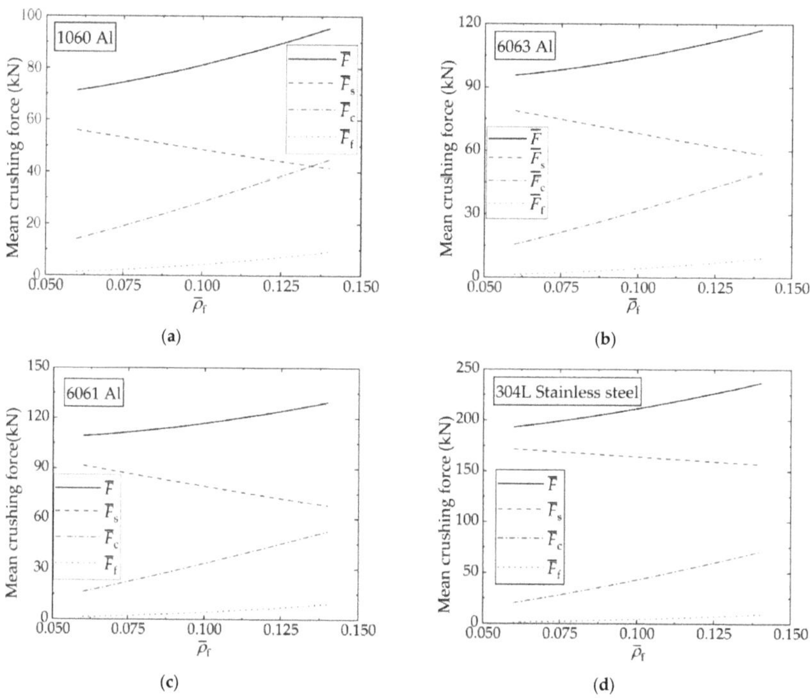

Figure 19. Theoretical predictions of mean crushing force of FFCSCSs with equal mass: (**a**) 1060 Al face sheets with different $\overline{\rho}_f$; (**b**) 6063 Al face sheets with different $\overline{\rho}_f$; (**c**) 6061 Al face sheets with different $\overline{\rho}_f$; (**d**) 304L stainless steel face sheets with different $\overline{\rho}_f$.

Figure 20 presents the variation characteristics of the proportion of $\overline{F}_s$, $\overline{F}_f$, and $\overline{F}_c$ in $\overline{F}$ under equal mass conditions as a function of $\overline{\rho}_f$. Figure 20a shows the proportional contribution of $\overline{F}_s$ in $\overline{F}$. For a given material, the proportion of $\overline{F}_s$ decreases as $\overline{\rho}_f$ increases. Conversely, for a given $\overline{\rho}_f$, higher material flow stress results in a more significant proportion of $\overline{F}_s$ in $\overline{F}$. Figure 20b displays the proportional contribution of $\overline{F}_f$ in $\overline{F}$. It can be observed that for a given material, the proportion of $\overline{F}_f$ increases with increasing $\overline{\rho}_f$, while for a given $\overline{\rho}_f$, higher material flow stress leads to a lower proportion of $\overline{F}_f$. Figure 20c demonstrates the proportional contribution of $\overline{F}_c$ in $\overline{F}$. It can be seen that for a given material, the proportion of $\overline{F}_c$ increases as $\overline{\rho}_f$ increases, whereas for a given $\overline{\rho}_f$, higher material flow stress results in a lower proportion of $\overline{F}_c$ in $\overline{F}$.

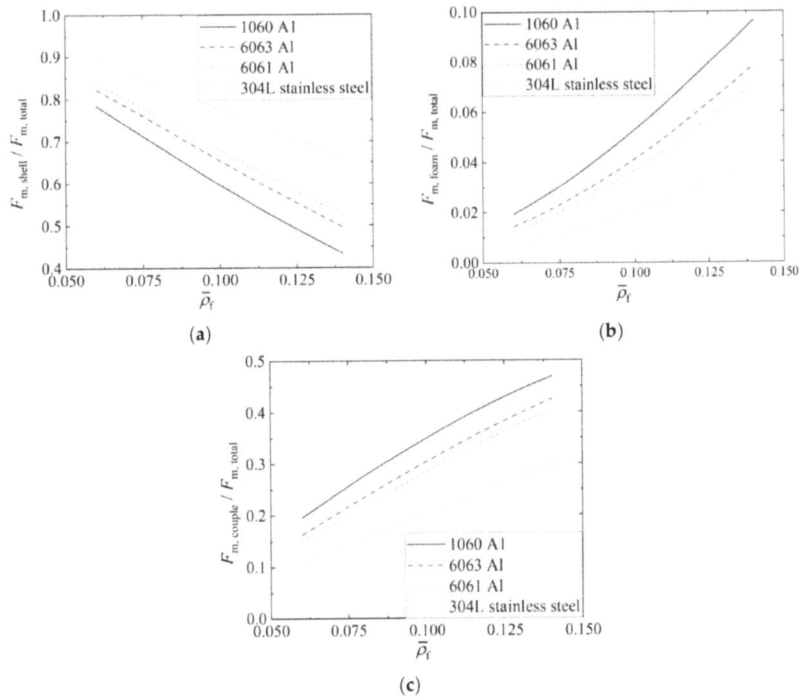

Figure 20. Proportion of mean crushing force of FFSCSCs: (**a**) $\overline{F}_s$; (**b**) $\overline{F}_f$; (**c**) $\overline{F}_c$.

In conclusion, based on equal mass conditions, a higher $\overline{\rho}_f$ contributes to greater energy absorption in FFCSCSs. This contribution can be attributed to both the foam itself and the coupling strengthening effect. When $\overline{\rho}_f$ is held constant, a higher shell flow stress leads to a larger energy absorption, with a significant contribution from the shell itself but a smaller contribution from the coupling effect and foam.

7. Conclusions

The concept of material hybrid design was introduced to incorporate aluminum foam into the corrugated channels of the corrugated core sandwich cylindrical shell, thereby creating a novel foam-filled corrugated sandwich cylindrical shell (FFCSCSs). The energy absorption characteristics of FFCSCSs were systematically investigated through a combination of simulations and theoretical analysis. The main conclusions are summarized as follows:

1. The FFCSCS demonstrates significantly enhanced energy absorption performance under axial compression, primarily due to the foam filling, resulting in maximum specific energy absorption of 60 kJ/kg. Furthermore, the coupling strengthening effect is notably pronounced, as evidenced by the maximum value of $\overline{F}_c/\overline{F}$, which reaches up to 40%.
2. The coupling strengthening effect is primarily observed in two aspects. Firstly, the intrusion of folds into the foam leads to a more comprehensive compression of the foam insertions. Secondly, influenced by foam insertions, the folds bend along the

compression direction and compress against each other, thereby expanding the plastic deformation zone.
3. In FFCSCSs, as the foam relative density, shell-wall thickness, and material flow stress increase, the coupling strengthening effect among the components strengthens, resulting in improved energy absorption performance, enhanced crushing efficiency, and increased mean crushing force.
4. The theoretical predictions strongly agree with the results of the finite element simulations. A parametric analysis based on the theoretical model shows that an increase in foam density leads to an increase in $\overline{F}$. Simultaneously, the proportion of $\overline{F}_s$ decreases, while the proportions of $\overline{F}_f$ and $\overline{F}_c$ increase.

Author Contributions: Methodology, P.S.; Software, B.H.; Validation, Y.W.; Investigation, P.S.; Data curation, P.S.; Writing—original draft, P.S.; Writing—review & editing, B.H., H.W., B.G. and T.J.L.; Supervision, B.H. and T.J.L.; Project administration, B.H. and T.J.L. All authors have read and agreed to the published version of the manuscript.

Funding: This work was supported by the National Key R&D Program of China (2022YFB4601804, 2022YFB4603103).

Institutional Review Board Statement: Not applicable.

Informed Consent Statement: Not applicable.

Data Availability Statement: The data presented in this study are available on request from the references and the corresponding author.

Conflicts of Interest: The authors declare no conflict of interest.

Nomenclature

H_s	initial height of cylindrical shell
R_o	outer radius of the sandwich cylindrical shell
R_i	inter radius of the sandwich cylindrical shell
t	thickness of the face sheets and corrugated core
w	width of the corrugated cell
N	number of the corrugated core
ρ_f	density of the filling foam
$\overline{\rho}_f$	relative density of the filling foam
σ_0	flow stress of the metal material
e	engineering strain of the foam
e_D	engineering compaction strain of the foam
σ_p	yield strength of the foam
E_p	modulus of elasticity of the foam
d	compressing displacement
F	crushing force
F_{max}	maximal value of F during compression
$\overline{F}$	mean crushing force of the entire structure
$\overline{F}_s$	mean crushing force of the corrugated sandwich cylindrical shell
$\overline{F}_f$	mean crushing force of the filling foam
$\overline{F}_c$	mean crushing force sourced from the coupling effect
E	energy absorption by the structure
A_E	crushing force efficiency
T_E	energy absorption efficiency
SEA	specific energy absorption
H	half-length of the fold
b	radius of toroidal surface in the super folding elements
ξ	effective crush distance coefficient
C_{avg}, α and β	dimensionless parameters which described the coupling enhancement effect

References

1. Lu, G.; Yu, T. *Energy Absorption of Structures and Materials*; Woodhead Publishing Ltd.: Cambridge, UK, 2003.
2. Jensen, Ø. Behaviour of Aluminium Extrusions Subjected to Axial Loading. Ph.D. Thesis, Norwegian University of Science and Technology, Trondheim, Norway, 2005.
3. Abramowicz, W.; Jones, N. Transition from initial global bending to progressive buckling of tubes loaded statically and dynamically. *Int. J. Impact Eng.* **1997**, *19*, 415–437. [CrossRef]
4. Guillow, S.R.; Lu, G.; Grzebieta, R.H. Quasi-static axial compression of thin-walled circular aluminium tubes. *Int. J. Mech. Sci.* **2001**, *43*, 2103–2123. [CrossRef]
5. Alexander, J.M. An approximate analysis of the collapse of thin cylindrical shells under axial loading. *Q. J. Mech. Appl. Math.* **1960**, *13*, 10–15. [CrossRef]
6. Abramowicz, W.; Jones, N. Dynamic progressive buckling of circular and square tubes. *Int. J. Impact Eng.* **1986**, *4*, 243–270. [CrossRef]
7. Abramowicz, W.; Jones, N. Dynamic axial crushing of square tubes. *Int. J. Impact Eng.* **1984**, *2*, 179–208. [CrossRef]
8. Abramowicz, W.; Wierzbicki, T. Axial Crushing of Multicorner Sheet Metal Columns. *J. Appl. Mech.* **1989**, *56*, 113–120. [CrossRef]
9. Hong, W.; Jin, F.; Zhou, J.; Xia, Z.; Xu, Y.; Yang, L.; Zheng, Q.; Fan, H. Quasi-static axial compression of triangular steel tubes. *Thin-Walled Struct.* **2013**, *62*, 10–17. [CrossRef]
10. Fan, Z.; Lu, G.; Yu, T.; Liu, K. Axial crushing of triangular tubes. *Int. J. Appl. Mech.* **2013**, *5*, 1350008. [CrossRef]
11. Alavi Nia, A.; Hamedani Haddad, J. Comparative analysis of energy absorption and deformations of thin walled tubes with various section geometries. *Thin-Walled Struct.* **2010**, *48*, 946–954. [CrossRef]
12. Zhang, X.; Zhang, H. Experimental and numerical investigation on crush resistance of polygonal columns and angle elements. *Thin-Walled Struct.* **2012**, *57*, 25–36. [CrossRef]
13. Umeda, T.; Mimura, K.; Morisaka, T. Study of energy absorption efficiency for a few thin-walled tubes in axial crushing. *J. Solid Mech. Mater. Eng.* **2010**, *4*, 875–890. [CrossRef]
14. Alavi Nia, A.; Parsapour, M. Comparative analysis of energy absorption capacity of simple and multi-cell thin-walled tubes with triangular, square, hexagonal and octagonal sections. *Thin-Walled Struct.* **2014**, *74*, 155–165. [CrossRef]
15. Fan, Z.; Lu, G.; Liu, K. Quasi-static axial compression of thin-walled tubes with different cross-sectional shapes. *Eng. Struct.* **2013**, *55*, 80–89. [CrossRef]
16. Mamalis, A.G.; Manolakos, D.E.; Ioannidis, M.B.; Kostazos, P.K.; Dimitriou, C. Finite element simulation of the axial collapse of metallic thin-walled tubes with octagonal cross-section. *Thin-Walled Struct.* **2003**, *41*, 891–900. [CrossRef]
17. Tang, Z.; Liu, S.; Zhang, Z. Energy absorption properties of non-convex multi-corner thin-walled columns. *Thin-Walled Struct.* **2012**, *51*, 112–120. [CrossRef]
18. Reddy, S.; Abbasi, M.; Fard, M. Multi-cornered thin-walled sheet metal members for enhanced crashworthiness and occupant protection. *Thin-Walled Struct.* **2015**, *94*, 56–66. [CrossRef]
19. Deng, X.; Liu, W.; Lin, Z. Experimental and theoretical study on crashworthiness of star-shaped tubes under axial compression. *Thin-Walled Struct.* **2018**, *130*, 321–331. [CrossRef]
20. Hanssen, A.G.; Langseth, M.; Hopperstad, O.S. Static and dynamic crushing of circular aluminium extrusions with aluminium foam filler. *Int. J. Impact Eng.* **2000**, *24*, 475–507. [CrossRef]
21. Reid, S.R.; Reddy, T.Y. Axial crushing of foam-filled tapered sheet metal tubes. *Int. J. Mech. Sci.* **1986**, *28*, 643–656. [CrossRef]
22. Reid, S.R.; Reddy, T.Y.; Gray, M.D. Static and dynamic axial crushing of foam-filled sheet metal tubes. *Int. J. Mech. Sci.* **1986**, *28*, 295–322. [CrossRef]
23. Seitzberger, M.; Rammerstorfer, F.G.; Gradinger, R.; Degischer, H.P.; Blaimschein, M.; Walch, C. Experimental studies on the quasi-static axial crushing of steel columns filled with aluminium foam. *Int. J. Solids Struct.* **2000**, *37*, 4125–4147. [CrossRef]
24. Santosa, S.P.; Wierzbicki, T.; Hanssen, A.G.; Langseth, M. Experimental and numerical studies of foam-filled sections. *Int. J. Impact Eng.* **2000**, *24*, 509–534. [CrossRef]
25. Zarei, H.R.; Kröger, M. Optimization of the foam-filled aluminum tubes for crush box application. *Thin-Walled Struct.* **2008**, *46*, 214–221. [CrossRef]
26. Børvik, T.; Hopperstad, O.S.; Reyes, A.; Langseth, M.; Solomos, G.; Dyngeland, T. Empty and foam-filled circular aluminium tubes subjected to axial and oblique quasistatic loading. *Int. J. Crashworthiness* **2003**, *8*, 481–494. [CrossRef]
27. Kavi, H.; Toksoy, A.K.; Guden, M. Predicting energy absorption in a foam-filled thin-walled aluminum tube based on experimentally determined strengthening coefficient. *Mater. Des.* **2006**, *27*, 263–269. [CrossRef]
28. Yan, W.; Durif, E.; Yamada, Y.; Wen, C. Crushing simulation of foam-filled aluminium tubes. *Mater. Trans.* **2007**, *48*, 1901–1906. [CrossRef]
29. Wang, W.; Wang, Y.; Zhao, Z.; Tong, Z.; Xu, X.; Lim, C.W. Numerical Simulation and Experimental Study on Energy Absorption of Foam-Filled Local Nanocrystallized Thin-Walled Tubes under Axial Crushing. *Materials* **2022**, *15*, 5556. [CrossRef]
30. Hussein, R.D.; Ruan, D.; Lu, G.; Guillow, S.; Yoon, J.W. Crushing response of square aluminium tubes filled with polyurethane foam and aluminium honeycomb. *Thin-Walled Struct.* **2017**, *110*, 140–154. [CrossRef]
31. Yin, H.; Wen, G.; Hou, S.; Chen, K. Crushing analysis and multiobjective crashworthiness optimization of honeycomb-filled single and bitubular polygonal tubes. *Mater. Des.* **2011**, *32*, 4449–4460. [CrossRef]

32. Zhang, Y.; Lu, M.; Sun, G.; Li, G.; Li, Q. On functionally graded composite structures for crashworthiness. *Compos. Struct.* **2015**, *132*, 393–405. [CrossRef]
33. Fang, J.; Gao, Y.; An, X.; Sun, G.; Chen, J.; Li, Q. Design of transversely-graded foam and wall thickness structures for crashworthiness criteria. *Compos. Part B Eng.* **2016**, *92*, 338–349. [CrossRef]
34. Suethao, S.; Shah, D.U.; Smitthipong, W. Recent Progress in Processing Functionally Graded Polymer Foams. *Materials* **2020**, *13*, 4060. [CrossRef] [PubMed]
35. Wang, A.; Yu, X.; Wang, H.; Li, Y.; Zhang, J.; Fan, X. Dynamic Response of Sandwich Tubes with Continuously Density-Graded Aluminum Foam Cores under Internal Explosion Load. *Materials* **2022**, *15*, 6966. [CrossRef]
36. Mohsenizadeh, S.; Alipour, R.; Shokri Rad, M.; Farokhi Nejad, A.; Ahmad, Z. Crashworthiness assessment of auxetic foam-filled tube under quasi-static axial loading. *Mater. Des.* **2015**, *88*, 258–268. [CrossRef]
37. Duarte, I.; Krstulović-Opara, L.; Dias-de-Oliveira, J.; Vesenjak, M. Axial crush performance of polymer-aluminium alloy hybrid foam filled tubes. *Thin-Walled Struct.* **2019**, *138*, 124–136. [CrossRef]
38. Linul, E.; Pietras, D.; Sadowski, T.; Marşavina, L.; Rajak, D.K.; Kovacik, J. Crashworthiness performance of lightweight Composite Metallic Foams at high temperatures. *Compos. Part A Appl. Sci. Manuf.* **2021**, *149*, 106516. [CrossRef]
39. Linul, E.; Movahedi, N.; Marsavina, L. On the Lateral Compressive Behavior of Empty and Ex-Situ Aluminum Foam-Filled Tubes at High Temperature. *Materials* **2018**, *11*, 554. [CrossRef]
40. Li, M.; Li, J.; Barbat, S.; Baccouche, R.; Lu, W. Enhanced filler-tube wall interaction in liquid nanofoam-filled thin-walled tubes. *Compos. Struct.* **2018**, *200*, 120–126. [CrossRef]
41. Li, Z.; Chen, R.; Lu, F. Comparative analysis of crashworthiness of empty and foam-filled thin-walled tubes. *Thin-Walled Struct.* **2018**, *124*, 343–349. [CrossRef]
42. Zhang, Y.; Sun, G.; Li, G.; Luo, Z.; Li, Q. Optimization of foam-filled bitubal structures for crashworthiness criteria. *Mater. Des.* **2012**, *38*, 99–109. [CrossRef]
43. Zheng, G.; Wu, S.; Sun, G.; Li, G.; Li, Q. Crushing analysis of foam-filled single and bitubal polygonal thin-walled tubes. *Int. J. Mech. Sci.* **2014**, *87*, 226–240. [CrossRef]
44. Gao, Q.; Wang, L.; Wang, Y.; Guo, F.; Zhang, Z. Optimization of foam-filled double ellipse tubes under multiple loading cases. *Adv. Eng. Softw.* **2016**, *99*, 27–35. [CrossRef]
45. Djamaluddin, F.; Abdullah, S.; Ariffin, A.K.; Nopiah, Z.M. Optimization of foam-filled double circular tubes under axial and oblique impact loading conditions. *Thin-Walled Struct.* **2015**, *87*, 1–11. [CrossRef]
46. Goel, M.D. Deformation, energy absorption and crushing behavior of single-, double-and multi-wall foam filled square and circular tubes. *Thin-Walled Struct.* **2015**, *90*, 1–11. [CrossRef]
47. Bai, Z.; Chen, C.; Yang, X.; Zhong, Y.; Liu, R. Dynamic Characteristics and Effective Stiffness Properties of Sandwich Panels with Hierarchical Hexagonal Honeycomb. *Materials* **2023**, *16*, 5741. [CrossRef]
48. Valdevit, L.; Hutchinson, J.W.; Evans, A.G. Structurally optimized sandwich panels with prismatic cores. *Int. J. Solids Struct.* **2004**, *41*, 5105–5124. [CrossRef]
49. Hung, V.T.; Dong, D.T.; Thi Phuong, N.; Ly, L.N.; Minh, T.Q.; Trung, N.-T.; Nam, V.H. Nonlinear Buckling Behavior of Spiral Corrugated Sandwich FGM Cylindrical Shells Surrounded by an Elastic Medium. *Materials* **2020**, *13*, 1984. [CrossRef]
50. Ren, L.; Yang, H.; Liu, L.; Zhai, C.; Song, Y. Sound Insulation of Corrugated-Core Sandwich Panels: Modeling, Optimization and Experiment. *Materials* **2021**, *14*, 7785. [CrossRef]
51. Yao, J.; Cui, W.; Man, X.; Du, S. New Satellite Structure Design Technology. *Chin. Space Sci. Technol.* **2010**, *7*, 70–75, 83.
52. Song, S.; Xu, M.; Shao, S.; Jiang, W.; Liang, Y. Structural Design of Mechanically Reconfigurable Reflector Antennas. *Space Electron. Technol.* **2016**, *6*, 49–53.
53. Su, P.; Han, B.; Yang, M.; Wei, Z.; Zhao, Z.; Zhang, Q.; Zhang, Q.; Qin, K.; Lu, T.J. Axial compressive collapse of ultralight corrugated sandwich cylindrical shells. *Mater. Des.* **2018**, *160*, 325–337. [CrossRef]
54. Su, P.; Han, B.; Yang, M.; Zhao, Z.; Li, F.; Zhang, Q.; Zhang, Q.; Lu, T.J. Energy Absorption of All-Metallic Corrugated Sandwich Cylindrical Shells Subjected to Axial Compression. *J. Appl. Mech.* **2020**, *87*, 121008. [CrossRef]
55. Yan, L.; Yu, B.; Han, B.; Chen, C.; Zhang, Q.; Lu, T.J. Compressive strength and energy absorption of sandwich panels with aluminum foam-filled corrugated cores. *Compos. Sci. Technol.* **2013**, *86*, 142–148. [CrossRef]
56. Han, B.; Yan, L.; Yu, B.; Zhang, Q.; Chen, C.; Lu, T.J. Collapse mechanisms of metallic sandwich structures with aluminum foam-filled corrugated cores. *J. Mech. Mater. Struct.* **2014**, *9*, 397–425. [CrossRef]
57. Han, B.; Qin, K.; Yu, B.; Wang, B.; Zhang, Q.; Lu, T.J. Honeycomb–corrugation hybrid as a novel sandwich core for significantly enhanced compressive performance. *Mater. Des.* **2016**, *93*, 271–282. [CrossRef]
58. Wadley, H.N.G.; Dharmasena, K.P.; O'Masta, M.R.; Wetzel, J.J. Impact response of aluminum corrugated core sandwich panels. *Int. J. Impact Eng.* **2013**, *62*, 114–128. [CrossRef]
59. O'masta, M.; Compton, B.; Gamble, E.; Zok, F.; Deshpande, V.; Wadley, H. Ballistic impact response of an UHMWPE fiber reinforced laminate encasing of an aluminum-alumina hybrid panel. *Int. J. Impact Eng.* **2015**, *86*, 131–144. [CrossRef]
60. Han, B.; Zhang, Z.; Zhang, Q.; Zhang, Q.; Lu, T.J.; Lu, B.H. Recent advances in hybrid lattice-cored sandwiches for enhanced multifunctional performance. *Extrem. Mech. Lett.* **2017**, *10*, 58–69. [CrossRef]
61. Han, B.; Qin, K.; Yu, B.; Zhang, Q.; Chen, C.; Lu, T.J. Design optimization of foam-reinforced corrugated sandwich beams. *Compos. Struct.* **2015**, *130*, 51–62. [CrossRef]

62. Han, B.; Yu, B.; Xu, Y.; Chen, C.; Zhang, Q.; Lu, T.J. Foam filling radically enhances transverse shear response of corrugated sandwich plates. *Mater. Des.* **2015**, *77*, 132–141.
63. Han, B.; Yue, Z.; Wu, H.; Zhang, Q.; Lu, T.J. Superior compressive performance of hierarchical origami-corrugation metallic sandwich structures based on selective laser melting. *Compos. Struct.* **2022**, *300*, 116181. [CrossRef]
64. Yang, M.; Han, B.; Mao, Y.; Zhang, J.; Lu, T.J. Crashworthiness of foam filled truncated conical sandwich shells with corrugated cores. *Thin-Walled Struct.* **2022**, *179*, 109677. [CrossRef]
65. Lst, C. *LS-DYNA Keyword User's Manual*; Livermore Software Technology Corporation: Livermore, UK, 2007.
66. Henkel, C. *Technical Data Sheet of Loctite E-120HP Hysol Epoxy Adhesive*; Henkel Corporation: Dusseldorf, Germany, 2008.
67. Stout, M.G.; Follansbee, P.S. Strain rate sensitivity, strain hardening, and yield behavior of 304L stainless steel. *J. Eng. Mater. Technol.* **1986**, *108*, 344–353. [CrossRef]
68. Hanssen, A.G.; Hopperstad, O.S.; Langseth, M.; Ilstad, H. Validation of constitutive models applicable to aluminium foams. *Int. J. Mech. Sci.* **2002**, *44*, 359–406. [CrossRef]
69. Hanssen, A.G.; Langseth, M.; Hopperstad, O.S. Static and dynamic crushing of square aluminium extrusions with aluminium foam filler. *Int. J. Impact Eng.* **2000**, *24*, 347–383. [CrossRef]

materials

MDPI

Article

Crashworthiness of 3D Lattice Topologies under Dynamic Loading: A Comprehensive Study

Autumn R. Bernard and Mostafa S. A. ElSayed *

Mechanical and Aerospace Engineering, Carleton University, Ottawa, ON K1S 5B6, Canada; autumnbernard@cmail.carleton.ca
* Correspondence: mostafa.elsayed@carleton.ca

Abstract: Periodic truss-based lattice materials, a particular subset of cellular solids that generally have superior specific properties as compared to monolithic materials, offer regularity and predictability that irregular foams do not. Significant advancements in alternative technologies—such as additive manufacturing—have allowed for the fabrication of these uniquely complex materials, thus boosting their research and development within industries and scientific communities. However, there have been limitations in the comparison of results for these materials between different studies reported in the literature due to differences in analysis approaches, parent materials, and boundary and initial conditions considered. Further hindering the comparison ability was that the literature generally only focused on one or a select few topologies. With a particular focus on the crashworthiness of lattice topologies, this paper presents a comprehensive study of the impact performance of 24 topologies under dynamic impact loading. Using steel alloy parent material (manufactured using Selective Laser Melting), a numerical study of the impact performance was conducted with 16 different impact energy–speed pairs. It was possible to observe the overarching trends in crashworthiness parameters, including plateau stress, densification strain, impact efficiency, and absorbed energy for a wide range of 3D lattice topologies at three relative densities. While there was no observed distinct division between the results of bending and stretching topologies, the presence of struts aligned in the impact direction did have a significant effect on the energy absorption efficiency of the lattice; topologies with struts aligned in that direction had lower efficiencies as compared to topologies without.

Keywords: energy absorption; finite element analysis; dynamic compression; 316L stainless steel; truss lattice materials

Citation: Bernard, A.R.; ElSayed, M.S.A. Crashworthiness of 3D Lattice Topologies under Dynamic Loading: A Comprehensive Study. *Materials* **2024**, *17*, 1597. https://doi.org/10.3390/ma17071597

Academic Editors: Madhav Baral and Charles Lu

Received: 9 February 2024
Revised: 7 March 2024
Accepted: 9 March 2024
Published: 31 March 2024

1. Introduction

Cellular solids are materials that can be found directly in nature, such as wood or coral, but have also been developed and manufactured in the industrial capacity for a wide range of applications in biological and medical sciences, aviation and aerospace, and defense and automotive industries [1–8]. They are useful for such a variety of applications due to their highly customizable nature; it is possible to tailor one or multiple attributes to obtain a unique and desired set of properties, including high specific stiffness and strength and high energy absorption [1,9,10]. The customization capability of these materials means their tailored set of properties may not be achievable by other existing monolithic materials nor a fully dense solid of the same parent material, and, as such, cellular solids extend the material property selection design space into areas once inaccessible [1,3,11].

Cellular solids are generally defined as solid materials made of cells—which are themselves an assembly of connected struts or plates—which are tessellated in a random (stochastic) or periodic manner to fill a design space [2]. The subset of cellular solids, where the cells are made up of an assembly of plate-like faces and arranged in a stochastic fashion, are generally called *foams* [12]. Conversely, the result of the periodic tessellation of cells is generally called a *lattice material* [3,13–15]. This subset of cellular solids—*lattice materials*—are of interest as they typically exhibit superior properties when compared to stochastic cellular solids and, due to their periodic nature, are more controllable and offer greater repeatability and predictability with regards to overall properties and performance [12,13,16–18]. Indeed, the advancement in additive manufacturing technology allows for the ability to investigate and characterize these complex geometric materials and for the repeated fabrication of samples and parts, with key features that can be on the micro- or nano-scale [8,15,19–23].

As mentioned, lattice materials are known for their high energy absorption capabilities, which allow them to be excellent candidates for applications such as protective packaging, shock absorption, and crash and blast mitigation, particularly within the aviation and aerospace industries [2,10,12,24,25]. In the literature, typical methods of approach for characterizing and quantifying the behavior of lattice materials—whether subjected to quasi-static or dynamic loading—involve utilizing numerical models, generally developed from or validated by experimental data and/or experimental testing with lattices produced using additive manufacturing.

Nasrullah et al. [26] investigated the dynamic response of eleven topologies (cube, cube open-cell, Kagome, octahedron, octet, pyramid, rhombicuboctahedron, rhombic-dodecahedron, tetrahedron, truncated pyramid, and a twisted octet) modeled for LS-DYNA and designed with material data based on AlSi-12 manufactured using Selective Laser Melting (SLM). Their numerical model results reveal that the topology-optimized octet topology—the "twisted octet"—had the highest specific energy absorption when the deformation mode was bending-dominated (at a relative density of less than 20%) versus when it is stretching-dominated (at a relative density above 25%). This twisted octet topology was then utilized in an aircraft subfloor structure and showed better specific energy absorption than the original design, indicating that the new configuration could be utilized in this type of system.

Mueller et al. [10] utilized ABAQUS/Explicit to characterize the impact performance of four aluminum alloy-based topologies (cube, Delauney, octet, and Voronoi) undergoing quasi-static and dynamic ($< 10^4$/s) strain rates. They concluded that while relative density was the most important parameter for determining the response of stochastic foams, the deformation mode (bending or stretching) was the most important for periodic structures. By investigating the effect of unit cell rotation with respect to impact direction, they also concluded that this orientation was important for energy absorption properties since the same topology at different orientations might exhibit different deformation modes.

Wang et al. [27] used theoretical and numerical methods to characterize the deformation mode and energy absorption potential of an aluminum alloy FCC lattice at different orientations and impact velocities. The theoretical analysis calculated plateau stresses for the unit cell at different orientations and validated the numerical model, which used the solver ABAQUS. They found that unit cell orientation and impact velocity had an effect on energy absorption, plateau stress, densification strain, and deformation mode. At high impact velocities, however, the deformation mode "I"—defined by the concentration of the deformation in a band perpendicular to the load direction—prevailed regardless of unit cell orientation. For the FCC lattice, an orientation of 45° resulted in the highest energy absorption and plateau stress but the lowest densification strain (and vice versa for a 20° orientation).

Ozdemir et al. in [24,28] observed the energy absorption properties of three titanium alloy-based (Ti6Al4V) topologies (cubic, diamond, and re-entrant cube) under quasi-static and dynamic loading, manufactured using the additive manufacturing technique Electron Beam Melting. First, they performed quasi-static experiments under compression using a universal test machine, while the dynamic tests were performed using a Hopkinson pressure bar. Then, in [28], they created a numerical non-linear finite element model based on the experimental results of the previous paper, finding good accuracy between experimental and numerical results. It is believed that those were the first papers to investigate the dynamic response of the diamond and re-entrant topologies.

Jin et al. [29] additively manufactured samples of four different topologies (two novel diamond lattices,—Dfcc and Dhex—FCC, and BCC) made of titanium alloy Ti6Al4V manufactured with SLM. The dynamic response was tested on a split Hopkinson pressure bar, and the numerical model, created in LS-DYNA, showed a good correlation with the experimental results. They found that the Dfcc and Dhex lattices showed a mixture of stretching- and bending-dominated deformation modes, while the FCC and BCC topologies were stretching- and bending-dominated, respectively. Under dynamic loading, they found that the lattices exhibiting a stretching-dominated deformation mode had better mechanical properties.

Lei et al. [30] fabricated aluminum alloy multi-layer BCC and BCC-Z lattice samples using SLM and, prior to experimental testing, examined the specimens using X-ray micro-computed tomography (μ-CT). This data allowed the novel finite element model they developed to factor in printing imperfections, which are not generally a consideration. Using such a design approach for the lattice within the numerical model, they found that predicted compressive modulus and initial crushing strength were consistent with experimental results.

It is obvious that the design of lattice materials requires the consideration of many multiscale variables, including parent material, geometry and topology, and relative density, among others [1,31]. However, the literature identified previously is generally limited in the variety of topologies examined, a limitation which was also described by Helou and Kara in [3] after a review of over 45 publications in the realm of lattice structures. Helou and Kara also critiqued the difficulty in comparing data from one paper to the next due to the variations in materials and investigative approaches and having no standardized test methods for analyzing lattice structures. Indeed, there are many modeling strategies available for the investigation of lattice structures; Giorgio et al. [32] defined the mechanical behavior of pantographic lattices using a second-grade elasticity model, while Tran and Niiranen [33] and Dong et al. [34] formulated a non-linear Euler–Bernoulli beam model that maintained high accuracy while saving computational costs and developed a numerical homogenization method for 3D cellular materials in MATLAB, respectively.

To begin to close the gap caused by differences in methods between publications, further exacerbated by limited topologies being investigated per publication, this paper analyzes over 20 different topologies using the same base FEA model. To quantify impact performance, we look specifically at energy absorption capabilities—calculating energy absorption efficiency and densification strain—across multiple impact strain rate levels. This work was accomplished by validating the developed FEA model with experimental data and utilizing FEA model material data published and validated in [16,17].

This paper is organized into three main sections following this introduction. In Section 2, the lattice topology design, finite element model, design of experiment approach, and validation are described. In Section 3, the results collected are presented and discussed. Finally, the conclusion is in Section 4, with acknowledgments and references following.

2. Materials and Methods

2.1. Lattice Geometry and Design

For this study, 24 topologies were investigated. Each topology was modeled as a single unit cell with an original height of 10 mm, where the radius of the struts was varied to investigate three different relative densities, namely 0.10, 0.20, and 0.30. While it is theoretically possible to utilize Equation (1) from [26] to determine the design radius (R) for a given topology based on unit cell height (h) and relative density ($\overline{\rho}$), since k and c are correction coefficients dependent on the topology (k—correction of total strut length of a unit cell, c—correction of total strut length of geometrical cubic), in practice, it is much more difficult [26].

$$\overline{\rho} = k\left(\frac{R}{h}\right)^2 - c\left(\frac{R}{h}\right)^3 \quad (1)$$

To determine those coefficients, one can derive them by observing the geometry of a single strut within the given topology as per the method in [35,36]. For simpler topologies, this derivation is straightforward. Additionally, for certain topologies, such as octet, the procedure and results have been well-documented in the literature [11,31,36–38]. However, the same data for other topologies is quite limited or non-existent: [36,38] for octet; [36] for truncated octahedron; [39] for BCC-Z; [39,40] for rhombic dodecahedron; and [41,42] for BCC. Additionally, sources may use a modified version of Equation (1), where instead of the height of the cell, the radius (or even diameter) is normalized against the length of one strut. The differences between the literature and a lack of values for the topologies of interest in this study led to determining the design radius for each of the 24 topologies using a combination of what was already identified in the literature and the homogenization code written in MATLAB from [34]. While the main purpose of that code is to determine the homogenized constitutive matrix of 3D cellular materials, it also outputs a relative density calculated using topology, cell size, and radius during the calculation process. This value was used to determine an approximate radius for geometry creation, which was ultimately confirmed after creating the geometry in ANSYS SpaceClaim (2020 R2) using a modified version of the code from [43].

Radius values and the corresponding relative density for each topology are provided in Table 1, as well as a visual representation of each topology. This data was utilized to compute a curve fit and the coefficients k and c from Equation (1) for the 24 studied topologies, making it easier to determine the radius for a given relative density (or vice versa) in the future. The plot of relative density vs radius–height ratio and the coefficients are provided in Figure 1 and Table 2. The plot is similar to the one presented in [14], though it expands on the number of topologies presented and provides numerical values for the coefficients for use in Equation (1).

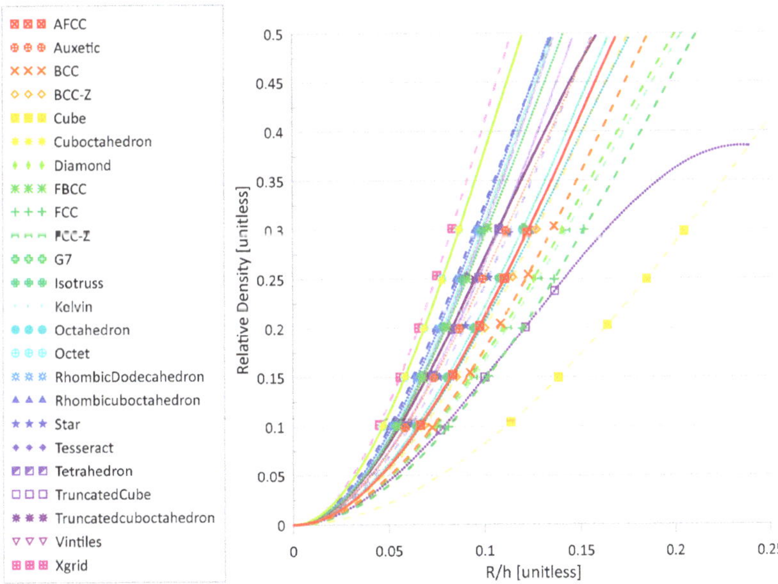

Figure 1. Relative density versus the ratio of radius to unit cell height. Line types distinguish between stretching (solid), bending (dotted), and mixed (dashed) deformation modes, discussed in Section 3.1. Line opacity indicates whether there is at least one strut aligned in the loading direction: opaque—no, semi-translucent—yes.

Table 1. Geometry and design data for lattice structures.

Topo.	Geometry	$\bar{\rho}$	R [mm]	Topo.	Geometry	$\bar{\rho}$	R [mm]
AFCC		0.1 0.2 0.3	0.663 0.973 1.218	Kelvin		0.1 0.2 0.3	0.665 0.985 1.250
Auxetic		0.1 0.2 0.3	0.583 0.867 1.108	Octahedron		0.1 0.2 0.3	0.645 0.945 1.195
BCC		0.1 0.2 0.3	0.723 1.080 1.357	Octet	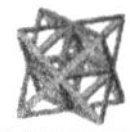	0.1 0.2 0.3	0.466 0.684 0.867
BCC-Z		0.1 0.2 0.3	0.683 1.000 1.267	Rhombic Dodecahedron		0.1 0.2 0.3	0.519 0.760 0.965
Cube		0.1 0.2 0.3	1.130 1.636 2.047	Rhombi-cuboctahedron		0.1 0.2 0.3	0.500 0.742 0.950
Cuboctahedron		0.1 0.2 0.3	0.467 0.683 0.867	Star		0.1 0.2 0.3	0.615 0.900 1.125
Diamond		0.1 0.2 0.3	0.740 1.095 1.400	Tesseract		0.1 0.2 0.3	0.551 0.792 1.000
FBCC		0.1 0.2 0.3	0.542 0.800 1.017	Tetrahedron		0.1 0.2 0.3	0.585 0.850 1.075
FCC		0.1 0.2 0.3	0.808 1.192 1.517	Truncated Cube [1]		0.1 0.2	0.767 1.208
FCC-Z		0.1 0.2 0.3	0.750 1.117 1.425	Truncatedcuboctahedron	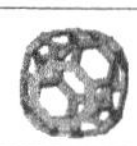	0.1 0.2 0.3	0.558 0.825 1.075
G7		0.1 0.2 0.3	0.642 0.942 1.200	Vintiles	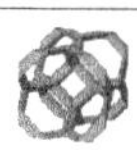	0.1 0.2 0.3	0.667 0.983 1.250
IsoTruss		0.1 0.2 0.3	0.533 0.783 0.983	Xgrid	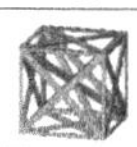	0.1 0.2 0.3	0.450 0.655 0.830

[1] The truncated cube cannot be built beyond a relative density of ~24%.

Table 2. Coefficients k and c for the equation $\bar{\rho} = k\left(\frac{R}{h}\right)^2 - c\left(\frac{R}{h}\right)^3$, where $\bar{\rho}$ is unitless, R is radius, and h is unit cell height, and both have the same units.

Topology	k	c	Topology	k	c
AFCC	26.656	54.618	Kelvin	26.657	60.331
Auxetic	34.381	90.751	Octahedron	28.049	59.170
BCC	21.765	39.187	Octet	53.313	154.493
BCC-Z	24.907	49.179	Rhombic Dodecahedron	43.530	117.566
Cube	9.425	11.311	Rhombicuboctahedron	46.847	144.491
Cuboctahedron	53.313	154.493	Star	31.190	69.157
Diamond	21.765	46.361	Tesseract	39.156	90.673
FBCC	39.542	102.212	Tetrahedron	33.635	71.375
FCC	17.774	31.282	Truncated Cube	20.919	59.354
FCC-Z	20.913	43.497	Truncatedcuboctahedron	39.391	123.597
G7	28.048	59.169	Vintiles	26.657	60.330
IsoTruss	40.614	99.127	Xgrid	170.547	57.846

2.2. Finite Element Model

2.2.1. Initial Model Creation

The finite element model was developed in Altair's HyperMesh (v2020, Altair Engineering Inc., Troy, MI, USA) for the explicit finite element solver Radioss (v2020, Altair Engineering Inc., Troy, MI, USA). The initial finite element model was based on that model from the literature containing the material model (further described in Section 2.2.4.), which provided quasi-static results for a 3 × 3 ×3 rhombic dodecahedron lattice with overall dimensions of 24 mm × 24 mm × 24 mm and a relative density of 11.68% [17]. As in [40], a one-quarter model was created, reducing the complexity and computation time for the model. Boundary conditions for the one-quarter model along the planes of symmetry were additionally applied following [40]. Figures 2 and 3 illustrate the stress–strain and deformation results, respectively, of the quasi-static (0.001/s) experimental results from Cao et al. [17] and the developed numerical model for this work. Note that stress and strain values are for the lattice as a whole; load and displacement data were utilized, along with original model dimensions, to calculate stress and strain values. These figures suggested a good correlation to the experimental results, initially validating the model. Further requirements to reduce the model complexity (due to the large number of numerical simulations to be completed, detailed in Section 2.3) meant further validation for a model containing a single unit cell, described in the following subsections.

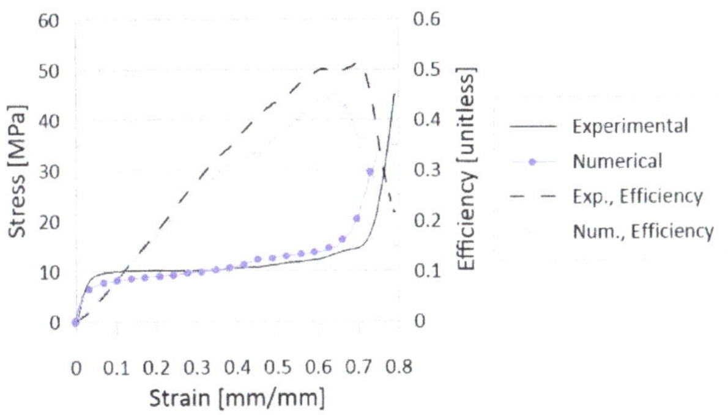

Figure 2. Stress–strain and efficiency–strain results for the quasi-static experiments (from Cao et al. [17]) and the corresponding numerical model as designed for this work. Select deformation behavior illustrated in Figure 3.

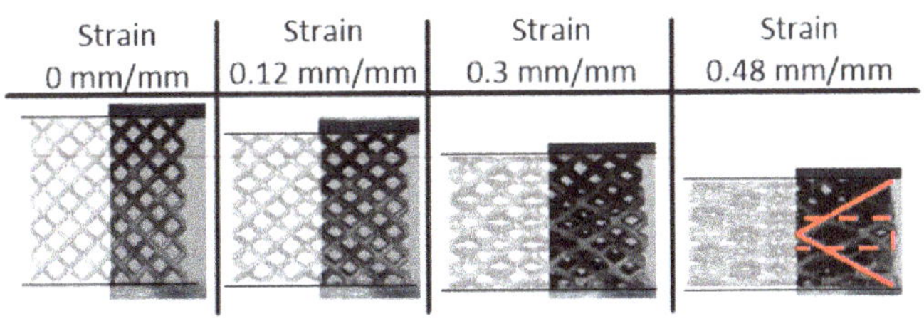

Figure 3. Deformation behavior of rhombic dodecahedron cluster from experiments (half-image on the right, from Cao et al. [17]) and corresponding numerical model for this work (half-images on the left) for given strain values. Corresponding stress and efficiency results are provided in Figure 2.

2.2.2. Unit Cell Model

The model was constructed such that a single unit cell made of solid tetrahedron elements was impacted at a given initial velocity by a flat plate impactor ("impactor") made of 4-node shell elements. An additional non-moving, fixed flat plate ("base") is modeled beneath the single unit cell, illustrated in Figure 4. This model has boundary conditions, as detailed in Section 2.2.3., which represent a single-layer lattice; multi-layer lattice behavior is considered out-of-scope for this work.

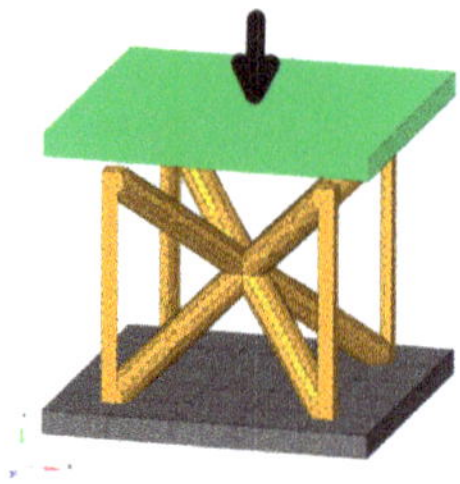

Figure 4. General finite element model components, using a BCC-Z unit cell (orange) for illustrative purposes. Base plate (grey) is fixed and not permitted to translate or rotate. Impactor (green) is given an initial velocity in the downward y-direction as indicated by the black arrow.

Both the impactor and the base have side lengths of 14 mm and thicknesses of 1 mm. They are modeled as rigid, and the mass and initial velocity of the impactor are parametrized, allowing for variations to the initial kinetic energy ("impact energy"). The initial velocity was applied to the primary node of the rigid body, in the middle of the impactor, and only in the direction of axial compression.

For unit cell topologies, it should be noted that while impact only occurs in the global y-direction, for non-isotropic orientations (such as BCC-Z, octahedron, etc.), the results may change depending on the orientation of the unit cell; the investigation of unit cell orientation was beyond the scope of this work.

There are three interfaces defined in the model: (i) the interaction between the impactor and lattice cell, (ii) the self-contact of the lattice cell during compression, and (iii) the interaction between the fixed plate and lattice cell. The former and latter are both controlled by a solid contact interface between a primary surface (impactor or plate) and secondary nodes (lattice cell). The self-contact of the lattice cell was controlled by a single-surface interface.

2.2.3. Boundary Conditions

Considering the Cartesian coordinate system defined in Figure 4, all translational and rotational degrees of freedom of the rigid plate impactor are restricted except the vertical translational motion, as the impactor was only allowed to move along the y-axis. For the lattice unit cell, periodic boundary conditions (BCs) were applied such that they mimicked the periodicity of the lattice.

For validation of these BCs, four separate models were created and tested: a model containing a unit cell whose nodes on the side surfaces were restricted to permit only motion in the y-axis direction; a model containing a unit cell whose nodes on the side surfaces were not restricted; and models containing a single-layer of either 3×3 or 5×5 clustered unit cells lattice whose nodes on the side surfaces were not restricted. The interest in these clustered lattices was for the comparison of internal energy, force, and general deformation behavior of the middle unit cell to those resulting from the single unit cell models. These results are depicted in Figure 5 for internal energy and force over the course of the compression event and Figure 6 for general deformation behavior characteristics. It should be noted that all models were all compressed at the same rate and the lattices were all defined by the same material parameters presented in Section 2.2.4.

From Figure 5, it can be seen that fixed BCs applied to a single unit cell represent the response of the middle unit cell in a clustered (3×3 or 5×5) layer most accurately, as compared to the results of the unit cell with free BCs. In general, across the compression event, the percent difference between fixed BC results and the middle unit cell of the cluster is below 20% (absolute value), whereas the free BC model reaches a percent difference of almost 70% and 90% for internal energy and force, respectively. Even the deformation behavior, illustrated in Figure 6, is best captured by the fixed BCs as opposed to the free

BCs; the vertical struts of the cubic unit cell in the fixed BC model "squish" in the same manner as those same struts of the middle unit cell in the 3 × 3 cluster while the vertical struts of the unit cell in the free BC model instead buckles outwards.

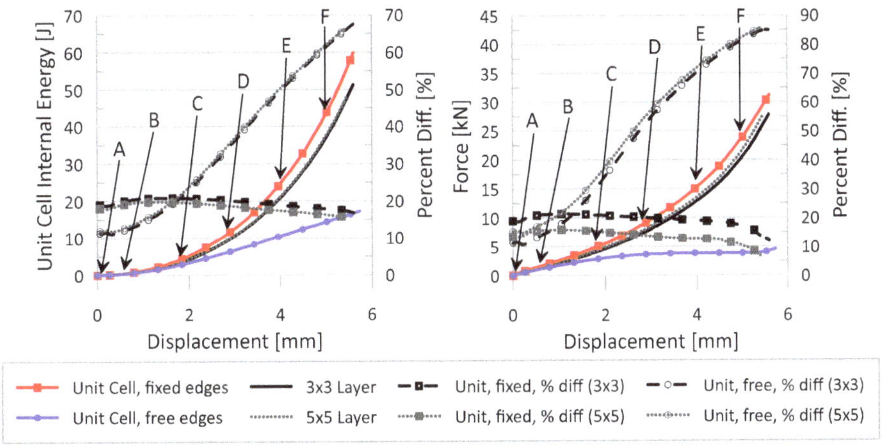

Figure 5. (**Left**) Unit cell internal energy over compression of lattice. (**Right**) Interface force over compression of lattice. Lettered displacement locations (A, B, C, D, E, and F) correspond to images in Figure 6 for the two single-unit cells and the 3 × 3 layer.

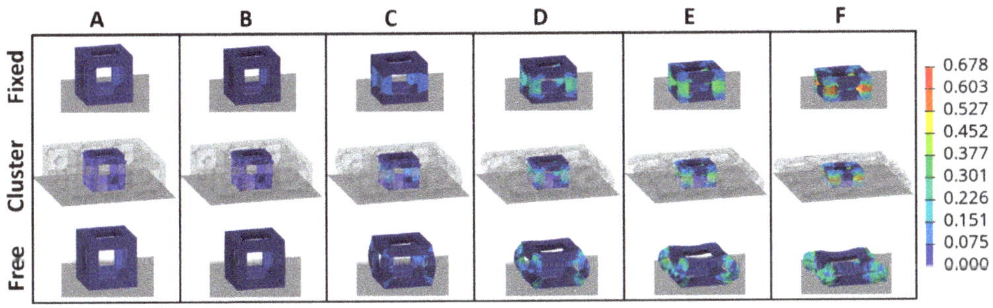

Figure 6. Deformation behavior of two unit cells with different boundary conditions applied to the sides (fixed in top row and free in bottom row), as compared to the middle unit cell of a single-layer 3 × 3 lattice cluster (middle row). Color contour is of plastic strain, units [mm/mm]. Letters correspond to displacement locations in Figure 5.

Such validation testing was also used to justify the use of a single unit cell as opposed to a cluster of unit cells, reducing the overall computation time and memory requirements that would come with the larger lattice model. In Table 3, the number of elements in the lattice, the output file size (animation file and results file), and the overall run time (all simulations were run using parallelization across 32 cores) are provided for the three simulation runs. Being able to utilize one unit cell with the appropriate BCs allows for a minimum 70% to 89% reduction of those measures. As such, all topologies were modeled as a single lattice unit cell for the remainder of this work.

Table 3. Number of elements in model, output file size, and run time for unit cells and single-layer cluster, along with associated percent reductions.

Model	Num. Elem.	Output File Size [MB]	Run Time [min]	Percent Reduction [1] [%]		
				Num. Elem.	Output File Size	Run Time
Free BC	3139	74	15	89	85	72
Fixed BC	3139	74	16	89	84	70
3 × 3 Single-Layer Cluster	27,520	498	53	-	-	-
5 × 5 Single-Layer Cluster	77,547	1010	339	-	-	-

[1] As compared to the 3 × 3 cluster.

2.2.4. Material and Failure Model

The lattice unit cell was modeled as a 316L stainless steel alloy with properties as produced by the additive manufacturing technique SLM, and, in Radioss, the material itself was defined using a Johnson–Cook strength model [44,45].

The Johnson–Cook strength model considers the plastic stress to be made up of the product of three terms, each considering a different mechanism, namely: strain hardening; strain rate hardening; and temperature softening. Equation (2) presents this product, where A, B, C, m, and n are all material constants, determined from experimental testing, and $\bar{\varepsilon}_p$, $\dot{\varepsilon}^*$, and T^* are the equivalent plastic strain, the normalized equivalent plastic strain rate, and the dimensionless homologous temperature, respectively. Equations (3) and (4) define the normalized equivalent plastic strain rate and the dimensionless homologous temperature, respectively, where $\dot{\varepsilon}$ is the plastic strain rate, $\dot{\varepsilon}_0$ is the reference strain rate, T_0 is room temperature, and T_m is melting temperature.

$$\bar{\sigma} = \left[A + B\bar{\varepsilon}_p^n\right]\left[1 + C\ln\dot{\varepsilon}^*\right]\left[1 - T^{*m}\right] \quad (2)$$

$$\dot{\varepsilon}^* = \frac{\dot{\varepsilon}}{\dot{\varepsilon}_0} \quad (3)$$

$$T^* = \frac{T - T_0}{T_m - T_0} \quad (4)$$

For this model, no temperature effects were considered. All other material constants required to define the Johnson–Cook strength model for the 316L stainless steel, as manufactured by SLM, are provided in Table 4, including material density (ρ_0), Young's modulus (E), and Poisson's ratio (ν). These values come from the work of Cao et al. in [16,17], where a numerical model employed the Johnson–Cook strength and failure models for the lattice structures and was validated by experimental testing.

Table 4. Mechanical properties of 316L stainless steel alloy (manufactured by SLM) for use in Johnson–Cook strength and failure models.

	ρ_0	E	ν	
	7960 kg/m^3	93 GPa	0.3	
A	B	C	n	$\dot{\varepsilon}_0$
310 MPa	622	0.1	0.8	0.001
D_1	D_2	D_3	D_4	D_5
0.1152	1.0116	−1.7684	−0.05279	0.5256

As in [16,17], the material card definition for this FEA model also utilized the Johnson-Cook failure model. This failure model splits the plastic portion of the stress–strain behavior into *damage initiation* and *damage evolution* up until the complete failure of the material.

Damage initiation can be described by Equation (5), specifically that when $\omega = 1$, damage is initiated (ω can vary between 0 and 1).

$$\omega = \sum (\Delta\bar{\varepsilon}_p / \bar{\varepsilon}_f) \tag{5}$$

In this equation, $\Delta\bar{\varepsilon}_p$ is the increment of the equivalent plastic strain and $\bar{\varepsilon}_f$ is the equivalent plastic strain at failure. The equivalent plastic strain at failure can be defined by Equation (6), where D_1 through D_5 are material constants and σ^* is the ratio of the mean stress (σ_{mean}) to the Mises equivalent stress (σ_{eff}). The material constants D_1 through D_5 are presented in Table 4. As with the constants for the Johnson–Cook strength model, these values come from the work of Cao et al. in [16,17], and temperature effects are not considered.

$$\bar{\varepsilon}_f = \left[D_1 + D_2 \exp\left(D_3 \sigma^*\right)\right]\left[1 + D_4 \ln \dot{\varepsilon}^*\right]\left[1 + D_5 T^*\right] \tag{6}$$

2.3. Design of Experiments

The energy absorption capabilities of each lattice topology unit cell and relative density were observed through four variations to impactor initial kinetic energy (KE) and four variations to impactor initial velocity for a total of 16 different initial velocity-KE combinations per a given topology and relative density set. An "energy matrix" is presented in Table 5, which outlines the different speed and impactor energy values (and corresponding impactor mass) for each of the 16 different runs. The 16 Radioss files per topology and relative density pair were created from the "parent" HyperMesh models by utilizing commands in .tcl files to modify the mass and velocity parameters of the impactor. Once passed through the solver, the animation and time history files were processed, and result data was collected using a developed .oml file in Altair's Compose (v2021, Altair Engineering Inc., Troy, MI, USA). The final post-processing of the collected data was performed in MATLAB (R2020b, MathWorks, Natick, MA, USA) and Golden Software's Grapher (20.1.251, Golden Software LLC, Golden, CO, USA).

Table 5. Speed-kinetic energy matrix for the 16 runs per topology-relative density, with mass [kg] in the center.

Strain Rate [1] [1/s]	Speed [m/s]	Initial Kinetic Energy [J] 1	5	50	100
100	1	2	10	100	200
1000	10	0.02	0.1	1	2
10,000	100	0.0002	0.001	0.01	0.02
100,000	1000	0.000002	0.00001	0.0001	0.0002

[1] Strain rate is calculated by dividing speed by the height of the unit cell (10 mm).

3. Results

3.1. Data Analysis and Crashworthiness Parameters

Data was collected from the finite element solver so that an analysis of the crashworthiness of the lattices could be performed. The determination of the performance of those lattices involved calculating the following crashworthiness parameters:

Energy Absorption Efficiency. Also, "EA efficiency" or "efficiency", this parameter, η, is calculated from the stress–strain (σ–ε) curve as

$$\eta(\varepsilon) = \frac{1}{\sigma(\varepsilon)} \int_0^{\varepsilon} \sigma(\varepsilon) d\varepsilon \tag{7}$$

where, in the numerical model, the stress is obtained from the impact force between the impactor and lattice divided by the surface area of the upper surface of the lattice unit cell envelope, and the strain is obtained from the displacement of the uppermost surface of the lattice divided by the lattice unit cell height, a method commonly used in the literature [25,46,47].

Densification Strain. Using the energy absorption efficiency method, the densification strain, ε_D, is

$$\left.\frac{d\eta(\varepsilon)}{d\varepsilon}\right|_{\varepsilon=\varepsilon_D} = 0 \quad (8)$$

which is simply the strain at the maximum efficiency point (η_{max}) [25,48,49].

Plateau Stress. Plateau stress, σ_{pl}, along with densification strain, are considered to be the most important values with regard to the energy absorption of materials [50]. This stress is determined using [25]

$$\sigma_{pl} = \frac{\int_0^{\varepsilon_D} \sigma(\varepsilon) d\varepsilon}{\varepsilon_D} \quad (9)$$

Energy Absorption. Energy absorption, EA or IE (internal energy), is calculated as

$$EA = V \int_0^{\varepsilon_D} \sigma(\varepsilon) d\varepsilon \quad (10)$$

It should be noted that while Equation (10) utilizes the densification strain in the integral for calculation, the actual selected strain values do vary in the literature [10,25,26,31,51].

Specific Energy Absorption. Specific Energy Absorption (SEA) is calculated either per unit volume—SEA_V—or per unit mass—SEA_m. Equations for both of these specific energy absorptions are provided as follows [25,51,52]

$$SEA_V = \frac{EA}{V} = \int_0^{\varepsilon_D} \sigma(\varepsilon) d\varepsilon \quad (11)$$

$$SEA_m = \frac{EA}{m} = \frac{\int_0^{\varepsilon_D} \sigma(\varepsilon) d\varepsilon}{\bar{\rho}\rho_0} \quad (12)$$

where V is the volume of the lattice structure, and m is the mass of the lattice structure. It should be noted that these values are also calculated up to the densification strain.

Bending- versus Stretching-Dominated Behavior. Maxwell's stability criterion [53] can be utilized to help understand the deformation behavior of a topology, which can be helpful in understanding appropriate applications for a lattice made of that topology. However, it should be noted that even if the stability criterion designates the unit cell of a certain topology as bending-dominated, it is still possible that the actual behavior of the topology is stretching-dominated (or mixed-mode), particularly since the criterion does not account for the loading direction, including the presence of struts aligned directly in the loading direction [54,55]. Indeed, Calladine [56] described *tensegrity* structures as "[constituting] a paradoxical exception to Maxwell's rule," going on to examine the conditions required to "break" Maxwell's rule. In 1986, Peregrino and Calladine [57] discussed a modified version of this criterion, and it has been noted the criterion is only a necessary condition, and not a sufficient condition, for determining truss stiffness [1,58]. However, the determination of the variables for the modified criterion is not trivial [57] and is considered to be beyond the scope of this work. Instead, with this knowledge of the criterion and its limitations, the determination of the compression behavior of the topologies in this work considers several characteristics: Maxwell's criterion for a single unit cell, strut orientation with respect to loading, and classification in the literature. A summary of these quantitative and qualitative characteristics is provided in Table 6. As can be seen from Table 6, Maxwell's stability criterion may not accurately reflect the compressive behavior of a topology if it is predicted to be bending-dominated by the criterion. However, those topologies that are determined to be stretching-dominated by the criterion will still exhibit predominantly stretching-dominated behavior during compression.

Table 6. Topologies as bending- or stretching-dominated and whether the topology has struts aligned directly in the loading direction.

Topology		b [1]	j [2]	M [3]	S/B [4]	S/B [5]	Ref.	LD? [6]
AFCC		24	12	−6	B	S	[59]	N
Auxetic		36	18	−12	B	B	[60,61]	Y
BCC		8	9	−13	B	S/B	[14,29,54,55]	N
BCC-Z		12	9	−9	B	S/B	[54,55]	Y
Cube		12	8	−6	B	S/B	[10,14,26,62]	Y
Cuboctahedron		36	13	3	S	S	[60]	N
Diamond		16	14	−20	B	S/B	[14,24,28]	N
FBCC		24	13	−9	B	B	[60]	N
FCC		16	12	−14	B	S/B	[54,55]	N
FCC-Z		20	12	−10	B	S/B	[54,55]	Y
G7		16	9	−5	B	B	[60,63]	Y
IsoTruss		26	15	−13	B	S	[10,59]	Y
Kelvin		36	24	−30	B	B	[10,14,64,65]	N
Octahedron		12	6	0	S	S	[26,59]	Y
Octet		36	14	0	S	S	[10,26,29,59,65]	N
Rhombic Dodecahedron		32	20	−22	B	B	[16,40]	N
Rhombicuboctahedron		48	24	−18	B	S/B	[14,26,64]	N
Star		20	9	−1	B	S/B	[14,59]	Y

Table 6. *Cont.*

	Topology		b [1]	j [2]	M [3]	S/B [4]	S/B [5]	Ref.	LD? [6]
◆	Tesseract		32	16	−10	B	S	[60,66,67]	Y
◩	Tetrahedron		22	9	1	S	S	[68]	Y
□	Truncated Cube	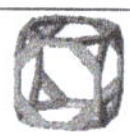	36	24	−30	B	B	[14]	N
✸	Truncatedcuboctahedron		72	48	−66	B	S	[60]	N
▽	Vintiles	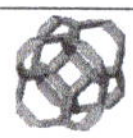	36	28	−42	B	B	[10]	N
⊞	Xgrid		44	15	5	S	S/B	[59]	Y

[1] Number of struts. [2] Number of nodes. [3] Maxwell's number. [4] Stretching (S, blue color) or Bending (B, orange color) based on Maxwell's number. [5] Stretching (S, blue color) or Bending (B, orange color) based on the literature references in the next column. (S/B) indicates a mix of Stretching and Bending modes (red color). [6] Strut(s) in loading direction? Yes (Y, green color) or No (N, yellow color).

3.2. *Numerical Model Results and Discussion*

All performance parameters mentioned in the previous section, as well as time-history internal and kinetic energies, force, and displacement data, were carefully analyzed for all 24 lattice topology unit cells at three relative densities for each of the 16 speed-impact energy scenarios of the Design of Experiments. There are four main independent variables in this study, namely, (i) impact KE, (ii) impact speed, (iii) relative density, and (iv) topology—the effects of which are further discussed in the following subsections.

3.2.1. Effect of Initial KE

Based on internal energy variations over the compression event and stress–strain results for all topologies, the following observations were made:

- It was noted that simulations with lower initial impactor kinetic energies mimicked the initial behavior of the larger initial KE, particularly for simulations with initial impactor speeds of 100 m/s or less. That is, the results suggest that running simulations at a larger initial KE could still predict the behavior of the simulations at a lower initial KE. A set of internal energy over compression displacement curves for the AFCC topology (at all three relative densities) is provided in Figure 7 to illustrate such a phenomenon. In the top graphs of this figure, it is evidently seen that the lower initial impact energy curves (solid lines) follow the path of the higher initial internal energy curves (dashed lines) up until the system limit is reached and the higher energy curves continue.
- In looking at stress–strain results, it was revealed that initial impactor KE of 1 J and 5 J was not sufficiently large to reveal complete elastic-plateau-densification stress–strain curves and a demonstratory set of stress–strain curves for the AFCC topology is provided in Figure 8. As a result, calculated crashworthiness parameters—such as densification strain or energy absorption—for the simulations with those lower initial KE would not accurately reflect the behavior and capabilities of the topology. It should be noted that for some topologies, even an initial impactor KE of 50 J was not always sufficient to reveal the complete elastic-plateau-densification characteristics expected of a stress–strain curve for lattice materials, particularly at higher relative densities.

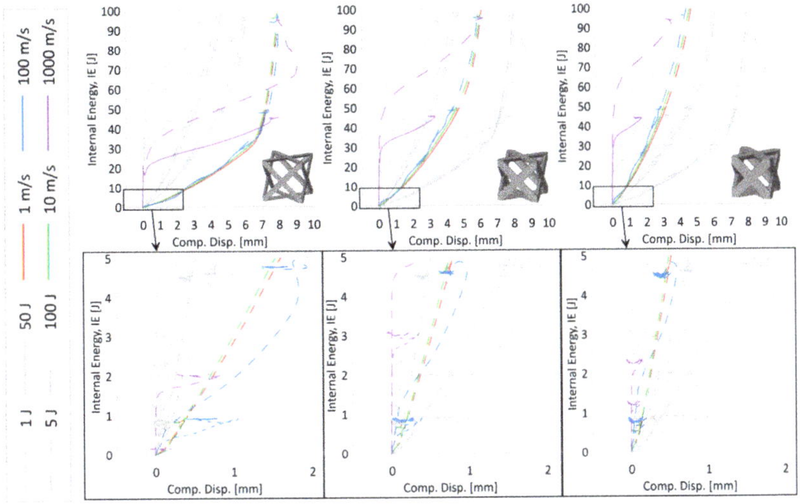

Figure 7. Internal energy over compression displacement for AFCC topology at three relative densities, four speeds, and four initial kinetic energies. Top graphs are for initial KE of 50 J and 100 J, and bottom graphs are for initial KE of 1 J and 5 J (identified by line type per legend). Variations in initial impact speeds are distinguished using the line color specified in legend. From left to right: relative density 10%, 20%, 30%; images of unit cell provided for reference. Grey lines are used to help compare relative densities.

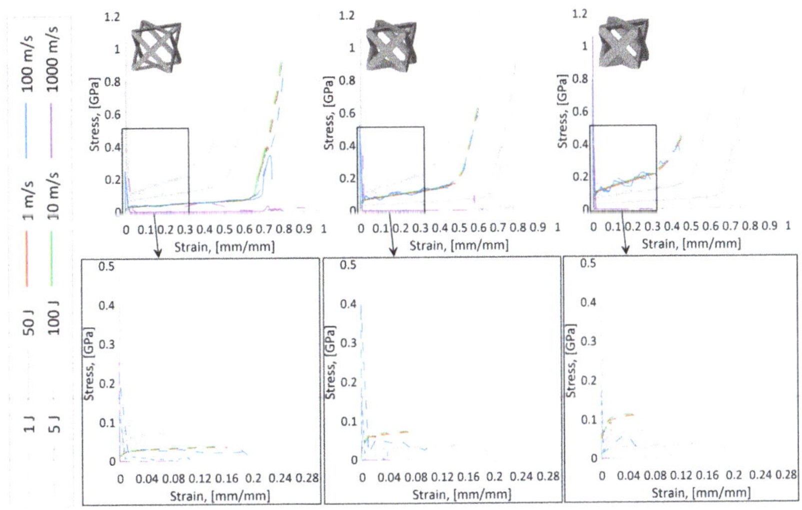

Figure 8. Stress over strain for AFCC topology at three relative densities, four speeds, and four initial kinetic energies. Top graphs are for initial KE of 50 J and 100 J, and bottom graphs are for initial KE of 1 J and 5 J (identified by line type per legend). Variations in initial impact speeds are distinguished using the line color specified in legend. From left to right: relative density 10%, 20%, 30%; images of unit cell provided for reference. Grey lines are used to help compare relative densities.

Based on such observations—which suggest that the full potential of the topology is not being utilized if the impactor initial KE is lower and that the curves for low KE mimicked the behavior of the larger KE anyway—the focus for further investigations was narrowed to include only results from the 100 J initial KE simulations.

3.2.2. Effect of Speed

Based on internal energy variations over the compression event and stress–strain results for all topologies, the following observations were made:

- As illustrated in Figure 7, despite the strain-sensitivity of the material, up to initial impactor speeds of approximately 10 m/s, there is very little difference in results, suggesting that at those low speeds, the sensitivity of the material to strain rate variations is negligible. While it also appears that data from simulations with initial impactor speeds of 100 m/s is also insignificantly different from the two lower speed levels in those plots, stress–strain plots for certain topologies (as an example, in Figure 8) did suggest that results from 10 m/s could not always be used to accurately predict results for a speed of 100 m/s.
- Also seen from stress–strain results was the indication that the current model setup was insufficient to appropriately reveal topology behavior at an initial impactor speed of 1000 m/s. While the internal energy over the compression event curves (Figure 7) and associated simulation animations (Figure 9) for numerical models at initial impactor speeds at 1000 m/s appear to illustrate the phenomenon of layer-by-layer collapse at high impact speeds (versus all layers deforming in the same or similar manner during a quasi-static or low-speed compression event), the jump in magnitude from 100 m/s to 1000 m/s results in erratic time-history behavior, bringing into question the reliability and credibility of the data. Additionally, with only one unit cell contained in this model, such a conclusion would realistically require multiple layers. Coupled with the stress–strain result observations, data for numerical models with impact speeds of 1000 m/s were perceived to have a high level of uncertainty.

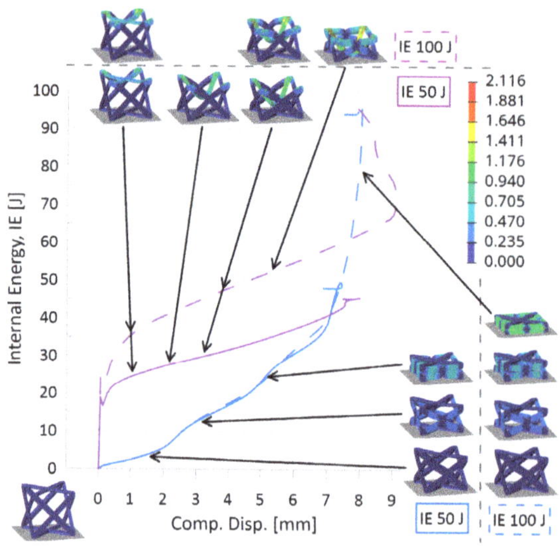

Figure 9. Internal energy over compression displacement for AFCC topology at a relative density of 10%. Variations in initial KE are 50 J and 100 J (distinguished by line type). Variations in initial impact speeds are 100 m/s and 1000 m/s (distinguished by line color). Color contour for images of compression of unit cells is for plastic strain [mm/mm].

Based on the observations for the effect of speed and initial KE, the focus for additional data analysis was placed on simulations with an initial KE of 100 J at speeds of 10 m/s and 100 m/s.

3.2.3. Effect of Relative Density

To first get a sense of the static mechanical properties of the topologies of interest in this work, the homogenized properties—such as the constitutive matrix and Young's modulus (in x-, y-, and z-axes directions)—were computed for all topologies at the three relative densities investigated (10% through 30%, at 10% increments) using the homogenization code for cellular materials in MATLAB presented in Dong et al. [34] and a modified version of the MATLAB code for visualizing elastic anisotropy from Nordmann et al. [69]. In analyzing the dynamic response of these topologies, this homogenized data was useful in understanding their relationships and behavior.

Plateau stress results have been plotted in Figure 10 alongside the homogenized Young's modulus results (in the compression direction). It is seen that those topologies, such as cubic (■) and FCC-Z (↔), that have higher homogenized strengths for a given relative density also have higher resulting plateau stresses, and understandably, as relative density increases, the plateau stress increases as well. Notably, the rate of increase for the tesseract topology (◆) is lower in comparison to other topologies. While the data for the 10 m/s speed and initial impact energy of 100 J is presented in Figure 10, it is noted that, except for actual numerical result values, the trends and relations between topologies are generally the same for a speed of 100 m/s.

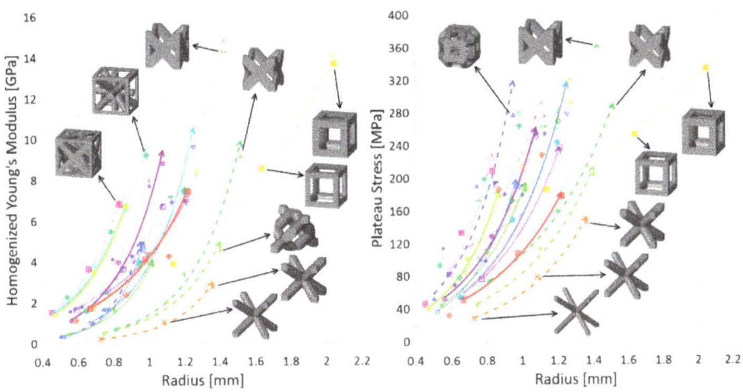

Figure 10. (**Left**) static homogenized Young's modulus (i.e., in compression direction) versus strut radius and (**right**) plateau stress versus strut radius (impact energy 100 J, speed 10 m/s). Legend as shown in Figure 1. Line types distinguish stretching (solid), bending (dotted), and mixed (dashed) deformation modes, discussed in Section 3.1. Line opacity indicates whether there is at least one strut aligned in the loading direction: opaque—no, semi-translucent—yes (see Table 6 for clear classification on whether strut(s) are aligned in loading direction or not). The arrow at the end of the line indicates increasing relative density.

In addition to an increase in plateau stress for an increase in relative density, the following trends were also generally observed, regardless of topology:

- *Decrease in densification strain*. Shown in Figure 11 (right), this statement is true for all topologies, except auxetic (⊕). The expectation for this decrease with an increase in relative density is understandable if one considers densification strain to represent the point at which the lattice has been compressed and begins behaving like the solid parent material; at lower relative densities, the lattice must be compressed *more* before it self-contacts and can behave as a monolithic material. Thus, the auxetic behavior is intriguing and will be discussed further in the following subsection.
- *Increase in absorbed energy up to densification strain*. Such a trend can be seen in Figure 12 (left). Since absorbed energy is related to both stress and strain, and it has been observed that the plateau stress increased with increasing relative density, this result is also unsurprising. However, given that the rate at which the plateau stress increased for the tesseract topology was notably lower in comparison to other topologies, the

absorbed energy actually *decreased* with an increase in relative density from 0.1 to 0.2; its decrease in densification strain was enough to cause a shift in direction. Additionally, the two highest-strength (and highest plateau-stress) topologies—cubic and FCC-Z—had a consistent internal energy at densification regardless of relative density; had the initial impact energy been increased beyond 100 J, it may have been possible for the trend of *increasing IE for increasing relative density* to be observed.

- *Decrease in energy absorption efficiency*. This decreasing trend can be seen in Figure 13, and more discussion regarding the EA efficiency is provided in the following subsection with regards to whether there are struts aligned in the loading direction.

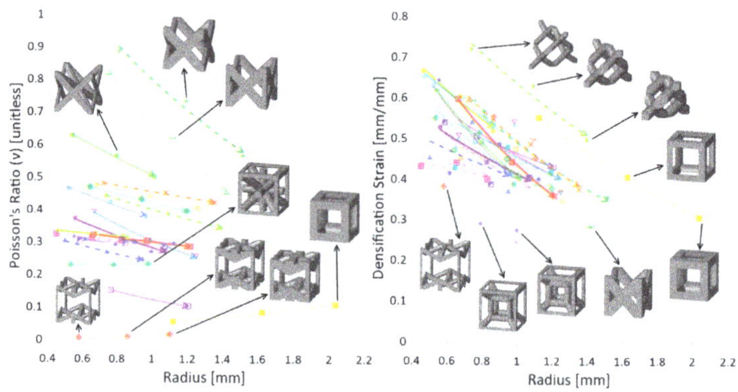

Figure 11. (**Left**) homogenized Poisson's ratio (in compression direction) versus strut radius and (**right**) densification strain versus strut radius (impact energy 100 J, speed 10 m/s). Legend as shown in Figure 1. Line types distinguish stretching (solid), bending (dotted), and mixed (dashed) deformation modes. Line opacity indicates whether there is at least one strut aligned in the loading direction: opaque—no, semi-translucent—yes (see Table 6 for clear classification on whether strut(s) are aligned in the loading direction or not). The arrow at the end of the line indicates increasing relative density.

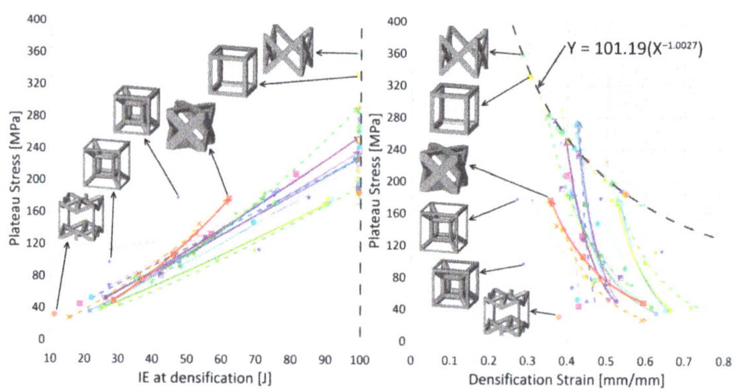

Figure 12. (**Left**) plateau stress versus IE at densification strain and (**right**) plateau stress versus densification strain. Both sets of data are for an impact energy of 100 J, speed of 10 m/s. Legend as shown in Figure 1. Line types distinguish stretching (solid), bending (dotted), and mixed (dashed) deformation modes. Line opacity indicates whether there is at least one strut aligned in the loading direction: opaque—no, semi-translucent—yes (see Table 6 for clear classification on whether strut(s) are aligned in the loading direction or not). The arrow at the end of the line indicates increasing relative density.

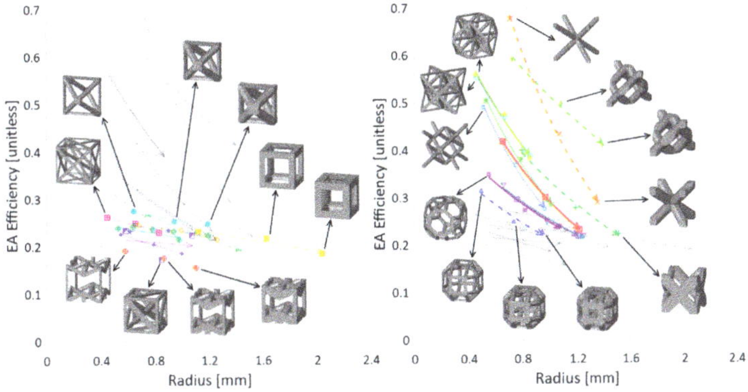

Figure 13. EA efficiency versus strut radius (impact energy 100 J, speed 10 m/s)—left: struts in loading direction; right—no struts in loading direction (see Table 6 for clear classification on whether strut(s) are aligned in loading direction or not). Legend as shown in Figure 1. Line types distinguish stretching (solid), bending (dotted), and mixed (dashed) deformation modes. The arrow at the end of the line indicates increasing relative density. Grey lines are for ease of comparison between the data in the two plots (grey data is found as color data in the other plot).

3.2.4. Effect of Topology

For a mass comparison of the crashworthiness parameters of the 24 topologies, Figures 12 and 13 illustrate some of the interesting relationships observed within the data.

- *Bending versus Stretching.* While bending-dominated cellular materials are known to be better energy absorbers and be compliant (lower strength and stiffness), and stretching-dominated cellular materials have higher stiffness and strength, the dynamic data collected for this work using the numerical model set up described suggests no noticeable delineation between these two categories of topologies (nor mixed-mode topologies), though it is noted that absorbed energy is measured up to the densification of the given topology, not up to a stationary strain point. Additionally, as has been noted by others in the literature, it appears that there is a non-negligible influence of the strut orientation with respect to the loading direction when it comes to deformation modes, which is not accounted for in Maxwell's criterion and can increase the strength of otherwise bending-dominated topologies.
- *Strut(s) aligned in loading direction.* With regards to energy absorption efficiency (Figure 13), there is, quite interestingly, a clear separation of performance when it comes to a topology having strut(s) aligned in the loading direction versus a topology not having any struts aligned in that direction. For those topologies with struts aligned in the loading direction— ⊕ ◇ ■ ⊢ ✣ ⊕ ● ★ ♦ ◩ ⊞ , Figure 13 (left)—the resulting maximum energy absorption efficiency during impact is lower than for topologies that do not have struts aligned in that direction (Figure 13 (right), ⊠ × ✱ ♦ ✳ + ⊕ ✲ ▲ □ ✵ ▽). For plateau stress (Figure 10 (right)), it is obvious that the cubic unit cell—which has struts aligned in the compression direction and fewer struts, leading to a larger overall strut radius to reach a given relative density—has a greater strength than other topologies when considering only a given relative density. Indeed, the topology that has the next highest plateau stress for a given relative density is FCC-Z, which also has struts aligned in the impact direction and has a similarly large radius for struts given the fewer total number of struts compared to other topologies (not including cubic). The BCC (×) topology, which has no struts aligned in the compression direction and is generally bending-dominated, has the lowest strength.
- *Topology Highlight: Auxetic (⊕).* Unlike other topologies, the densification strain increased with an increase in relative density for the auxetic topology. The Poisson's

ratio of this topology in relation to other topologies—Figure 11 (left)—is notably significantly smaller, almost equal to zero (generally, auxetic materials are those materials with negative Poisson's ratio, unlike other types of topologies).

- *Topology Highlight: Cubic* (■) *and FCC-Z* (⊢). These two topologies had noticeably larger plateau stresses for a given relative density than other topologies. For a constant relative density and unit cell height, these two topologies have the largest strut radius as compared to other topologies with struts aligned in the impact direction. Thus, the combination of large strut radius and struts aligned in the loading direction seems to have allowed for an increase in overall strength and higher plateau stress during impact. Notably, observing the response of the FCC-Z, FCC (+), diamond (♦), and BCC (×) topologies at a constant strut radius of approximately 1.4 mm in Figure 10 (left or right), it is the topology with struts aligned directly in the loading direction (FCC-Z) that has the highest Young's modulus and plateau stress, whereas the others reach less than 70% of those properties of the FCC-Z topology. Also interesting is that the cubic topology requires a larger strut radius to reach a similar strength and plateau stress as compared to the FCC-Z topology; at a radius of approximately 1.4 mm, it yields similar results to the FCC, diamond, and BCC topologies.
- *Topology Highlight: Tesseract* (◆). The tesseract topology was noted to have a lower rate of increase in plateau stress for an increase in strut radius (thus, relative density) (Figure 10 (right)). Together with a decrease in densification strain from a relative density of 0.1 to 0.2, the absorbed energy up to densification actually decreased with the increase in relative density (71.2 J vs. 27.6 J). However, with an additional 10% relative density from 0.2 to 0.3, the absorbed energy also increased (to 47.6 J), the results of which are plotted in Figure 12 (left).
- *Topology Highlight: Octahedron* (●), *Truncated Cube* (□), *Truncatedcuboctahedron* (✱), *Rhombicuboctahedron* (▲). Of the topologies with struts aligned in the impact direction, the octahedron topology is the only one that has two half-struts aligned in that direction, as opposed to four-quarter struts. Interestingly, the energy absorption efficiency results from Figure 13 (left) place the octahedron topology along the *strut-no strut* division. Also close to this split are the truncated cube, truncatedcuboctahedron, and rhombicuboctahedron (Figure 13 (right)), which do not have full struts aligned in the impact direction (i.e., a strut with a length equal to the unit height) but do have smaller-length struts aligned in the impact direction. Such an observation seems to suggest that while it is possible to observe a generally distinct division between the energy absorption efficiency results of Figure 13 based on whether or not there are struts aligned directly in the loading direction, the amount and length of those struts can shift the efficiency.

3.2.5. Discussion Concluding Remarks

Figure 12 plots plateau stress versus both absorbed energy up to densification and the actual densification strain. As expected, Figure 12 (left) shows a positive relationship between increasing plateau stress and absorbed energy—a higher plateau stress means more energy is absorbed during deformation. However, the energy of the system was limited by the initial impact energy of the impactor—100 J for Figure 12 —and, as such, there were some topologies/relative densities that, at densification, had absorbed the impact energy. During analysis of the data, there appears to be a relationship between the impact energy, plateau stress, and densification strain, shown in Figure 12 (right) with the limiting grey dashed curve. For a given initial impact energy, there were densification-plateau stress pairings that were unattainable; for an initial impact energy of 100 J, this area has been highlighted in Figure 12 (right), and the limiting curve equation has been provided. Other limiting curves for the lower initial impact energies—1 J, 5 J, 50 J—were also observed; they had similar power exponents (approximately equal to −1) but lower coefficients (due to their lower initial impact energies).

While this work dealt with results predicted by a developed numerical model, the material data itself came from additively manufactured samples. Since additive manufacturing technologies have greatly expanded the ability to research lattice materials, it is important to note their limitations, particularly in regard to minimum geometrical dimensions, such as the strut radii. From the results for these 24 topologies across the range of relative densities, it can be seen that for the range of strut diameters (approximately 0.8 mm to 4.4 mm), there is a variety of mechanical property sets that could be obtained simply by modifying the topology: Figure 10 (right) indicates that the plateau stress of a cube (■) at a relative density of 10% is almost equivalent to the plateau stress of an AFCC topology (⊠) at a relative density of 30%, allowing for a reduction in structure weight; Figure 11 (right) indicates that there is a minimal change in densification strain for changing relative density for the Xgrid topology (⊞); and Figure 13 shows that by simply removing the struts aligned in the impact direction, the energy absorption increases.

4. Conclusions

In an attempt to close the gap in the literature caused by the difficulty in comparing data between papers due to material, model, and setup differences, this work analyzed the dynamic impact performance of over 1,000 lattice simulations, which included over 24 topologies at three relative densities. To ensure the data captured represented a wide dynamic range, four different impact speeds and initial impact energy levels (each) were simulated using a developed numerical model containing a single unit cell with a material model sensitive to strain rate and validated boundary conditions. Several conclusions based on careful analysis of time-history-based data as well as crashworthiness parameters, including energy absorption and densification strain, were drawn:

- Despite the strain-rate-sensitive material properties of the SLM steel alloy, changes in impact performance from 1 m/s to 10 m/s were negligible (for the same initial impact energy). There were some changes between speeds of 10 m/s and 100 m/s, but general trends were similar regardless of the data set observed. At speeds of 1000 m/s, significant performance changes were observed, but it was determined that there was a large amount of uncertainty in results given the current numerical model setup—single unit cell with periodic boundary conditions—and that a multi-layered, larger cluster of cells would probably be better suited to illustrate the layer-by-layer collapse observed for high impact speeds with lattice materials.
- Increasing initial impact energy allowed for a better representation of the elastic-plateau-densification stress–strain curve characteristic of lattice materials. Lower impact energies whose stress–strain curves did not reach densification simply revealed a portion of the stress–strain curve, higher energies were better for performance comparison, given more representative values for densification strain and energy absorption up to densification.
- For increasing relative density, it was generally seen that the plateau stress increased, the densification strain decreased, the energy absorbed at densification increased, and the energy absorption efficiency decreased. A few exceptions include the auxetic and tesseract topologies.
- It was clear that topologies with struts aligned in the impact direction had lower energy absorption efficiencies as compared to topologies that had no struts aligned in that direction.
- Based on the numerical model setup, there was no clear separation in performance based on bending, stretching, or mixed deformation modes.

Author Contributions: Conceptualization, M.S.A.E.; methodology, A.R.B.; software, A.R.B.; validation, A.R.B.; formal analysis, A.R.B. and M.S.A.E.; investigation, A.R.B.; resources, M.S.A.E.; data curation, A.R.B.; writing—original draft preparation, A.R.B.; writing—review and editing, A.R.B. and M.S.A.E.; visualization, A.R.B.; supervision, M.S.A.E.; project administration, M.S.A.E.; funding acquisition, M.S.A.E. All authors have read and agreed to the published version of the manuscript.

Funding: This research was funded by Bombardier Inc. Montreal in collaboration with MITACS Canada, grant number IT29280.

Institutional Review Board Statement: Not applicable.

Informed Consent Statement: Not applicable.

Data Availability Statement: Data are contained within the article.

Conflicts of Interest: The authors declare no conflicts of interest. The funders had no role in the design of the study, in the collection, analyses, or interpretation of data, in the writing of the manuscript, or in the decision to publish the results.

References

1. Ashby, M.F. The Properties of Foams and Lattices. *Philos. Trans. R. Soc. A Math. Phys. Eng. Sci.* **2006**, *364*, 15–30. [CrossRef] [PubMed]
2. Gibson, L.J.; Ashby, M.F. *Cellular Solids: Structure and Properties*, 2nd ed.; Cambridge University Press: Cambridge, UK, 1997; ISBN 9780521499118.
3. Helou, M.; Kara, S. Design, Analysis and Manufacturing of Lattice Structures: An Overview. *Int. J. Comput. Integr. Manuf.* **2018**, *31*, 243–261. [CrossRef]
4. Tsopanos, S.; Mines, R.A.W.; McKown, S.; Shen, Y.; Cantwell, W.J.; Brooks, W.; Sutcliffe, C.J. The Influence of Processing Parameters on the Mechanical Properties of Selectively Laser Melted Stainless Steel Microlattice Structures. *J. Manuf. Sci. Eng.* **2010**, *132*, 0410111–04101112. [CrossRef]
5. Parthasarathy, J.; Starly, B.; Raman, S.; Christensen, A. Mechanical Evaluation of Porous Titanium (Ti6Al4V) Structures with Electron Beam Melting (EBM). *J. Mech. Behav. Biomed. Mater.* **2010**, *3*, 249–259. [CrossRef] [PubMed]
6. Banhart, J. Manufacture, Characterisation and Application of Cellular Metals and Metal Foams. *Prog. Mater. Sci.* **2001**, *46*, 559–632. [CrossRef]
7. Ashby, M.F. The Mechanical Properties of Cellular Solids. *Metall. Trans.* **1983**, *14A*, 1755–1769. [CrossRef]
8. Bhate, D.; Penick, C.A.; Ferry, L.A.; Lee, C. Classification and Selection of Cellular Materials in Mechanical Design: Engineering and Biomimetic Approaches. *Designs* **2019**, *3*, 19. [CrossRef]
9. Fleck, N.A.; Deshpande, V.S.; Ashby, M.F. Micro-Architectured Materials: Past, Present and Future. *Proc. R. Soc. A* **2010**, *466*, 2495–2516. [CrossRef]
10. Mueller, J.; Matlack, K.H.; Shea, K.; Daraio, C. Energy Absorption Properties of Periodic and Stochastic 3D Lattice Materials. *Adv. Theory Simul.* **2019**, *2*, 1900081. [CrossRef]
11. Deshpande, V.S.; Fleck, N.A.; Ashby, M.F. Effective Properties of the Octet-Truss Lattice Material. *J. Mech. Phys. Solids* **2001**, *49*, 1747–1769. [CrossRef]
12. Evans, A.G.; Hutchinson, J.W.; Fleck, N.A.; Ashby, M.F.; Wadley, H.N.G. The Topological Design of Multifunctional Cellular Metals. *Prog. Mater. Sci.* **2001**, *46*, 309–327. [CrossRef]
13. Yan, C.; Hao, L.; Hussein, A.; Young, P.; Raymont, D. Advanced Lightweight 316L Stainless Steel Cellular Lattice Structures Fabricated via Selective Laser Melting. *Mater. Des.* **2014**, *55*, 533–541. [CrossRef]
14. Park, K.-M.; Min, K.-S.; Roh, Y.-S. Design Optimization of Lattice Structures under Compression: Study of Unit Cell Types and Cell Arrangements. *Materials* **2022**, *15*, 97. [CrossRef] [PubMed]
15. Bhate, D. Four Questions in Cellular Material Design. *Materials* **2019**, *12*, 1060. [CrossRef] [PubMed]
16. Cao, X.; Ren, X.; Zhao, T.; Li, Y.; Xiao, D.; Fang, D. Numerical and Theoretical Analysis of the Dynamic Mechanical Behaviour of a Modified Rhombic Dodecahedron Lattice Structure. *Int. J. Mech. Mater. Des.* **2020**, *17*, 271–283. [CrossRef]
17. Cao, X.; Xiao, D.; Li, Y.; Wen, W.; Zhao, T.; Chen, Z.; Jiang, Y.; Fang, D. Dynamic Compressive Behavior of a Modified Additively Manufactured Rhombic Dodecahedron 316L Stainless Steel Lattice Structure. *Thin-Walled Struct.* **2020**, *148*, 106586. [CrossRef]
18. Wadley, H.N.G. Cellular Metals Manufacturing. *Adv. Eng. Mater.* **2002**, *4*, 726–733. [CrossRef]
19. Rashed, M.G.; Ashraf, M.; Mines, R.A.W.; Hazell, P.J. Metallic Microlattice Materials: A Current State of the Art on Manufacturing, Mechanical Properties and Applications. *Mater. Des.* **2016**, *95*, 518–533. [CrossRef]
20. Yuan, S.; Chua, C.K.; Zhou, K. 3D-Printed Mechanical Metamaterials with High Energy Absorption. *Adv. Mater. Technol.* **2019**, *4*, 1800419. [CrossRef]
21. Al-Saedi, D.S.J.; Masood, S.H.; Faizan-Ur-Rab, M.; Alomarah, A.; Ponnusamy, P. Mechanical Properties and Energy Absorption Capability of Functionally Graded F2BCC Lattice Fabricated by SLM. *Mater. Des.* **2018**, *144*, 32–44. [CrossRef]
22. Schaedler, T.A.; Carter, W.B. Architected Cellular Materials. *Annu. Rev. Mater. Res.* **2016**, *46*, 187–210. [CrossRef]
23. Mieszala, M.; Hasegawa, M.; Guillonneau, G.; Bauer, J.; Raghavan, R.; Frantz, C.; Kraft, O.; Mischler, S.; Michler, J.; Philippe, L. Micromechanics of Amorphous Metal/Polymer Hybrid Structures with 3D Cellular Architectures: Size Effects, Buckling Behavior, and Energy Absorption Capability. *Small* **2017**, *13*, 1602514. [CrossRef] [PubMed]
24. Ozdemir, Z.; Hernandez-Nava, E.; Tyas, A.; Warren, J.A.; Fay, S.D.; Goodall, R.; Todd, I.; Askes, H. Energy Absorption in Lattice Structures in Dynamics: Experiments. *Int. J. Impact Eng.* **2016**, *89*, 49–61. [CrossRef]

25. Yang, C.; Li, Q.M. Advanced Lattice Material with High Energy Absorption Based on Topology Optimisation. *Mech. Mater.* **2020**, *148*, 103536. [CrossRef]
26. Nasrullah, A.I.H.; Santosa, S.P.; Dirgantara, T. Design and Optimization of Crashworthy Components Based on Lattice Structure Configuration. *Structures* **2020**, *26*, 969–981. [CrossRef]
27. Wang, P.; Bian, Y.; Yang, F.; Fan, H.; Zheng, B. Mechanical Properties and Energy Absorption of FCC Lattice Structures with Different Orientation Angles. *Acta Mech.* **2020**, *231*, 3129–3144. [CrossRef]
28. Ozdemir, Z.; Tyas, A.; Goodall, R.; Askes, H. Energy Absorption in Lattice Structures in Dynamics: Nonlinear FE Simulations. *Int. J. Impact Eng.* **2017**, *102*, 1–15. [CrossRef]
29. Jin, N.; Wang, F.; Wang, Y.; Zhang, B.; Cheng, H.; Zhang, H. Failure and Energy Absorption Characteristics of Four Lattice Structures under Dynamic Loading. *Mater. Des.* **2019**, *169*, 107655. [CrossRef]
30. Lei, H.; Li, C.; Meng, J.; Zhou, H.; Liu, Y.; Zhang, X.; Wang, P.; Fang, D. Evaluation of Compressive Properties of SLM-Fabricated Multi-Layer Lattice Structures by Experimental Test and μ-CT-Based Finite Element Analysis. *Mater. Des.* **2019**, *169*, 107685. [CrossRef]
31. Tancogne-Dejean, T.; Spierings, A.B.; Mohr, D. Additively-Manufactured Metallic Micro-Lattice Materials for High Specific Energy Absorption under Static and Dynamic Loading. *Acta Mater.* **2016**, *116*, 14–28. [CrossRef]
32. Giorgio, I.; dell'Isola, F.; Steigmann, D.J. Second-Grade Elasticity of Three-Dimensional Pantographic Lattices: Theory and Numerical Experiments. *Contin. Mech. Thermodyn.* **2023**, 1–13. [CrossRef]
33. Tran, L.V.; Niiranen, J. A Geometrically Nonlinear Euler–Bernoulli Beam Model within Strain Gradient Elasticity with Isogeometric Analysis and Lattice Structure Applications. *Math. Mech. Complex Syst.* **2020**, *8*, 345–371. [CrossRef]
34. Dong, G.; Tang, Y.; Zhao, Y.F. A 149 Line Homogenization Code for Three-Dimensional Cellular Materials Written in MATLAB. *J. Eng. Mater. Technol.* **2019**, *141*, 011005. [CrossRef]
35. Tancogne-Dejean, T.; Mohr, D. Stiffness and Specific Energy Absorption of Additively-Manufactured Metallic BCC Metamaterials Composed of Tapered Beams. *Int. J. Mech. Sci.* **2018**, *141*, 101–116. [CrossRef]
36. Qi, D.; Yu, H.; Liu, M.; Huang, H.; Xu, S.; Xia, Y.; Qian, G.; Wu, W. Mechanical Behaviors of SLM Additive Manufactured Octet-Truss and Truncated-Octahedron Lattice Structures with Uniform and Taper Beams. *Int. J. Mech. Sci.* **2019**, *163*, 105091. [CrossRef]
37. Dong, L.; Deshpande, V.; Wadley, H. Mechanical Response of Ti-6Al-4V Octet-Truss Lattice Structures. *Int. J. Solids Struct.* **2015**, *60*, 107–124. [CrossRef]
38. Kaur, M.; Yun, T.G.; Han, S.M.; Thomas, E.L.; Kim, W.S. 3D Printed Stretching-Dominated Micro-Trusses. *Mater. Des.* **2017**, *134*, 272–280. [CrossRef]
39. Zhang, M.; Yang, Z.; Lu, Z.; Liao, B.; He, X. Effective Elastic Properties and Initial Yield Surfaces of Two 3D Lattice Structures. *Int. J. Mech. Sci.* **2018**, *138–139*, 146–158. [CrossRef]
40. Cao, X.; Duan, S.; Liang, J.; Wen, W.; Fang, D. Mechanical Properties of an Improved 3D-Printed Rhombic Dodecahedron Stainless Steel Lattice Structure of Variable Cross Section. *Int. J. Mech. Sci.* **2018**, *145*, 53–63. [CrossRef]
41. Ushijima, K.; Cantwell, W.J.; Chen, D.H. Prediction of the Mechanical Properties of Micro-Lattice Structures Subjected to Multi-Axial Loading. *Int. J. Mech. Sci.* **2013**, *68*, 47–55. [CrossRef]
42. Ushijima, K.; Cantwell, W.; Mines, R.; Tsopanos, S.; Smith, M. An Investigation into the Compressive Properties of Stainless Steel Micro-Lattice Structures. *J. Sandw. Struct. Mater.* **2011**, *13*, 303–329. [CrossRef]
43. Parisien, A. *Characterization and Design Optimization of Periodic Cellular Solids for Optimal Bone Ingrowth in Osseointegration Implants*; Carleton University: Ottawa, ON, Canada, 2021.
44. Altair Engineering Inc. *Altair Radioss 2022: Theory Manual*; Altair Engineering Inc.: Troy, MI, USA, 2022.
45. Altair Engineering Inc. *Altair Radioss 2022: Reference Guide*; Altair Engineering Inc.: Troy, MI, USA, 2022.
46. Labeas, G.N.; Sunaric, M.M. Investigation on the Static Response and Failure Process of Metallic Open Lattice Cellular Structures. *Strain* **2010**, *46*, 195–204. [CrossRef]
47. Bernard, A.R.; Elsayed, M.S.A. Impact Resistance and Specific Energy Absorption Efficiency Limits of Periodic Cellular Solids. In *AIAA SCITECH 2022 Forum*; AIAA Inc.: Reston, VA, USA, 2022. [CrossRef]
48. Chen, Z.; Wang, Z.; Zhou, S.; Shao, J.; Wu, X. Novel Negative Poisson's Ratio Lattice Structures with Enhanced Stiffness and Energy Absorption Capacity. *Materials* **2018**, *11*, 1095. [CrossRef] [PubMed]
49. Li, Q.M.; Magkiriadis, I.; Harrigan, J.J. Compressive Strain at the Onset of Densification of Cellular Solids. *J. Cell. Plast.* **2006**, *42*, 371–392. [CrossRef]
50. Lu, G.; Yu, T. *Energy Absorption of Structures and Materials*; Woodhead Publishing Limited: Cambridge, UK, 2003; ISBN 0849317681.
51. Xu, P.; Yang, C.; Peng, Y.; Yao, S.; Xing, J.; Li, B. Cut-out Grooves Optimization to Improve Crashworthiness of a Gradual Energy-Absorbing Structure for Subway Vehicles. *Mater. Des.* **2016**, *103*, 132–143. [CrossRef]
52. Ha, N.S.; Lu, G. Thin-Walled Corrugated Structures: A Review of Crashworthiness Designs and Energy Absorption Characteristics. *Thin-Walled Struct.* **2020**, *157*, 106995. [CrossRef]
53. Maxwell, J.C. On the Calculation of the Equilibrium and Stiffness of Frames. *Lond. Edinb. Dublin Philos. Mag. J. Sci.* **1864**, *27*, 294–299. [CrossRef]
54. Leary, M.; Mazur, M.; Elambasseril, J.; McMillan, M.; Chirent, T.; Sun, Y.; Qian, M.; Easton, M.; Brandt, M. Selective Laser Melting (SLM) of AlSi12Mg Lattice Structures. *Mater. Des.* **2016**, *98*, 344–357. [CrossRef]

55. Leary, M.; Mazur, M.; Williams, H.; Yang, E.; Alghamdi, A.; Lozanovski, B.; Zhang, X.; Shidid, D.; Farahbod-Sternahl, L.; Witt, G.; et al. Inconel 625 Lattice Structures Manufactured by Selective Laser Melting (SLM): Mechanical Properties, Deformation and Failure Modes. *Mater. Des.* **2018**, *157*, 179–199. [CrossRef]
56. Calladine, C.R. Buckminster Fuller's "Tensegrity" Structures and Clerk Maxwell's Rules for the Construction of Stiff Frames. *Int. J. Solids Struct.* **1978**, *14*, 161–172. [CrossRef]
57. Pellegrino, S.; Calladine, C.R. Matrix Analysis of Statically and Kinematically Indeterminate Frameworks. *Int. J. Solids Struct.* **1986**, *22*, 409–428. [CrossRef]
58. Tarnai, T. Simultaneous Static and Kinematic Indeterminacy of Space Trusses with Cyclic Symmetry. *Int. J. Solids Struct.* **1980**, *16*, 347–359. [CrossRef]
59. Altamimi, S.; Lee, D.W.; Barsoum, I.; Rowshan, R.; Jasiuk, I.M.; Abu Al-Rub, R.K. On Stiffness, Strength, Anisotropy, and Buckling of 30 Strut-Based Lattices with Cubic Crystal Structures. *Adv. Eng. Mater.* **2022**, *24*, 2101379. [CrossRef]
60. Parisien, A.; ElSayed, M.S.A.; Frei, H. Mechanoregulation Modelling of Stretching versus Bending Dominated Periodic Cellular Solids. *Mater. Today Commun.* **2022**, *33*, 104315. [CrossRef]
61. Wagner, M.A.; Lumpe, T.S.; Chen, T.; Shea, K. Programmable, Active Lattice Structures: Unifying Stretch-Dominated and Bending-Dominated Topologies. *Extreme Mech. Lett.* **2019**, *29*, 100461. [CrossRef]
62. Bauer, J.; Hengsbach, S.; Tesari, I.; Schwaiger, R.; Kraft, O. High-Strength Cellular Ceramic Composites with 3D Microarchitecture. *Proc. Natl. Acad. Sci. USA* **2014**, *111*, 2453–2458. [CrossRef] [PubMed]
63. Li, S.J.; Xu, Q.S.; Wang, Z.; Hou, W.T.; Hao, Y.L.; Yang, R.; Murr, L.E. Influence of Cell Shape on Mechanical Properties of Ti–6Al–4V Meshes Fabricated by Electron Beam Melting Method. *Acta Biomater.* **2014**, *10*, 4537–4547. [CrossRef] [PubMed]
64. Habib, F.N.; Iovenitti, P.; Masood, S.H.; Nikzad, M. Fabrication of Polymeric Lattice Structures for Optimum Energy Absorption Using Multi Jet Fusion Technology. *Mater. Des.* **2018**, *155*, 86–98. [CrossRef]
65. Zheng, X.; Lee, H.; Weisgraber, T.H.; Shusteff, M.; DeOtte, J.; Duoss, E.B.; Kuntz, J.D.; Biener, M.M.; Ge, Q.; Jackson, J.A.; et al. Ultralight, Ultrastiff Mechanical Metamaterials. *Science* **2014**, *344*, 1373–1377. [CrossRef]
66. Zhang, J.; Huang, H.; Liu, G.; Zong, H.; Zhang, C. Stiffness and Energy Absorption of Additive Manufactured Hybrid Lattice Structures. *Virtual Phys. Prototyp.* **2021**, *16*, 428–443. [CrossRef]
67. Athanker, P.; Singh, A.K. Elastic and Elasto-Plastic Analysis of Ti6Al4V Micro-Lattice Structures under Compressive Loads. *Math. Mech. Solids* **2021**, *26*, 591–615. [CrossRef]
68. Arabnejad, S.; Burnett Johnston, R.; Pura, J.A.; Singh, B.; Tanzer, M.; Pasini, D. High-Strength Porous Biomaterials for Bone Replacement: A Strategy to Assess the Interplay between Cell Morphology, Mechanical Properties, Bone Ingrowth and Manufacturing Constraints. *Acta Biomater.* **2016**, *30*, 345–356. [CrossRef] [PubMed]
69. Nordmann, J.; Aßmus, M.; Altenbach, H. Visualising Elastic Anisotropy: Theoretical Background and Computational Implementation. *Contin. Mech. Thermodyn.* **2018**, *30*, 689–708. [CrossRef]

materials

Article

A Study on the Critical Saturation Response Characteristics of Simple and Sandwich Cylindrical Shells under Long-Duration Blast Loading

Mao Yang [1], Jun Zhang [1], Yunfei Mu [1], Hanjun Huang [1], Bin Han [2] and Yongjian Mao [1,*]

1 Institute of Systems Engineering, China Academy of Engineering Physics, Mianyang 621999, China; 15050822354@163.com (M.Y.); hjzhangj@caep.cn (J.Z.); muyunfei@caep.cn (Y.M.); huanghj@caep.cn (H.H.)
2 School of Mechanical Engineering, Xi'an Jiaotong University, Xi'an 710049, China; hanbinghost@xjtu.edu.cn
* Correspondence: maoyj@caep.cn

Abstract: Experimental research and numerical simulations of the structural response to shock waves with pulse durations of hundreds of milliseconds, or even seconds, are extremely challenging. This paper takes typical single-layer and sandwich cylindrical shells as the research objects. The response rules of cylindrical shells under long-duration blast loadings were studied. The results show that when the pulse duration is greater than or equal to 4~5 times the first-order period of the structure, the maximum response of the structure tends to be consistent, that is, the maximum response of the cylindrical shells with different vibration shapes shows a saturation effect as the pulse duration increases. This study established the relationship between the saturation loading time and the inherent characteristics of the structure. It was found that the saturation effect was applicable under the following conditions, including different load waveforms, elastic–plastic deformation of the structure, and the loading object being a sandwich shell. This will help transform the long-duration explosion wave problem into a finite pulse-duration shock wave problem that can be realized by both experiments and numerical simulations.

Keywords: critical saturation response; cylindrical shells; long-duration blast loading; inherent characteristics; sandwich structures

Citation: Yang, M.; Zhang, J.; Mu, Y.; Huang, H.; Han, B.; Mao, Y. A Study on the Critical Saturation Response Characteristics of Simple and Sandwich Cylindrical Shells under Long-Duration Blast Loading. *Materials* **2024**, *17*, 1990. https://doi.org/10.3390/ma17091990

Academic Editor: Aniello Riccio

Received: 14 March 2024
Revised: 16 April 2024
Accepted: 20 April 2024
Published: 25 April 2024

1. Introduction

Shock wave parameters are closely related to the explosion yield and distance. Small- and medium-yield shock waves have generally short pulse durations of microseconds to milliseconds, while the shock waves of large-yield explosions exhibit long durations (tens to hundreds of milliseconds or more) in the medium and far fields [1,2]. Wars and large-scale explosion disasters all involve long-duration blast wave loads. However, it is very difficult to directly conduct experimental research and numerical simulations of structural responses under the action of long-duration blast loading of hundreds of milliseconds or even seconds. On the one hand, long-pulse explosion impact experiments are huge in scale and extremely low in cost-effectiveness [3,4]; on the other hand, numerical simulations require huge computing resources [5,6].

Previous research has found that if the impact load lasts for a long time, the saturation impulse phenomenon will appear in the dynamic response of the structure [7], that is, when the plate is subjected to a strong transverse pressure pulse load, it will produce large deformation, and the membrane force induced by the large deformation will enhance the load-bearing capacity of the plate. If the plate is subjected to a long enough rectangular pressure pulse, after the pulse load reaches the saturation time, the deformation mode of the structure will no longer continue to change with the increasing loading time. Regarding the research on saturation impulse theory, Zhao et al. [7,8] discovered, for the first time, the saturation impulse phenomenon that occurs in the large-deflection plastic dynamic response of structures under moderate-intensity pulse loads. Moreover, a reasonable explanation was given for the saturation impulse phenomenon, and the dimensionless saturation impulse values of a simply supported beam or fixed beam were given. On the basis of the above, Zhu et al. [9] further studied the elastic–plastic dynamic response of a square plate and proposed two types of saturation impulses corresponding to the maximum deformation and permanent deformation based on the elastic–plastic analysis. Zhu et al. [10] verified that the saturation impulse phenomenon satisfies the scale rate by analyzing the saturation impulse phenomenon of square plates of geometrically similar bilinear elastic–plastic materials. In addition, the concept of a saturation impulse is not limited to structures such as beams, circular plates, and square plates, but is also applicable to other complex structures, such as stiffened plates [7]. Xi et al. [11] studied the rigid-plastic response of a hinged circular plate under uniform pulse loading and discovered the saturation impulse phenomenon in the high load range. Xi et al. [12] took steel beams as their research object and pointed out that for strain rate-sensitive structures that undergo plastic deformation under pulse loads, the occurrence of the saturation impulse phenomenon depends on two necessary parameters: the pulse amplitude and length. In addition, Zhu et al. [13] also studied the effects of the aspect ratio and boundary conditions on the saturation impulse of rectangular plates; the effects of material strain rate sensitivity and strain hardening on the saturation impulse [14]; the saturation impulse of a square plate under different pulse loads [15]; the saturation impulse of a square plate considering a moving hinge [16]; and the saturation impulse of a beam taking into account the transient response stage and the accurate yield surface [17]. The above saturation impulse method reflects the relationship between the load impulse and structural characteristics and provides a simple and reliable calculation method for structural deformation calculations under an explosion load. In general, the research on saturation impulse theory has become relatively mature.

The above saturation impulse method is mainly used in the dynamic response analysis of structures under conventional air burst loads. Conventional air burst loads are characterized by a high shock wave pressure attenuation rate and a short action time, which are far lower than the response time of the structure. Therefore, the dynamic deformation and failure of the structure are often only related to the impulse of the shock wave load. Under the principle of impulse equivalence, the structural response has nothing to do with the shape of the shock wave load [18,19].

However, the shock wave of a large-yield explosion exhibits a long duration (tens to hundreds of milliseconds or more) in the mid-to-far field, and the duration of a nuclear explosion shock wave can even reach the order of seconds. The above impulse equivalence method may no longer be applicable. However, it is very difficult to directly conduct experimental research and numerical simulations of structural responses under the action of long-duration blast loading of hundreds of milliseconds or even seconds. This paper draws on the research ideas of the saturation impulse method to establish the relationship between the saturation loading time of the maximum response of a cylindrical shell under long-duration blast loading and the first-order period of the structure and discusses its applicability. This will help transform the long-duration shock wave problem into a finite pulse-duration (saturation loading time) shock wave problem that can be more easily realized by both experiments and numerical simulations.

2. Finite Element Model

In subsequent studies, four types of single-layer cylindrical shells and one corrugated sandwich cylindrical shell (CSCS) were used as the research objects. The specific dimensions of the single-layer cylindrical shells are shown in Table 1, wherein the structure numbers are S1, S2, S3, and S4, respectively. The geometric dimensions of the CSCS are shown in Figure 1, and the specific geometric dimensions are shown in Table 2.

Table 1. Geometric dimensions of cylindrical shells.

Num.	Diameter (mm)	Height (mm)	Wall Thickness (mm)
S1	1000	1500	40
S2	1000	1500	80
S3	3000	1000	40
S4	2000	1500	40

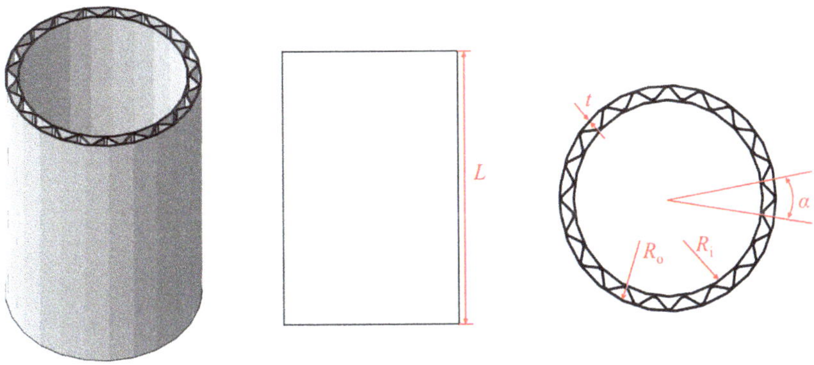

Figure 1. Geometric schematic diagram of corrugated sandwich cylindrical shell.

Table 2. Geometric dimensions of corrugated sandwich cylindrical shell.

Num.	L (mm)	R_i (mm)	R_o (mm)	α (°)	t (mm)
CSCS	1500	450	500	18	10

The finite element analysis of the mechanical response of the above-mentioned shell under the action of dynamic pressure is realized by the nonlinear explicit FE algorithm. The finite element model of S1 is described in detail below. The model settings of the other structures are basically the same as those of S1. The finite element model of S1 is shown in Figure 2. The bottom end of the cylindrical shell is fixed, and one side of the shell (the red area in Figure 1) is subjected to dynamic-pressure loads inward along the normal direction. In addition, the vibration characteristics of the cylindrical shell are obtained via the linear perturbation analysis method based on a Lanczos eigensolver. The finite element model uses a 1:1 3D model. In this study, Q235, a common metal material in engineering structures, is used as a representative material. In Sections 3.1 and 3.2, only elastic deformation occurs, the material's constitutive model is a linear elastic constitutive model, and the material properties are a density of 7.8 g/cm^3, an elastic modulus of 2×10^5 MPa, and Poisson's ratio of 0.33. In Sections 3.3 and 3.4, the structure undergoes plastic deformation, the material's constitutive model in the finite element model is the Johnson–Cook model, and the specific parameters are shown in Table 3 [20], among which A is the initial yield stress, B is the strain hardening modulus of the material, n is the hardening index of the material, and m is the thermal softening index of the material. A four-node shell element is used, and upon checking the convergence of the numerical solutions, the optimal mesh size is selected as 20 mm × 20 mm.

Table 3. Parameters of constitutive model for steel.

ρ (g/cm^{-3})	E (MPa)	ν	A (MPa)	B (MPa)	n	m
7.8	2×10^5	0.33	293.8	230.2	0.578	0.706

Considering that the main purpose of this article is to verify the saturation effect, when this effect is satisfied for a certain strain component, other strain values will also satisfy the corresponding law. Therefore, the typical strain (axial strain component LE11) of the shell is extracted herein as a representative for this study. The response of the structure is to extract the LE11 of the "Response point", and the specific location of the "Response point" is shown in Figure 2.

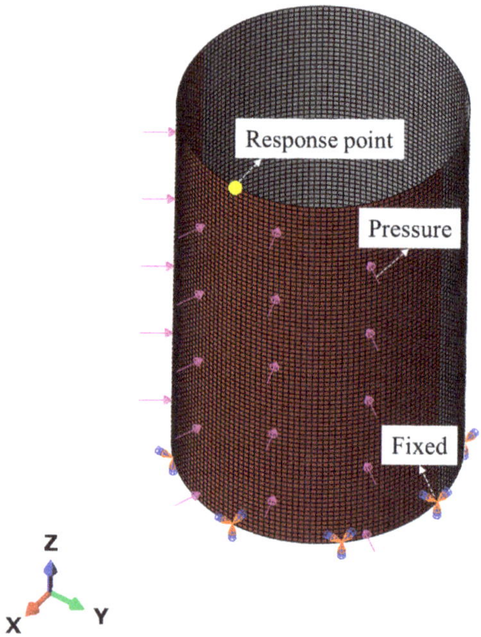

Figure 2. Finite element model of S1.

3. Results and Discussion

3.1. A Preliminary Exploration of the Saturation Effect

Analyzing the free vibration characteristics of the cylindrical shells, it can be seen in Figure 3 that the first-order vibration modes of structures S1 and S2 are both breathing vibration modes. In addition, the first-order natural frequency of S1 is 168.91 Hz, that is, the first-order period T = 5.9 ms. The first-order natural frequency of S2 is 256.45 Hz, that is, the first-order period T = 3.9 ms.

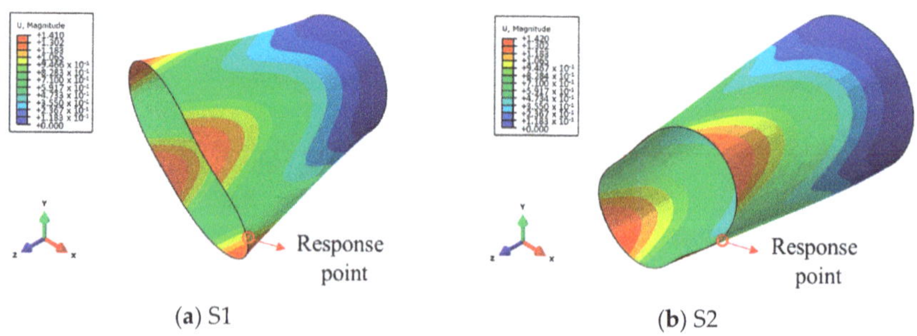

Figure 3. The first-order vibration shapes of the structures of S1 and S2.

In order to analyze the saturation response characteristics of the structure under the action of long-duration, dynamic-pressure loads with different pulse durations are applied on one side of the outer surface of the cylindrical shell. The load waveform is a rectangular wave with an amplitude of 1 MPa. The pulse duration (t_d) is N times the first-order period of the structure (that is, $t_d = NT$, whereby for S1, when $N = 1$, $t_d = 5.9$ ms). In subsequent studies, dynamic loads with different pulse durations, different waveforms, and different amplitudes will be applied. In order to more intuitively compare the differences between the different loads, Figure 4 shows the pressure–time curves of the dynamic loads of rectangular waves and sawtooth waves with different pulse widths when $t_d = 5.9$ ms.

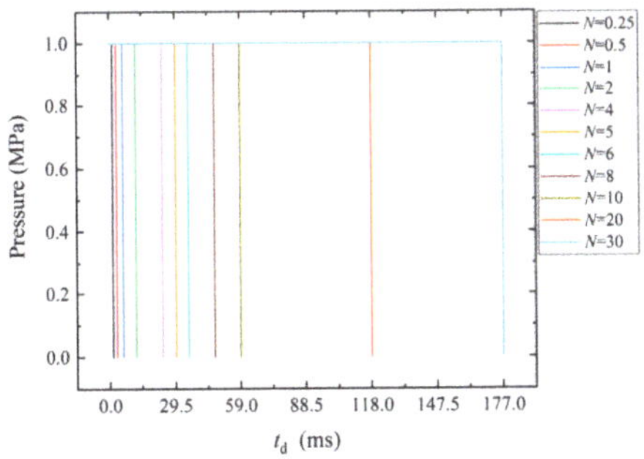

(a)

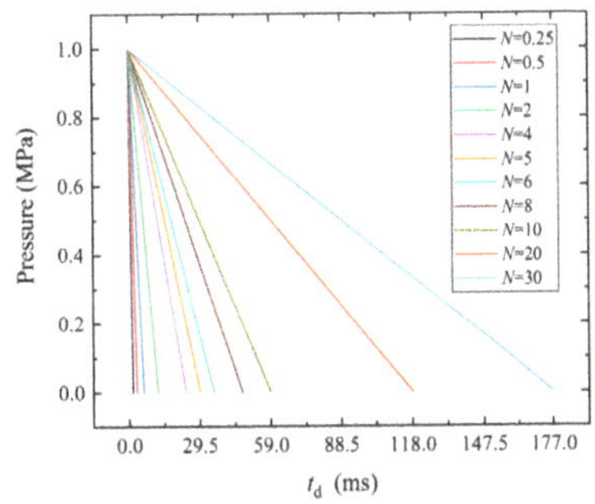

(b)

Figure 4. The pressure–time curves of dynamic loads with different pulse widths when $t_d = 5.9$ ms. (**a**) Rectangular waves; (**b**) sawtooth waves.

Combining the data of the maximum strain and its occurrence time, as shown in Figure 5, it can be seen that when $N < 5$, the maximum strain and its occurrence time fluctuate as N increases. When $N \geq 5$, the maximum strain and its occurrence time tend to be stable. Further analysis of the data shows that when $N \geq 5$; the error between the maximum value (0.41384×10^{-4}) and the minimum value (0.4006×10^{-4}) of the maximum strain of S1 is only 3.2%; and the error between the maximum value (0.14454×10^{-3}) and the minimum value (0.144×10^{-3}) of the maximum strain of S2 is only 0.4%. Such an error can obviously satisfy subsequent practical engineering applications. The structural response shows a saturation effect as the load pulse duration increases.

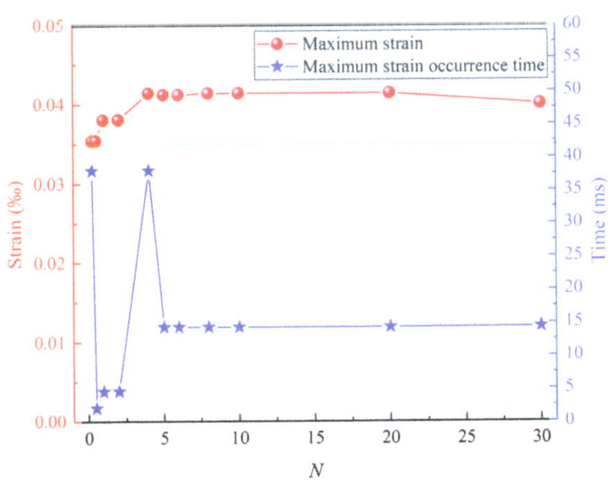

(**a**) S1

Figure 5. *Cont.*

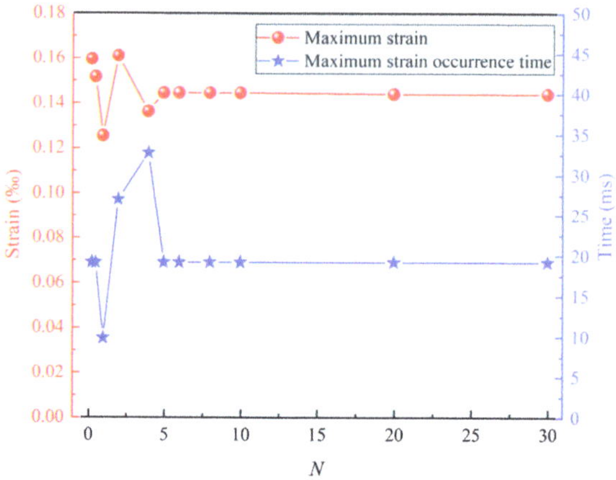

(**b**) S2

Figure 5. The maximum strain and its occurrence time.

The above research found that under the action of rectangular wave loads with different pulse durations, the maximum strain response of the cylindrical shell showed a saturation effect as the pulse duration increased. However, the above results are satisfied under the following conditions: the load is a rectangular wave, the free vibration mode of the structure is the breathing mode, the structural deformation is elastic deformation, and the structure is a simple single-layer cylindrical shell. In subsequent sections, the above conditions will be studied to explore the applicability of this saturation effect.

3.2. *Influence of Structural Vibration Shape and Load Waveform*

The vibration shape of a structure is related to its structural form, material properties, etc. In this study, in order to explore whether the saturation effect is applicable to structures with different vibration modes, three structures (S1, S2, and S3) with different geometric sizes and different vibration modes were selected as representatives for this research. The specific geometric dimensions are shown in Table 1. The free vibration modes of S1, S3, and S4 are the breathing mode, bending mode, and triangle mode, respectively, as shown in Figure 6. In addition, the first-order natural frequency of S1 is 168.91 Hz, that is, the first-order period T = 5.9 ms; the first-order natural frequency of S3 is 97.01 Hz, that is, the first-order period T = 10.3 ms; and the first-order natural frequency of S4 is 137.0 Hz, that is, the first-order period T = 7.3 ms.

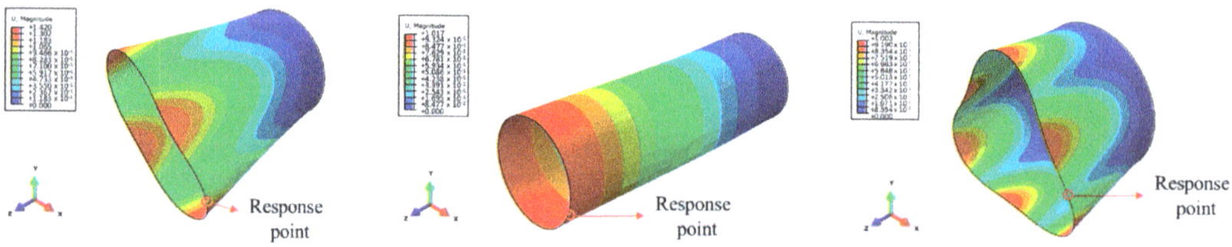

(**a**) S1 (**b**) S3 (**c**) S4

Figure 6. The first-order vibration shapes of the structures of S1, S3 and S4S..

Dynamic-pressure loads with different pulse durations are applied to the outer surfaces of the above-mentioned structures S1, S3, and S4. The load waveform is a rectangular wave or a sawtooth wave, the amplitude is 1 MPa, and the pulse duration (t_d) is N times the first-order period of the structure. Figure 6 shows the maximum strain data of S1, S3, and S4 under different pulse durations. It can be seen in Figure 7 that when $N < 5$, the maximum strain fluctuates with the increase in N. When $N \geq 5$, the maximum strain tends to be stable. Further analysis of the data shows that when the loads are rectangular waves and $N \geq 5$, the error between the maximum value (0.41384×10^{-4}) and the minimum value (0.4006×10^{-4}) of the maximum strain of S1 is only 3.2%; the error between the maximum value (0.1957×10^{-3}) and the minimum value (0.189×10^{-3}) of the maximum strain of S3 is only 3.4%; and the error between the maximum value (0.1401×10^{-3}) and the minimum value (0.137×10^{-3}) of the maximum strain of S4 is only 2.2%. When the loads are sawtooth waves and $N \geq 5$, the error between the maximum value (0.378×10^{-4}) and the minimum value (0.355×10^{-4}) of the maximum strain of S1 is only 6.1%; the error between the maximum value (0.172×10^{-3}) and the minimum value (0.167×10^{-3}) of the maximum strain of S3 is only 2.9%; and the error between the maximum value (0.134×10^{-4}) and the minimum value (0.125×10^{-4}) of the maximum strain of S4 is only 6.7%. Such an error can obviously satisfy subsequent practical engineering applications. In summary, the above saturation effects still exist for cylindrical shells with different vibration shapes under rectangular-wave and sawtooth-wave loads.

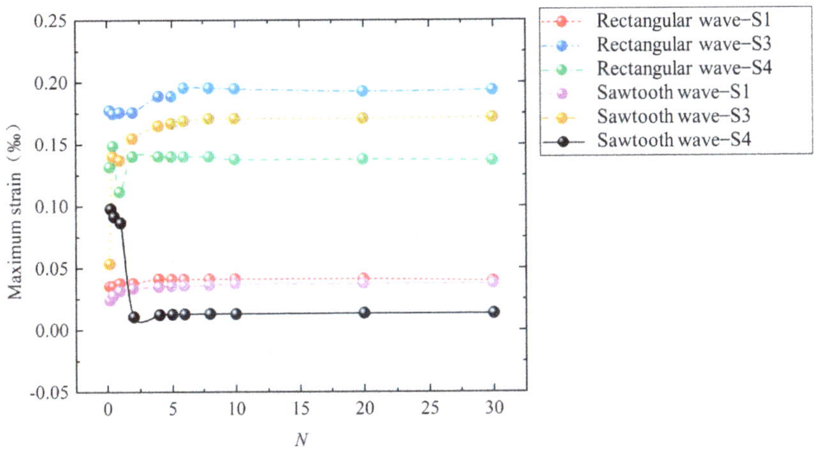

Figure 7. Maximum strain of the structures under different loading pulse durations and different loading waveforms.

3.3. *Influence of Load Amplitude*

Dynamic-pressure loads with different amplitudes and different pulse durations are applied to the outer surface of the above-mentioned structure S1. The load waveform is a rectangular wave, the amplitude is 1 MPa~10 MPa, and the pulse duration (t_d) is N times the first-order period of the structure. Figure 8 shows the maximum strain data of S1 under different pulse durations. It can be seen in Figure 8 that when $N < 4$, the maximum strain fluctuates with the increase in N. When $N \geq 5$, the maximum strain tends to be stable. Figure 9 shows the deformation mode of the structure when the amplitude is 10 MPa. It can be seen in Figure 9 that when $N > 1$, the maximum stress of the structure exceeds the yield strength of Q235, that is, the structure begins to undergo plastic deformation. Moreover, when $N \geq 5$, the deformation mode of the structure basically no longer changes. Further analysis of the data shows that when the load amplitude is 1 MPa and $N \geq 5$, the error between the maximum value (0.41384×10^{-4}) and the minimum value (0.4006×10^{-4}) of the maximum strain of S1 is only 3.2%; when the load amplitude is 2 MPa and $N \geq 5$, the error between the maximum value (0.832×10^{-4}) and the minimum value (0.811×10^{-4}) of the maximum strain of S1 is only 2.5%; when the load amplitude is 4 MPa and $N \geq 5$, the error between the maximum value (0.173×10^{-3}) and the minimum value (0.168×10^{-3}) of the maximum strain of S1 is only 2.9%; when the load amplitude is 6 MPa and $N \geq 5$, the error between the maximum value (0.263×10^{-3}) and the minimum value (0.257×10^{-3}) of the maximum strain of S1 is only 2.3%; and when the load amplitude is 8 MPa and $N \geq 5$, the error between the maximum value (5.7×10^{-3}) and the minimum value (5.7×10^{-3}) of the maximum strain of S1 is only 0.0%. Such an error can obviously satisfy subsequent practical engineering applications. In summary, it can be seen that for different amplitude loads, whether the structure undergoes elastic deformation or large plastic deformation, the above-mentioned saturation effect still exists.

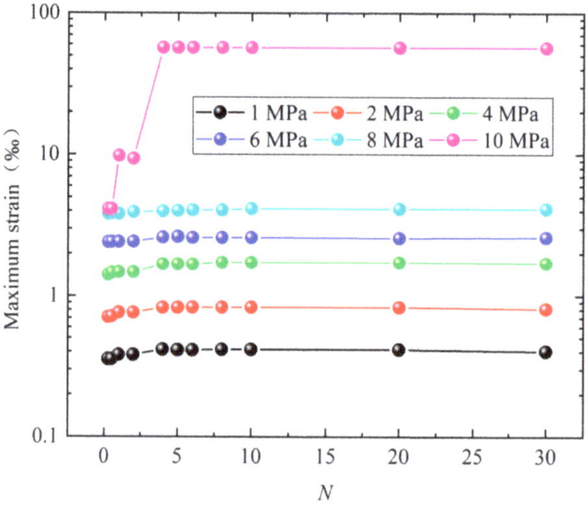

Figure 8. Maximum strain of structure under different load amplitudes.

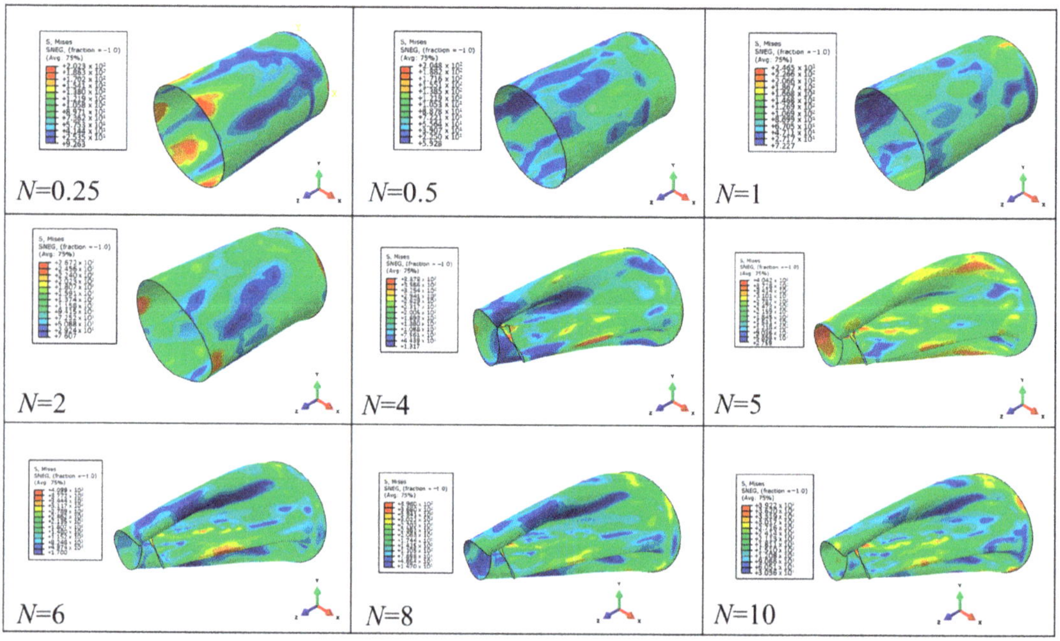

Figure 9. Deformation modes of the structure under different pulse durations when the amplitude is 10 MPa.

3.4. Sandwich Shells

The structures studied above were all single-layer cylindrical shells. This section will study more complex cylindrical shells. As a novel structure, sandwich cylindrical shells have excellent mechanical properties and have been widely researched in recent years [21–25]. This section takes a corrugated sandwich cylindrical shell as the research object to explore the applicability of the above saturation effect.

When large plastic deformation occurs, mutual extrusion and friction will occur between the components of the sandwich shell. This is a significant difference between a sandwich shell and a single-layer shell. However, there is no such difference during elastic deformation. Therefore, this paper discusses the situation of large plastic deformation of a sandwich shell. The load waveform is a rectangular wave, the amplitude is 8 MPa, and the pulse duration (t_d) is N times the first-order period of the structure.

Figure 10 shows the maximum strain data of the corrugated sandwich cylindrical shell under different pulse durations. It can be seen in Figure 10 that when $N < 4$, the maximum strain fluctuates with the increase in N. When $N \geq 4$, the maximum strain tends to be stable. Figure 10 shows the deformation mode of the structure. It can be seen that the structure has undergone large plastic deformation, but when $N \geq 4$, the deformation mode basically no longer changes. Further analysis of the data shows that when $N \geq 4$, the error between the maximum value (0.27×10^{-3}) and the minimum value (0.26×10^{-3}) of the maximum strain of S1 is only 3.7%. Such an error can obviously satisfy subsequent practical engineering applications. In summary, the above saturation effects apply to both simple single-layer cylindrical shells and corrugated sandwich cylindrical shells.

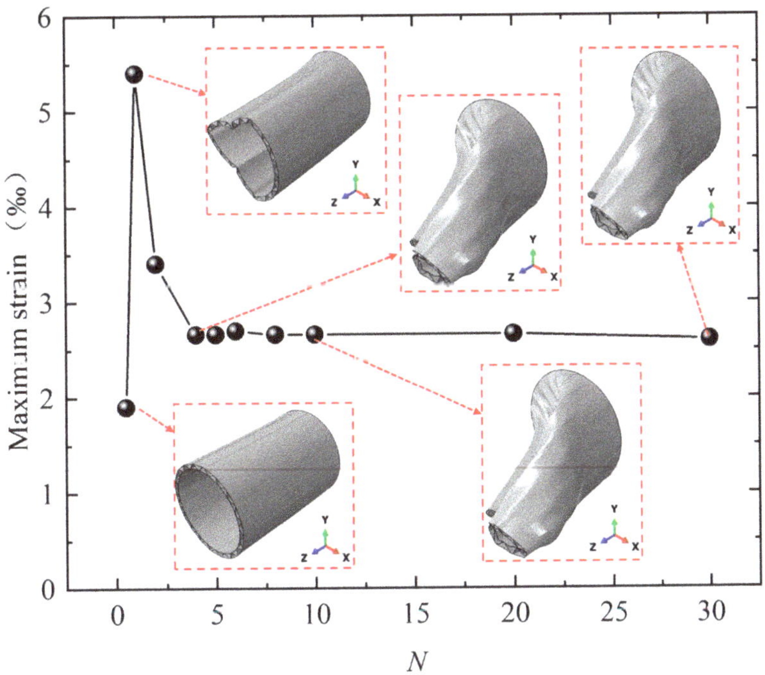

Figure 10. Maximum strain and deformation modes of corrugated sandwich cylindrical shells under different load amplitudes.

4. Conclusions

This article focused on the difficult problem of experimental research and numerical simulations of structural responses under the action of shock waves with long durations of hundreds of milliseconds or even seconds. This study took the typical cylindrical shell structure as the research object. The response rules of cylindrical shells under long-duration blast loadings were studied, and the following conclusions were found:

1. When the pulse duration of the load is greater than or equal to four to five times the first-order period of the structure, the maximum response of the structure tends to be consistent. That is, the maximum response of the cylindrical shell of different vibration modes (the breathing mode, bending mode, and triangle mode) shows a saturation effect as the pulse duration increases.
2. When the load waveform is a typical rectangular wave or sawtooth wave, the above saturation effect is applicable.
3. Whether the structure undergoes elastic deformation or large plastic deformation, the above saturation effect is applicable.
4. The above saturation effect applies to both simple single-layer cylindrical shells and relatively complex structures (i.e., corrugated sandwich cylindrical shells).

This study established the relationship between the saturation loading time of the maximum response of the cylindrical shell under long-duration blast loading and the inherent characteristics of the structure. After the load reaches the saturation loading time, the maximum response of the structure no longer changes as the pulse duration increases. This will help transform the long-duration shock wave problem into a finite pulse-duration (saturation loading time) shock wave problem that can be more easily realized by both experiments and numerical simulations. In the future, the mechanism will be further studied, and a unified theory will be established together with the short pulse duration saturation effect mentioned above.

Author Contributions: Methodology, M.Y., J.Z., B.H. and Y.M. (Yongjian Mao); Software, M.Y.; Validation, M.Y.; Investigation, M.Y., J.Z., Y.M. (Yunfei Mu), H.H. and Y.M. (Yongjian Mao); Resources, Y.M. (Yongjian Mao); Writing—original draft, M.Y.; Writing—review & editing, Y.M. (Yongjian Mao). All authors have read and agreed to the published version of the manuscript.

Funding: This work was supported by the National Natural Science Foundation of China (12072334).

Institutional Review Board Statement: Not applicable.

Informed Consent Statement: Not applicable.

Data Availability Statement: Data are contained within the article.

Conflicts of Interest: The authors declare no conflict of interest.

Nomenclature

T	First-order period of cylindrical shells
t_d	The pulse duration
N	$t_d = NT$
L	The height of the sandwich cylindrical shell
R_i	The radius of the inner panel of the sandwich cylindrical shell
R_o	The radius of the outer panel of the sandwich cylindrical shell
α	A single corrugation corresponds to the central angle
t	The wall thickness of the sandwich cylindrical shell

References

1. Clubley, S.K. Non-linear long duration blast loading of cylindrical shell structures. *Eng. Struct.* **2014**, *59*, 113–126. [CrossRef]
2. Clubley, S.K. Long duration blast loading of cylindrical shell structures with variable fill level. *Thin-Walled Struct.* **2014**, *85*, 234–249. [CrossRef]
3. Børvik, T.; Burbach, A.; Langberg, H.; Langseth, M. On the ballistic and blast load response of a 20ft ISO container protected with aluminum panels filled with a local mass-Phase II: Validation of protective system. *Eng. Struct.* **2008**, *30*, 1621–1631. [CrossRef]
4. Keys, R.A.; Clubley, S.K. Experimental analysis of debris distribution of masonry panels subjected to long duration blast loading. *Eng. Struct.* **2017**, *130*, 229–241. [CrossRef]

5. Clough, L.G.; Clubley, S.K. Steel column response to thermal and long duration blast loads inside an air blast tunnel. *Struct. Infrastruct. Eng.* **2019**, *15*, 1510–1528. [CrossRef]
6. Denny, J.W.; Clubley, S.K. Evaluating long-duration blast loads on steel columns using computational fluid dynamics. *Struct. Infrastruct. Eng.* **2019**, *15*, 1419–1435. [CrossRef]
7. Zhao, Y.P.; Yu, T.X.; Fang, J. Large dynamic plastic deflection of a simply supported beam subjected to rectangular pressure pulse. *Arch. Appl. Mech.* **1994**, *64*, 223–232. [CrossRef]
8. Zhao, Y.P.; Yu, T.; Fang, J.; Zheng, Z.M.; Tan, Q.M. Dynamic plastic response of structures with finite deflection and saturation impulse. In Proceedings of the Iutam Symposium on Impact Dynamics, Iutam Symposium on Impact Dynamics, Beijing, China, 11–15 October 1993.
9. Zhu, L.; Yu, T.X. Saturated impulse for pulse-loaded elastic-plastic square plates. *Int. J. Solids Struct.* **1997**, *34*, 1709–1718. [CrossRef]
10. Zhu, L.; He, X.; Yu, T.X.; Chen, F.L. Scaling effect on saturated impulse for square plates under rectangular pulse loading. In Proceedings of the 35th International Conference on Ocean, Offshore and Arctic Engineering (OMAE 2016), Busan, Republic of Korea, 19–24 June 2016.
11. Xi, F.; Yang, J.L. Dynamic Response Analysis of Elastic-plastic Thin Circular Plates under Impulse Loading with Considenration of Large Deflection. *Explos. Shock. Waves* **2000**, *20*, 379–384.
12. Xi, F.; Zhang, Y. The effects of strain rate on the dynamic response and abnormal behavior of steel beams under pulse loading. *Explos. Shock. Waves* **2012**, *32*, 34–42.
13. Bai, X.Y.; Zhu, L.; Yu, T.X. Saturated impulse for pulse-loaded rectangular plates with various boundary conditions. *Thin-Walled Struct.* **2017**, *119*, 166–177. [CrossRef]
14. Zhu, L.; He, X.; Chen, F.L.; Bai, X. Effects of the strain rate sensitivity and strain hardening on the saturated impulse of plates. *Lat. Am. J. Solids Struct.* **2017**, *14*, 1273–1292. [CrossRef]
15. Zhu, L.; Bai, X.Y.; Yu, T.X. The saturated impulse of fully clamped square plates subjected to linearly decaying pressure pulse. *Int. J. Impact Eng.* **2017**, *110*, 198–207. [CrossRef]
16. Bai, X.Y.; Zhu, L.; Yu, T.X. Saturated impulse for fully clamped square plates under blast loading. *Int. J. Mech. Sci.* **2018**, *146–147*, 417–431. [CrossRef]
17. Tian, L.R.; Chen, F.L.; Zhu, L.; Yu, T.X. Saturated analysis of pulse-loaded beams based on membrane factor method. *Int. J. Impact Eng.* **2019**, *131*, 17–26. [CrossRef]
18. Nurick, G.N.; Martin, J.B. Deformation of thin plates subjected to impulsive loading: A review Part II: Experimental studies. *Int. J. Impact Eng.* **1989**, *8*, 171–186. [CrossRef]
19. Nurick, G.N.; Martin, J.B. Deformation of thin plates subjected to impulsive loading: A review: Part I: Theoretical considerations. *Int. J. Impact Eng.* **1989**, *8*, 159–170. [CrossRef]
20. Guo, Z.; Gao, B.; Guo, Z.; Zhang, W. Dynamic constitutive relation based on J-C model of Q235steel. *Explos. Shock. Waves* **2018**, *38*, 804–810.
21. Pham, T.T.; Changsoo, K.; Hyounseung, J.; Kim, T.; Kim, J. Buckling behavior analysis of hybrid-honeycomb sandwich cylindrical shells. *Ocean. Eng.* **2023**, *276*, 114214.
22. Shahgholian-Ghahfarokhi, D.; Rahimi, G. A sensitivity study of the free vibration of composite sandwich cylindrical shells with grid cores. *Iran. J. Sci. Technol. Trans. Mech. Eng.* **2020**, *44*, 149–162. [CrossRef]
23. Ghahfarokhi, D.S.; Rahimi, G. An analytical approach for global buckling of composite sandwich cylindrical shells with lattice cores. *Int. J. Solids Struct.* **2018**, *146*, 69–79. [CrossRef]
24. Zarei, M.; Rahimi, G.H. Buckling resistance of joined composite sandwich conical–cylindrical shells with lattice core under lateral pressure. *Thin-Walled Struct.* **2022**, *174*, 109027. [CrossRef]
25. Tho Hung, V.; Thuy Dong, D.; Thi Phuong, N.; Ngoc Ly, L.; Quang Minh, T.; Trung, N.-T.; Hoai Nam, V. Nonlinear buckling behavior of spiral corrugated sandwich FGM cylindrical shells surrounded by an elastic medium. *Materials* **2020**, *13*, 1984. [CrossRef] [PubMed]

materials

MDPI

Article

Effective Mechanical Properties of Periodic Cellular Solids with Generic Bravais Lattice Symmetry via Asymptotic Homogenization

Padmassun Rajakareyar, Mostafa S. A. ElSayed *, Hamza Abo El Ella and Edgar Matida

Department of Mechanical and Aerospace Engineering, Carleton University, Ottawa, ON K1S 5B6, Canada; padmassun.rajakareyar@carleton.ca (P.R.); hamzaella@cunet.carleton.ca (H.A.E.E.); edgar.matida@carleton.ca (E.M.)

* Correspondence: mostafa.elsayed@carleton.ca (M.S.A.E.); Tel.: +1-613-520-2600 (ext. 4138)

Abstract: In this paper, the scope of discrete asymptotic homogenization employing voxel (cartesian) mesh discretization is expanded to estimate high fidelity effective properties of any periodic heterogeneous media with arbitrary Bravais's lattice symmetry, including those with non-orthogonal periodic bases. A framework was developed in Python with a proposed fast–nearest neighbour algorithm to accurately estimate the periodic boundary conditions of the discretized representative volume element of the lattice unit cell. Convergence studies are performed, and numerical errors caused by both voxel meshing and periodic boundary condition approximation processes are discussed in detail. It is found that the numerical error in periodicity approximation is cyclically dependent on the number of divisions performed during the meshing process and, thus, is minimized with a refined voxel mesh. Validation studies are performed by comparing the elastic properties of 2D hexagon lattices with orthogonal and non-orthogonal bases. The developed methodology was also applied to derive the effective properties of several lattice topologies, and variation of their anisotropic macroscopic properties with relative densities is presented as material selection charts.

Keywords: asymptotic homogenization; cartesian mesh; voxel mesh; non-orthogonal periodic basis; lattice material; Bravais lattice symmetry

Citation: Rajakareyar, P.; ElSayed, M.S.A.; Abo El Ella, H.; Matida, E. Effective Mechanical Properties of Periodic Cellular Solids with Generic Bravais Lattice Symmetry via Asymptotic Homogenization. *Materials* **2023**, *16*, 7562. https://doi.org/10.3390/ma16247562

Academic Editors: Madhav Baral and Charles Lu

Received: 23 October 2023
Revised: 19 November 2023
Accepted: 23 November 2023
Published: 8 December 2023

1. Introduction

Periodic cellular solids, also known as lattice materials, are formed by tessellating a unit cell in an infinite periodicity to fill a design space. The unit cell, the representative volume element (RVE), is the smallest depiction of the lattice that has the most accurate statistical representation of its physical macroscale properties. The dimensions of the unit cell are at least an order of magnitude smaller than the characteristic length of the macroscopic structure.

While having complicated geometries that cannot be fabricated using conventional manufacturing methods, recent advances in additive manufacturing have permitted the fabrication of cellular materials with increasingly intricate structures and new topologies [1–4], opening the door for their adoption in many applications. Over the last few decades, lattice materials have been used in a wide range of applications, including acoustic and vibrational damping and reflections [5–7], thermal and heat exchangers [8–10], actuation for shape morphing structures [11,12], energy absorption [1,13–15], and multiscale optimization [16,17].

Multiscale numerical simulations to evaluate lattice material characteristics considering structural attributes at all length scales are normally challenging and computationally expensive. Thus, homogenization methods are developed to analyze the effective qualities of a multiscale material such that its macroscale properties may be represented statistically.

A number of analytical, numerical, and experimental studies have been proposed in the literature [18–26] to characterize the effective properties of lattice materials. Some of the proposed homogenization techniques include energy methods [18–21], the wavelet-reduced order model [22], the Cauchy–Born hypothesis [23–26], discrete homogenization [27–29], and asymptotic homogenization [30–34].

Some notable works [35–37] have provided closed-form representations of the effective mechanical characteristics of lattice materials. Their methods, however, were only applicable to simple topology with straightforward arrangement of lattice cells.

The characteristics of planar lattice materials have also been homogenized using matrix-based methods based on Bloch's theorem and the Cauchy–Born hypothesis [24,25]. In order to derive the material's macroscopic stiffness parameters, Hutchinson and Fleck [24] first converted the microscopic nodal deformations of a lattice in terms of the macroscopic strain field. A method was developed to describe cell topologies, such as the Kagome lattice and the Triangular–Triangular lattice, which have a particular degree of symmetry. This approach was expanded by ElSayed and Pasini [25] and ElSayed [38] to handle planar topologies that can have any arbitrary cell geometry. A more generic matrix-based method for the analysis of arbitrary bi-dimensional and tri-dimensional cell topologies with open and closed cells was described by Vigliotti and Pasini [23,26,39].

The characterization of cellular materials has also been effectively accomplished using discrete homogenization approaches [27–29]. These methods simulate the lattice cell walls using discrete elements like beam or rod elements. The homogenized characteristics are produced by converting the discrete sum of equilibrium equations into a continuous relation of stress and strain.

The asymptotic homogenization (AH) theory has been effectively used among other numerical methods for predicting the effective mechanical characteristics of periodic lattice materials [30–34,40–44]. Their results have been validated by experimental tests, demonstrating the effectiveness of the AH method [45–48]. The main discrepancy between the experimental and the theoretical analyses was mainly due to defects and deviations present in the manufactured part. The defects invalidated the periodic boundary condition and caused deviations in the measured experimental values. A standout benefit of AH in comparison to other homogenization systems is its ability to precisely identify the stress distribution in the lattice unit cell, which can then be utilized for an in-depth analysis of the strength and damage of heterogeneous periodic materials [29,49].

The double-scale AH method, considering the use of 2D iso-parametric plate elements, was employed by Andreassen and Andreasen [33] to determine the elastic properties, thermal expansion, thermal conductivity, and fluid permeability of 2D periodic cellular lattices and composite materials. Andreassen and Andreasen [33] were able to analyze RVE's cell envelope with monoclinic, orthorhombic, tetragonal, and hexagonal periodicities by changing the shape of the 2D iso-parametric quadrilateral plate element.

Considering the use of solid elements for discrete homogenization techniques, the homogenized elastic tensor could be obtained by applying unit strains and performing volume averaging of the stresses and strains in the discretized elements. This approach has been performed using popular commercial finite element packages, such as Abaqus [50] and ANSYS Material Designer [51]. The main limitation for these commercial applications adopting this approach is that they are only developed for RVEs with orthorhombic, tetragonal, or cubic periodicity [50–52] mainly due to the limitations imposed by the meshing process, leading to an incompatibility while applying the periodic boundary conditions. When discretizing the RVE, the locations and the number of nodes along a periodic boundary face would not necessarily match while applying periodic boundary conditions.

Dong et al. [34] expanded the work of Andreassen and Andreasen [33] by considering the use of 3D solid elements (iso-parametric hexahedral element shown in Figure 1, also hereafter referred to as 3D voxel). The code developed by Dong et al. [34] was limited to analyzing the periodic cell envelope with the orthogonal periodic basis.

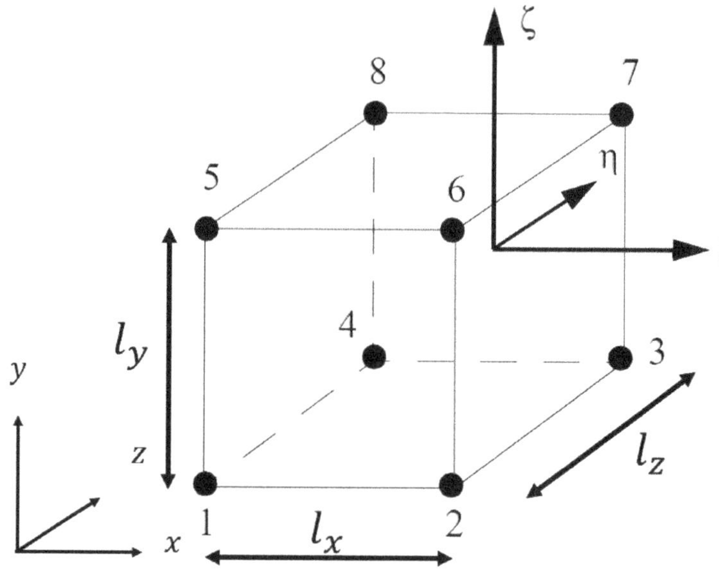

Figure 1. Iso-parametric hexahedral voxel element, with the global coordinate system (x, y, z), local coordinate system (ξ, ζ, η), and voxel edges aligned with a global cartesian coordinate system with lengths (l_x, l_y, l_z), respectively.

This paper expands on the work performed by Dong et al. [34]. The prior work only considered cell envelopes with orthogonal periodicity, whereas the current work expands upon this by (a) proposing methods to discretize unit cells with non-orthogonal bases and (b) applying approximated periodic boundary conditions for non-orthogonal cell envelopes to evaluate the homogenized elastic properties.

The paper is organized into five sections. After the introduction, in Section 2, the methodology for discretizing the geometry and approximating the periodic boundary condition, along with the numerical homogenization process for determining high fidelity homogenized elastic properties, are presented. In Section 3, the validation for the numerical analysis is detailed. The developed methodology is then applied to triclinic and monoclinic Bravais grid lattice topologies, and the results are documented in Section 4. The paper is concluded in Section 5.

The developed methodology was also used to derive the effective properties of 35 lattice topologies, and variation of their anisotropic macroscopic properties with relative densities which is presented in the appendices included in the Supplementary File.

2. Methodology

The periodic cellular solids can have an open or a closed cell construction. The former can be modelled as a micro-truss-like structure, while the latter is commonly represented with shells and plates. In this paper, an open-cell micro-truss with circular cross-section elements is considered.

The numerical homogenization performed in this paper is based on the asymptotic double-scale homogenization [30–33,53,54] with a workflow written in Python [55]. Here, the periodic cellular lattices are categorized, based on the shape of the cell envelope, as either orthogonal or non-orthogonal. Cell envelopes with orthogonal periodic bases could be part of an orthorhombic, tetragonal, or cubic Bravais lattice system, as shown in Figure 2. On the other hand, non-orthogonal cell envelopes have non-orthogonal periodic bases and could be part of triclinic, monoclinic, and hexagonal systems.

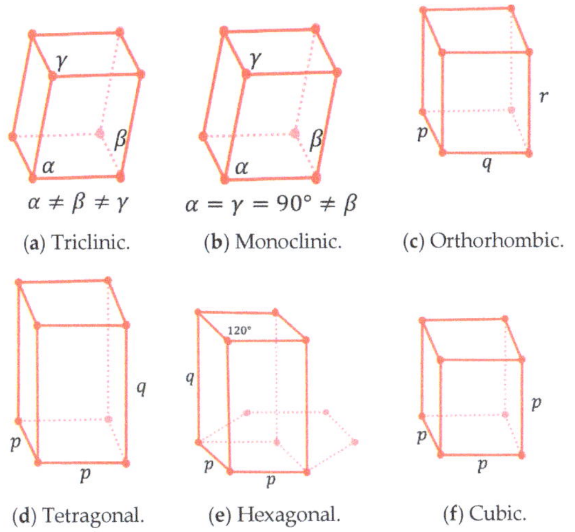

Figure 2. Primitive Bravais lattices. α, β, and γ are angles ($\alpha \neq \beta \neq \gamma$), whereas p, q and r are lengths ($p \neq q \neq r$).

2.1. *Representative Volume Element*

The RVE of a lattice is its unit cell represented by a cell envelope and the lattice structure, as shown in Figure 3. This figure shows the honeycomb lattice topology with different possibilities of cell envelope representation, including both orthogonal and non-orthogonal bases. Figure 3b shows a discretized 2D hexagon geometry with a non-orthogonal cell envelope. Three different types of voxels are shown in this figure. The green voxels represent the voids within the RVE's cell envelope, and they have a non-zero volume and zero material property. The green voxels are only considered when calculating total volume but ignored when finding periodic node pairs and calculating the stiffness matrix. The red voxels represent the discretized geometry, with both non-zero material property and non-zero volume. The blue voxels are assigned a zero volume and zero material property because they are outside the RVE's cell envelope. The cell envelope is defined as a face with a normal pointing outside the cell envelope, as shown in Figure 4. Based on the cell envelope's plane definition, any voxel elements that are outside the cell envelope (same side as the normal vector) are removed during the analysis.

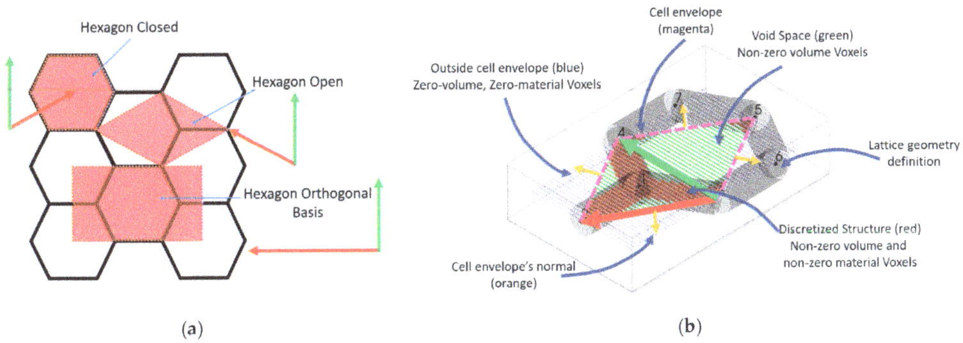

Figure 3. (**a**) Honeycomb lattice RVE with multiple cell envelope definitions. Periodic bases are shown with red and green arrows for the three proposed envelopes. (**b**) Visualization of the 2D Open Hexagon RVE, RVE's envelope, voxels, and periodic basis. Voxels' shape is not to scale.

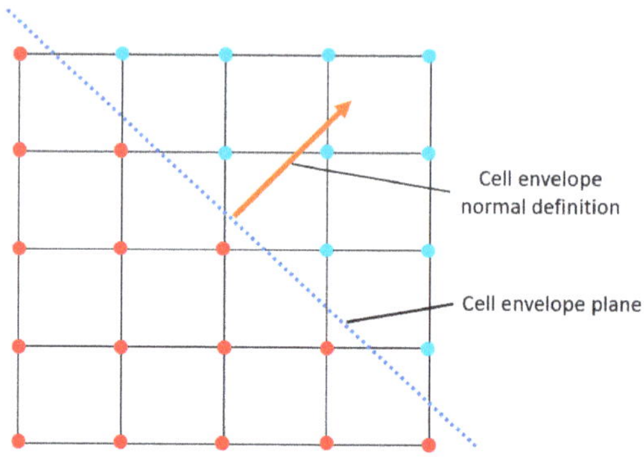

Figure 4. Cell envelope definition for filtering voxels with zero and non-zero volume. Red voxel centers are considered to be inside the cell envelope, whereas blue-coloured voxel centers are outside the cell envelope based on the normal direction of the cell envelope (orange arrow).

The parent material property for the individual voxel is defined using the two Lame's parameters, namely, λ and μ, which are expressed as:

$$\begin{aligned} \lambda &= \frac{vE}{(1+v)(1-2v)} \\ \mu &= \frac{E}{2(1+v)} \end{aligned} \tag{1}$$

where E and v are the Young's modulus and the Poisson's ratio, respectively. The analysis presented in this paper employs isotropic steel with $E_{iso} = 2.0 \times 10^{11}$ Pa and $v_{iso} = 0.3$, which correspond to lame parameters of $\lambda = 1.153 \times 10^{11}$ Pa and $\mu = 7.692 \times 10^{11}$ Pa.

The voxel element is formulated as an iso-parametric brick element whose edges are orthogonal with lengths of l_x, l_y and l_z aligned in the global cartesian coordinate system. The individual voxels with non-zero volume, which are inside the cell envelope, are treated as having an isotropic material property, and the element's stiffness material matrix ($\mathbf{C}^{(e)}$) is formulated as:

$$\mathbf{C}^{(e)} = \lambda^{(e)} \begin{bmatrix} 1 & 1 & 1 & 0 & 0 & 0 \\ 1 & 1 & 1 & 0 & 0 & 0 \\ 1 & 1 & 1 & 0 & 0 & 0 \\ 0 & 0 & 0 & 0 & 0 & 0 \\ 0 & 0 & 0 & 0 & 0 & 0 \\ 0 & 0 & 0 & 0 & 0 & 0 \end{bmatrix} + \mu^{(e)} \begin{bmatrix} 2 & 0 & 0 & 0 & 0 & 0 \\ 0 & 2 & 0 & 0 & 0 & 0 \\ 0 & 0 & 2 & 0 & 0 & 0 \\ 0 & 0 & 0 & 1 & 0 & 0 \\ 0 & 0 & 0 & 0 & 1 & 0 \\ 0 & 0 & 0 & 0 & 0 & 1 \end{bmatrix} \tag{2}$$

2.2. Discretization and Approximation of the Periodic Boundary Condition

The process to determine the translational periodicity of the node pairs is to translate the coordinates of a node using the periodic basis vector and then find the closest node within a prescribed search radius.

A periodic physical quantity of a structure can be represented as:

$$\mathcal{F}(\boldsymbol{x} + \boldsymbol{N}\boldsymbol{Y}) = \mathcal{F}(\boldsymbol{x}) \tag{3}$$

where $\boldsymbol{x} = [x_1, x_2, x_3]^T$ is the position vector of a point where the physical quantity $\mathcal{F}$ is evaluated, $\boldsymbol{N} = \lceil \mathrm{n}_1, \mathrm{n}_2, \mathrm{n}_3 \rfloor$ is a 3×3 diagonal matrix which consists of arbitrary integer values, and $\boldsymbol{Y} = [Y_1,\ Y_2,\ Y_3]^T$ is a vector that denotes the period of the structure, where this value could be a scalar, vector, or tensor function of $\boldsymbol{x}$ [30].

The static response of a periodic structure can then be represented as:

$$\begin{aligned} \sigma_{ij} &= C_{ijkl}(x)\varepsilon_{kl} = C_{ijkl}(x + NY)\varepsilon_{kl} \\ i, j, k, l &\in \{1, 2, 3\} \end{aligned} \tag{4}$$

where C_{ijkl} is the stiffness tensor, ε_{kl} is the strain tensor, and σ_{ij} is the stress tensor. By definition of $\boldsymbol{x}, \boldsymbol{N}$, and $\boldsymbol{Y}$, the stiffness tensor C_{ijkl} can be expanded as

$$\begin{aligned} C_{ijkl}(\mathrm{x}_1 + \mathrm{n}_1\mathrm{Y}_1, \mathrm{x}_2 + \mathrm{n}_2\mathrm{Y}_2, \mathrm{x}_3 + \mathrm{n}_3\mathrm{Y}_3) &= C_{ijkl}(\mathrm{x}_1, \mathrm{x}_2, \mathrm{x}_3) \\ i, j, k, l \ &\in \{1, 2, 3\} \end{aligned} \tag{5}$$

The voxel nodes have been discretized such that the nodal coordinates (x_1^i, x_2^i, x_3^i) of the ith node are defined as:

$$\begin{pmatrix} x_1^i \\ x_2^i \\ x_3^i \end{pmatrix} = \begin{pmatrix} A_1 l_x + O_x \\ A_2 l_y + O_y \\ A_3 l_z + O_z \end{pmatrix}, \ where \begin{array}{l} A_1 = \{0, 1,\ 2,\ \ldots,\ x_{divs}\} \\ A_2 = \{0,\ 1,\ 2,\ \ldots,\ y_{divs}\}, \\ A_3 = \{0,\ 1,\ 2,\ \ldots,\ z_{divs}\} \end{array}$$
$$i = \{1,\ 2,\ \ldots,\ (x_{divs} + 1) * (y_{divs} + 1) * (z_{divs} + 1)\} \tag{6}$$

where, A_1, A_2, and A_3 are integers and $\{O_x,\ O_y,\ O_z\}$ is the origin. The i superscript indicates node numbers; l_x, l_y, and l_z are the lengths of the voxel along the global x, y, and z axis, respectively; and the x_{divs}, y_{divs} and z_{divs} are integers denoting the number of voxel discretization along the global x, y, and z axes, respectively. Typically, l_x, l_y, and l_z are the same for all the voxels and can be calculated as:

$$\begin{pmatrix} l_x \\ l_y \\ l_z \end{pmatrix} = \begin{pmatrix} \frac{\max(\mathrm{x}_1^i) - \min(\mathrm{x}_1^i)}{x_{divs}} \\ \frac{\max(\mathrm{x}_2^i) - \min(\mathrm{x}_2^i)}{y_{divs}} \\ \frac{\max(\mathrm{x}_3^i) - \min(\mathrm{x}_3^i)}{z_{divs}} \end{pmatrix} \text{ and } \begin{pmatrix} O_x \\ O_y \\ O_z \end{pmatrix} = \begin{pmatrix} \min(\mathrm{x}_1^i) \\ \min(\mathrm{x}_2^i) \\ \min(\mathrm{x}_3^i) \end{pmatrix} \tag{7}$$

The periodicity vector $\boldsymbol{Y}$ can be decomposed such that:

$$\boldsymbol{Y} = \begin{pmatrix} \mathrm{Y}_1 \\ \mathrm{Y}_2 \\ \mathrm{Y}_3 \end{pmatrix} = \begin{pmatrix} B_1 l_x \\ B_2 l_y \\ B_3 l_z \end{pmatrix} + \begin{pmatrix} \mathrm{G}_1 \\ \mathrm{G}_2 \\ \mathrm{G}_3 \end{pmatrix} : \begin{array}{c} B_1,\ B_2,\ B_3 \in \mathbb{Z} \\ \mathrm{G}_1,\ \mathrm{G}_2,\ \mathrm{G}_3 \in \mathbb{R} \\ \mathrm{G}_1 < l_x,\ \mathrm{G}_2 < l_y,\ \mathrm{G}_3 < l_z, \end{array} \tag{8}$$

where B_1, B_2, and B_3 are integers and G_1, G_2, and G_3 are the residual or the remainder which are smaller than the lengths of the voxel l_x, l_y, and l_z, respectively.

For each node of the voxel defined by Equation (6) that is part of the RVE's cell envelope, there exists a periodic node pair that is also part of the discretized RVE's cell envelope.

If any of the residuals G_1, G_2, or G_3 are non-zero, then it would not be possible to find the exact node that is part of the voxels, as shown in Figure 5, (which leads to approximating the periodic node pairs, causing some numerical errors which are discussed in Sections 3.1 and 3.4). Approximating the location of the periodic node pair involves ignoring the residuals (G_i). A perfect voxel mesh can be obtained by minimizing the residual term. For an orthogonal periodic basis, aligned with the global cartesian coordinate system, the residuals are automatically zero. But for an RVE with a non-orthogonal periodic basis, the x_{divs}, y_{divs}, and z_{divs} must be altered to minimize the residuals.

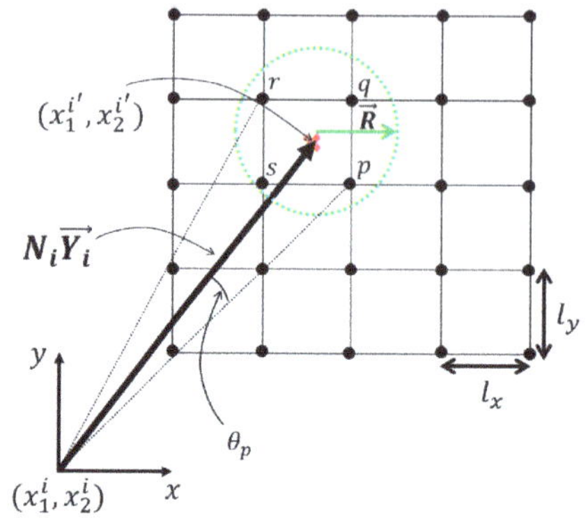

Figure 5. Determination of a periodic node pair for node i located at (x_1^i, x_2^i) with its periodic pair $i\prime$ at $N_i\vec{Y}_i$ and the search radius $\vec{R}$ (green arrow) required to search the approximate periodic node pair due to the difference between the voxel element basis and RVE's periodic basis. This search radius can be used with a KD-Tree algorithm [56] to query the closest point.

Thus, to find the periodic node pair for an RVE with a non-orthogonal periodic basis, a local search must be conducted within a radius of R, such that at least one point lies within the defined search radius, as shown in Figure 5. If the periodic node is not found for the node in the cell boundary, or the node pair is incorrectly identified, it can introduce small numerical errors. A suggested search radius is proposed as follows:

$$R = \max\begin{pmatrix} 0.51 \max(l_x, l_y, l_z) \\ 0.51\left(\sqrt{l_x^2 + l_y^2 + l_z^2}\right) \end{pmatrix} \tag{9}$$

For a coarse mesh (where l_x, l_y, l_z values are similar or larger relative to the truss radius), it is suggested to use angular constraints, where the angle θ_p, shown in Figure 5, is more than $5°$, rather than the suggested radius constraint in Equation (9). The angle θ_p, as shown in Figure 5, can be calculated by performing the dot product operation between the two normalized vectors, $\vec{Y}$ and $\overrightarrow{x^i p}$. The misalignment of the periodic basis with the voxel's element natural coordinate system introduces slight numerical errors, which are discussed in Section 3.

To determine the periodicity of the lattice, the nodal coordinates of all the voxels are used to create a k-d tree from [56] to implement a fast–nearest neighbour algorithm. Then, similar to Equation (5), an offset from Equation (8) is applied to all of the voxel coordinates based on all of the unique combinations of the basis, and the closest point within a certain minimum distance R and a minimum angle θ is chosen as its periodic nodal pair. Unit cells with the orthogonal periodic basis that align with the voxel's natural coordinate system do not need the minimum distance and the angle restrictions. But for a lattice geometry with a periodic basis that is not an integer multiple of the voxel's natural coordinate system, the process of finding the periodic nodal pair requires an additional step because the offset coordinate is not coincident with the second nodal pair. Thus, a local search must be conducted to find the closest point that is subject to minimum distance and angle restrictions described by (9).

2.3. Numerical Homogenization

The asymptotic homogenization process is based on the double-scale expansion theory [30–33,53,54]. The homogenized macroscopic elasticity tensor can be written as:

$$\boldsymbol{E}_{ijkl}^{H} = \frac{1}{|V|}\int_V \boldsymbol{E}_{pqrs}\left(\varepsilon_{pq}^{0(ij)} - \varepsilon_{pq}^{(ij)}\right)\left(\varepsilon_{rs}^{0(kl)} - \varepsilon_{rs}^{(kl)}\right)\mathrm{d}V, \quad i,j,k,l,p,q,r,s \in \{1,2,3\} \tag{10}$$

where V is the volume of the base RVE, E_{pqrs} is the locally varying stiffness tensor, $\varepsilon_{pq}^{0(ij)}$ represents the macroscopic strain, and $\varepsilon_{pq}^{(ij)}$ is the microscopic strain (locally periodic). The superscript H denotes the homogenized quantity. More details for 3D and 2D voxels can be found at [33,34]. $\varepsilon_{pq}^{(ij)}$ is defined as:

$$\varepsilon_{pq}^{(ij)} = \varepsilon_{pq}\left(\chi^{ij}\right) = \frac{1}{2}\left(\chi_{p,q}^{ij} + \chi_{q,p}^{ij}\right), \quad p,q \in \{1,2,3\} \tag{11}$$

Based on displacement fields χ^{kl}, which are found by solving the elasticity equations with prescribed macroscopic strains:

$$\begin{gathered}\int_V E_{ijpq}\varepsilon_{ij}(v)\varepsilon_{pq}\left(\chi^{kl}\right)dV = \int_V E_{ijpq}\varepsilon_{ij}(v)\varepsilon_{pq}^{0(kl)}dV \\ i,j,k,l,p,q \in \{1,2,3\}\end{gathered} \tag{12}$$

where v is the virtual displacement field. Homogenization is normally performed numerically using discretization and finite element method to solve Equation (12).

3. Validation

In this section, the analysis of a voxelized RVE with a non-orthogonal base is validated against results of a commercial finite element software for the same lattice topology but represented by a different RVE with orthogonal bases, as reported by Gibson and Ashby [36], Vigliotti and Pasini [26], and other homogenization codes by Andreassen and Andreasen [33], as well as Dong et al. [34]. For low relative density lattice, the results of the voxelized non-orthogonal hexagon are compared against similar non-orthogonal hexagons but formulated using beam theory [27–29]. Grid convergence studies have been performed in this section, and the numerical errors caused by both the voxel meshing and the periodic boundary condition approximation processes are discussed in detail. The material used in the validation study is isotropic steel with $E_{iso} = 2.0 \times 10^{11}$ Pa and $\nu_{iso} = 0.3$.

3.1. Grid Convergence Study

For the grid convergence study, the lattice materials' truss radius is held constant, and the number of divisions along the global axis is varied. For the first mesh sensitivity study, a 3D primitive cubic lattice with an orthogonal cell envelope was analyzed, and the results are shown in Figure 6. The primitive cubic lattice, which had a unit length of 1 with a truss radius of 0.157, was targeted, which converged to a volume fraction of 0.18. The convergence of elastic properties was satisfied around a discretization of 60 voxels in each direction.

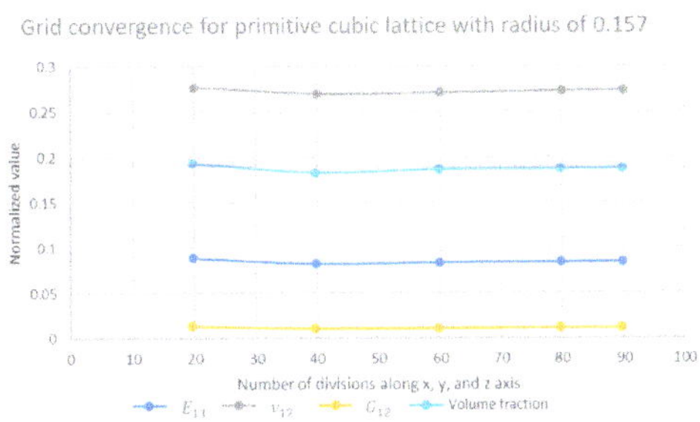

Figure 6. Grid convergence study.

Then, a 2D closed hexagon lattice was analyzed, and a volume fraction of 0.25 was targeted using a radius of 0.117 units. The length of the truss was 1 unit, and the cell angle was 60°. The elastic constants of the stiffness matrix and the elastic properties are presented in Figure 7 It was observed that increasing the number of divisions affects the volume fraction of the discretized geometry; this is due to the voxel meshing process, where at some critical point, the volume fraction increases due to the voxel center coordinate suddenly being included as part of the truss radius (refer to Figure 8). In the plots below, the other quantities generally follow a similar trend as the volume fraction (black dashed lines). However, the observed discrepancy is generally due to the errors in approximating periodicity, which generally converges while refining the mesh.

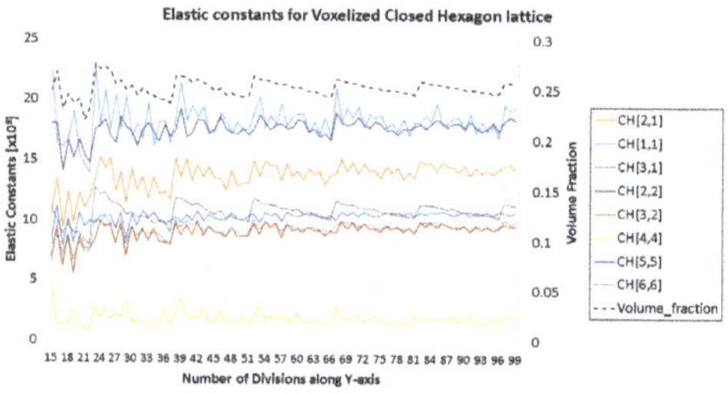

(**a**)

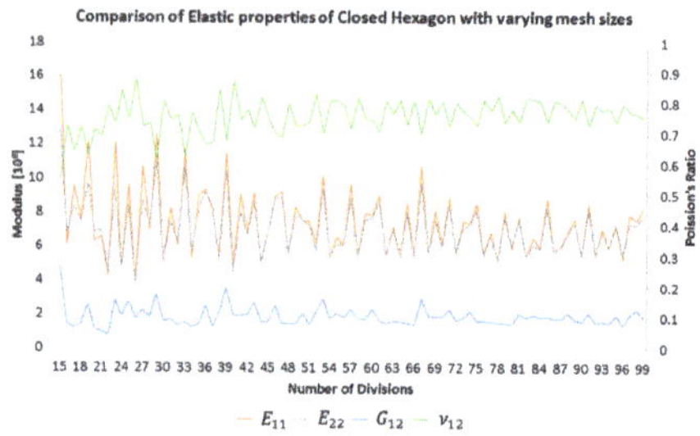

(**b**)

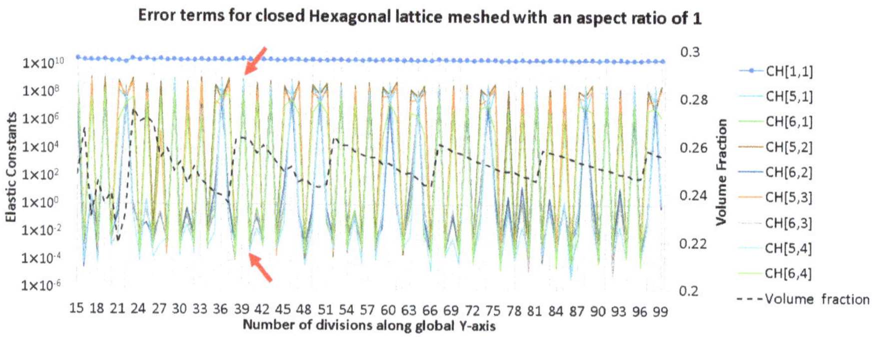

(**c**)

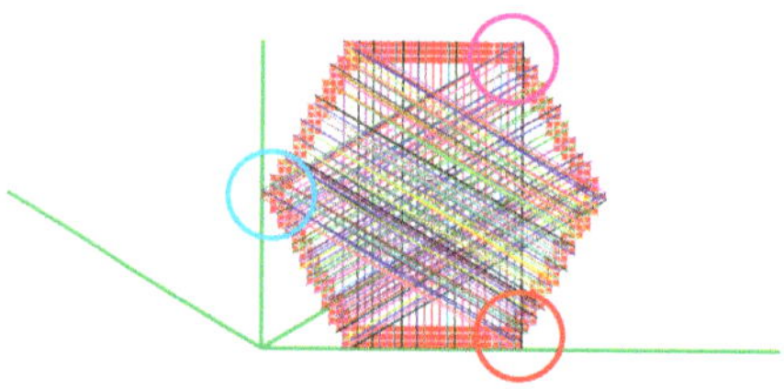

(**d**)

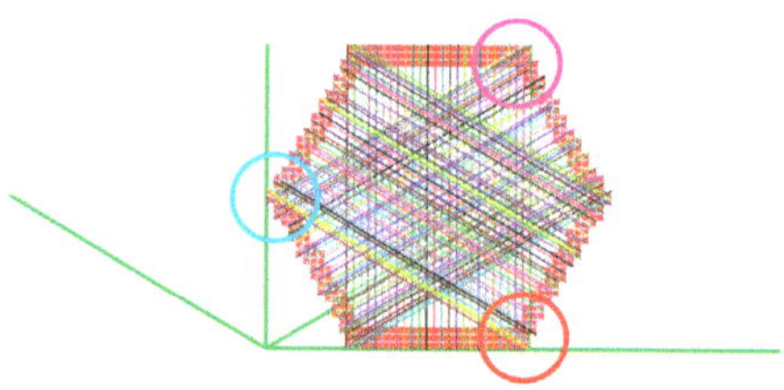

(**e**)

Figure 7. Numerical error analysis of a closed hexagon with voxel aspect ratios of unity. (**a**) The large non-zero elastic constants. For a constant truss radius, the volume fraction changes due to the voxel meshing process, and the volume fraction converges for a smaller voxel mesh. (**b**) The homogenized effective properties of the hexagon lattice, where the fluctuations are caused by the voxel meshing process and approximating the periodic boundary conditions. (**c**) The residual terms in a log plot, where the residuals fluctuate between a high and a low value, corresponds to the error caused by approximating the periodic boundary conditions. (compare colored circles in (**d**,**e**)) The periodic boundary conditions for the voxelized hexagon lattice are marked in (**c**). (**d**) Large error (AR $\approx$ 1) for 46 by 39 divisions, VF = 0.263. This voxelized hexagon corresponds to the upper red arrow shown in (**c**). (**e**) Small error (AR $\approx$ 1) for 47 by 40 divisions, VF = 0.262. This voxelized hexagon corresponds to the lower red arrow shown in (**c**).

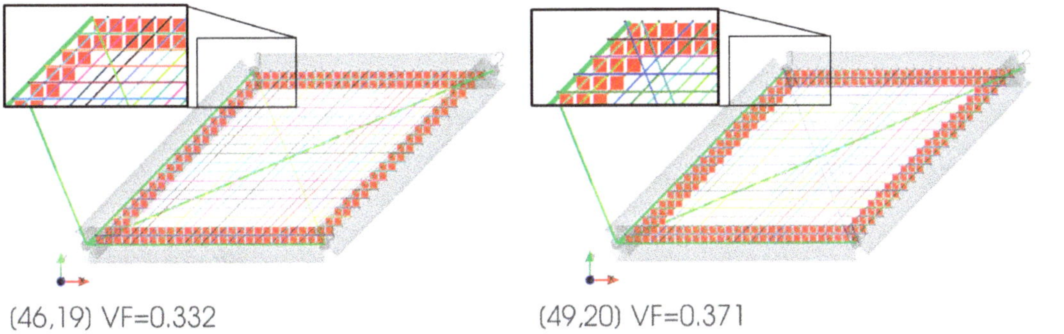

Figure 8. Variation of volume fraction due to increasing the number of voxel divisions in the y-direction while maintaining an aspect ratio of unity. The voxel's center coordinate has been represented as smaller voxels. The periodic basis is plotted as a bold green line.

The theoretical zero terms presented in an orthotropic stiffness matrix are plotted in Figure 7 above. It can be observed that the supposed zero terms fluctuate between a high and low value. The higher values are at least two orders of magnitude smaller than the main diagonals of the matrix, whereas the lower error terms are at least eight orders of magnitude smaller than the main diagonals. The main cause for this is because the periodicity is discretized where the remainder term (G_i) mentioned in Equation (8) is minimized; this means that the periodicity basis vector could be resolved as an integer multiple of the voxel lengths.

The effect of reduced remainder term leads to a case where the periodic nodes are matched in an orderly fashion, as shown in Figures 8 and 7b. When comparing the periodicity lines from Figure 7a's blue circle to the magenta circle, the periodicity lines are not parallel, as shown in Figure 7b. Furthermore, comparing the periodicity from the blue circle to the red circle, a single periodic node is shared with two other nodes; this can lead to the periodicity being mismatched. Several of these scenarios are visualized in Figure 9. Based on the visualization, the planes of symmetry are disrupted when periodicity is not matched properly, which causes the orthotropic cellular lattices to have smaller magnitude non-zero terms like a monoclinic or a triclinic material. An in-depth analysis is presented in Section 3.4.

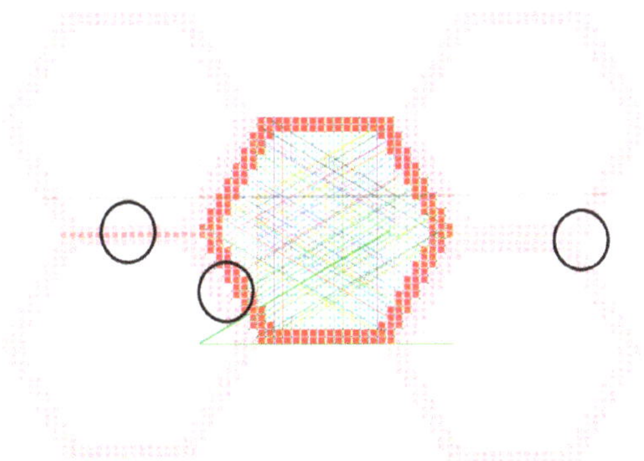

Figure 9. Visualization of mismatched periodicity (black circles) and its effect on lattice symmetry. The periodic node pairs are connected using randomly colored lines.

3.2. Comparison with ANSYS

The results obtained from the voxelization process are compared with the results from ANSYS Material Designer, which can perform homogenization of elastic and thermal properties for RVE with an orthogonal periodic basis. The results obtained from ANSYS are comparable with the results obtained with the voxelized method, with differences of less than 2%. The anisotropy plot for the primitive cubic lattice and hexagon lattice is shown in Figure 10. The primitive cubic lattice in ANSYS was meshed using tetrahedrons, and the meshing strategy made sure that the meshes in all of the faces were similar, such that the periodicity of the lattice could be found easily. The hexagon lattice in ANSYS follows a similar meshing strategy, but the Honeycomb was meshed with a mix of hexahedral (brick) and tetrahedral elements. However, the cell envelope's faces are meshed using brick elements, which ensures that the cell envelope's nodes can be matched easily to apply the periodic boundary conditions. ANSYS also provides an option to calculate the homogenized properties by applying symmetric boundary conditions for some of the simpler lattices.

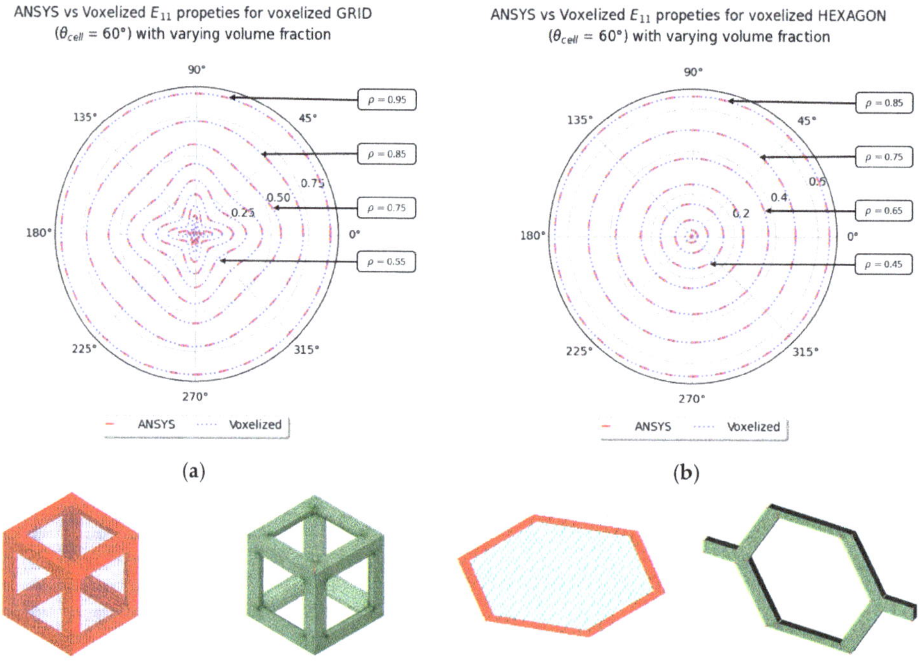

Figure 10. Comparison of normalized E_{11} anisotropy plot at multiple volume fractions for (**a**) voxelized grid with ANSYS' cubic lattice model, and (**b**) voxelized hexagon with non-orthogonal periodic basis and ANSYS' Honeycomb lattice cell with orthogonal periodic basis. (**c**) Voxelized grid and ANSYS' cubic lattice model. (**d**) Voxelized hexagon with non-orthogonal periodic basis and ANSYS' Honeycomb model with orthogonal periodic basis.

Three planes of symmetry exist in the x, y, and z-planes for both the primitive cubic lattice in 3D and the hexagon lattice in 2D (with periodic conditions imposed in the out-of-plane). Thus, the homogenized elastic properties for these two lattices can be considered as an orthotropic. Theoretically, the homogenized elasticity tensor in the Voigt notation for an orthotropic lattice cell takes the form of:

$$\left[C_{ij}^{H}\right] = \begin{bmatrix} C_{11} & C_{12} & C_{13} & 0 & 0 & 0 \\ C_{21} & C_{22} & C_{23} & 0 & 0 & 0 \\ C_{31} & C_{32} & C_{33} & 0 & 0 & 0 \\ 0 & 0 & 0 & C_{44} & 0 & 0 \\ 0 & 0 & 0 & 0 & C_{55} & 0 \\ 0 & 0 & 0 & 0 & 0 & C_{66} \end{bmatrix} \quad i,j \in \{1,2,3\} \tag{13}$$

Due to some small numerical residuals, the zero terms shown in Equation (13) were not zeros when the hexagon and the cubic lattice were homogenized. A comparison of the terms for C_{41}^{H} for the hexagon and cubic lattice is shown in Figure 11. The data exported from ANSYS, which is marked as "D [1,1]" and "D [1,4]" in the figures, are from the grid and the hexagon lattice cell, respectively.

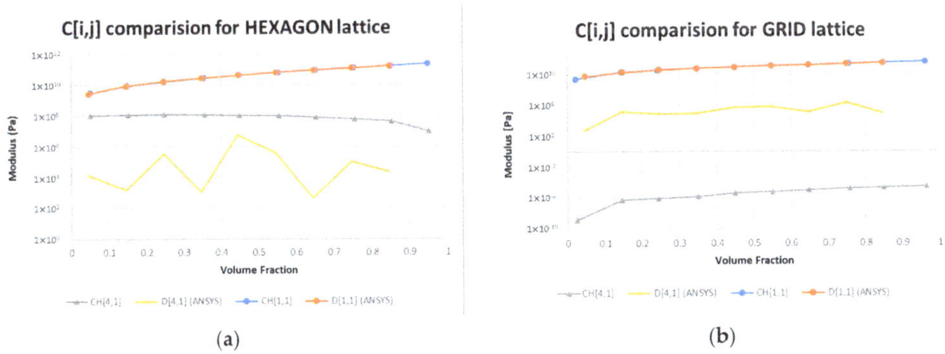

Figure 11. Comparison of the [1,4] term of the homogenized elastic tensor for (**a**) hexagon lattice and (**b**) grid lattice with the [1,1] term for a material of $E_{iso} = 2 \times 10^{11}$ and $nu_{iso} = 0.3$. The CH and D labels correspond to the data from the voxelization and ANSYS material modeller, respectively.

Based on the data presented in Figure 11, it can be observed that results from ANSYS have a lower C_{41}^{H} term for the hexagon lattice, and a higher error for the grid lattice, in comparison to the voxelized homogenization process developed in this paper. This small numerical error is caused by the meshing scheme, as the grid lattice was meshed using tetrahedrons, while the hexagons were meshed mostly using brick elements, as shown in Figure 10c,d, respectively. In the next section, numerical errors caused by approximating periodicity are discussed.

3.3. RVE Rotation

The anisotropic behaviour of the elastic properties of the lattice cell is investigated by rotating the microstructure of the unit cell with respect to the applied macroscopic strain field; this is carried out by applying the rotation tensor as follows:

$$E_{ijkl}^{H\,\prime} = Q_{ip}Q_{jq}Q_{kr}Q_{ls}E_{pqrs}^{H} \quad i,j,k,l,p,q,r,s \in \{1,2,3\} \tag{14}$$

where Q is the orthogonal tensor, which corresponds to an orthogonal transformation from x_i' to x_i basis. The C_{ij}^{H} in the Voigt notation is expanded to E_{ijkl}^{H}, and an Einstein summation [57] was performed, where $Q_{ip} = Q_{jq} = Q_{kr} = Q_{ls} = Q$.

The 2D anisotropy plot produced in this paper (shown in Figure 10) corresponds to a rotation in the x_3 the axis of the global coordinate system, where:

$$Q_{ip} = \begin{bmatrix} \cos(\alpha) & -\sin(\alpha) & 0 \\ \sin(\alpha) & \cos(\alpha) & 0 \\ 0 & 0 & 1 \end{bmatrix}, \; 0 \le \alpha \le 2\pi \tag{15}$$

The 3D anisotropy plot produced in this paper corresponds to a rotation in the x_3 axis and the x_2 axis, where:

$$Q_{ip} = \begin{bmatrix} \cos(\alpha) & -\sin(\alpha) & 0 \\ \sin(\alpha) & \cos(\alpha) & 0 \\ 0 & 0 & 1 \end{bmatrix} \begin{bmatrix} \cos(\beta) & 0 & \sin(\beta) \\ 0 & 1 & 0 \\ -\sin(\beta) & 0 & \cos(\beta) \end{bmatrix}, \; \begin{matrix} 0 \le \alpha \le 2\pi \\ 0 \le \beta \le 2\pi \end{matrix} \tag{16}$$

Using the compliance matrix by finding the inverse of the elasticity matrix, the following elastic properties can be determined:

$$\begin{gathered} S_{ij} = E_{ij}^{-1} \\ E_x^H = \frac{1}{S_{11}},\; E_y^H = \frac{1}{S_{22}},\; E_z^H = \frac{1}{S_{33}}, \\ G_{xy}^H = \frac{1}{S_{44}},\; G_{yz}^H = \frac{1}{S_{55}},\; G_{xz}^H = \frac{1}{S_{66}}, \\ v_{xy}^H = -\frac{S_{12}}{S_{11}},\; v_{xz}^H = -\frac{S_{13}}{S_{11}},\; v_{yz}^H = -\frac{S_{23}}{S_{22}} \end{gathered} \tag{17}$$

where S_{ij} is the compliance matrix, and E_{ij} is the elastic stiffness matrix. The sensitivity of the primitive cubic lattice's elastic property to the cell rotation was performed and presented in Figure 12. The radius for each plot was adjusted such that a volume fraction of 0.3 could be achieved. The RVE was rotated along the z-axis by an angle of α, and the E_{11} was plotted in the anisotropy plot, which was plotted for $\alpha = 0°$. For a volume fraction of 0.3, it was noted that 50 voxel discretization was sufficient to prove that applying the rotation matrix to the homogenized elastic tensor is equivalent to rotating the RVE.

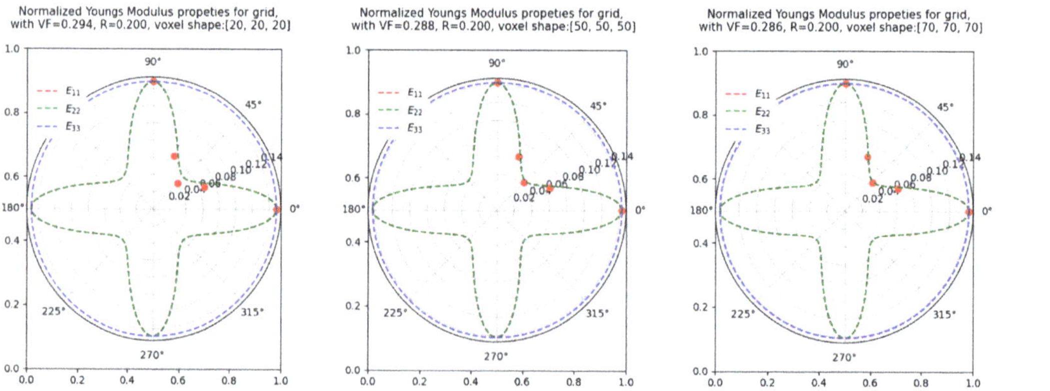

Figure 12. Anisotropy plot of the Youngs Modulus Elastic property of a 3D primitive cubic lattice cell, where the red circles mark the E_{11} values at multiple rotation cell angles similar to the rotated cells shown in Figure 13. The dashed E_{11} values are overlapped with the E_{22} values due to geometrical symmetry.

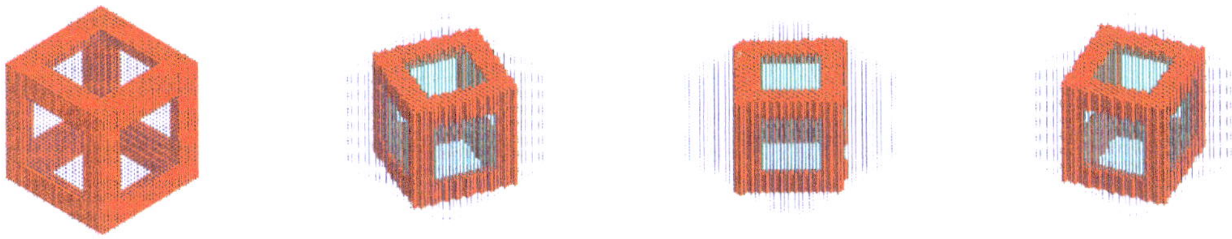

Figure 13. A grid lattice cell discretized by 30 voxels along the x, y, and z axis, where the unit cell rotated at 0°, 22°, 45°, and 68° along the z-axis.

3.4. Numerical Errors Due to Approximating Periodicity

In this section, a hexagon was meshed with an equal number of divisions along the global x and y axes and one element along the z-axis. The following term (τ_e) has been visualized in Figure 14:

$$\tau_e^{(i)(j)} = \frac{1}{C_{ij}^H} \int_{V_e} \left(\chi_e^{0(i)} - \chi_e^{(i)}\right)^T \mathbf{k}_e \left(\chi_e^{0(j)} - \chi_e^{(j)}\right) dV_e \tag{18}$$

where $\chi_e^{0(i)}$ and $\chi_e^{(i)}$ are the element's macroscopic and microscopic displacement due to unit strain i, and τ_e represents the contribution of element e to the homogenized elastic tensor C_{ij}^H normalized by the total C_{ij}^H. By visualizing τ_e for each element, it is possible to observe the elements that could cause the numerical errors shown in Figure 11. The colour scheme used in Figure 11 is such that the maximum absolute value dictates the positive (red) and negative (blue) limits of the plot. Values close to zero are given a green colour.

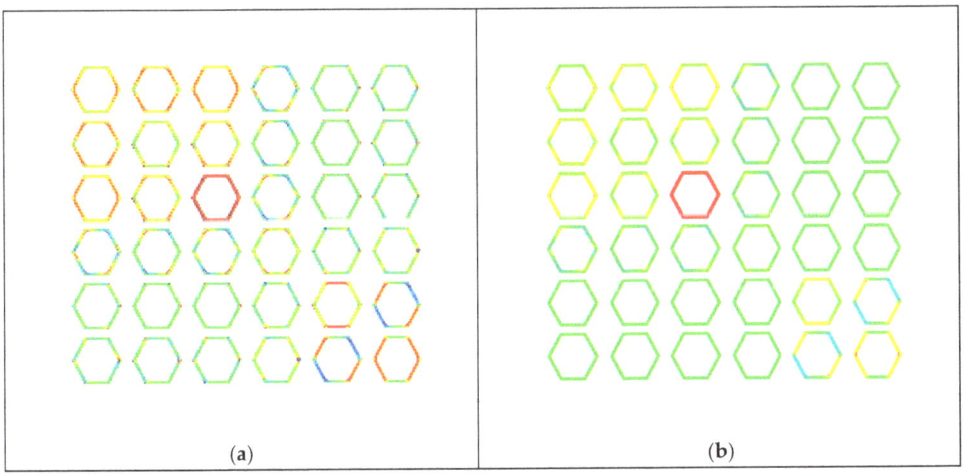

Figure 14. Visualization of the integral in Equation (18) for individual voxel elements, a hexagon with $l_x/l_y = 1.155$ and a voxel shape of (**a**) 30 × 30 × 1 and (**b**) 150 × 150 × 1. Each of the individual hexagon visualizations were normalized by the value of the C_{ij}^H term.

For the hexagon closed topology with a volume fraction of 0.29 using isotropic steel with $E_{iso} = 2.0 \times 10^{11}$ and $\nu_{iso} = 0.3$, which has been discretized as $30 \times 30 \times 1$ voxels ($l_x/l_y = 1.155$), the homogenized elastic tensor was computed to be:

$$\left[C_{ij}^{H}\right] = \begin{bmatrix} 23.8422 & 18.3845 & 12.5175 & -0.0127 & -0.0589 & -0.1269 \\ 18.3845 & 23.0329 & 12.3423 & -0.0116 & -0.0204 & 0.0223 \\ 12.5175 & 12.3423 & 72.047 & -0.0278 & -0.0421 & -0.0378 \\ -0.0127 & -0.0116 & -0.0278 & 3.3903 & -0.0154 & 0.0016 \\ -0.0589 & -0.0204 & -0.0421 & -0.0154 & 13.0261 & -0.0147 \\ -0.1269 & 0.0223 & -0.0378 & 0.0016 & -0.0147 & 14.7842 \end{bmatrix} \times 10^{10} \quad (19)$$

After the mesh resolution was increased to $150 \times 150 \times 1$ voxels ($l_x/l_y = 1.155$), with a volume fraction of 0.30, the refined, homogenized elastic tensor was computed to be:

$$\left[C_{ij}^{H}\right] = \begin{bmatrix} 24.6521 & 18.4319 & 12.9212 & 0.0001 & 0.0001 & -0.0009 \\ 18.4319 & 24.5085 & 12.8797 & -0.0001 & 0.0002 & 0.0014 \\ 12.9212 & 12.8797 & 71.0627 & -0.0001 & 0.0004 & 0.0006 \\ 0.0001 & -0.0001 & -0.0001 & 3.2672 & -0.0004 & 0.0001 \\ 0.0001 & 0.0002 & 0.0004 & -0.0004 & 14.020 & -0.0002 \\ -0.0009 & 0.0014 & 0.0006 & 0.0001 & -0.0002 & 14.1044 \end{bmatrix} \times 10^{10} \quad (20)$$

The contributions of the voxels that contribute the small non-zero terms for the $30 \times 30 \times 1$ and the $150 \times 150 \times 1$ are shown in Figure 14a b, respectively. The comparison of the eigen values with and without the zero terms is tabulated in Table 1. Comparing the hexagon corresponding to C_{51}^{H} in Figure 14a with Figure 15a, it can be seen that the voxels that have the highest contributions belong to the group whose periodicity has not been matched properly due to discretizing the periodicity basis vector; this can be seen in Figure 15a—the line that marks the periodic node pair for the nodes near the marked area of a_1 to a_5, is not parallel with the periodic basis (bold green line), and a single node from a_1 is periodic, with two nodes from a_5 zone. This behaviour causes over-stiffening of the nodal degree of freedom because the stiffness corresponding to another node would be accumulated into one node during the global stiffness formulation. Similarly, if the nodes on the cell envelope are not periodically matched with another node on the envelope, it would cause the stiffness of the nodal degree of freedom to be smaller than the ones that have been matched. The over-stiffening and the under-stiffening impact the microscopic displacements, which contribute to values that are above or below the actual threshold. The mismatch in the periodic degrees of freedom would remove planes of symmetry and would cause the RVE with hexagonal symmetry to be represented as a monoclinic or a triclinic material, as shown in Figure 9.

Table 1. Comparison of eigenvalues of lower resolution (19) and higher resolution (20) hexagon lattice for an imperfectly voxelized hexagon with a voxel aspect ratio of 1.155.

Eigenvalues of (19) (10^{11})	Eigenvalues of (19) with Zeros (10^{11})	% Diff.	Eigenvalues of (20) (10^{11})	Eigenvalues of (20) with Zeros (10^{11})	% Diff.	% Diff. Between (19) and (20)
8.0118	8.0117	0.0002	8.0050	8.0050	0	0.08
3.3757	3.3757	0.0006	3.4025	3.4025	0	−0.79
1.4785	1.4784	0.0070	1.4104	1.4104	0	4.8
1.3026	1.3026	−0.0009	1.4020	1.4020	0	−7.1
0.5047	0.5048	−0.0241	0.6148	0.6148	0	−17.94
0.3390	0.3390	−0.0012	0.3267	0.3267	0	3.77

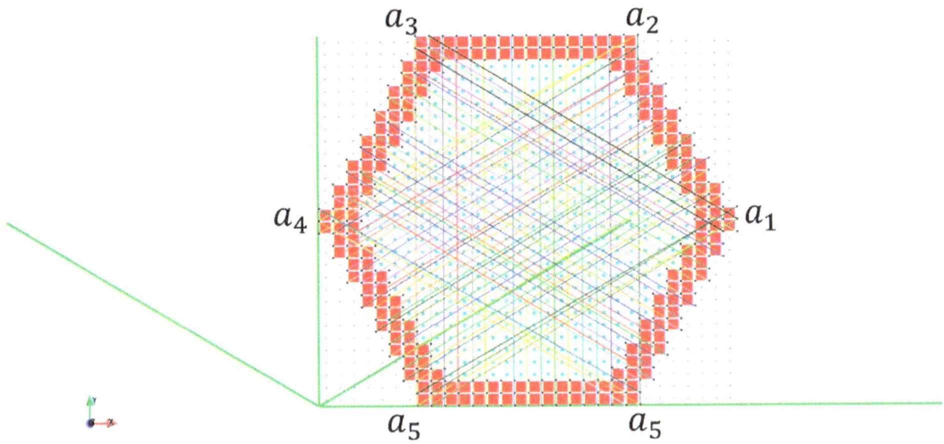

(a) Approximated periodic boundary condition for a coarsely meshed hexagon lattice.

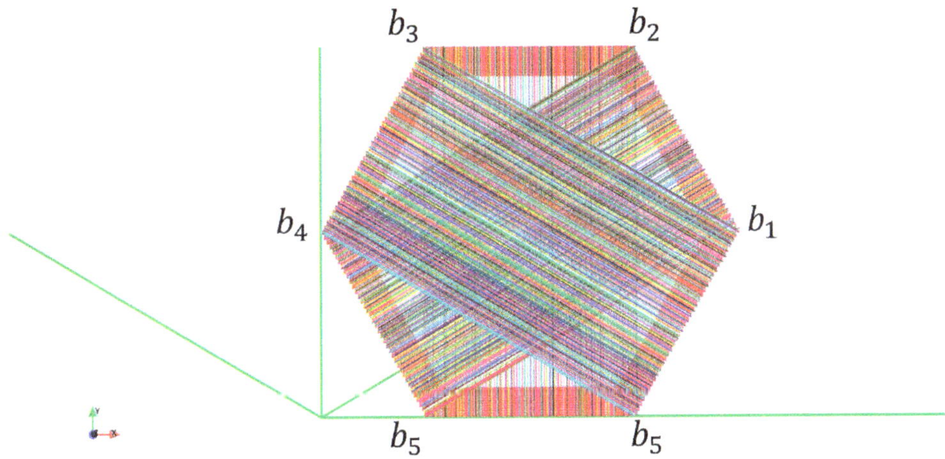

(**b**) Approximated periodic boundary condition for a finely meshed hexagon lattice.

Figure 15. Visualization of the periodicity node pairs. Each line of a random colour represents a coupling of the degrees of freedom for that node pair. The periodic basis has been highlighted as thick green lines. The approximated periodic boundary condition can be observed while comparing how the periodic boundary condition is applied at (**a**) between regions a_1 to a_5 and a_2 to a_4. Due to the finer mesh size, the approximation of the periodic boundary condition at (**b**) is not affected.

Furthermore, it can be observed from Figures 14 and 16 that increasing the voxel discretization reduces the small non-zero components in the $\left[C_{ij}^{H}\right]$ tensor because each voxel contributes a smaller value due to the smaller voxel volume. In Figure 14, for the hexagon with a larger voxel size, the residual terms are contributed by the voxels along the edges caused by imperfect periodic boundary conditions; this is also the case for the refined hexagon lattice, but the ratio of voxels in the edge is smaller than the hexagon with larger voxel size, so the contribution of the residual error is decreased.

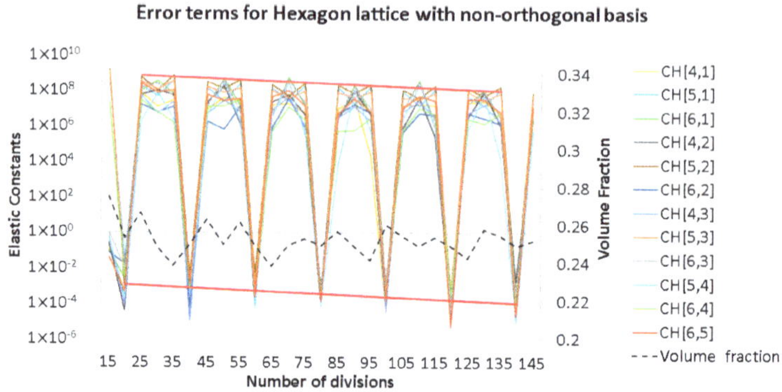

Figure 16. Convergence of error terms for the hexagon lattice with non-unity aspect ratio. The datum shown above is a reduced subset of Figure 7c to emphasize the decreasing residual errors (thick red line).

3.5. Comparison of Young's Modulus with Volume Fraction

The E_{11} elastic property for the hexagon lattice that is obtained from the voxelization code and ANSYS are compared with different homogenization procedures obtained from Gibson and Ashby [36] and Vigliotti and Pasini [26]. In Figure 17, the E_{11}, properties calculated from the voxelized closed cell hexagon, open cell hexagon, and hexagon cell envelope (all shown in Figure 18) with orthogonal basis are compared. No significant deviations were observed when comparing it to the results from ANSYS, which were obtained from a hexagon cell envelope with an orthogonal basis.

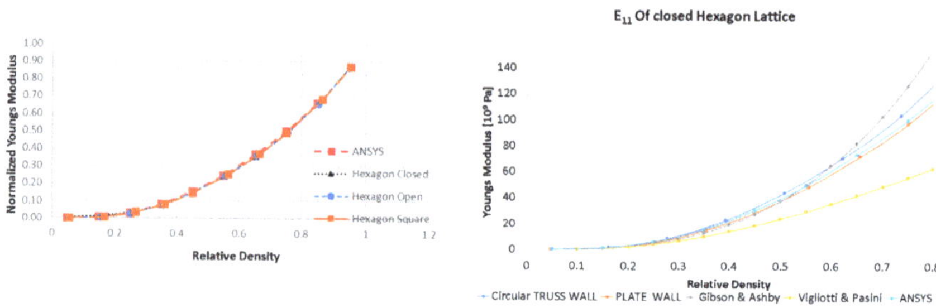

Figure 17. Comparison of normalized E_{11} for the hexagon lattice at multiple volume fractions across multiple homogenization schemes present in the literature [26,36].

Figure 18. Multiple representations of the Honeycomb hexagon lattice with different representative volume elements with their corresponding periodic basis. The representative volume element of the unit cell is shown as red voxels.

A hexagon lattice made from a circular truss (i.e., without considering periodicity in the out-of-plane direction) and a hexagon with a plate wall (3D voxel with a single layer and periodicity in the z-direction) are considered below. Gibson and Ashby's model [36] starts to overpredict the elasticity, whereas Vigliotti and Pasini's [26] model, which uses a beam model, underpredicts the elasticity at higher relative density.

3.6. Comparison of Elastic Properties for Hexagon Lattice with Varying Cell Angle

In this section, the cell angle is changed, and Young's modulus and shear modulus are plotted for a hexagon with a relative density of 0.2. The results obtained from the voxel homogenization process match very well, with some slight discrepancies; this is caused by the geometry discretization process, where a volume fraction of 0.2 was not achieved due to the linear interpolation used with the optimizer to define the relative density as a function of the radius of the truss elements. The variation of the homogenized elastic properties for the hexagon lattice w.r.t cell angle is presented in Figure 19.

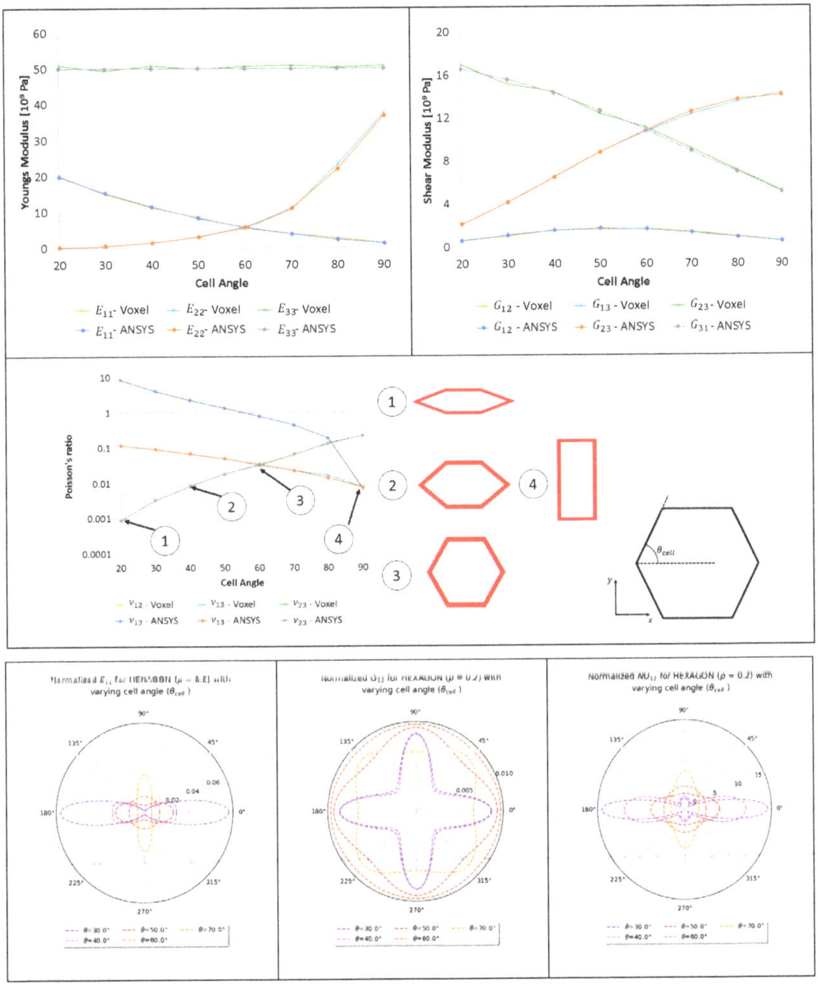

Figure 19. Elastic properties of the 2D hexagon lattice with a relative density of 0.25 made from isotropic material with Young's modulus of 2×10^{11} Pa and Poisson's ratio of 0.3. Two-dimensional anisotropic diagram of elastic properties for the hexagon with a volume fraction of 0.2 due to the variation of the hexagon cell angle.

3.7. Comparison of Elastic Constants for a 2D Monoclinic RVE Lattice with Varying Cell Angle

In this section, a square-like 2D lattice is considered; the angle shown varies from 45° to 90°. The results are compared with code published by Andreassen and Andreasen [33] as shown in Figure 20. Andreassen's work is considered more accurate because the 2D voxel elements change shape as the cell angle changes, and there are no approximations performed when applying periodicity boundary conditions to the lattice. The voxel code underestimates the C_{22} component of the stiffness matrix because of the periodicity approximations and the voxel meshing algorithm. The main cause is the voxel meshing, where the truss elements are jagged, as shown in Figure 8.

Homogenized elastic properties of a square lattice with varying cell angles.

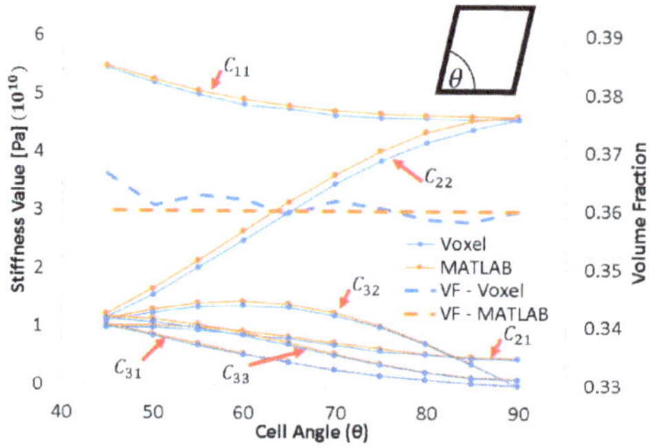

Figure 20. Comparison of homogenized elastic properties of a square lattice with varying cell angles between 2D MATLAB voxel and 3D voxel code.

4. Application

In this section, application of the developed voxel code is applied to different lattice topologies with non-orthogonal bases. The material used in this study is isotropic steel with $E_{iso} = 2.0 \times 10^{11}$ Pa and $\nu_{iso} = 0.3$.

4.1. Three-Dimensional Triclinic and a Monoclinic Bravais Grid Lattice

In this section, a triclinic and a monoclinic Bravais grid lattice are analyzed by applying the approximated periodicity boundary condition. The geometry was discretized using voxels, as shown in Figure 21.

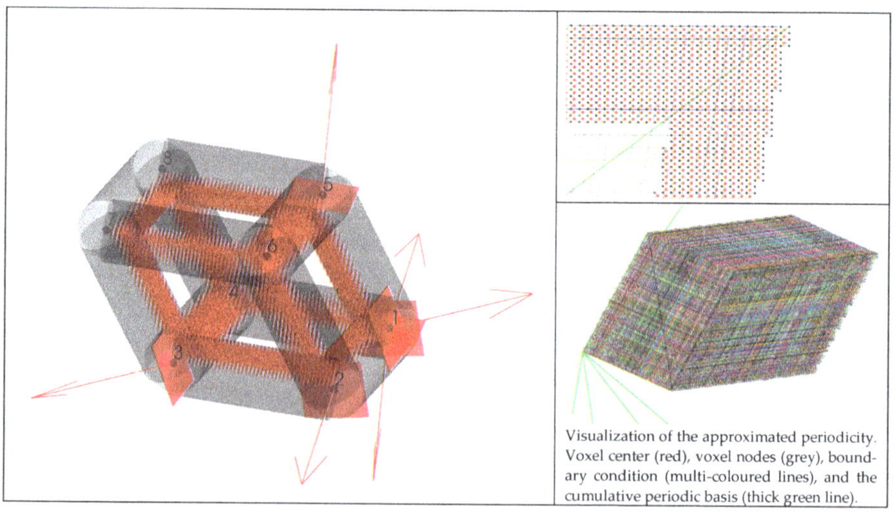

Figure 21. Discretized triclinic Bravais grid lattice. The Voxel center is represented as a small sphere for a volume fraction of 0.3. The planes that represent the cell envelope definition and its normal have also been plotted.

The lattice definition used for the current analysis to create the triclinic and monoclinic Bravais grid lattice is shown in Figure 22. The lattice is defined such that the red trusses are in the global x–y plane. The trusses T_{14} and T_{23} are aligned parallel to the global x-axis. The other red trusses T_{12} and T_{43}, are rotated from the global y-axis using angle gamma (γ) in the x–y plane. The green trusses are created by offsetting the red trusses using the blue trusses. The blue trusses are defined using a spherical coordinate system with ϕ and θ rotation angles, as shown in Figure 22. The periodic basis for the lattice was defined using the trusses T_{12}, T_{14}, and T_{15}. The angular restrictions for ϕ, θ, and γ for the Bravais lattice system are tabulated in Table 2.

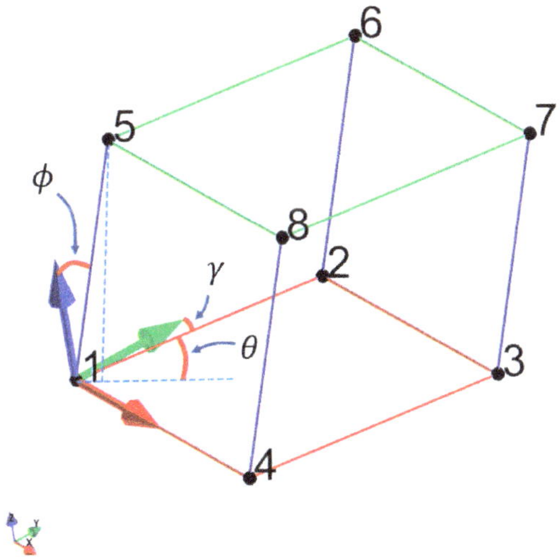

Figure 22. Base lattice definition for monoclinic and triclinic Bravais lattice.

Table 2. Angular constraints for the Bravais lattice are based on the angle definitions used in this paper.

	Phi (ϕ)	Theta (θ)	Gamma (γ)
Triclinic ($\phi \neq \theta \neq \gamma$)	$\neq 0°$	$\neq 0°$	$\neq 0°$
Monoclinic	$\neq 0°$	$= 0°$	$= 0°$
Cubic, Orthorhombic, and Tetragonal	$= 0°$	$= 0°$	$= 0°$

The key difference between the triclinic and the monoclinic Bravais grid lattice used for the current analysis is the slight change in the truss angles and their corresponding periodicity bases. It can be noted that for the monoclinic lattice, the trusses that make up the red and the green frames are orthogonal to each other ($\gamma = 0°$). Furthermore, the point set {1,4,5,8} and {2,3,6,7} are co-planar with the z–x plane.

For the monoclinic Bravais grid lattice shown in Figure 23b, which had a volume fraction of 0.29 and was discretized as 30 $\times$ 30 $\times$ 1 voxels, the homogenized elastic tensor was computed to be:

$$\left[C_{ij}^{H}\right] = \begin{bmatrix} 31.0421 & 3.8754 & 5.0174 & 0.0000 & 0.0000 & 1.9074 \\ & 30.0581 & 2.9965 & 0.0000 & 0.0000 & 1.2017 \\ & & 24.1917 & 0.0000 & 0.0000 & 6.9772 \\ & & & 2.7767 & 0.5141 & 0.0000 \\ & & & & 2.4535 & 0.0000 \\ & & & & & 5.2315 \end{bmatrix} \times 10^{9} \quad (21)$$

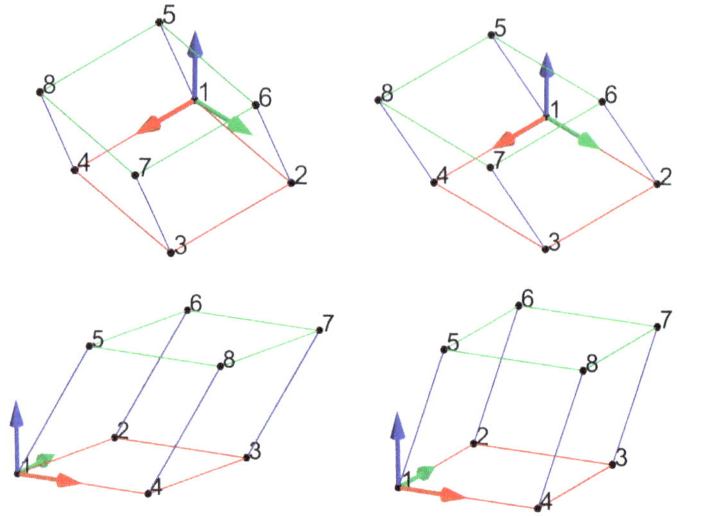

(**a**) Triclinic
$\phi = 30°, \theta = 20°, \gamma = 10°$

(**b**) Monoclinic
$\phi = 20°, \theta = 0°, \gamma = 0°$

Figure 23. Geometric definitions for the monoclinic and triclinic grid lattice are used in this paper.

Similarly, for the triclinic Bravais grid lattice shown in Figures 23a and 23, which had a volume fraction of 0.29 and was discretized as 30 × 30 × 1 voxels, the homogenized elastic tensor was computed to be

$$\left[C_{ij}^{H}\right] = \begin{bmatrix} 31.8431 & 4.3748 & 5.2469 & 1.3106 & 1.1570 & 2.8994 \\ & 27.8371 & 2.3521 & 4.1020 & 0.3740 & 1.3446 \\ & & 18.1782 & 1.6143 & 2.6998 & 7.1696 \\ & & & 3.5984 & 0.8139 & 1.2457 \\ & & & & 2.5514 & 1.3011 \\ & & & & & 6.3248 \end{bmatrix} \times 10^{9} \quad (22)$$

The elastic properties of the monoclinic and the triclinic lattice are tabulated in Table 3. The evolution of the elastic properties visualized as 2D and 3D anisotropic plots is shown in Figures 24 and 25 for the triclinic and monoclinic Bravais grid lattice, respectively. The key difference between the 3D anisotropy plot between the triclinic and monoclinic Bravais grid lattice is how the peaks of the anisotropy plots are warped.

Table 3. Elastic properties of the sample monoclinic and triclinic Bravais lattice.

Elastic Properties	Monoclinic	Triclinic
E_{11} [Pa]	2.96×10^{10}	2.96×10^{10}
E_{22} [Pa]	2.93×10^{10}	2.27×10^{10}
E_{33} [Pa]	1.46×10^{10}	9.28×10^{9}
G_{12} [Pa]	2.67×10^{9}	2.68×10^{9}
G_{13} [Pa]	2.36×10^{9}	2.03×10^{9}
G_{23} [Pa]	3.21×10^{9}	3.36×10^{9}
ν_{12}	0.108	0.1247
ν_{13}	0.156	0.1668
ν_{23}	0.076	0.0778

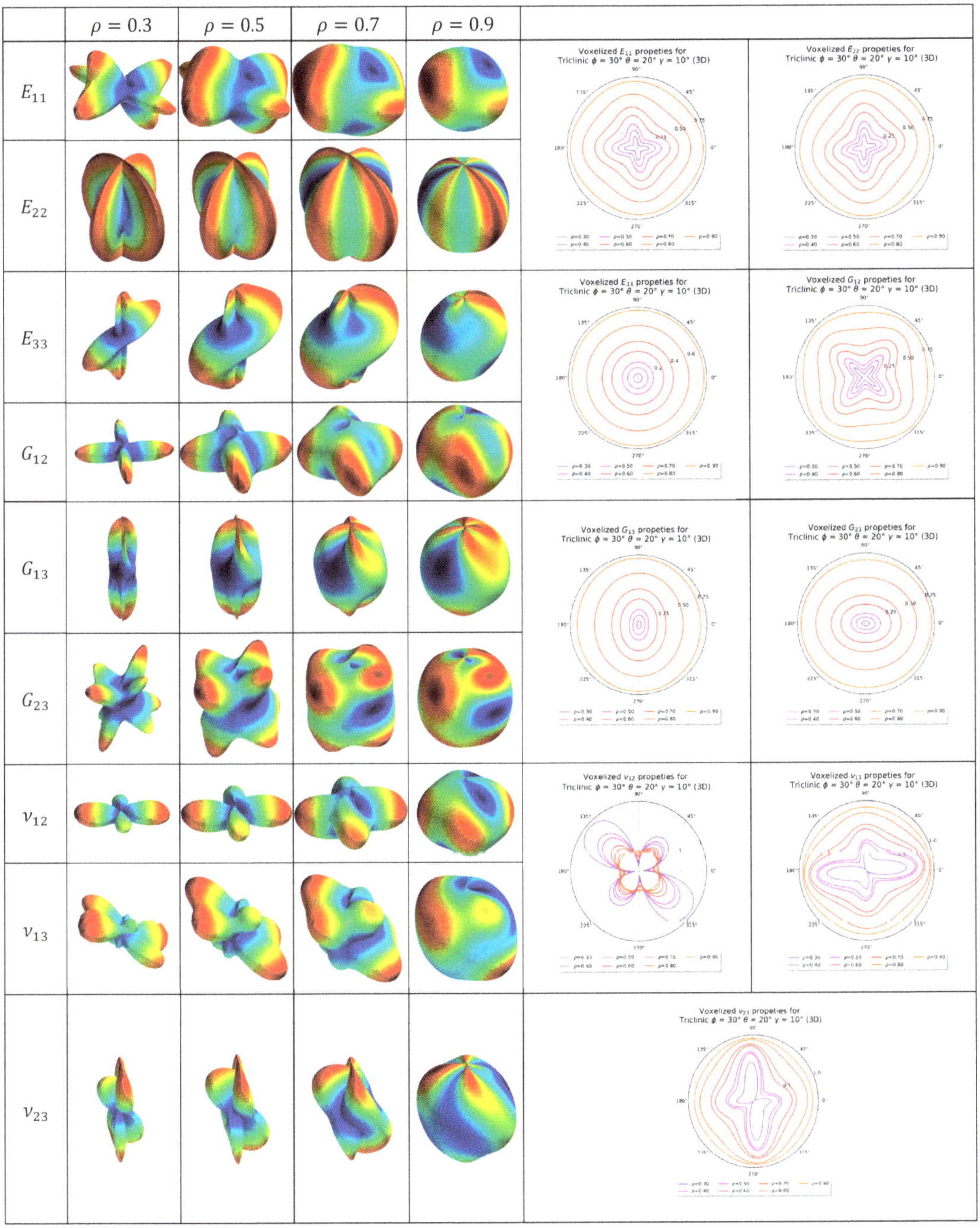

Figure 24. Two-dimensional and three-dimensional anisotropy plots for triclinic Bravais grid lattice (3D). Three-dimensional anisotropy plots are not to scale.

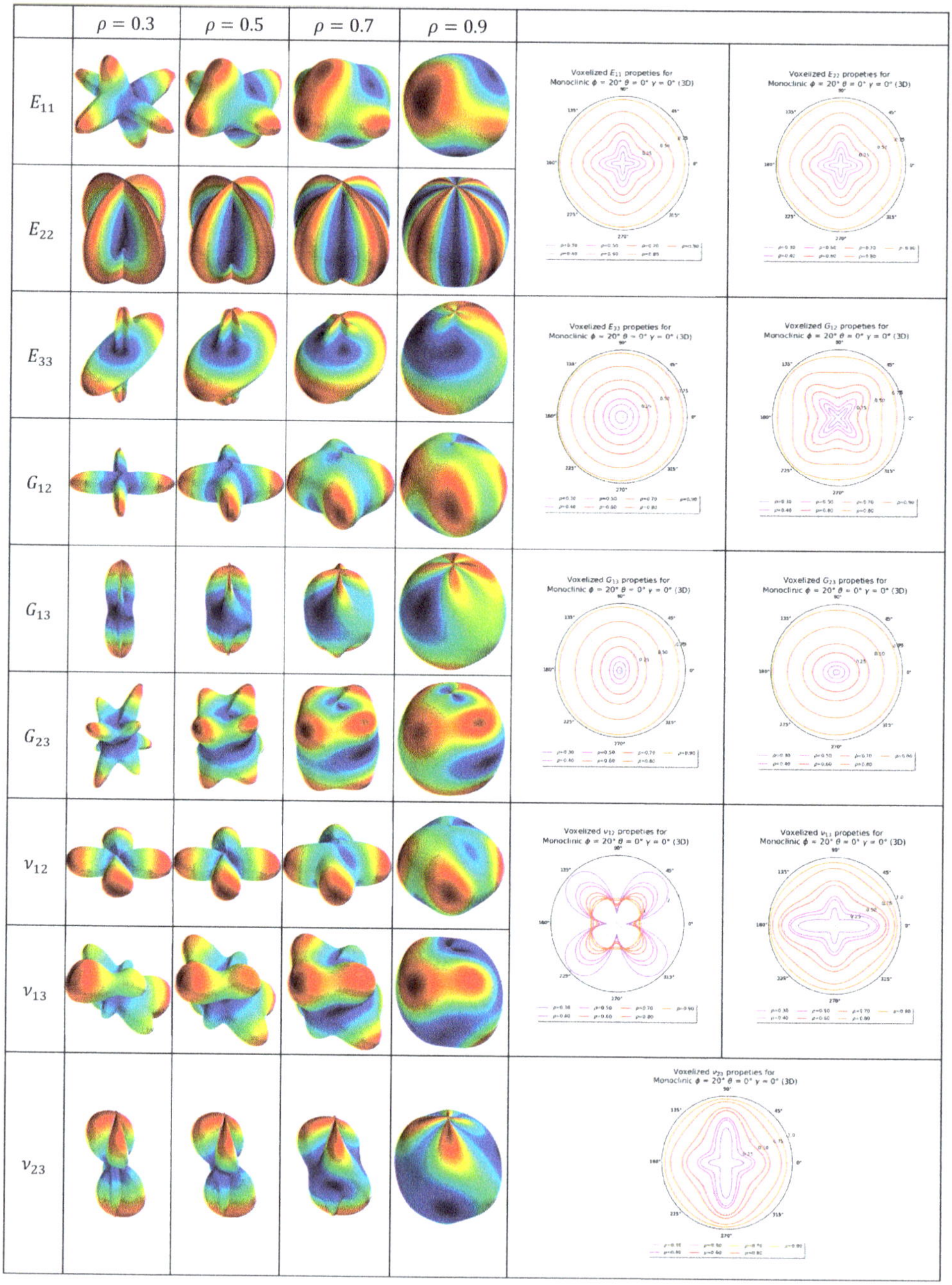

Figure 25. Two-dimensional and three-dimensional anisotropy plots for monoclinic Bravais grid lattice (3D). Three-dimensional anisotropy plots are not to scale.

4.2. Two-Dimensional Non-Orthogonal Lattice

In this section, the elastic properties of the 3.4.6.4 2D lattice are plotted. The homogenized elastic tensor of the 3.4.6.4 lattices shown in Figure 26 is tabulated in Table 3. The tabulated tensor is based on voxels that are generated as a single layer with an additional periodicity added out of a plane (z-axis). Considering this periodicity means that the single layer is equivalent to a fully tessellated geometry that has infinite depth in the out-of-plane direction. The tabulated tensors are obtained by fitting a third-order polynomial for each of the tensor entries, where the volume fraction varies from 0.1 to 0.9.

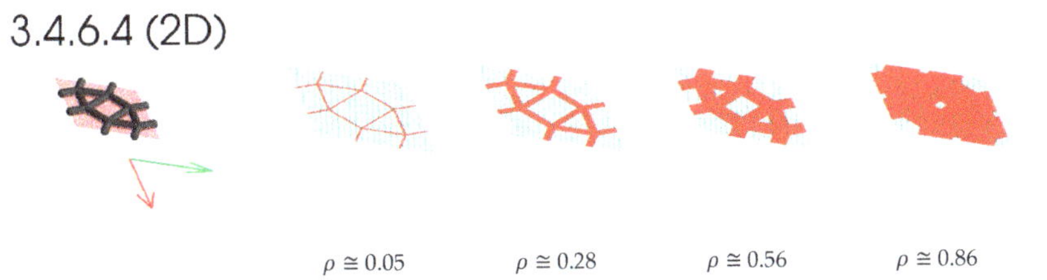

Figure 26. 3.4.6.4 lattice cells with a non-orthogonal periodic basis. The representative volume element of the unit cell is shown as red voxels. Evolution of selected 2D latices as the volume fraction increases.

Figure 26 depicts the 3.4.6.4 lattice for multiple volume fractions. This figure shows how the geometry of the lattice changes as the volume fraction increases. This sort of filling behaviour is one of the reasons for the Euler–Bernoulli beam formulation being invalid for larger relative density, as it cannot predict the interactions of the geometry within the lattice. The 2D and 3D anisotropy plots for the 3.4.6.4 lattice are shown in Figure 27. The coefficients for the homogenized elastic tensor as a function of relative density is tabulated in Table 4.

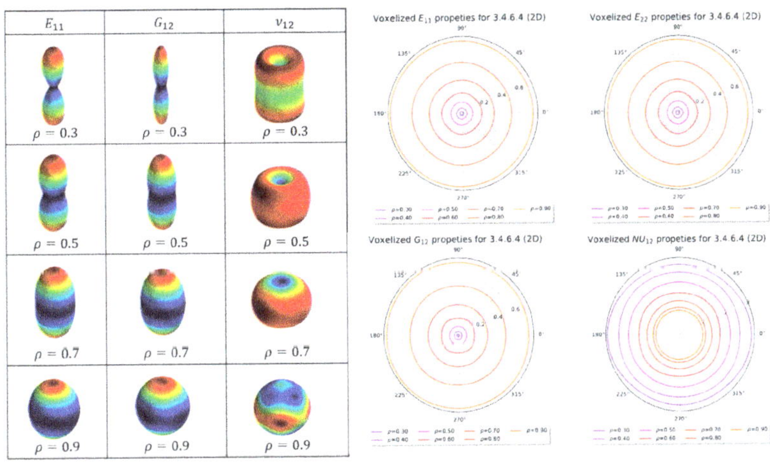

Figure 27. Two-dimensional and three-dimensional anisotropy plots for 3.4.6.4 lattice.

4.3. *Three-Dimensional Sandwich Panel Lattice*

In this section, the elastic properties of the sandwiched X (body-centred cubic) lattice are plotted. For the sandwich panel analysis, periodicity in the z-direction (sandwich plate normal) is assumed; thus, the elastic properties that are plotted in this section are for sandwich panels that are assumed to be stacked. For all the sandwich panel geometries analyzed in this paper, the thickness of the sandwich panel is held constant at a unit of 0.05 of the cell length. The homogenized elastic tensor of the Sandwich X lattice is shown and tabulated in Table 3. The 2D and 3D anisotropy plots for the Sandwich X lattice are shown in Figure 28. The coefficients for the homogenized elastic tensor as a function of relative density is tabulated in Table 5.

Table 4. Homogenized elastic tensor coefficients (based on Equation (13)) and correlation coefficient (R) for the polynomial fit for the 3.4.6.4 lattice.

		C_{11}	C_{22}	C_{33}	C_{44}	C_{55}	C_{66}	C_{12}	C_{13}	C_{23}
	ρ^3	1.591	1.644	0.377	0.538	0.210	0.178	0.476	0.620	0.636
	ρ^2	−0.662	−0.731	−0.189	−0.136	−0.014	0.026	−0.351	−0.304	−0.325
	ρ	0.380	0.390	1.132	0.011	0.190	0.180	0.347	0.218	0.221
3.4.6.4 (2D)	R	0.999	0.999	1.000	1.000	1.000	1.000	0.995	0.999	0.999

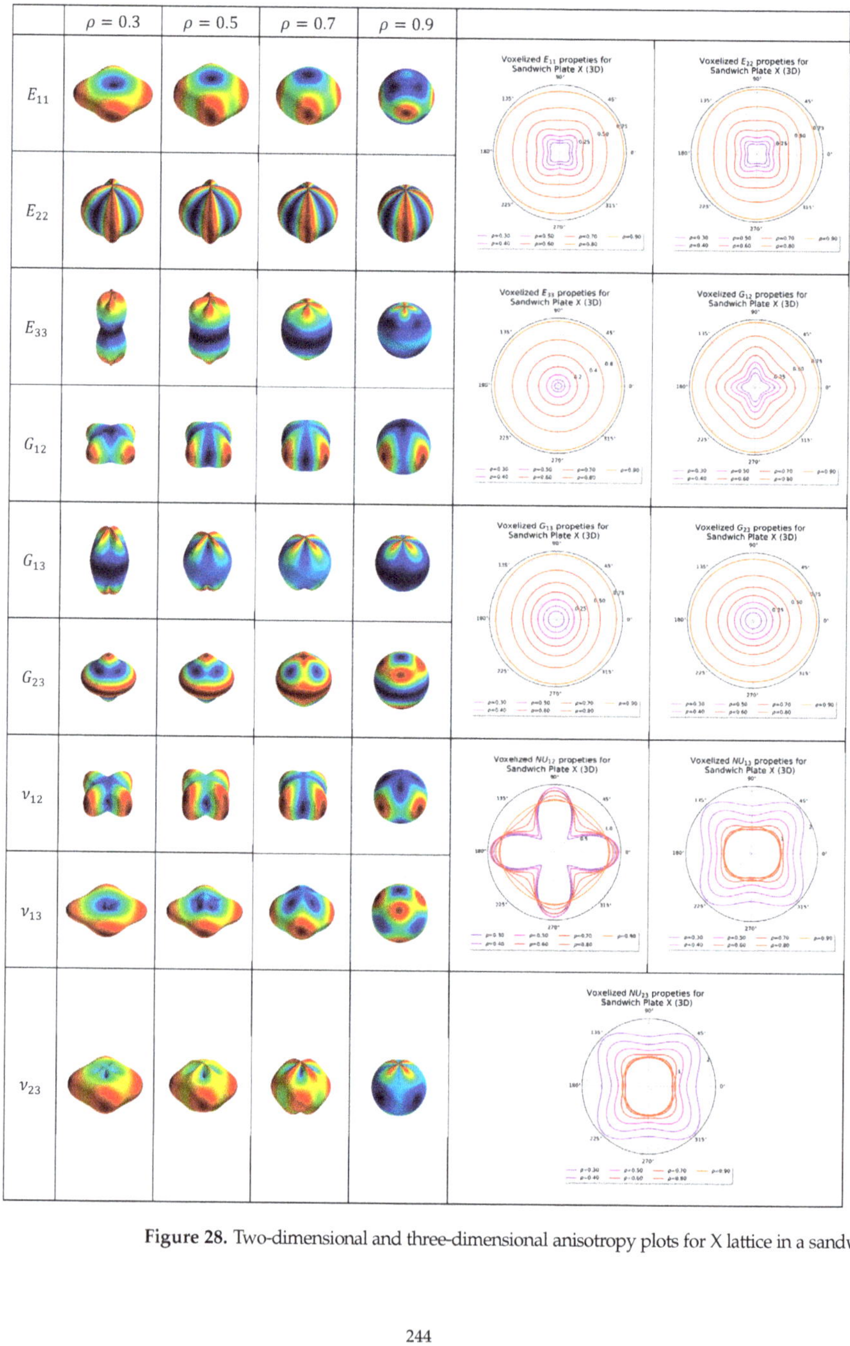

Figure 28. Two-dimensional and three-dimensional anisotropy plots for X lattice in a sandwich panel (3D).

Table 5. Homogenized elastic tensor coefficients (based on Equation (13)) and correlation coefficient (R^2) for the polynomial fit for the Sandwich X lattice.

		C_{11}	C_{22}	C_{33}	C_{44}	C_{55}	C_{66}	C_{12}	C_{13}	C_{23}
	ρ^3	2.318	2.318	1.499	0.456	0.127	0.127	0.78	0.607	0.607
	ρ^2	−1.915	−1.915	−0.296	−0.387	0.193	0.193	−0.646	−0.265	−0.265
	ρ	0.941	0.941	0.118	0.32	0.045	0.045	0.365	0.14	0.14
Sandwich X Lattice (3D)	R^2	0.999	0.999	1.0	0.999	0.999	0.999	0.998	0.995	0.995

5. Conclusions

The homogenized elastic properties of several 2D and 3D lattices can be performed using the voxel mesh approach regardless of whether the periodicity is orthogonal or non-orthogonal. Approximating the periodicity on an imperfect voxel mesh misaligns the periodic node pairs; this alters or removes the planes of symmetry within the RVE and introduces errors in the stiffness matrix. These errors are at least two orders of magnitude smaller than the main diagonals for the imperfect voxel mesh. The mesh is considered perfectly voxelized if the translated nodes of the cell envelope, using the periodicity basis, align themselves with another node. For this perfectly voxelized mesh, the numerical errors are at least eight orders of magnitude smaller than the main diagonals; this can be achieved by altering the number of discretizations performed along the global coordinate system. Furthermore, the numerical error caused by approximating the non-orthogonal periodicity decreases as the voxel size is reduced. We have shown that it is possible to evaluate the elastic properties of periodic cellular materials whose periodic basis is of any Bravais lattice system.

Supplementary Materials: The following supporting information can be downloaded at: https://www.mdpi.com/article/10.3390/ma16247562/s1.

Author Contributions: Conceptualization, P.R., M.S.A.E., H.A.E.E. and E.M.; methodology, P.R. and M.S.A.E.; software, P.R.; validation, P.R.; formal analysis, P.R.; investigation, P.R., M.S.A.E., H.A.E.E. and E.M.; resources, P.R. and M.S.A.E.; data curation, P.R.; writing—original draft preparation, P.R.; writing—review and editing, P.R., M.S.A.E., H.A.E.E. and E.M.; visualization, P.R.; supervision, M.S.A.E., H.A.E.E. and E.M.; project administration, M.S.A.E. and H.A.E.E.; funding acquisition, M.S.A.E. All authors have read and agreed to the published version of the manuscript.

Funding: This research was funded by BOMBARDIER INC., Montreal, in collaboration with MITACS Canada, grant number IT29280.

Institutional Review Board Statement: Not applicable.

Informed Consent Statement: Not applicable.

Data Availability Statement: The data that support the findings of this study are available on request from the corresponding author.

Acknowledgments: Geometry data from [34,38] was used to construct most of the 2D and the 3D lattice geometries. The following Python packages are used in-house code to create the homogenization module: Numpy [58], Scipy [56] (sparse matrices, cKd-Tree [59]), matplotlib [60], Mayavi [61], ELATE [62], matscipy [57], numpy-indexed [63] and numba [64]. The authors would like to acknowledge the financial support from BOMBARDIER INC., Montreal, in collaboration with MITACS Canada.

Conflicts of Interest: The authors declare no conflict of interest.

References

1. Tancogne-Dejean, T.; Spierings, A.B.; Mohr, D. Additively-Manufactured Metallic Micro-Lattice Materials for High Specific Energy Absorption under Static and Dynamic Loading. *Acta Mater.* **2016**, *116*, 14–28. [CrossRef]
2. Ling, B.; Wei, K.; Wang, Z.; Yang, X.; Qu, Z.; Fang, D. Experimentally Program Large Magnitude of Poisson's Ratio in Additively Manufactured Mechanical Metamaterials. *Int. J. Mech. Sci.* **2020**, *173*, 105466. [CrossRef]
3. Wang, C.; Gu, X.; Zhu, J.; Zhou, H.; Li, S.; Zhang, W. Concurrent Design of Hierarchical Structures with Three-Dimensional Parameterized Lattice Microstructures for Additive Manufacturing. *Struct. Multidiscip. Optim.* **2020**, *61*, 869–894. [CrossRef]

4. Chen, L.-Y.; Liang, S.-X.; Liu, Y.; Zhang, L.-C. Additive Manufacturing of Metallic Lattice Structures: Unconstrained Design, Accurate Fabrication, Fascinated Performances, and Challenges. *Mater. Sci. Eng. R Rep.* **2021**, *146*, 100648. [CrossRef]
5. Chen, Y.; Zhao, B.; Liu, X.; Hu, G. Highly Anisotropic Hexagonal Lattice Material for Low Frequency Water Sound Insulation. *Extrem. Mech. Lett.* **2020**, *40*, 100916. [CrossRef]
6. Yin, S.; Wu, L.; Yang, J.; Ma, L.; Nutt, S. Damping and Low-Velocity Impact Behavior of Filled Composite Pyramidal Lattice Structures. *J. Compos. Mater.* **2014**, *48*, 1789–1800. [CrossRef]
7. Murray, G.; Gandhi, F.; Hayden, E. Polymer-Filled Honeycombs to Achieve a Structural Material with Appreciable Damping. *J. Intell. Mater. Syst. Struct.* **2012**, *23*, 703–718. [CrossRef]
8. Yang, R.; Yang, Q.; Niu, B. Design and Study on the Tailorable Directional Thermal Expansion of Dual-Material Planar Metamaterial. *Proc. Inst. Mech. Eng. Part C J. Mech. Eng. Sci.* **2020**, *234*, 837–846. [CrossRef]
9. Xu, H.; Farag, A.; Pasini, D. Routes to Program Thermal Expansion in Three-Dimensional Lattice Metamaterials Built from Tetrahedral Building Blocks. *J. Mech. Phys. Solids J. Homepage* **2018**, *117*, 54–87. [CrossRef]
10. Wei, K.; Chen, H.; Pei, Y.; Fang, D. Planar Lattices with Tailorable Coefficient of Thermal Expansion and High Stiffness Based on Dual-Material Triangle Unit. *J. Mech. Phys. Solids* **2016**, *86*, 173–191. [CrossRef]
11. Nelissen, W.E.D.; Ayas, C.; Tekõ Glu, C. 2D Lattice Material Architectures for Actuation. *J. Mech. Phys. Solids J. Homepage* **2019**, *124*, 83–101. [CrossRef]
12. McHale, C.; Telford, R.; Weaver, P.M. Morphing Lattice Boom for Space Applications. *Compos. Part B Eng.* **2020**, *202*, 108441. [CrossRef]
13. Li, M.Z.; Stephani, G.; Kang, K.J. New Cellular Metals with Enhanced Energy Absorption: Wire-Woven Bulk Kagome (WBK)-Metal Hollow Sphere (MHS) Hybrids. *Adv. Eng. Mater.* **2011**, *13*, 33–37. [CrossRef]
14. Murray, G.J.; Gandhi, F. Auxetic Honeycombs with Lossy Polymeric Infills for High Damping Structural Materials. *J. Intell. Mater. Syst. Struct.* **2013**, *24*, 1090–1104. [CrossRef]
15. Tao, Y.; Duan, S.; Wen, W.; Pei, Y.; Fang, D. Enhanced Out-of-Plane Crushing Strength and Energy Absorption of in-Plane Graded Honeycombs. *Compos. Part B Eng.* **2017**, *118*, 33–40. [CrossRef]
16. Dinovitzer, M.; Miller, C.; Hacker, A.; Wong, G.; Annen, Z.; Rajakareyar, P.; Mulvihill, J.; El Sayed, M. Structural Development and Multiscale Design Optimization of Additively Manufactured UAV with Blended Wing Body Configuration Employing Lattice Materials. In Proceedings of the AIAA Scitech 2019 Forum, San Diego, CA, USA, 7–11 January 2019; American Institute of Aeronautics and Astronautics: Reston, VA, USA, 2019.
17. Zhu, J.H.; Zhang, W.H.; Xia, L. Topology Optimization in Aircraft and Aerospace Structures Design. *Arch. Comput. Methods Eng.* **2016**, *23*, 595–622. [CrossRef]
18. Askar, A.; Cakmak, A.S.S. A Structural Model of a Micropolar Continuum. *Int. J. Eng. Sci.* **1968**, *6*, 583–589. [CrossRef]
19. Chen, J.Y.; Huang, Y.; Ortiz, M. Fracture Analysis of Cellular Materials: A Strain Gradient Model. *J. Mech. Phys. Solids* **1998**, *46*, 789–828. [CrossRef]
20. Bažant, Z.P.; Christensen, M. Analogy between Micropolar Continuum and Grid Frameworks under Initial Stress. *Int. J. Solids Struct.* **1972**, *8*, 327–346. [CrossRef]
21. Kumar, R.S.; McDowell, D.L. Generalized Continuum Modeling of 2-D Periodic Cellular Solids. *Int. J. Solids Struct.* **2004**, *41*, 7399–7422. [CrossRef]
22. van Tuijl, R.A.; Remmers, J.J.C.; Geers, M.G.D. Multi-Dimensional Wavelet Reduction for the Homogenisation of Microstructures. *Comput. Methods Appl. Mech. Eng.* **2019**, *359*, 112652. [CrossRef]
23. Vigliotti, A.; Pasini, D. Mechanical Properties of Hierarchical Lattices. *Mech. Mater.* **2013**, *62*, 32–43. [CrossRef]
24. Hutchinson, R.G.; Fleck, N.A. The Structural Performance of the Periodic Truss. *J. Mech. Phys. Solids* **2006**, *54*, 756–782. [CrossRef]
25. Elsayed, M.S.A.; Pasini, D. Analysis of the Elastostatic Specific Stiffness of 2D Stretching-Dominated Lattice Materials. *Mech. Mater.* **2010**, *42*, 709–725. [CrossRef]
26. Vigliotti, A.; Pasini, D. Linear Multiscale Analysis and Finite Element Validation of Stretching and Bending Dominated Lattice Materials. *Mech. Mater.* **2012**, *46*, 57–68. [CrossRef]
27. Florence, C.; Sab, K. A Rigorous Homogenization Method for the Determination of the Overall Ultimate Strength of Periodic Discrete Media and an Application to General Hexagonal Lattices of Beams. *Eur. J. Mech. A/Solids* **2006**, *25*, 72–97. [CrossRef]
28. Assidi, M.; Dos Reis, F.; Ganghoffer, J.-F. Equivalent Mechanical Properties of Biological Membranes from Lattice Homogenization. *J. Mech. Behav. Biomed. Mater.* **2011**, *4*, 1833–1845. [CrossRef]
29. Dos Reis, F.; Ganghoffer, J.F. Equivalent Mechanical Properties of Auxetic Lattices from Discrete Homogenization. *Comput. Mater. Sci.* **2012**, *51*, 314–321. [CrossRef]
30. Hassani, B.; Hinton, E. A Review of Homogenization and Topology Optimization I—Homogenization Theory for Media with Periodic Structure. *Comput. Struct.* **1998**, *69*, 707–717. [CrossRef]
31. Hollister, S.J.; Kikuehi, N. A Comparison of Homogenization and Standard Mechanics Analyses for Periodic Porous Composites. *Comput. Mech.* **1992**, *10*, 73–95. [CrossRef]
32. Guedes, J.J.M.J.; Kikuchi, N. Preprocessing and Postprocessing for Materials Based on the Homogenization Method with Adaptive Finite Element Methods. *Comput. Methods Appl. Mech. Eng.* **1990**, *83*, 143–198. [CrossRef]

33. Andreassen, E.; Andreasen, C.S. How to Determine Composite Material Properties Using Numerical Homogenization. *Comput. Mater. Sci.* **2014**, *83*, 488–495. [CrossRef]
34. Dong, G.; Tang, Y.; Zhao, Y.F. A 149 Line Homogenization Code for Three-Dimensional Cellular Materials Written in MATLAB. *J. Eng. Mater. Technol. Trans. ASME* **2019**, *141*, 011005. [CrossRef]
35. Masters, I.G.; Evans, K.E. Models for the Elastic Deformation of Honeycombs. *Compos. Struct.* **1996**, *35*, 403–422. [CrossRef]
36. Gibson, L.J.; Ashby, M.F.; Gibson, L.J.; Ashby, M.F. *Cellular Solids: Structure and Properties*; Cambridge University Press: Cambridge, UK, 1997; pp. 1–510. [CrossRef]
37. Christensen, R.M. Mechanics of Cellular and Other Low-Density Materials. *Int. J. Solids Struct.* **2000**, *37*, 93–104. [CrossRef]
38. Elsayed, M.S.A. *Multiscale Mechanics and Structural Design of Periodic Cellular Materials*; McGill University: Montreal, QC, Canada, 2010.
39. Vigliotti, A.; Pasini, D. Stiffness and Strength of Tridimensional Periodic Lattices. *Comput. Methods Appl. Mech. Eng.* **2012**, *229–232*, 27–43. [CrossRef]
40. Kalamkarov, A.L.; Andrianov, I.V.; Danishevs'kyy, V.V. Asymptotic Homogenization of Composite Materials and Structures. *Appl. Mech. Rev.* **2009**, *62*, 030802. [CrossRef]
41. Masoumi, E.; Abad, K.; Khanoki, S.A.; Pasini, D. Fatigue Design of Lattice Materials via Computational Mechanics: Application to Lattices with Smooth Transitions in Cell Geometry. *Int. J. Fatigue* **2012**, *47*, 126–136. [CrossRef]
42. Cheng, G.D.; Cai, Y.W.; Xu, L. Novel Implementation of Homogenization Method to Predict Effective Properties of Periodic Materials. *Acta Mech. Sin. Xuebao* **2013**, *29*, 550–556. [CrossRef]
43. Liu, P.; Liu, A.; Peng, H.; Tian, L.; Liu, J.; Lu, L. Mechanical Property Profiles of Microstructures via Asymptotic Homogenization. *Comput. Graph.* **2021**, *100*, 106–115. [CrossRef]
44. Arabnejad, S.; Pasini, D. Mechanical Properties of Lattice Materials via Asymptotic Homogenization and Comparison with Alternative Homogenization Methods. *Int. J. Mech. Sci.* **2013**, *77*, 249–262. [CrossRef]
45. Peng, X.; Cao, J. A Dual Homogenization and Finite Element Approach for Material Characterization of Textile Composites. *Compos. Part B Eng.* **2002**, *33*, 45–56. [CrossRef]
46. Guinovart-Daz, R.; Rodrguez-Ramos, R.; Bravo-Castillero, J.; Sabina, F.J.; Dario Santiago, R.; Martinez Rosado, R. Asymptotic Analysis of Linear Thermoelastic Properties of Fiber Composites. *J. Thermoplast. Compos. Mater.* **2007**, *20*, 389–410. [CrossRef]
47. Visrolia, A.; Meo, M. Multiscale Damage Modelling of 3D Weave Composite by Asymptotic Homogenisation. *Compos. Struct.* **2013**, *95*, 105–113. [CrossRef]
48. Takano, N.; Zako, M.; Kubo, F.; Kimura, K. Microstructure-Based Stress Analysis and Evaluation for Porous Ceramics by Homogenization Method with Digital Image-Based Modeling. *Int. J. Solids Struct.* **2003**, *40*, 1225–1242. [CrossRef]
49. Matsui, K.; Terada, K.; Yuge, K. Two-Scale Finite Element Analysis of Heterogeneous Solids with Periodic Microstructures. *Comput. Struct.* **2004**, *82*, 593–606. [CrossRef]
50. Omairey, S.L.; Dunning, P.D.; Sriramula, S. Development of an ABAQUS Plugin Tool for Periodic RVE Homogenisation. *Eng. Comput.* **2019**, *35*, 567–577. [CrossRef]
51. Material Designer User's Guide. Available online: https://ansyshelp.ansys.com/account/secured?returnurl=/Views/Secured/corp/v194/acp_md/acp_md.html (accessed on 16 December 2020).
52. Barbero, E.J. *Finite Element Analysis of Composite Materials Using AbaqusTM.*; CRC Press: Boca Raton, FL, USA, 2013; ISBN 9781466516632.
53. Hassani, B.; Hinton, E. A Review of Homogenization and Topology Opimization II—Analytical and Numerical Solution of Homogenization Equations. *Comput. Struct.* **1998**, *69*, 719–738. [CrossRef]
54. Hassani, B.; Hinton, E. A Review of Homogenization and Topology Optimization III—Topology Optimization Using Optimality Criteria. *Comput. Struct.* **1998**, *69*, 739. [CrossRef]
55. Rajakareyar, P. *Thermo-Elastic Mechanics of Morphing Lattice Structures with Applications in Shape Optimization of BLI Engine Intakes*; Carleton University: Ottawa, ON, Canada, 2023.
56. Virtanen, P.; Gommers, R.; Oliphant, T.E.; Haberland, M.; Reddy, T.; Cournapeau, D.; Burovski, E.; Peterson, P.; Weckesser, W.; Bright, J.; et al. SciPy 1.0: Fundamental Algorithms for Scientific Computing in Python. *Nat. Methods* **2020**, *17*, 261–272. [CrossRef]
57. Kermode, J. Lars Pastewka Matscipy/Elasticity.Py at Master LibAtoms/Matscipy. Available online: https://github.com/libAtoms/matscipy/blob/master/matscipy/elasticity.py (accessed on 11 December 2020).
58. Van Der Walt, S.; Colbert, S.C.; Varoquaux, G. The NumPy Array: A Structure for Efficient Numerical Computation. *Comput. Sci. Eng.* **2011**, *13*, 22–30. [CrossRef]
59. SciPy. Scipy.Spatial.CKDTree—SciPy v1.5.4 Reference Guide. Available online: https://docs.scipy.org/doc/scipy/reference/generated/scipy.spatial.cKDTree.html (accessed on 11 December 2020).
60. Hunter, J.D. Matplotlib: A 2D Graphics Environment. *Comput. Sci. Eng.* **2007**, *9*, 90–95. [CrossRef]
61. Ramachandran, P.; Varoquaux, G. Mayavi: 3D Visualization of Scientific Data. *Comput. Sci. Eng.* **2011**, *13*, 40–51. [CrossRef]
62. Gaillac, R.; Pullumbi, P.; Coudert, F.-X. ELATE: An Open-Source Online Application for Analysis and Visualization of Elastic Tensors. *J. Phys. Condens. Matter* **2016**, *28*, 275201. [CrossRef] [PubMed]

63. Hoogendoorn, E.; van Golen, K. Numpy-Indexed PyPI. Available online: https://pypi.org/project/numpy-indexed/ (accessed on 11 December 2020).
64. Lam, S.K.; Pitrou, A.; Seibert, S. Numba: A LLVM-Based Python JIT Compiler. In Proceedings of the Second Workshop on the LLVM Compiler Infrastructure in HPC, New York, NY, USA, 15 November 2015; pp. 1–6.

materials

MDPI

Article

Using the Radial Shear Rolling Method for Fast and Deep Processing Technology of a Steel Ingot Cast Structure

Alexandr Arbuz [1,*], Anna Kawalek [2], Alexandr Panichkin [3], Kirill Ozhmegov [2], Fedor Popov [4] and Nikita Lutchenko [1]

[1] Core Facilities—Office the Provost, AEO Nazarbayev University, 53 Kabanbay Batyr Avenue, Astana 010000, Kazakhstan; nikita.lutchenko@nu.edu.kz
[2] Department of Production Management, Faculty of Engineering Production and Materials Technology, Częstochowa University of Technology, ul. J.H. Dąbrowskiego 69, 42-201 Częstochowa, Poland; kawalek.anna@wip.pcz.pl (A.K.); kvozhmegov@wp.pl (K.O.)
[3] Institute of Metallurgy and Ore Benefication JSC, Satbayev University, 29/133 Shevchenko Street, Almaty 050010, Kazakhstan; a.panichkin@satbayev.university
[4] Department of Metallurgy, Faculty of Metallurgy and Mechanical Engineering, Karaganda Industrial University, 30 Republic Avenue, Temirtau 101400, Kazakhstan; fedor_popoff@mail.ru
* Correspondence: alexandr.arbuz@nu.edu.kz

Abstract: In advancing special materials, seamless integration into existing production chains is paramount. Beyond creating improved alloy compositions, precision in processing methods is crucial to preserve desired properties without drawbacks. The synergy between alloy formulation and processing techniques is pivotal for maximizing the benefits of innovative materials. By focusing on advanced deep processing technology for small ingots of modified 12% Cr stainless steel, this paper delves into the transformation of cast ingot steel structures using radial shear rolling (RSR) processing. Through a series of nine passes, rolling ingots from a 32 mm to a 13 mm diameter with a total elongation factor of 6.02, a notable shift occurred. This single-operation process effectuated a substantial change in sample structure, transitioning from a coarse-grained cast structure (0.5–1.5 mm) to an equiaxed fine-grained structure with peripheral grain sizes of 1–4 μm and an elongated rolling texture in the axial part of the bar. The complete transformation of the initial cast dendritic structure validates the implementation of the RSR method for the deep processing of ingots.

Keywords: rheology; plastometry; radial-shear rolling; severe plastic deformation; FEM-simulation; ingots; casting structure; fine-grained structure

Citation: Arbuz, A.; Kawalek, A.; Panichkin, A.; Ozhmegov, K.; Popov, F.; Lutchenko, N. Using the Radial Shear Rolling Method for Fast and Deep Processing Technology of a Steel Ingot Cast Structure. *Materials* **2023**, *16*, 7547. https://doi.org/10.3390/ma16247547

Academic Editors: Madhav Baral and Charles Lu

Received: 4 November 2023
Revised: 2 December 2023
Accepted: 5 December 2023
Published: 7 December 2023

1. Introduction

In recent years, there has been a renaissance in the exploration of liquid-metal-cooled fast reactor (LMFR) fuels and fuel cycle possibilities [1]. This renewed interest has not only been observed in various national research and development initiatives, but has also extended to international collaborative efforts. Some of these global undertakings include the International Project on Innovative Reactors and Fuel Cycles (INPRO), the Generation IV International Forum (GIF), and the Global Nuclear Energy Partnership (GNEP) [2–4].

Austenitic nickel-based stainless steels in the 300s family were initially chosen because of their good long-term mechanical properties at high temperatures and resistance to sodium coolant [5,6]. However, nickel, which stabilizes the austenite phase and gives steel its heat resistance, so necessary in high-temperature Gen-IV, experiences problems with induced radioactivity [7,8]. Nickel transition under neutron irradiation from the long-lived radioactive isotope nickel-63 makes reactor core structure disposal very problematic [9]. It is also known that matrix precipitates γ', which gives nickel steels increased strength, are unstable under irradiation [10,11]. This increases brittleness and reduces the strength of grain boundaries of such steel, which were unacceptably high, leading to the development of alternative stainless steels [12–14].

Nickel can be replaced by particulate, fine dispersed inclusions such as oxides or carbides. As reinforcing and stabilizing elements, yttrium oxide (Y_2O_3) particles are most effective [15,16]. This is the main dispersion strengthened (ODS) steel idea. Moreover, yttrium oxide has increased resistance to radiation [17–20], thereby prolonging the maximum service life of structural materials, and solving labor-consumption issues and the dangerous processing and disposal of steel structures in decommissioned nuclear power plant (NPP) cores.

ODS steels are usually produced using powder metallurgy by various types of mechanically alloyed powder sintering [21,22]. These methods are associated with problems related to residual porosity, uneven composition, and large grain size of 80 μm or more [23,24]. In addition, billets, as well as ingots, need additional deformational and heat treatments to improve structures.

The production of such special materials, as a rule, is carried out in small-scale production, special enterprises [25,26]. The technological chain usually includes smelting an ingot using a vacuum of at least 10–3 mbar, hot deformation processing (forging, pressing, and rolling), and cold deformation processing (pilger rolling, drawing, etc.) [27–30]. Due to the high quality requirements of semi-finished and finished products, these technological schemes are characterized by significant metal losses and low yields [31]. As a result, finished products for the nuclear industry will have high costs, thus the development of technologies that reduce these negative consequences is highly relevant.

Even minor changes in chemical composition can affect the manufacturability of a material during product manufacturing, primarily at cold deformation processing stages [32,33]. In addition, as a rule, due to the high costs of smelting full-scale ingots as in mass production, ingots of smaller diameter are used here [34,35]. A decrease in diameter leads to a decrease in the deformation of the cast structure during hot processing, which can lead to structural inhomogeneities during subsequent processing operations [36]. Therefore, deformation processing studies of small ingots of special materials are of great relevance. This is the main focus of this work.

For the deep processing of a cast structure, severe plastic deformation (SPD) methods can be used [37,38]. For this initial purpose, the equal-channel angular pressing (ECAP) method was invented by Segal [39,40]. This method later became generally accepted and most widespread in the bulk ultrafine-grained (UFG) and nanostructured material field by SPD [41–43]. Various SPD process schemes are shown in Figure 1.

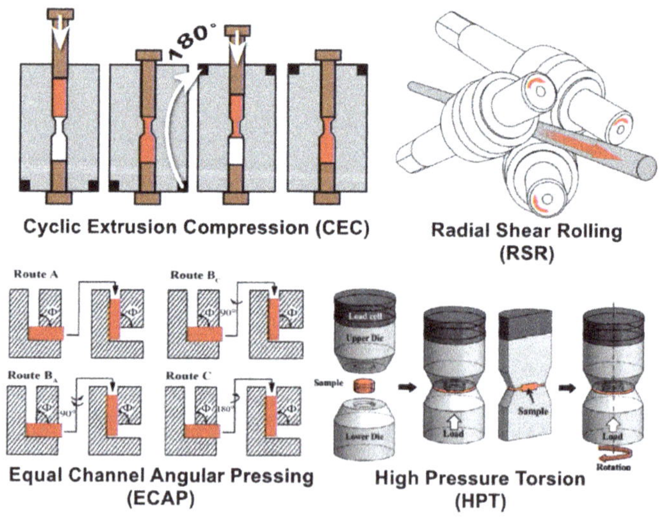

Figure 1. Scheme showing various SPD processes.

Generally, using plastic strain intensification to improve the quality of ingots makes a lot of sense and has a previous application history. The most common use of SPD methods is in the forging of ingots using special dies [44,45] or in all-round forging [46]. High-effective microstructure processing methods such as high-pressure torsion (HPT) [47–49] are unfortunately not applicable outside laboratories, and even more so for ingots, not applicable due to sample size limitations and method scaling factors. The same can be said for cyclic extrusion (CEC) [50,51], accumulated roll bonding (ARB) [52], and some other methods [53–59]. Of course, ECAP can be used for pressing ingots, but this is not technologically easy due to the need for several pressing cycles and considerable effort and equipment strength requirements. This general point can be seen clearly from the SPD process schemes in Figure 1.

One of the alternative and still not widely used methods of intensive structure refinement using the SPD method is Radial Shear Rolling (RSR) [60,61]. This method is outwardly similar to the classical Mannesmann method for producing seamless pipes [62,63] but differs significantly from it. Figure 1 shows the process scheme.

As can be seen from Figure 1, there are not two skew rolls, like in the pipe piercing scheme, but three skew rolls, and not a pipe is rolled, but a solid bar. A more important difference is not visible in the figure; due to the special combination of roll angles, all-around compressive stresses are realized in the deformation zone, without tensile stresses, as it works in pipe piercing processes. Due to increased roll skew angles of 18–21 degrees, metal flows in the deformation zone have an axisymmetric vortex character with a gradient along the radius [64]. Such a stress–strain state ensures intense deformation of the outer sample zone with the formation of an equiaxed ultrafine-grained structure [65]. Laminar metal flow along the rolling axis in the central third zone of the bar ensures the formation of a rolling texture [66].

This method has been successfully used for the significant refinement of titanium [67,68], zirconium [64,69], copper [70], and steel [71]. In many of these cases, a gradient structure is formed. All experiments with radial shear rolling involve the processing of semi-products of rolling production with a structure either after hot rolling or after recrystallization annealing.

One of the main requirements is to provide a structure. NPP core zone structural elements should be predominantly recrystallized with a grain size of at least 9–12 points according to ASTM E112 [72,73]. Such a structure type can be achieved by the described method, therefore, the use of RSR is justified.

Equally, the influence of radial shear rolling on the structure of ingots is of no less scientific interest. Implemented high levels of accumulated deformation and a stress state scheme, which are favorable for materials with reduced plasticity, promise significant effects with relatively small changes in workpiece dimensions.

The main purpose of this study was to investigate ingot initial structure behaviors under the influence of radial shear rolling. The possibilities of this method were realized by evaluating the processing microstructures of special alloy ingots in small-scale production conditions of NPP critical parts.

2. Materials and Methods

Yttrium-modified stainless steel small ingots were used as experimental materials for this study. Ingots were obtained by vacuum induction furnace melting as part of other research focused on generating oxide dispersion-strengthened (ODS) steel. The base material for melting was ordinary Fe-13%Cr steel (AISI 403). During the melting process, a small quantity of metallic yttrium and iron oxide (0.5% together) was added to steel. Metallic yttrium undergoes oxidation through the reduction of iron oxide, transforming it into yttrium oxide (Y_2O_3) particles in steel. The melting operation was carried out at a temperature of 1600 °C and a furnace pressure ranging from 1 to $5{\cdot}10^2$ Pa. The final composition percentages were Fe-86.50, Cr-12.47, Si-0.25, Y-0.12, and Other-0.65. The melting technology and its features, including oxide dispersion and composition fluctuations are other ongoing research subjects and will be published separately. Here, we studied casted structure deep processing. A typical casting dendrite ingot structure was reached for a 32 mm diameter ingot intended for RSR rolling experiments (Figure 2).

Figure 2. Initial 32 mm diameter ingots.

Casted ingots were examined for rheology properties using a Gleeble 3800 (Dynamic systems Inc., Austin, DX, USA) plastometer using the "Pocket Jaw" module. Plastometric tests were carried out by the uniaxial compression of cylindrical specimens at strain rates of 0.5 s^{-1}, 5 s^{-1}, and 15 s^{-1} at temperatures of 600 °C, 800 °C, 1000 °C, and 1200 °C. That is, 12 deformation cases of modified steel were studied. Strain rate and temperature ranges were selected based on all-round forging and radial shear rolling conditions of the resulting alloy, and accounted for the thermal effects of plastic deformation. Although the average strain rate during radial shear rolling corresponded to 5 s^{-1}, there were regions of localized metal flow where the strain rate was lower (central part of the bar) and higher (periphery). The testing process is shown in Figure 3.

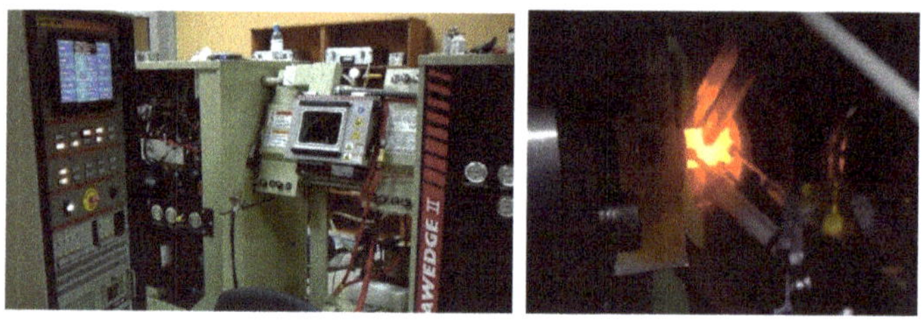

Figure 3. Gleeble 3800 testing machine (**left**) and a photo showing the testing of cast ODS-steel specimens in ISO-T model dies with mounted extensometers for longitudinal and transverse measurements of specimens during testing (**right**).

Cylindrical specimens with a working part diameter of 10 mm and a length of 12 mm from several initial ingots were made. The test temperature was controlled using a chromel–copel thermocouple welded to the central part of the specimen on the Therwocouple welder tool supplied with the Gleeble 3800 set. Graphite-based thin gaskets were used as a lubricant in tests. Working dies of the ISO-T model were additionally lubricated with OKS255 (OKS, Maisach, Germany) grease after each test.

Radial-Shear Rolling for ingot processing computer simulations by the Finite Element Method (FEM) using DEFORM-3D (SFTC, Columbus, OH, USA) software were conducted.

A verification experiment was conducted using the RSR-10/30 radial shear rolling mill at Karaganda Industrial University. The rolling process included 10 passes with a 2 mm step (in terms of diameter), commencing from an initial diameter of 32 mm and concluding at a diameter of 13 mm, with an initial heating temperature of 1200 °C. These represented the maximum and minimum achievable rolling diameters for the mill to attain the highest deformation level. Rolling diameters for each pass were the next route (mm): 32-30-28-26-24-22-20-18-16-13. Rolling reduction factor and temperature selections were guided by reference materials and previous studies [74–77]. The heating process was conducted using the Nabertherm LH-30/14 heating furnace.

The experimental rolling process is shown in Figure 4. The RSR mill the next rolling gape diameter setting process to the next pass took 2–5 min. The rolling process is shown in Figure 4.

Figure 4. Experimental rolling and the RSR-10/30 radial shear rolling mill.

After rolling, the bar was cut for microstructure characterization into short small bars, which were cut in half along the axis using a Brilliant-220 (QATM, Mammelzen, Germany) precision cutting machine with a cutting speed of 10 µm/s and intensive water cooling to minimize deformation–temperature damage to structures. From both halves of the section, two thin plates were cut—one for Transmission Electron Microscopy (Jeol, Tokyo, Japan) specimen preparation to characterize deformed peripheral zone fine structure zones. Another plate was used for Electron Backscatter Diffraction (EBSD) sample preparation by electrolytic polishing (jet-polishing). The longitudinal section was preliminarily polished on a Saphire-520 (QATM, Mammelzen, Germany) grinding and polishing machine, and then electrolytically polished on a LectroPol-5 (Struers, Copenhagen, Denmark) unit. A 3 mm diameter round TEM specimen was punched out with a Gatan puncher and subjected to electrolytic thinning on a TenuPol-5 (Struers, Copenhagen, Denmark) machine. In both cases, A2 electrolytes recommended by the manufacturer were used. Electrolytes forced cooling to −20 °C using attached Julabo 600F cryostats (Julabo, Seelbach, Germany). The cutting scheme is shown in Figure 5.

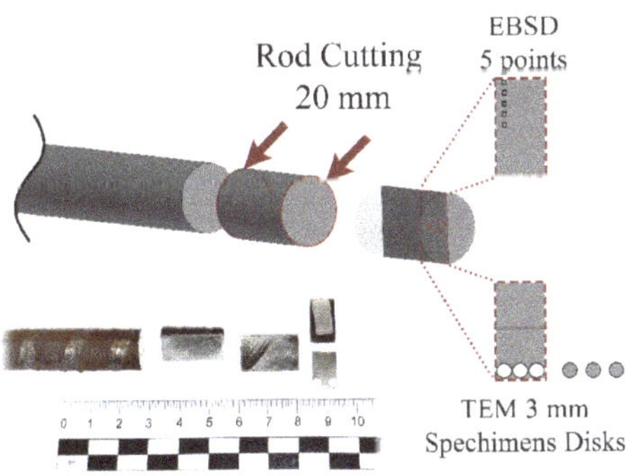

Figure 5. The rolled bar cutting scheme.

Electron microscopy methods were used to study microstructural changes. The microstructural characterization of grains, including crystallographic orientation, was performed using Electron backscatter diffraction (EBSD) on a Carl Zeiss Crossbeam-540 field emission scanning electron microscope (Carl Zeiss, Jena, Germany) with an NordlysNano EBSD(Oxford Instruments, Abingdon, UK) attachment. Diffraction pattern recognition and mapping were performed using HKL Channel-5 Tango software v.5.12.74.0 (Oxford Instruments, Abingdon, UK). Maps were built along the radius line from the center to the periphery at five points and at the same distance with steps of 1.625 mm. Once the grains were reconstructed, a variety of grain size parameters were automatically calculated, e.g., Circle Equivalent Diameter (CED) was better and more accurate in characterizing deformed grain sizes. Therefore, we used this parameter later for grain size. Major and minor axis ratios were used to characterize grain elongation parameters. The scheme according to the Channel-5 User Manual is shown in Figure 6.

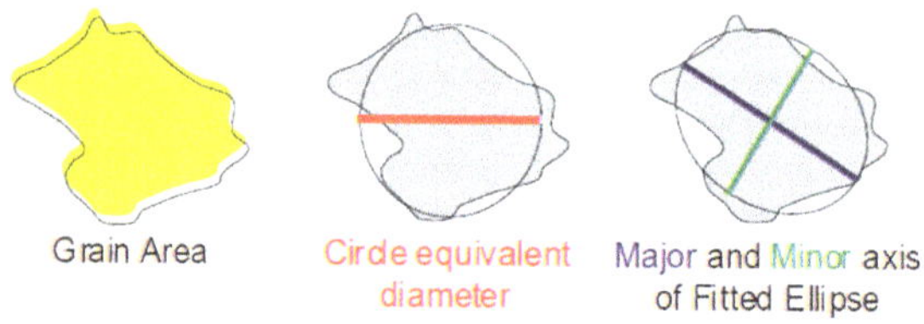

Figure 6. Scheme for calculating grain size parameters.

The fine structures of the most deformed peripheral zones were studied on a JEOL JEM-1400PLUS transmission electron microscope with a Gatan OneView camera (Gatan Inc., Pleasanton, CA, USA) in bright field mode. The original ingot sample microstructure, due to its very large grain size and beyond the minimum field of view, was performed only on a SEM JEOL JSM IT-200LA (Jeol, Tokyo, Japan) in back-scattered electron (BSE) mode.

A microhardness test was performed on the EBSD-ready-cross-section plate using the Shimadzu HMV-G31ST (Shimadzu, Kyoto, Japan) tester and the Vickers method (10 HV) using five times averaging with a load of (2.942) N and 10 s holding times.

3. Results and Discussion

3.1. Rheological Properties and Database Making

Based on Gleeble 3800 plastometer plastometric test results, ODS steel stress-strain flow curve graphs for 0.5–15 s^{-1} strain rate ranges and 600–1200 °C temperature ranges were plotted. To construct curves, three tests were conducted and a total of 30 tests were carried out. Flow curves are shown in Figures 7–10.

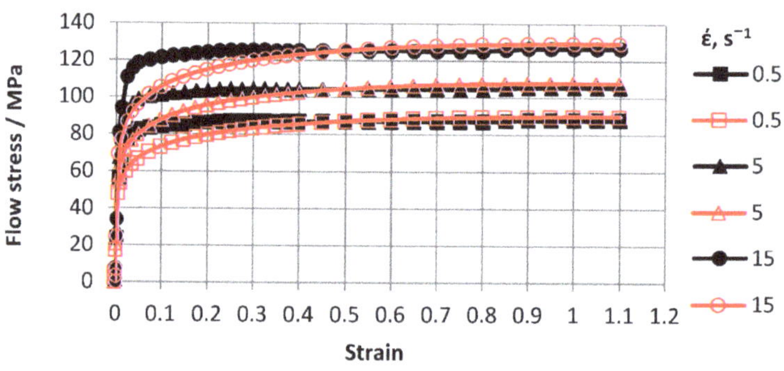

Figure 7. Material flow curves obtained at 600 °C: solid markers—indicate experimental data and transparent markers—indicate approximation results.

The advantage of the compression test scheme was the similarity of the load application scheme with many metal pressure treatment processes and data generation on the resistance of the metal to plastic deformation over a wide range of deformations. The Figures 7–10 showed that with increasing temperatures from 600 to 1200 °C, deformation resistance values decreased by more than 10 times. Increasing the strain rate from 0.5 s^{-1} to 15.0 s^{-1} increased strain resistance values, and at T = 600 °C, the increase was approximately 30%. At T = 800 °C, the increase was approximately 48%. At T = 1000 °C, the increase was approximately 57%, and at T = 1200 °C, the increase was 55%. Thus, the effects of increasing strain rates on strain resistance increased with increasing test temperature.

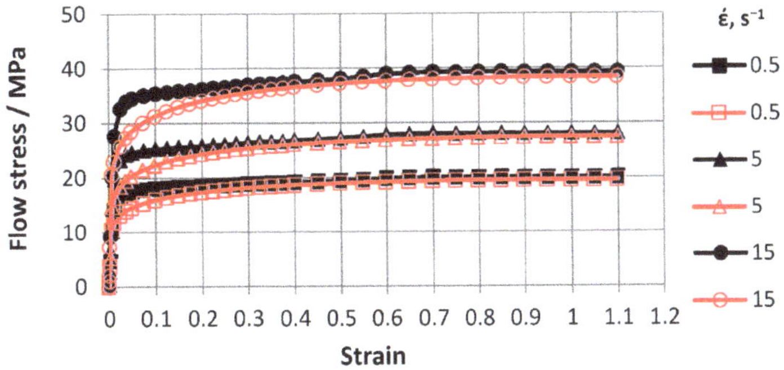

Figure 8. Material flow curves obtained at 800 °C: solid markers—indicate experimental data and transparent markers—indicate approximation results.

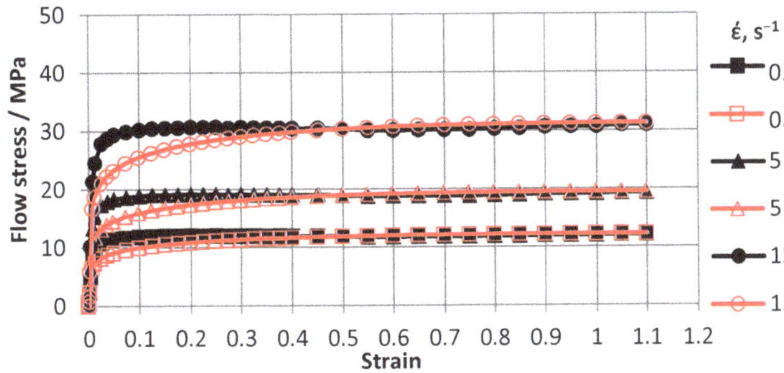

Figure 9. Material flow curves obtained at 1000 °C: solid markers—indicate experimental data and transparent markers—indicate approximation results.

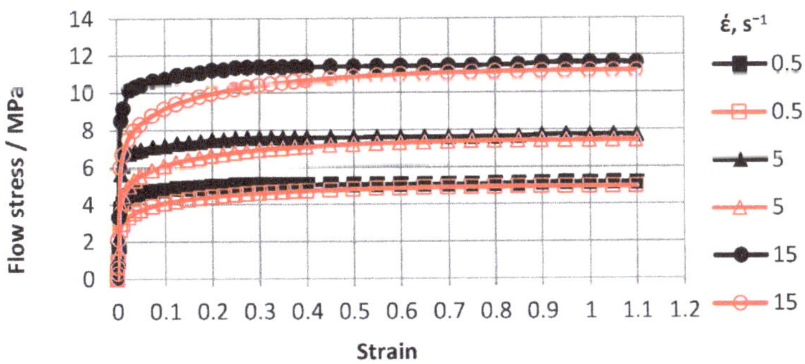

Figure 10. Material flow curves obtained at 1200 °C: solid markers—indicate experimental data and transparent markers—indicate approximation results.

Flow curve characters in temperature ranges coincided. In the strain range up to $\varepsilon = 0.5$, flow curves reached steady-state stages. Moreover, with increasing strain rates, eliminated stages occurred at higher strain values.

For the practical use plastometric study results, an approximation of flow curves was carried out using a generalized dependence—the function (1) of Hensel A., Spittel T. [33].

$$\sigma_p = A\exp(m_1 T)T^{m_9}\varepsilon^{m_2}\exp\left(m_4/\varepsilon\right)(1+\varepsilon)^{m_5 T}\exp(m_7\varepsilon)\dot{\varepsilon}^{m_3}\dot{\varepsilon}^{m_8 T} \quad (1)$$

where A and $m_1 \ldots m_9$—are unknown coefficients in the deformation resistance model.

To solve the regression, we used the built-in program of the Forge 2008® "RheologyDatabase" program, which used the Levenberg–Marquardt method. This method provides the automatic determination of coefficients based on an array of experimental points. The operation was performed at the Institute of Metal Forming Processes and Safety Engineering, Czestochowa Polytechnic (Poland). The approximating coefficients were as follows: $A = 2251$; $m_1 = -0.0052$; $m_2 = -0.01$; $m_3 = 0.2$; and $m_4 = -0.05$. An analysis of the verification results of approximated and real results showed an approximation error of no more than 10%. The approximation was considered optimal when the approximation error was not more than 15%. The difference between flow curves obtained experimentally and calculated by deformation $\varepsilon < 0.4$ was due to the fact that deformation resistance σ_p values were influenced by elastic deformation, which was not accounted for in Equation (1). Thus, in experimental flow curves, plastic deformation began at large deformation resistance σ_p values. As the degree of deformation increased, the error decreased. Due to the large magnitude of plastic deformation that occurred in simulated processes, the difference between the magnitude of the actual deformation and the value obtained after approximation in the range of smaller deformations of 0.4 did not affect the results.

After processing the data, a new material database was created for a new material base was created for Deform-3D suitable for modeling deformation processes of the cast ODS-steel in the temperature range of 600–1200 °C and strain rates of 0.5–15 s^{-1}.

The database can be used to simulate hot deformation processing conditions (hot pressing, hot forging, and hot rolling), which are used at the beginning of the technological cycle for the manufacture of fast reactors fuel assemblies, fuel element claddings, and plugs.

3.2. Computer Simulation of RSR Ingot Processing

Radial shear rolling processing was simulated for the sequential rolling of a 32 mm diameter ingot to a 13 mm diameter bar in nine passes with a 2–3 mm reduction in diameter on the RSR-10/30 (MISIS, Moscow, Russia) rolling mill. The total rolling elongation factor was 6.05. Simulation results are shown in Figure 11.

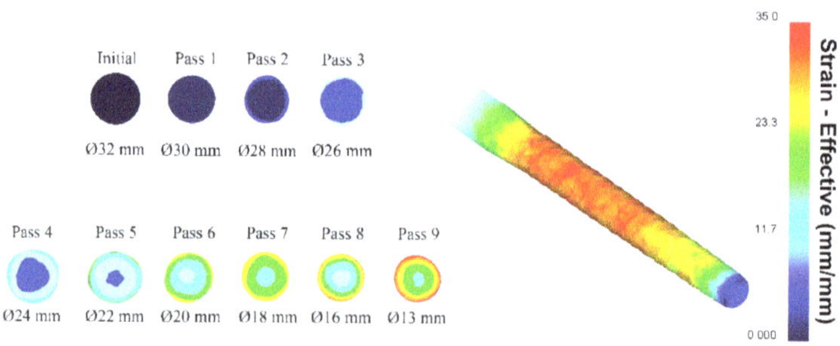

Figure 11. Computer simulation of RSR ingot processing.

The accumulated total strain degree in some areas of the final bar reached 30 mm/mm with a minimum value of 15 mm/mm. Such deformation levels should be sufficient for initial casted structure deep transformation. Strain levels had axisymmetric radial gradients regarding bar cross-sections. These differences can be used to interpret microstructural changes.

3.3. Microstructural Change Study

Initial ingot microstructures at various magnifications are shown in Figure 12. The images clearly show the light axes of dendrites and darker interdendritic spaces. The grain sizes fall within the range of 0.5–1.5 mm. Consistently, dendritic cast structures are prevalent throughout the steel volume, embodying typical ingot structures as expected. This microstructure was suitable for casted structure evolution research during RSR processing.

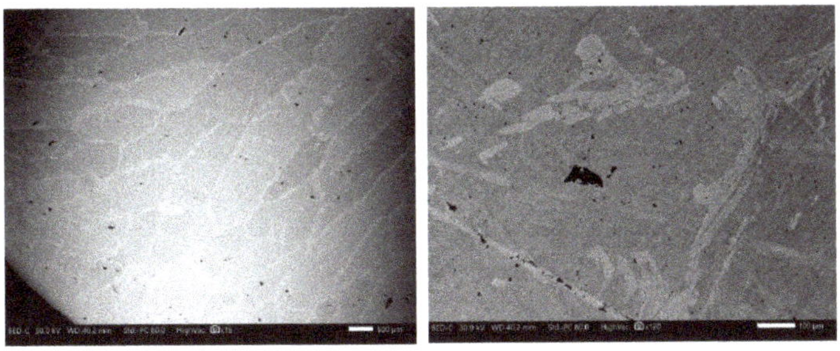

Figure 12. SEM (BSE) scale images of original ingot dendritic structures.

Following the rolling process, a gradient structure was expected. To characterize this, gradient EBSD mapping was used. The mapping spans along the longitudinal section radius from the axial zone to the peripheral zone at equal distance measurements at five specific points. The resulting EBSD maps, showcasing full-size microstructure images of the most differenced axial and peripheral zones of the bar (left and right images respectively), are presented in Figure 13.

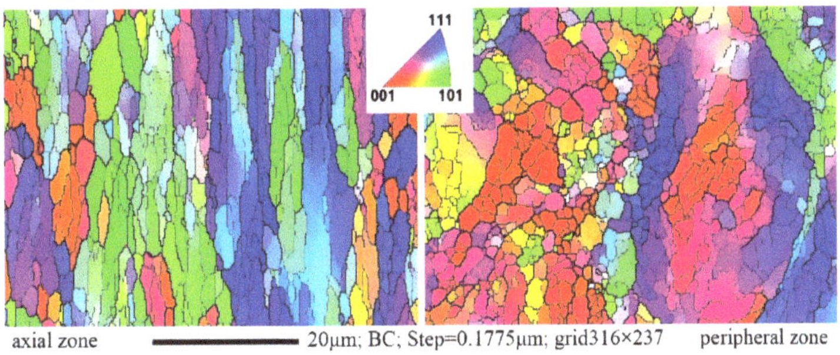

Figure 13. EBSD map of the axial zone (**left**) and peripheral zone (**right**) of a longitude bar section after final rolling.

The peripheral zone map of the bar revealed a fine-grained structure characterized by equiaxed grains with diverse orientations. Within structures, similar orientation grains to elongated along rolling direction chains aligned and interspersed randomly, exhibiting significantly different orientation grain chains. The occurrence of identical orientation grain chains suggested unfinished potential dynamic recrystallization processes. However, this fact generally correlated with well-known work on radial shear rolling [24–30,51,52] and SPD processing generally [4–6], and indicated significant processing of the cast structure with its complete transformation into fine-grained structures close to nuclear engineering needs [53,54].

The center of the bar microstructure exhibited predominantly similar orientations of large grains elongated in the rolling direction. However, there were notable regions with fine grain clusters and varying orientation zones, rendering the structure more intricate than anticipated based on existing research. It did not resemble a heavily deformed laminar shape typically observed in metals with a strong rolling texture, as previously seen [18,26]. Instead, it appeared to represent an intermediate state between the texture and structure of the bar's center after potential deformation using the RSR method, as outlined elsewhere [51].

This complex structure likely resulted from a combination of processes involving dynamic recrystallization, manifesting as clusters or colonies of fine grains within the matrix with different orientations, and static recrystallization, leading to the formation of large grains that retained their longitudinal orientation in the rolling direction. However, these large grains were now irregular and more rounded in shape, exhibiting evident misorientation. This phenomenon was probably induced by longer, smoother deformational heating in the central part, contrasting with rapid heating and cooling cycles experienced by the near-surface layer due to large shear deformations and subsequent swift cooling from cold rolls and the environment.

To conduct a comprehensive and quantitative analysis of microstructure changes across cross-sections, all EBSD maps were meticulously processed and subjected to statistical analysis. This analysis generated a graph featuring two primary indicators, as depicted in Figure 14. The gray columns, with a scale on the left, represented the average grain size by CED in the axial-peripheral direction. Concurrently, the red plot, with a scale on the right, illustrated the average ratio of the larger and smaller axes of grain dimensions, serving as a measure of equiaxiality. A value close to 1 indicated a more equiaxed grain, while a value closer to 0 suggested a more elongated grain, characteristic of rolling texture morphology. This indicator was introduced to address gradient structures across the rod's cross-section and to provide a more objective characterization of microstructures than the average grain size by CED alone.

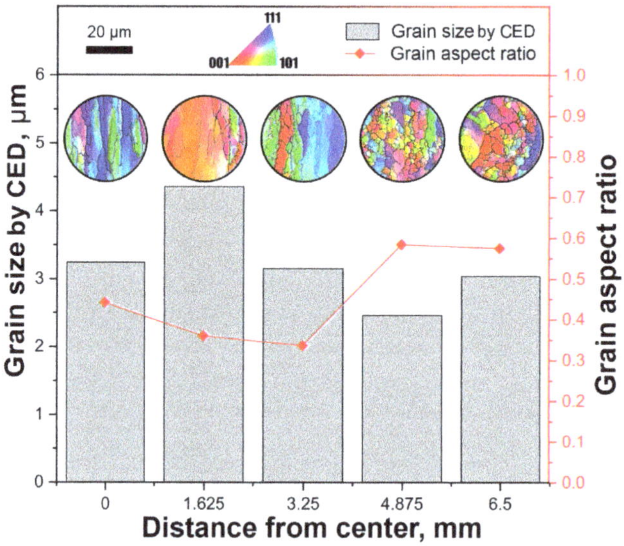

Figure 14. EBSD study of the gradient microstructures of rolled bar longitudinal sections.

In the axial zone of the rod, the grain ratio value was 0.58, showcasing a 23% change towards the peripheral zone. This indicated a decrease in grain equiaxiality from the peripheral to the axial zone. As illustrated in Figure 13, this aligned with the transition from a fine-grained structure in the peripheral zone to an elongated fine-grained structure along the axis of the bar.

In the provided miniatures, it was evident that grains tended to cluster in fine-grained groups with closely aligned orientations across almost all images. A notable shift in grain equiaxiality was observed around the halfway point of the rod's radius. The highest concentration of elongated grains started from this region. However, interestingly, the average grain size remained comparable to that of the periphery. Grain growth was initiated at a depth of approximately 2/3 of the radius from the surface. These data were comparable with FEM-simulation data (Figure 10). They had a general gradient similarity regarding the structure-changing radial distribution. A higher total strain level of 25 mm/mm and more for outer zones corresponded to a better structure. Medium 18–24 mm/mm levels corresponded to a transition structure. Although the average grain size at this stage was still notably smaller than the original sample's grains, which were larger than 1000 μm, the change was considerable. The average grain size across the entire cross-sectional area of the rolled sample was 3.2 μm, representing a reduction of over 300 times compared to initial grain size.

Overall, the cast structure of ingots had a complete transformation throughout the entire cross-section, leading to the successful application of the RSR method for ingot deep processing. The transformed structure of the periphery zone mainly corresponded to ASTM E112 requirements [72,73].

The fine structure TEM characterization of the longitudinal section peripheral part (2 mm from the edge along the radius of the rod) at various magnifications is shown in Figure 15. The TEM study affirmed grain refinement down to sizes of 1–4 μm in the peripheral part of the rod, accompanied by a similar misorientation level. Dislocations were primarily concentrated at grain boundaries. Within the grain body, dislocation clusters, if present in sufficient quantities, demonstrated a trend towards dislocation rearrangements, resulting in the formation of a cellular structure. Simultaneously, numerous grains displayed a small number of dislocations, suggesting increased material plasticity. High deformation heating probably triggered fine structure recovery processes with dislocation annihilation. Notably, these observations slightly deviated from previous data [52], which indicated a significantly higher number of dislocations in peripheral ultrafine-grained (UFG) zone grains.

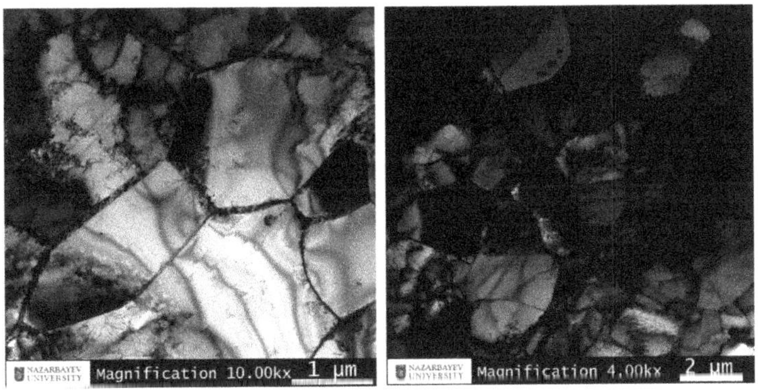

Figure 15. Fine structure of fine-grained zones in the peripheral bar part after final rolling.

In this case, despite substantial processing, the peripheral region did not transition to the Ultrafine-Grained (UFG) state. Probably, between passes, recovery and recrystallization processes had likely occurred. Specific thermal and rolling conditions played crucial roles in achieving desired microstructural outcomes. Further optimization of the processing parameters may be necessary to achieve a targeted UFG state in the peripheral zone.

Microhardness evaluations showed decreasing values from 212 HV to 183 HV. These changes corresponded to microstructural changes in the full transformation of non-plastic casted structures to normal. Also, a decrease in hardness was explained by dislocation-free grains according to the TEM study.

4. Conclusions

Based on our analyses, the following conclusions were formulated:

1. The use of radial shear rolling processing led to changes in the structure of a small ingot of stainless steel modified with yttrium.
2. The original coarse-grained dendritic structure of casted ingots was significantly refined to a fine-grained structure. The region farthest from the axis of the rod, constituting 1/3 of the radius of the rod, had a fine-grained structure with a grain size of 1–4 μm, with an equiaxial structure, without a clear dominant orientation structure.

The axial zone, which also occupied 1/3 of the radius of the rod, had a more elongated and coarse-grained structure with grain sizes from 4 to 10 μm. The predominant grain orientation in this region was oriented along the rolling axis. In the middle region, there was a transition-type structure between these two. The average cross-sectional grain size was approximately 3.2 microns, which was 300 times smaller than grain sizes before the rolling process.

3. Grains in the highly deformed peripheral zone had a relatively small number of dislocations. Based on fine structure and some EBSD features, we conclude that processing can be improved.
4. Resulting blanks were close to nuclear engineering needs and could be used as semi-product improvements for the manufacture of some structural elements for nuclear power plant cores.

Author Contributions: Conceptualization, writing, and funding acquisition, A.A.; methodology, A.P., A.K., K.O. and F.P.; writing and melting procedures, A.P. and F.P.; formal analysis and sample preparation, N.L. and F.P.; writing—review and editing A.A. and F.P. All authors have read and agreed to the published version of the manuscript.

Funding: This research was funded by the Science Committee of the Ministry of Education and Science of the Republic of Kazakhstan (Grant No. AP09259982).

Institutional Review Board Statement: Not applicable.

Informed Consent Statement: Not applicable.

Data Availability Statement: Data are contained within the article.

Acknowledgments: The work was performed under the state budget-funded theme № AP09259982 "Development and research of technology for obtaining and advance refinement of oxide dispersion-strengthened steel for use in nuclear engineering" of the program "Grant Financing of Scientific Research" (Customer— the Science Committee of the Ministry of Education and Science of the Republic of Kazakhstan). The authors would like to thank the Electron Microscopy Lab staff of Nazarbayev University Core Facilities team for their support with microscopy work (Nurgul Danieva for TEM and Alisher Rapikov for SEM) and the Core Facilities head, Aidos Baumuratov.

Conflicts of Interest: The authors declare no conflict of interest.

References

1. Yang, G.-L.; Liao, H.-L.; Ding, T.; Chen, H.-L. Development and Validation of a New Oxide Fuel Rod Performance Analysis Code for the Liquid Metal Fast Reactor. *Nucl. Sci. Tech.* **2022**, *33*, 66. [CrossRef]
2. IAEA. *Internationale Atomenergie-Organisation Structural Materials for Liquid Metal Cooled Fast Reactor Fuel Assemblies: Operational Behaviour*; IAEA Nuclear Energy Series; IAEA: Vienna, Austria, 2012; ISBN 978-92-0-127510-3.
3. GNEP organization. *An Overview of The Global Nuclear Energy Partnership (GNEP), Including Proposed Advanced Reactor Technologiesf Recycling Nuclear Waste*; Scholar's Choice: Los Angeles, CA, USA, 2015; ISBN 1-297-01179-1.
4. Sokolov, Y.; Omoto, A.; Facer, I.; Khoroshev, M. The IAEA international project on innovative nuclear reactors and fuel cycles (INPRO): Experience of assessments and development of approaches. Proceeding of the International Conference on Emerging Nuclear Energy Systems ICENES, Brussels, Belgium, 22–26 August 2005.
5. Dalle, F.; Blat-Yrieix, M.; Dubiez-Le Goff, S.; Cabet, C.; Dubuisson, P. Conventional Austenitic Steels as Out-of-Core Materials for Generation IV Nuclear Reactors. In *Structural Materials for Generation IV Nuclear Reactors*; Elsevier: Amsterdam, The Netherlands, 2017; pp. 595–633. ISBN 978-0-08-100906-2.
6. Kleykamp, H. Selection of Materials as Diluents for Burning of Plutonium Fuels in Nuclear Reactors. *J. Nucl. Mater.* **1999**, *275*, 1–11. [CrossRef]
7. Heyn, F.A. The Radioactivity of Cobalt, Nickel, Copper and Zinc Induced by Neutrons. *Physica* **1937**, *4*, 160–165. [CrossRef]
8. El Mokhtari, B.; Chetaine, A.; Amsil, H.; Embarch, K.; Benchrif, A.; Laraki, K.; Marah, H. Modeling and Simulating the Induced Effect on the Steel Collimator Plug Used in the PGAA Facility of the Moroccan TRIGA Mark-II Reactor under Different Neutron Irradiation Levels. *Appl. Radiat. Isot.* **2021**, *170*, 109620. [CrossRef] [PubMed]
9. Pavliuk, A.O.; Kotlyarevskiy, S.G.; Bespala, E.V.; Zakharova, E.V.; Ermolaev, V.M.; Volkova, A.G. Experience of On-Site Disposal of Production Uranium-Graphite Nuclear Reactor. *J. Environ. Radioact.* **2018**, *184–185*, 22–31. [CrossRef]
10. Zinkle, S.J.; Was, G.S. Materials Challenges in Nuclear Energy. *Acta Mater.* **2013**, *61*, 735–758. [CrossRef]
11. Chen, Y. Irradiation effects of HT-9 martensitic steel. *Nucl. Eng. Technol.* **2013**, *45*, 311–322. [CrossRef]

12. Henry, J.; Maloy, S.A. Irradiation-Resistant Ferritic and Martensitic Steels as Core Materials for Generation IV Nuclear Reactors. In *Structural Materials for Generation IV Nuclear Reactors*; Elsevier: Amsterdam, The Netherlands, 2017; pp. 329–355. ISBN 978-0-08-100906-2.
13. Zinkle, S.J.; Ghoniem, N.M. Prospects for Accelerated Development of High Performance Structural Materials. *J. Nucl. Mater.* **2011**, *417*, 2–8. [CrossRef]
14. Yvon, P.; Carré, F. Structural Materials Challenges for Advanced Reactor Systems. *J. Nucl. Mater.* **2009**, *385*, 217–222. [CrossRef]
15. Wilms, M.B.; Pirch, N.; Gökce, B. Manufacturing Oxide-Dispersion-Strengthened Steels Using the Advanced Directed Energy Deposition Process of High-Speed Laser Cladding. *Prog. Addit. Manuf.* **2023**, *8*, 159–167. [CrossRef]
16. Wang, J.; Liu, S.; Xu, B.; Zhang, J.; Sun, M.; Li, D. Research Progress on Preparation Technology of Oxide Dispersion Strengthened Steel for Nuclear Energy. *Int. J. Extrem. Manuf.* **2021**, *3*, 032001. [CrossRef]
17. Huet, J.-J.; Coheur, L.; Bremaecker, A.D.; Wilde, L.D.; Gedopt, J.; Hendrix, W.; Vandermeulen, W. Fabrication and Mechanical Properties of Oxide Dispersion Strengthening Ferritic Alloy Canning Tubes for Fast Reactor Fuel Pins. *Nucl. Technol.* **1985**, *70*, 215–219. [CrossRef]
18. Yamashita, S.; Akasaka, N.; Ukai, S.; Ohnuki, S. Microstructural Development of a Heavily Neutron-Irradiated ODS Ferritic Steel (MA957) at Elevated Temperature. *J. Nucl. Mater.* **2007**, *367–370*, 202–207. [CrossRef]
19. Ukai, S.; Nishida, T.; Okuda, T.; Yoshitake, T. R&D of Oxide Dispersion Strengthened Ferritic Martensitic Steels for FBR. *J. Nucl. Mater.* **1998**, *258–263*, 1745–1749. [CrossRef]
20. Little, E.A. *Effects of Radiation on Materials: 17th International Symposium*; Gelles, D.S., Nanstad, R.K., Kumar, A.S., Little, E.A., Eds.; ASTM: West Conshohocken, PA, USA, 1996; pp. 739–752.
21. Xu, H.; Lu, Z.; Jia, C.; Feng, D.; Liu, C. Influence of Mechanical Alloying Time on Morphology and Properties of 15Cr-ODS Steel Powders. *High Temp. Mater. Process.* **2016**, *35*, 473–477. [CrossRef]
22. Mao, X.; Kang, S.H.; Kim, T.K.; Kim, S.C.; Oh, K.H.; Jang, J. Microstructure and Mechanical Properties of Ultrafine-Grained Austenitic Oxide Dispersion Strengthened Steel. *Metall. Mater. Trans. A* **2016**, *47*, 5334–5343. [CrossRef]
23. Couvrat, M.; Chaffron, L.; Nunes, D.; Bonnaillie, P.; Mathon, M.H.; Perrut, M. Microstructure Evolution of Mechanically Alloyed ODS Ferritic Steels during Hot Extrusion. *Solid State Phenom.* **2011**, *172–174*, 721–726. [CrossRef]
24. Gao, R.; Ge, W.; Miao, S.; Zhang, T.; Wang, X.; Fang, Q. Hot Rolling and Annealing Effects on the Microstructure and Mechanical Properties of ODS Austenitic Steel Fabricated by Electron Beam Selective Melting. *Front. Mater. Sci.* **2016**, *10*, 73–79. [CrossRef]
25. Phillips, J.A.; Nagley, S.G.; Shaber, E.L. Fabrication of Uranium Oxycarbide Kernels and Compacts for HTR Fuel. *Nucl. Eng. Des.* **2012**, *251*, 261–281. [CrossRef]
26. Byalik, G.A.; Gontarenko, V.I.; National University "Zaporizhzhia Polytechnic"; Bazhmina, E.A. Features of the Production of High-Quality Carbon Cast Steel in Induction Crucible Furnaces of Small Capacity. *Metalozn. Obrobka Met.* **2020**, *95*, 76–84. [CrossRef]
27. Banaszek, G.; Ozhmegov, K.; Kawałek, A.; Sawicki, S.; Arbuz, A.; Naizabekov, A. Modeling of Closure of Metallurgical Discontinuities in the Process of Forging Zirconium Alloy. *Chem. Mater. Sci.* 2023, *online ahead of print*. [CrossRef] [PubMed]
28. McQueen, H.J.; Ryan, N.D. Constitutive Analysis in Hot Working. *Mater. Sci. Eng. A* **2002**, *322*, 43–63. [CrossRef]
29. Zhang, Y.; Zuo, T.T.; Tang, Z.; Gao, M.C.; Dahmen, K.A.; Liaw, P.K.; Lu, Z.P. Microstructures and Properties of High-Entropy Alloys. *Prog. Mater. Sci.* **2014**, *61*, 1–93. [CrossRef]
30. Saito, Y.; Utsunomiya, H.; Tsuji, N.; Sakai, T. Novel Ultra-High Straining Process for Bulk Materials—Development of the Accumulative Roll-Bonding (ARB) Process. *Acta Mater.* **1999**, *47*, 579–583. [CrossRef]
31. Song, R.; Ponge, D.; Raabe, D.; Speer, J.G.; Matlock, D.K. Overview of Processing, Microstructure and Mechanical Properties of Ultrafine Grained Bcc Steels. *Mater. Sci. Eng. A* **2006**, *441*, 1–17. [CrossRef]
32. Zhang, M.; Li, L.; Ding, J.; Wu, Q.; Wang, Y.-D.; Almer, J.; Guo, F.; Ren, Y. Temperature-Dependent Micromechanical Behavior of Medium-Mn Transformation-Induced-Plasticity Steel Studied by in Situ Synchrotron X-Ray Diffraction. *Acta Mater.* **2017**, *141*, 294–303. [CrossRef]
33. Kawalek, A.; Rapalska-Nowakowska, J.; Dyja, H.; Koczurkiewicz, B. Physical and numerical modelling of heat treatment the precipitation—Hardening complex—Phase steel (CP). *Metalurgija* **2013**, *52*, 23–26.
34. Ge, H.; Ren, F.; Li, J.; Hu, Q.; Xia, M.; Li, J. Modelling of Ingot Size Effects on Macrosegregation in Steel Castings. *J. Mater. Process. Technol.* **2018**, *252*, 362–369. [CrossRef]
35. Zhang, C.; Shahriari, D.; Loucif, A.; Jahazi, M.; Lapierre-Boire, L.-P.; Tremblay, R. Effect of Segregated Alloying Elements on the High Strength Steel Properties: Application to the Large Size Ingot Casting Simulation. In *TMS 2017 146th Annual Meeting & Exhibition Supplemental Proceedings*; The Minerals, Metals & Materials Series; Tms, T.M., Metals & Materials So, Eds.; Springer International Publishing: Cham, Switzerland, 2017; pp. 491–500. ISBN 978-3-319-51492-5.
36. Chelladurai, C.; Mohan, N.S.; Hariharashayee, D.; Manikandan, S.; Sivaperumal, P. Analyzing the Casting Defects in Small Scale Casting Industry. *Mater. Today Proc.* **2021**, *37*, 386–394. [CrossRef]
37. Valiev, R.Z.; Alexandrov, I.V.; Islamgaliev, R.K. Processing and Properties of Nanostructured Materials Prepared by Severe Plastic Deformation. In *Nanostructured Materials*; Chow, G.-M., Noskova, N.I., Eds.; Springer: Dordrecht, The Netherlands, 1998; pp. 121–142. ISBN 978-94-010-6100-1.
38. Valiev, R.Z.; Krasilnikov, N.A.; Tsenev, N.K. Plastic Deformation of Alloys with Submicron-Grained Structure. *Mater. Sci. Eng. A* **1991**, *137*, 35–40. [CrossRef]

39. Segal, V.M. The Method of Material Preparation for Subsequent Working. USSR Patent No. 575892, 1977.
40. Segal, V.M. Equal Channel Angular Extrusion: From Macromechanics to Structure Formation. *Mater. Sci. Eng. A* **1999**, *271*, 322–333. [CrossRef]
41. Cardoso, K.R.; Travessa, D.N.; Botta, W.J.; Jorge, A.M. High Strength AA7050 Al Alloy Processed by ECAP: Microstructure and Mechanical Properties. *Mater. Sci. Eng. A* **2011**, *528*, 5804–5811. [CrossRef]
42. Stepanov, N.D.; Kozin, A.N.; Salishchev, G.A.; Khlebova, N.E.; Pantsyrny, V.I. Effect of ECAP on Microstructure and Mechanical Properties of Cu-14Fe Microcomposite Alloy. *IOP Conf. Ser. Mater. Sci. Eng.* **2014**, *63*, 012098. [CrossRef]
43. Safari, M.; Joudaki, J. Effect of Temperature on Strength and Hardness in Multi-Pass Equal Channel Angular Pressing (ECAP) of Aluminum Alloys. *Trans. Indian Inst. Met.* **2020**, *73*, 619–627. [CrossRef]
44. Naizabekov, A.; Lezhnev, S.; Krivtsova, O.; Tolkushkin, A.; Panin, E. Simulation of the New Technology of Forging of Blanks Such as Slabs and Plates Made of Ferrous Metals and Alloys. *J. Achiev. Mater. Manuf. Eng.* **2014**, *62*, 75–80.
45. Tolkushkin, A.O.; Lezhnev, S.N.; Naizabekov, A.B. Computer Simulation and Study of the New Forging Technology of Billets in a Step-Wedge Dies. *Solid State Phenom.* **2020**, *299*, 699–704. [CrossRef]
46. Kryuchkov, D.I.; Nesterenko, A.V.; Smirnov, S.V.; Pugacheva, N.B.; Vichuzhanin, D.I.; Bykova, T.M. Influence of All-Round Forging under Short-Term Creep Conditions on the Structure and Mechanical Properties of the Al7075/10SiCp Composite with an Aluminum Matrix. Phys. *Metals Metallogr.* **2021**, *122*, 981–990. [CrossRef]
47. Lee, D.; Yoon, E.; Lee, S.; Kang, S.; Kim, H. Finite element analysis for compression behavior of high pressure torsion processing. *Rev. Adv. Mater. Sci.* **2012**, *31*, 25–30.
48. Strangward-Pryce, G.; Song, K.; Mizohata, K.; Hofmann, F. The Effect of High-Pressure Torsion on Irradiation Hardening of Eurofer-97. *Nucl. Mater. Energy* **2023**, *36*, 101468. [CrossRef]
49. Ni, H.; Ding, C.; Wang, H.; Lv, S.; Wang, X.; Liu, Y. The Evolutions of Microstructure, Texture and Hardness of A1050 Deformed by HPT at the Transition Area. *Materials* **2023**, *16*, 4686. [CrossRef]
50. Richert, M.W.; Richert, J.; Hotlos, A.; Mroczkowski, M.; Tokarski, T. The Effect of Reciprocating Extrusion (Cec) on the Consolidated Silver Powders Microstructure/Wpływ Dwustronnego Wyciskania (Cws) Na Mikrostrukture Konsolidowanych Proszków Srebra. *Arch. Metall. Mater.* **2013**, *58*, 73–75. [CrossRef]
51. Amani, S.; Faraji, G. Recrystallization and Mechanical Properties of WE43 Magnesium Alloy Processed via Cyclic Expansion Extrusion. *Int. J. Miner. Metall. Mater.* **2018**, *25*, 672–681. [CrossRef]
52. Abdelaziem, W.; Hamada, A.; Hassan, M.A.-N. Effect of the Cyclic Extrusion and Compression Processing on Microstructure and Mechanical Properties of Al-1%Cu Alloy. *Key Eng. Mater.* **2018**, *780*, 93–97. [CrossRef]
53. Alateyah, A.I.; El-Garaihy, W.H.; Alawad, M.O.; Sanabary, S.E.; Elkatatny, S.; Dahish, H.A.; Kouta, H. The Effect of ECAP Processing Conditions on Microstructural Evolution and Mechanical Properties of Pure Magnesium—Experimental, Mathematical Empirical and Response Surface Approach. *Materials* **2022**, *15*, 5312. [CrossRef]
54. Alateyah, A.I.; Alharbi, M.; El-Hafez, H.A.; El-Garaihy, W.H. The Effect of Equal-Channel Angular Pressing Processing on Microstructural Evolution, Hardness Homogeneity, and Mechanical Properties of Pure Aluminum. *SAE Int. J. Mater. Manf.* **2020**, *14*, 113–125. [CrossRef] [PubMed]
55. Galkin, S.P. Radial Shear Rolling as an Optimal Technology for Lean Production. *Steel Transl.* **2014**, *44*, 61–64. [CrossRef]
56. Karpov, B.V.; Patrin, P.V.; Galkin, S.P.; Kharitonov, E.A.; Karpov, I.B. Radial-Shear Rolling of Titanium Alloy VT-8 Bars with Controlled Structure for Small Diameter Ingots (≤200 Mm). *Metallurgist* **2018**, *61*, 884–890. [CrossRef]
57. Zhou, X.L.; Cao, J.C.; Cao, Y.; Fan, Y.J.; Chen, J.C. Effective Means of Microstructure Homogenization for AgMeO: Cumulative Extrusion Severe Plastic Deformation. *AMR* **2012**, *602–604*, 8–12. [CrossRef]
58. Zhou, X.L.; Cao, J.C.; Li, J.T.; Chen, J.C.; Peng, Y.Y.; Zhang, K.H.; Yu, J.; Feng, H.; Zhang, D. Effect of Severe Plastic Deformation on Microstructures and Properties of AgCuO Composites. *Adv. Mater. Res.* **2010**, *177*, 49–53. [CrossRef]
59. Hu, W.P.; Zhang, S.Y.; He, X.Y.; Liu, Z.Y.; Berghammer, R.; Gottstein, G. Investigations on Microstructure Evolution and Deformation Behavior of Aged and Ultrafine Grained Al-Zn-Mg Alloy Subjected to Severe Plastic Deformation. In *Materials Science Forum*; Trans Tech Publications Ltd.: Bäch, Switzerland, 2010; Volume 667, pp. 253–258. [CrossRef]
60. Naizabekov, A.; Dzamanbalin, K.; Volokitina, I.; Arbuz, A.; Lezhnev, S.; Panin, E.; Tolkushkin, A. Investigation of the Influence of Deformation Temperature on the Radial Shear Rolling Mill on the Microstructure Evolution of Copper. *Mach. Technol. Mater.* **2023**, *17*, 105–107.
61. Lezhnev, S.N.; Naizabekov, A.B.; Volokitina, I.E.; Panin, E.A.; Kuis, D.V. Recycling of stainless steel bar scrap by radial-shear rolling to obtain an ultrafine-grained gradient structure. *Lityo Metall.* **2021**, *2*, 61–67. [CrossRef]
62. Berazategui, D.A.; Cavaliere, M.A.; Montelatici, L.; Dvorkin, E.N. On the Modelling of Complex 3D Bulk Metal Forming Processes via the Pseudo-Concentrations Technique. Application to the Simulation of the Mannesmann Piercing Process. *Int. J. Numer. Meth. Eng.* **2006**, *65*, 1113–1144. [CrossRef]
63. Cho, J.M.; Kim, B.S.; Moon, H.K.; Lee, M.C.; Joun, M.S. Comparative Study on Mannesmann Roll Piercing Processes between Diescher's Guiding Disk and Stiefel's Guiding Shoe. In Proceedings of the 11th International Conference on Numerical Methods in Industrial Forming Processes, Shenyang, China, 6–10 July 2013.
64. Arbuz, A.; Kawalek, A.; Ozhmegov, K.; Dyja, H.; Panin, E.; Lepsibayev, A.; Sultanbekov, S.; Shamenova, R. Using of Radial-Shear Rolling to Improve the Structure and Radiation Resistance of Zirconium-Based Alloys. *Materials* **2020**, *13*, 4306. [CrossRef] [PubMed]

65. Lezhnev, S.N.; Naizabekov, A.B.; Panin, E.A.; Volokitina, I.E.; Arbuz, A.S. Graded Microstructure Preparation in Austenitic Stainless Steel during Radial-Shear Rolling. *Metallurgist* **2021**, *64*, 1150–1159. [CrossRef]
66. Sheremetyev, V.; Kudryashova, A.; Dubinskiy, S.; Galkin, S.; Prokoshkin, S.; Brailovski, V. Structure and Functional Properties of Metastable Beta Ti-18Zr-14Nb (at.%) Alloy for Biomedical Applications Subjected to Radial Shear Rolling and Thermomechanical Treatment. *J. Alloys Compd.* **2018**, *737*, 678–683. [CrossRef]
67. Galkin, S.P.; Gamin, Y.V.; Kin, T.Y. Analysis of Temperature Influence on Strain–Speed Parameters of Radial-Shear Rolling of Al-Zn-Mg-Ni-Fe Alloy. *Materials* **2022**, *15*, 7202. [CrossRef]
68. Naydenkin, E.V.; Mishin, I.P.; Ratochka, I.V.; Oborin, V.A.; Bannikov, M.V.; Bilalov, D.A.; Naydenkin, K.E. Fatigue and Fracture Behavior of Ultrafine-Grained near β Titanium Alloy Produced by Radial Shear Rolling and Subsequent Aging. *Mater. Sci. Eng. A* **2021**, *810*, 140968. [CrossRef]
69. Arbuz, A.; Kawalek, A.; Ozhmegov, K.; Panin, E.; Magzhanov, M.; Lutchenko, N.; Yurchenko, V. Obtaining an Equiaxed Ultrafine-Grained State of the Longlength Bulk Zirconium Alloy Bars by Extralarge Shear Deformations with a Vortex Metal Flow. *Materials* **2023**, *13*, 1062. [CrossRef]
70. Gamin, Y.V.; Muñoz Bolaños, J.A.; Aleschenko, A.S.; Komissarov, A.A.; Bunits, N.S.; Nikolaev, D.A.; Fomin, A.V.; Cheverikin, V.V. Influence of the Radial-Shear Rolling (RSR) Process on the Microstructure, Electrical Conductivity and Mechanical Properties of a Cu–Ni–Cr–Si Alloy. *Mater. Sci. Eng. A* **2021**, *822*, 141676. [CrossRef]
71. Naizabekov, A.; Lezhnev, S.; Arbuz, A.; Panin, E. The Effect of Radial-Shear Rolling on Microstructure and Mechanical Properties of Stainless Austenitic Steel AISI-321. *MATEC Web Conf.* **2018**, *190*, 11003. [CrossRef]
72. Bi, G.; Guan, Y.; Chen, X.; Tan, W.; Huang, W.; Gao, Y.; Wang, M. Acceptability Analysis for Determining the Average Grain Size by ASTM E112 Standard. *IOP Conf. Ser. Mater. Sci. Eng.* **2020**, *770*, 012023. [CrossRef]
73. Vander Voort, G.F. Is There Possible Bias in ASTM E112 Planimetric Grain Size Measurements? *Mater. Perform. Charact.* **2013**, *2*, 20120048. [CrossRef]
74. Napara-Volgina, S.G.; Orlova, L.N. Properties of Hot-Forged High Chromium Stainless Steels. *Sov. Powder Metall. Met. Ceram.* **1991**, *30*, 648–652. [CrossRef]
75. Kudryavtsev, A.S.; Okhapkin, K.A.; Markova, Y.M. Effect of Hot Rolling Production Parameters on 12% Chromium Steel Structure and Properties. *Metallurgist* **2019**, *62*, 1022–1029. [CrossRef]
76. Ilola, R.; Hänninen, H.; Kauppi, T. Hot and Cold Rolling of High Nitrogen Cr-Ni and Cr-Mn Austenitic Stainless Steels. *J. Mater. Eng. Perform.* **1998**, *7*, 661–666. [CrossRef]
77. Hawryluk, M.R.; Lachowicz, M.; Janik, M.; Gronostajski, Z.; Stachowicz, M. Effect of the Heating Temperature of a Nick-el-Chromium Steel Charge Material on the Stability of the Forging Process and the Durability of the Die. *Arch. Metall. Mater.* **2023**, *68*, 711–722. [CrossRef]

materials

MDPI

Article

Study on the Shear Strength and Erosion Resistance of Sand Solidified by Enzyme-Induced Calcium Carbonate Precipitation (EICP)

Gang Li [1], Qinchen Zhu [1], Jia Liu [1,*], Cong Liu [1] and Jinli Zhang [2]

[1] Shaanxi Key Laboratory of Safety and Durability of Concrete Structures, Xijing University, Xi'an 710123, China; t_bag945@126.com (G.L.); 15238796351@163.com (Q.Z.); liucong0147@outlook.com (C.L.)
[2] State Key Laboratory of Coastal and Offshore Engineering, Dalian University of Technology, Dalian 116024, China; jlzhang@dlut.edu.cn
* Correspondence: 15929935077@163.com

Abstract: Sand solidification of earth-rock dams is the key to flood discharge capacity and collapse prevention of earth-rock dams. It is urgent to find an economical, environmentally friendly, and durable sand solidification technology. However, the traditional grouting reinforcement method has some problems, such as high costs, complex operations, and environmental pollution. Enzyme-induced calcium carbonate precipitation (EICP) is an anti-seepage reinforcement technology emerging in recent years with the characteristics of economy, environmental protection, and durability. The erosion resistance and shear strength of earth-rock dams solidified by EICP need further verification. In this paper, EICP-solidified standard sand is taken as the research object, and EICP-cemented standard sand is carried out by a consolidated undrained triaxial test. A two-stage pouring method is adopted to pour samples, and the effects of dry density, cementation times, standing time, and confining pressure on the shear strength of cemented standard sand are emphatically analyzed. The relationship between cohesion, internal friction angle, and $CaCO_3$ formation was analyzed. After the optimal curing conditions are obtained through the triaxial shear strength test, the erosion resistance model test is carried out. The effects of erosion angle, erosion flow rate, and erosion time on the erosion resistance of EICP-solidified sand were analyzed through an erosion model test. The results of triaxial tests show that the standard sand solidified by EICP exhibits strain softening, and the peak strength increases with the increase in initial dry density, cementation times, standing time, and confining pressure. When the content of $CaCO_3$ increases from 2.84 g to 12.61 g, the cohesive force and internal friction angle change to 23.13 times and 1.18 times, and the determination coefficients reach 0.93 and 0.94, respectively. Erosion model test results indicate that the EICP-solidified sand dam has good erosion resistance. As the increase in erosion angle, erosion flow rate, and erosion time, the breach of solidified samples gradually becomes larger. Due to the deep solidification of sand by EICP, the development of breaches is relatively slow. Under different erosion conditions, the solidified samples did not collapse and the dam broke. The research results have important reference value and scientific significance for the practice of sand consolidation engineering in earth-rock dams.

Keywords: EICP; standard sand; earth-rock dam; erosion resistance; shear strength

Citation: Li, G.; Zhu, Q.; Liu, J.; Liu, C.; Zhang, J. Study on the Shear Strength and Erosion Resistance of Sand Solidified by Enzyme-Induced Calcium Carbonate Precipitation (EICP). *Materials* **2024**, *17*, 3642. https://doi.org/10.3390/ma17153642

Academic Editors: Madhav Baral and Charles Lu

Received: 26 May 2024
Revised: 20 July 2024
Accepted: 22 July 2024
Published: 24 July 2024

1. Introduction

The earth-rock dam is a traditional dam type with simple technology and low cost, which accounts for a large proportion of dam construction in the world. However, dam-break accidents occur frequently, and their safety has attracted much attention. There are more than 98,000 reservoirs built or under construction in China, including 756 large reservoirs, 3944 medium reservoirs, and 94,000 small reservoirs, with a total water storage capacity exceeding 900 billion cubic meters [1]. It is crucial to deal with dam failure to improve the utilization rate of resources. Traditional curtain grouting anti-seepage engineering uses cement slurry and chemical slurry, which have some shortcomings such as low permeability, high energy consumption, high costs, and environmental pollution. Therefore, it is an important task and necessary measure for geotechnical engineering to develop energy-saving, emission-reducing, eco-environmental, economical, and efficient dam seepage control and reinforcement technology. Plant-derived enzyme-induced calcium carbonate precipitation technology is a new solidification method. The mineralization principle is that urease is extracted from plants, and urea is decomposed into ammonium ions and carbonate ions under the catalytic action of urease. Subsequently, carbonate ions react with the calcium source to form a $CaCO_3$ precipitate, which bonds the soil particles to form a whole, achieving the purpose of solidification [2–7]. The EICP technology has the characteristics of high efficiency [8], environmental friendliness [9], economic benefit [10], durability [11], etc. This method not only has many applications in reinforcement but also involves the restoration and protection of ancient buildings [12], replacing some civil engineering materials [13], improving the thermal conductivity of backfill materials and soil [14], etc.

Under normal circumstances, the shear strength of soil is an important mechanical index of soil that is closely related to the bearing capacity of the foundation, the earth pressure of the retaining wall, and the stability of the slope. Alwalan et al. [15] conducted direct shear tests on consolidated sand spraying through EICP technology in four different ways: namely spraying, mix and compact, percolation, and injection. In the spraying method, EICP solution is directly sprayed on the top of the sample through a plastic spray bottle and then penetrates the sample. Compared with the untreated sample, the peak shear strength of sand increased by about 2.3 times. Meng et al. [16] found that the amount of calcium carbonate generated in the multiple-phase method is at least four times higher compared to the one-phase method. After 20 applications of cementation solution, the unconfined compressive strength of the EICP-treated sand exceeds 10 MPa, with a $CaCO_3$ content of 20%. He et al. [17] discovered that after EICP treatment of soil slopes was carried out four times in dry and humid environments, the permeability resistance increased to 92.1 N and 71.5 N in slope runoff erosion experiments, respectively, which were 7.7 times and 11.3 times higher than those in one treatment round. Wang et al. [18] found that there is a relationship between the cement solution concentration and the unconfined compressive strength of the EICP-treated sand soil. When grouting times are N = 4, as the cement solution concentration increases from 0.75 mol/L to 1.5 mol/L, the unconfined compressive strength continued to increase, reaching a maximum of 9.87 MPa, which is an improvement of 1.87 times. Gao et al. [19] found that soybeans crude extract-induced calcium carbonate precipitation (SICP) improved the strength of sandy soil. The experiment showed that the surface strength of SICP-treated sandy reached 306.2 kPa compared with untreated sandy soil, an increase of 1813.75%. Almajed et al. [20] found that compared to untreated samples, the samples treated with the EICP and sodium alginate (SA) combination had the lowest erosion rate. SA of 0.5% biopolymer and EICP solution significantly enhances the surface strength and the amount of carbon precipitation in sandy soil. Wu et al. [21] found that when the urease activity increased from 2.95 U/mL to 5.39 U/mL, the cementing solution increased from 0.25 M to 0.75 M, and the unconfined compressive strength of the EICP-solidified sand was enhanced from 180.58 kPa to 1850.70 kPa, which improved 9.25 times. Alotaibi et al. [22] found that in a life cycle assessment (LCA) experience, the abiotic depletion potential of EICP-treated sand in terms of capacity was reduced by nearly 90% compared with portland cement (PC), and the global warming potential of soil capacity was 3% lower than that of PC. Miao et al. [23] included that the EICP has higher production efficiency and production in a wide temperature range (10 $\pm$ 70 °C), significantly improving the water retention performance of the material, which is more suitable for desert environments. It is included that with the increase in spraying times, the urease mineralization method can better resist wind. Lee et al. [24] found that in EICP-solidified soil, when the soybean solution was 140 g/L and the urea–$CaCl_2$ solution was 3 M, the unconfined compressive strength increased from 273 kPa to 870 kPa when curing time was 7 days and 28 days, respectively. Baruah et al. [25] found that the influence of urea, calcium chloride, and urease enzyme on the development of sand strength was studied using unconfined compressive strength. A combination of 0.5 M urea, 1 M $CaCl_2$, and 6 g/L urease enzyme was used as the bonding medium to obtain the maximum calcium carbonate precipitation. Xu et al. [26] obtained that the optimal extraction time and dosage of urease solution in the EICP-treated sand cushion layer were 1 h and 100 g/L. As the number of EICP treatments increases, the content of calcium carbonate increases and the highest dynamic deformation modulus (Evd) reaches 50.55 MPa. Shen et al. [27] discovered that the addition of basal fiber (BF) and polyvinyl acetate emulsion (PVAC) in the EICP experiment resulted in high surface strength (SS) with the addition of 50 g/L PVAC. After treatment with EICP-0.5% BF, EICP-30 g/L PVAC, and EICP-50 g/L PVAC, the sandy soil exhibits high erosivity.

The mechanical properties of EICP-solidified sand have improved significantly, but the erosion resistance still needs to be studied. Through the triaxial shear strength test, the influence factors of dry density, cementation times, standing time, and confining pressure on the strength of solidified sand are discussed. The correlation between cohesion, internal friction angle, and $CaCO_3$ precipitation amount is established, and the optimum conditions for solidifying standard sand by EICP are determined. At the same time, the effects of erosion angle, flow rate, and time on erosion performance were analyzed by an erosion model test, and its durability under erosion was evaluated. These achievements provide an important reference for reinforcement and seepage control of earth-rock dams.

2. Materials and Methods

2.1. Test Materials

The test adopts (GB/T 17671-1999) [28] standard sand from Xiamen, China, and the parameters are shown in Table 1 below. According to its nonuniformity coefficient (C_u) and curvature coefficient (C_c), the sand is judged to be fine sand with poor particle gradation. According to the standard for geotechnical test methods (GB/T 50123-2019) [29], the sand particle size distribution curve is shown in Figure 1. Soybeans were purchased in the market, and their origin was Suihua City, Heilongjiang Province. The reagents used in the test are urea (Tianjin Hengxing Chemical Reagent Manufacturing Company, Tianjin, China) and calcium chloride (Suzhou Wuzhong District Luzhi Sheng Da Drying Reagent Company, Suzhou, China), and the purity is more than 99.0%.

Table 1. Xiamen ISO standard sand parameters.

ρ_{dmax} (g/cm^3)	ρ_{dmin} (g/cm^3)	d_{10} (mm)	d_{30} (mm)	d_{60} (mm)	C_u	C_c
1.890	1.480	0.136	0.300	0.660	4.853	1.003

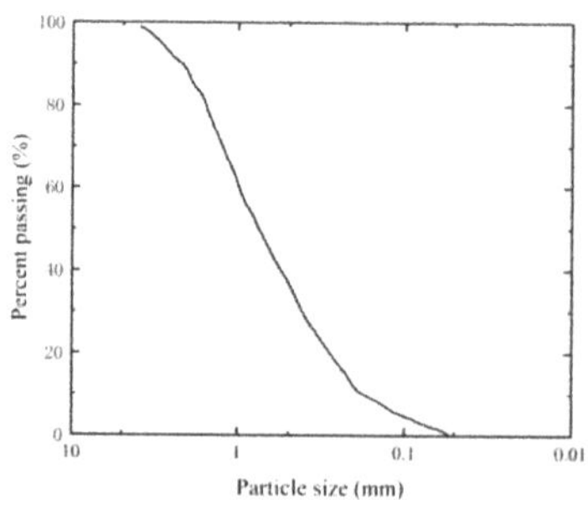

Figure 1. Sand particle size distribution curve.

2.2. Sample Preparation

The size of the standard sand sample is 39.1 mm in diameter and 80 mm in height. To carry out the triaxial shear strength test, a PVC pipe with an inner diameter of 39.1 mm and a height of 150 mm was selected as the die. The test steps should strictly follow the standard for geotechnical test methods (GB/T 50123-2019) [29]. Samples are made in four layers, and shaving treatment is needed between each layer. Before loading, lay three layers of filter paper on the bottom of the sample, and after loading, lay another layer of filter paper on the top. The bottom of the mold shall be plugged with a rubber plug with holes, and the reserved holes shall be connected to the liquid outlet. Urease is extracted from soybeans, urease activity is measured in a conductivity meter and the steps of extracting urease solution include the following: crushing dried soybean until the particle size is less than 0.10 mm, weighing an appropriate amount of soybean flour and adding it to deionized water to achieve a concentration of 100 g/L, stirring the solution for 30 min, and then refrigerating for 24 h at a temperature of 4 °C. The solution was centrifuged at high speed for 15 min, and the supernatant was taken to obtain the urease solution. The concentration of urea and calcium chloride solution was adjusted to 1.25 mol/L, and the pH value was 8.0. The sample was perfused by a two-stage method: first, 20 mL of urease solution, then 10 of mL calcium chloride solution, and 10 mL of urea solution, with an interval of 24 h each time. After completion, we rinsed with purified water three times to stop the reaction. Finally, the sample is saturated in an air-pumping vacuum saturation cylinder. Separation and extraction of soybean urease are shown in Figure 2.

(a) (b) (c) (d)

Figure 2. Separation and extraction process of soybean urease: (**a**) crushed; (**b**) stirring sample powder; (**c**) centrifugation of sample; (**d**) urease.

2.3. Test Methods

According to the standard for geotechnical test methods (GB/T 50123-2019) [29], the triaxial shear strength test adopts the KTL-LDF 50 soil static triaxial instrument (Xi'an Kang tuoli Instrument and Equipment Co., Ltd., Xi'an, China) for the consolidated undrained test. During the test, the loading mode was set to be controlled by a strain rate of 0.10%/min, and the test was stopped when the axial strain reached 15%. Based on relevant research [30–33], dry density, cementation times, standing times, and confining pressure were determined, and the test scheme is shown in Table 2. Carry out an erosion model test, analyze the influence of erosion angle, erosion flow rate, and erosion time on erosion resistance, and verify the erosion resistance of EICP-cemented sand. See Table 3 for the test scheme.

Table 2. Triaxial shear strength test scheme of EICP-cemented standard sand.

Dry Density ρ_d (g/cm^3)	Cementation Times n	Standing Times t_1 (d)	Confining Pressure σ_3 (kPa)
1.55, 1.60, 1.65	2, 4, 6	1, 3, 5	25
1.55, 1.60, 1.65	2, 4, 6	1, 3, 5	50
1.55, 1.60, 1.65	2, 4, 6	1, 3, 5	100

Table 3. EICP model test scheme for erosion resistance.

Erosion Angle α (°)	Erosion Flow Q (L/min)	Erosion Times t_2 (min)
15	1, 3, 5	5, 10, 15
30	1, 3, 5	5, 10, 15
45	1, 3, 5	5, 10, 15

Figure 3 shows a model box for erosion resistance in which plain sand is presented in the form of natural accumulation with a particle density of 1.65 g/cm^3, which is used for the erosion model test. The size of the dam model is 160 mm in length and 100 mm in width. This kind of experiment was carried out in a model box with a length of 50 cm, a width of 10 cm, and a depth of 12 cm. As shown in Figure 3, a self-circulating water pump is used for water supply, the dam model is placed at a position 2/3 away from the water inlet to prevent the uneven erosion of water flow, and a static water grid is set at a position 1/3 away from the water inlet. During the dam cementation process, urease solution (100 g/L), calcium chloride solution (1.25 mol/L), and urea solution (1.25 mol/L) were uniformly sprayed on the surface of the dam sample using T_70103405A adjustable spray cleaning bottle (Thermo Fisher Scientific, Massachusetts, USA). To ensure that the solidified dam samples have good shaping performance, a small amount of deionized water was added for mixing before spraying the solidification liquid.

Figure 3. Erosion resistance model test chamber.

3. Results and Discussion

3.1. Analysis of the Influence of Dry Density on the Strength of Solidified Sand

Figure 4 shows a stress–strain diagram of a cemented specimen under a confining pressure of 50 kPa. It can be seen from the figure that under the same confining pressure, with an increase in initial dry density, the peak strength of deviator stress increases with different standing times and cementation times. When the cementation times increased from two to six times and the dry density increased from ρ_d = 1.55 g/cm^3 to ρ_d = 1.65 g/cm^3, the peak strength of deviator stress increased significantly and the corresponding axial strain decreased gradually. The peak strength of cemented specimens is 518.39 kPa, 520.3 kPa, and 569.72 kPa when the dry density is 1.55 g/cm^3, 1.60 g/cm^3, and 1.65 g/cm^3, respectively, when the cementation times are six times, the confining pressure is 50 kPa, and the standing time is 1 d. The main reason for the above phenomenon is that during the EICP solidification process of sandy soil, urease accelerates the urea reaction and combines with the calcium source to produce $CaCO_3$ [2]. The produced $CaCO_3$ crystals mainly cement loose sand particles into a whole through bonding and filling. With the increase in sample dry density, the pore size between sand particles decreases, which improves the cementation efficiency of $CaCO_3$ and then makes the peak strength of the sample increase accordingly [34]. Under confining pressure, the cohesion and internal friction angle between sand particles increase, which makes the shear strength of cemented samples improve.

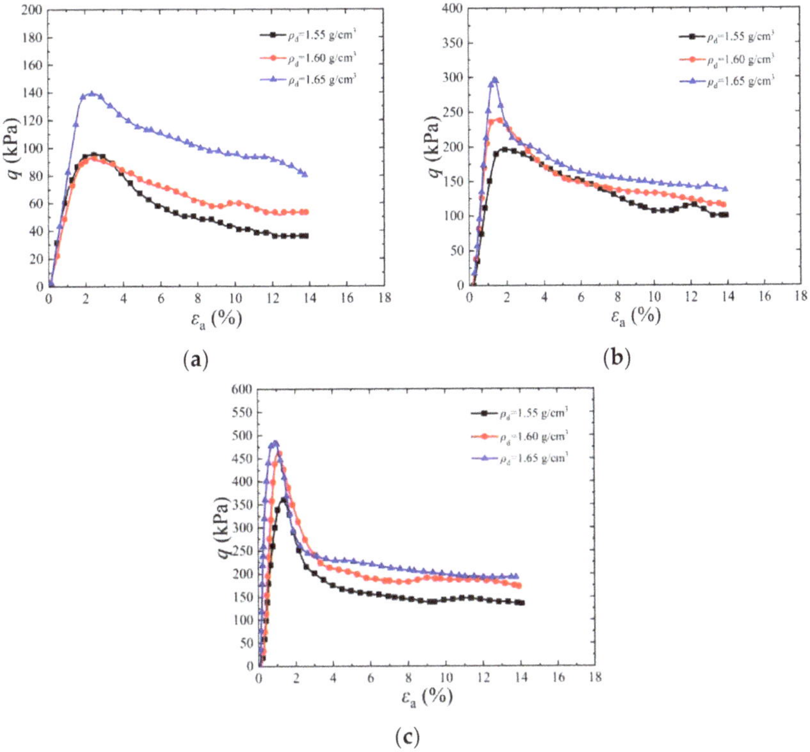

Figure 4. Effect of dry density on stress–strain curve of EICP-cemented standard sand: (**a**) t_1 = 1 d, n = 2; (**b**) t_1 = 1 d, n = 4; (**c**) t_1 = 1 d, n = 6.

3.2. Analysis of the Influence of Cementation Times on the Strength of Solidified Sand

Figure 5 shows the stress–strain diagram of the cemented specimen under a confining pressure of 25 kPa. It can be seen from the figure that under the same confining pressure, with the increase in cementation times, the amount of $CaCO_3$ generated gradually increases and the peak intensity of deviator stress increases continuously. When ρ_d is 1.65 g/cm^3, the confining pressure is 25 kPa, the standing time is 5 d, and cementation times are two, four, and six, respectively, the corresponding deviator stress peaks are 186.06 kPa, 385.43 kPa, and 548.10 kPa, respectively. The axial strain corresponding to the peak value of deviator stress is 1.85~2.51% when the cementation times are two times, and decreases to 0.78~1.84% when the cementation times increase to four times and six times. The main reason for the above phenomenon is that with the increase in cementation times from 2 times to 6 times, $CaCO_3$ crystals produced by the reaction between urease and cementing solution are also increasing, so the consolidation strength of samples is also increasing. In cemented specimens, when the deviator stress reaches its peak value, the brittle failure phenomenon will appear. At this time, the deviator stress decreases rapidly, and the more cementation times, the more obvious the brittle failure characteristics are, and this is consistent with the findings of reference [35]. Consequently, the axial strain of cemented specimens decreases accordingly.

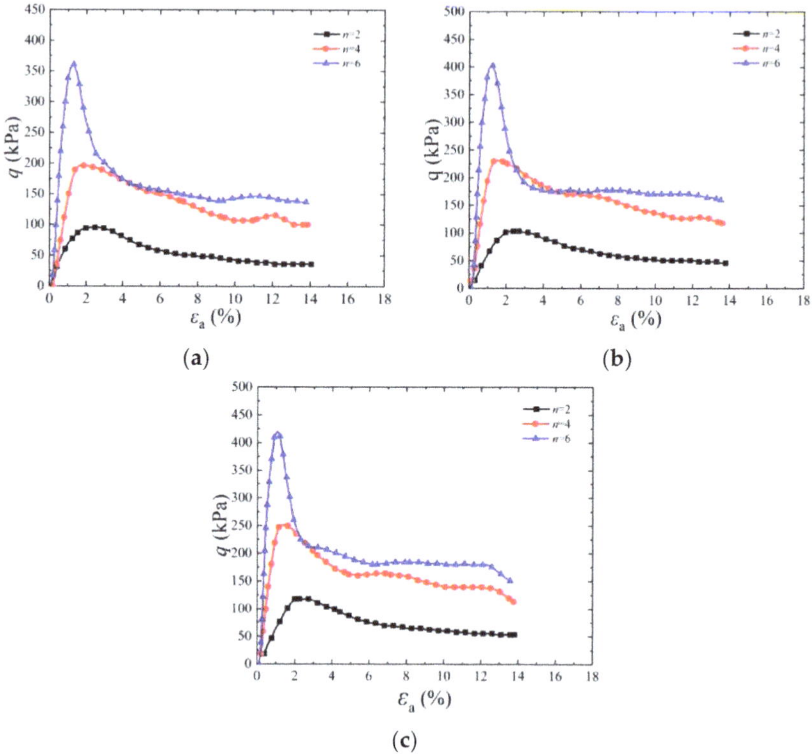

Figure 5. Effect of cementation times on stress–strain curve of EICP-cemented standard sand (σ_3 = 25 kPa): (**a**) ρ_d = 1.65 g/cm^3, t_1 = 1 d; (**b**) ρ_d = 1.65 g/cm^3, t_1 = 3 d; (**c**) ρ_d = 1.65 g/cm^3, t_1 = 5 d.

3.3. Analysis of the Influence of Standing Time on the Strength of Solidified Sand

Figure 6 is a stress–strain diagram of a cemented specimen at a dry density of 6. It can be seen from the figure that, under the same dry density, the peak strength of deviator stress of cemented specimens increases with the increase in confining pressure and standing time. When ρ_d is 1.65 g/cm^3, the confining pressure is 100 kPa, cementation times are 6, standing time is 5 days, and the maximum deviator stress of the specimen reaches 988.20 kPa. Compared with the peak strength of deviator stress under the same condition, when the standing time is 1 d, it increases by 12.52%. During the solidification process of EICP, urease catalyzes the production of carbonate ions from urea. The content of $CaCO_3$ precipitate produced by the combination of carbonate ions and calcium source ions has a significant impact on the strength of the sample [3]. This phenomenon is mainly due to the continuous reaction between the urease solution and cementing solution in the sample with the passage of time. The longer the standing time, the more sufficient the curing reaction, and then more $CaCO_3$ precipitates are generated from cement sand particles, thus increasing the peak value of the deviator stress of the sample [36]. The test results show that when the standing time is 5 days, the deviator stress of the cemented specimen reaches its highest value. However, when the standing time is 1 day, the peak value of deviator stress is low due to the insufficient reaction between urease and cementing solution.

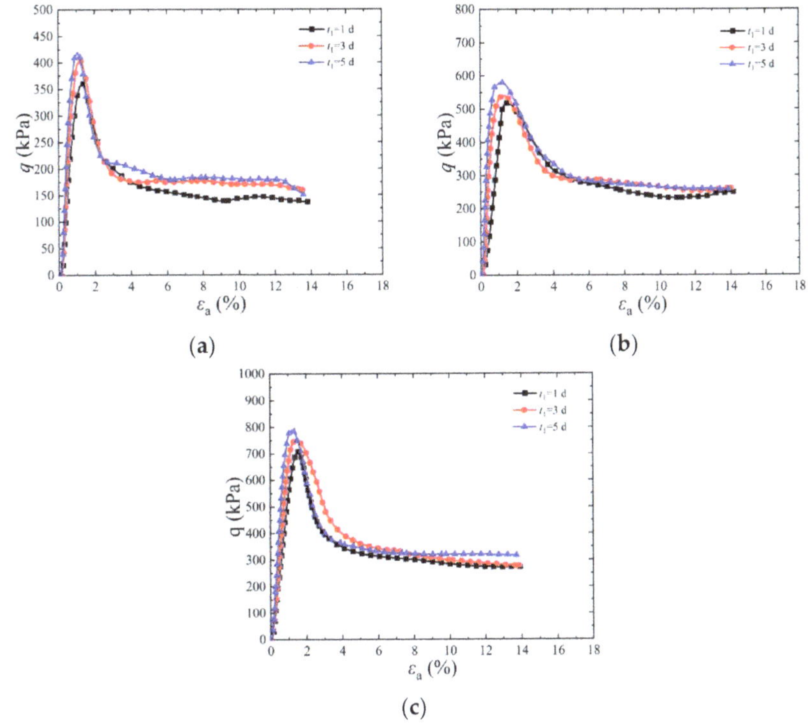

Figure 6. Effect of standing time on stress–strain curve of EICP–cemented standard sand (n = 6): (a) ρ_d = 1.65 g/cm^3, σ_3 = 25 kPa; (b) ρ_d = 1.65 g/cm^3, σ_3 = 50 kPa; (c) ρ_d = 1.65 g/cm^3, σ_3 = 100 kPa.

3.4. Influence of Confining Pressure on Strength of Solidified Sand

Figure 7 shows the stress–strain diagram of cemented specimens under different confining pressures. With an increase in the in situ depth of soil, its confining pressure increases accordingly. The figure shows the influence of different confining pressures on the shear strength characteristics of EICP-solidified specimens. It can be seen from the figure that when the cementation times are six, the residual strength of cemented samples under low confining pressure is significantly lower than the peak strength, and the stress–strain curve shows an obvious strain softening phenomenon. The peak strength of the cemented specimens at 25 kPa, 50 kPa, and 100 kPa is 482.12 kPa, 569.92 kPa, and 880.04 kPa, respectively, when the cementation times are 6 times and the standing time is 1 d. The main reason for the above phenomenon is that with the increase in confining pressure, the closer the contact between $CaCO_3$ crystal and soil particles, the more contact points there are, which leads to the enhancement of adhesion and friction between $CaCO_3$ crystal and soil particles, and this is consistent with the findings of reference [37]. When the specimen is subjected to an external load, the force conversion and transfer effect occur between the $CaCO_3$ crystal and soil particles, and they will share the external load, thus further improving the shear strength of EICP-solidified standard sand.

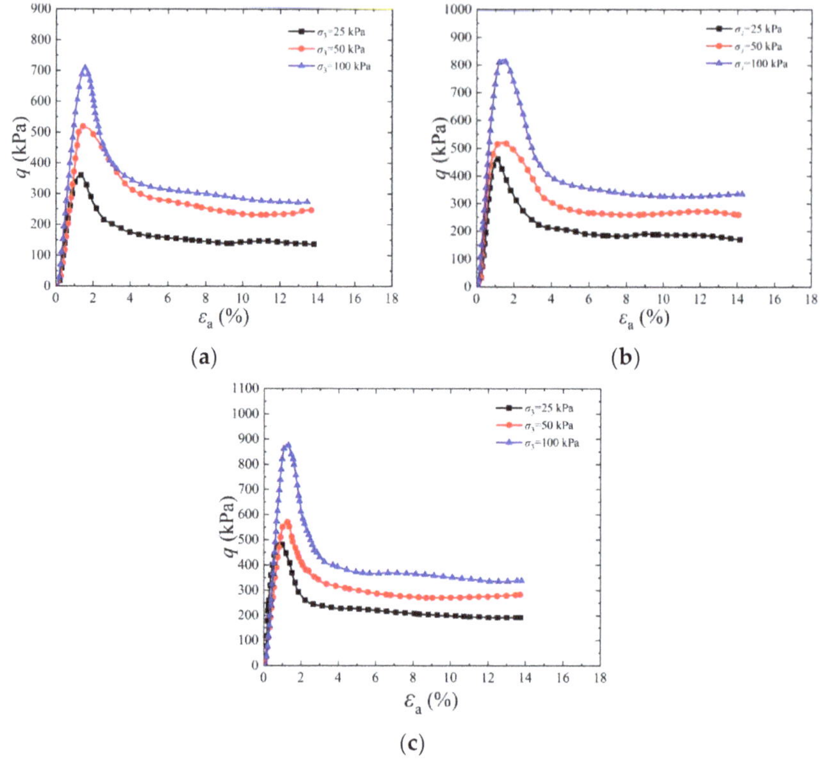

Figure 7. Effect of confining pressure on stress–strain curve of EICP-cemented standard sand (n = 6): (**a**) ρ_d = 1.55 g/cm^3, t_1 = 1 d; (**b**) ρ_d = 1.60 g/cm^3, t_1 = 1 d; (**c**) ρ_d = 1.65 g/cm^3, t_1 = 1 d.

3.5. Failure Mode Analysis of Samples

Figure 8 shows the failure modes of EICP-cemented samples under different cementation times and confining pressures. As shown in Figure 8a, when the cementation times are 2, with the continuous increase in strain, the specimen state first appears as shear shrinkage and then dilatancy failure, forming an indistinct shear band. With the increase in strain, the shear band becomes clear gradually, and the specimen is destroyed at the shear band. As shown in Figure 8b, the shear phenomenon is basically consistent with that in Figure 8a, but with the increase in cementation times, the surface strength of the sample is significantly improved compared with that in Figure 8a. However, with the continuous increase in confining pressure times, the surface shear failure phenomenon is gradually aggravated, and the shear band is becoming more and more obvious. As shown in Figure 8c, the shear phenomenon is basically consistent with Figure 8a, and the shear failure of the specimen decreases with the increase in cementation times. In Figure 8d, the continuous strain growth rate is faster than that in Figure 8a. At this time, the phenomenon is not local shear failure, but the cementation between sand particles and $CaCO_3$ is destroyed first. With the increase in cementation times and confining pressure, the time of shear back formation is relatively early, and the degree of shear failure is more thorough. Experiments indicate that local deformation occurs earlier at low confining pressure than at high confining pressure. Strain softening and obvious shear bands are formed in the specimens, and then the strength decreases obviously. This shows that EICP-solidified standard sand has poor plasticity under low confining pressure and a high cementation degree and is more prone to brittle failure, and this is consistent with the findings in [38].

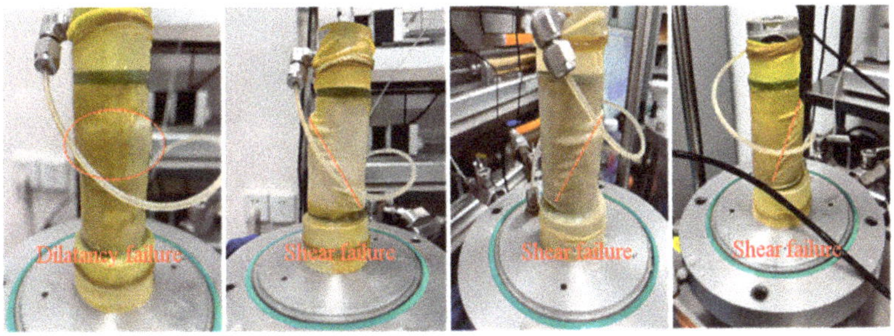

(**a**) (**b**) (**c**) (**d**)

Figure 8. Failure modes of EICP-cemented specimens under different cementation times and different confining pressures: (**a**) $n = 2$, $\sigma_3 = 25$ kPa; (**b**) $n = 4$, $\sigma_3 = 50$ kPa; (**c**) $n = 6$, $\sigma_3 = 25$ kPa; (**d**) $n = 6$, $\sigma_3 = 100$ kPa.

3.6. Analysis of the Relation between Cohesion, Internal Friction Angle, and $CaCO_3$ Formation

Figure 9 is the regression analysis curve of cohesion, internal friction angle, and $CaCO_3$ formation of the EICP-solidified standard sand sample. After EICP treatment of standard sand samples, it was found that the increase in cohesion brought about an improvement in shear strength. As shown in the figure, when the content of $CaCO_3$ increases from 2.84 g to 12.61 g, the cohesive force and internal friction angle change to 23.13 times and 1.18 times, respectively. It can be seen that there is a positive correlation between them, and the determination coefficient reaches 0.93 and 0.94, respectively, indicating that there is a significant correlation between cohesion and $CaCO_3$ production. The main reason for this phenomenon is that $CaCO_3$ crystals formed during the EICP process can fill and cement sand particles and enhance the adhesion and friction between sand particles. Therefore, the loose sand particles can be cemented into a whole and bear the external load together, thus improving its shear strength. Therefore, there is a significant positive correlation between the amount of $CaCO_3$ produced and the cohesion and internal friction angle of the specimen, so the main way to improve the shear strength of standard sand using EICP technology is to increase its cohesion.

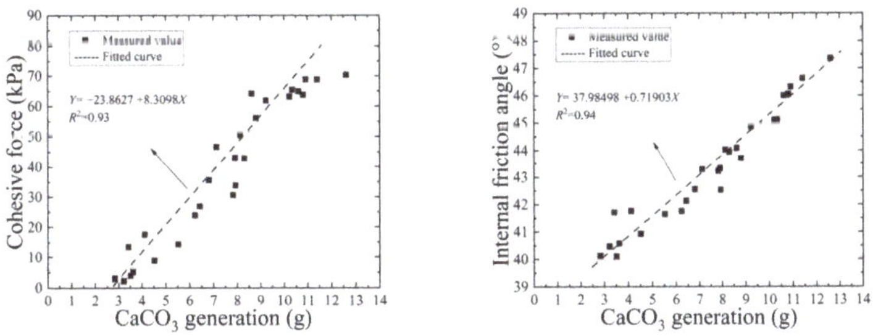

(a) (b)

Figure 9. Relationship between cohesion and internal friction angle and calcium carbonate production: (**a**) cohesion; (**b**) angle of internal friction.

4. Analysis of Erosion Characteristics of EICP-Solidified Sand

4.1. Erosion Performance Analysis of Uncemented Dam

Figure 10 is a standard sand dam with an uncemented $\alpha = 30°$ and $Q = 1$ L/min, and the erosion resistance model test is carried out. As shown in the figure, when the water level overflows the height of the dam crest, a narrow breach at the dam crest can be observed immediately, and the breach will spread and expand rapidly to both sides [39].

The erosion process is mainly affected by water flow, and the continuous expansion of the breach will also cause the collapse of the soil on both sides. The eroded particulate matter will be transported to the downstream slope toe for deposition along with the water flow, which leads to the slope of the slope toe gradually slowing down until it reaches a certain stable state. The whole erosion process is very short, lasting about 6 s, and its failure speed is extremely rapid, which cannot be directly compared with the dam samples after EICP cementation treatment. When the water flows over the top of the embankment, it will cause erosion on the surface of the embankment and produce local shear stress. Once this shear stress reaches or exceeds its critical value, it will lead to erosion failure, which will scour the soil particles to the foot of the slope and deposit them. The breach usually occurs in the weak link of the dam because the shear stress at the breach is higher than in the surrounding area, so the breach will expand rapidly.

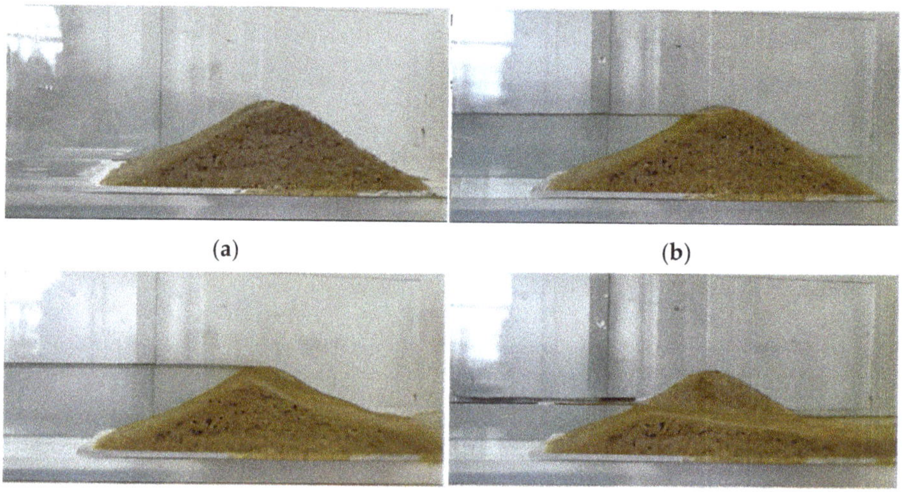

(a) (b) (c) (d)

Figure 10. Model test of standard sand dam: (**a**) uncemented specimen; (**b**) the water flow reaches the dam crest; (**c**) water erosion; (**d**) end of erosion.

4.2. Erosion Angle Analysis

Figure 11 shows the relationship between erosion angle, breach width, and breach depth under erosion flow rates (Q) of 5 L/min and 7 L/min. It can be seen from the figure that within 3 min, the depth of breach changes little in the erosion test, and the dam is relatively stable. When the erosion flow rate is 33 L/min and the erosion angles are 15°, 30°, and 45°, there is no breach in the dam specimen. As the angle increases to 30° and 45°, the breach appears, and the depth and width of the breach gradually expand with the passage of erosion time. The reason for the above phenomenon is that on the surface of the dam treated with EICP, urease accelerates the combination of carbonate ions and calcium ions to form $CaCO_3$ crystal shells [5]. The EICP test shows that the smaller the erosion angle, the lower the shear stress caused by overtopping erosion, the higher the dam stability, and the longer the dam break time. However, the larger the erosion angle is, the higher the top shear stress of water flow [40], and with the increase in erosion time, the breach will appear in the fragile part of surface cementation, but the breach develops relatively slowly due to the larger cementation depth.

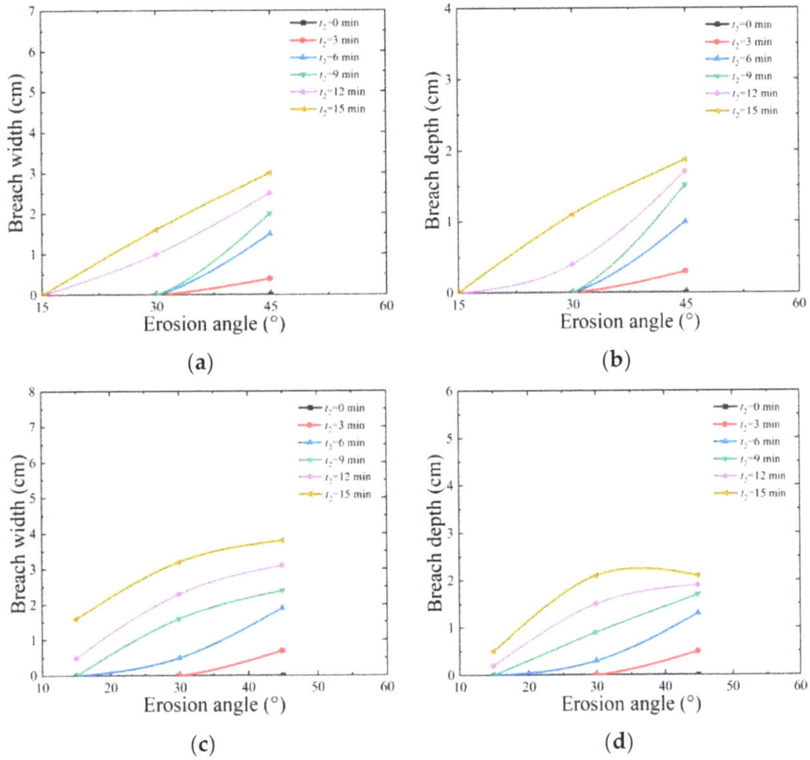

Figure 11. Influence of erosion angle on the top depth and widening of dam central axis breach: (**a**) Q = 5 L/min, breach width; (**b**) Q = 5 L/min, breach depth; (**c**) Q = 7 L/min, breach width; (**d**) Q = 7 L/min, breach depth.

4.3. Erosion Flow Analysis

Figure 12 shows the curves of erosion flow rate (Q) versus the breadth and depth of the breach at 15, 30, and 45 erosion angles. It can be seen from the figure that no obvious breach phenomenon is observed when the erosion angle is 15 and the erosion flow rate is 3 L/min. However, when the erosion flow rate is increased to 7 L/min, a significant breach phenomenon appears. The experimental results show that as the increase in erosion angle, the breadth and depth of the breach show an obvious increasing trend. At the same time, with the continuous increase in erosion flow, the width and depth of the breach are gradually expanded, which is consistent with the findings of reference [41]. The appearance of this phenomenon is due to the corresponding enhancement of water impact force caused by the increase in erosion flow rate, which in turn increases the shear stress for overtopping failure. Once the dam breaks, its energy release will be more intense. If the water flow rate continues to increase, the time of rapid erosion will be further shortened in the process of overtopping erosion. This will accelerate the development of breaches and the scale of the secondary collapse, which will lead to an increase in breach flow.

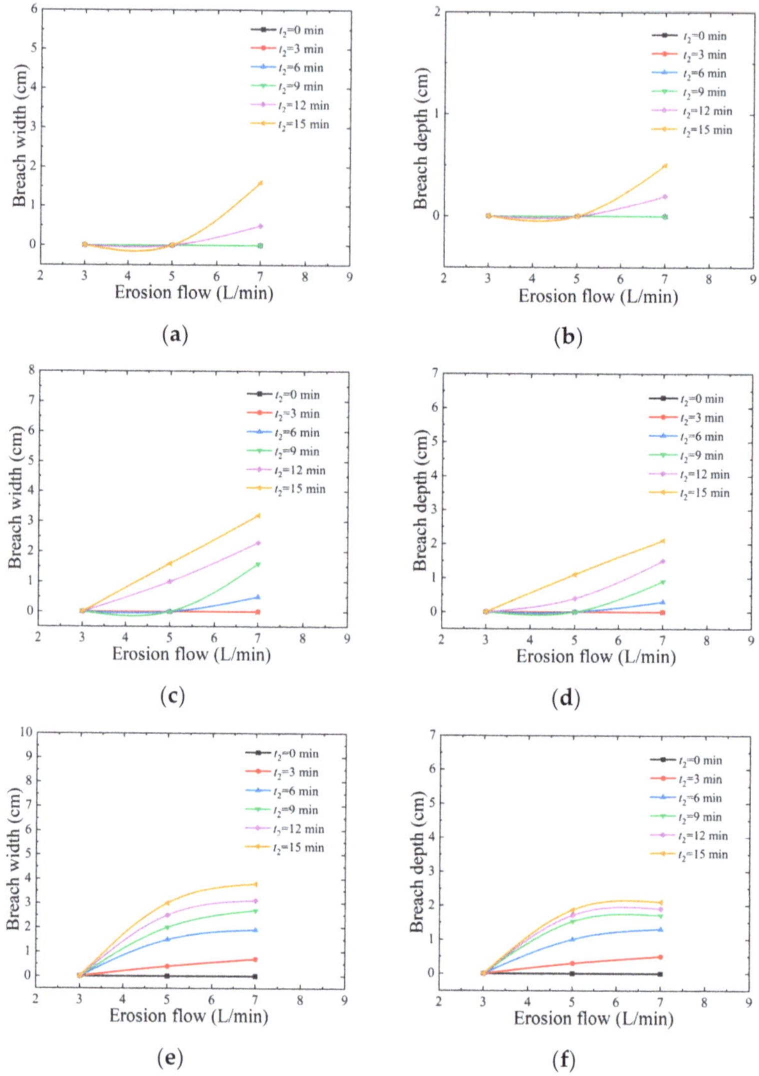

Figure 12. Influence of erosion discharge on the top depth and widening of dam central axis breach: (**a**) α = 15°, breach width; (**b**) α = 15°, breach depth; (**c**) α = 30°, breach width; (**d**) α = 30°, breach depth; (**e**) α = 45°, breach width; (**f**) α = 45°, breach depth.

4.4. Erosion Time Analysis

Figure 13 shows the relationship between erosion time, breach width, and breach depth at 15, 30, and 45 erosion angles. It can be seen from the figure that after the dam breach occurs, the depth and width of the breach will gradually expand with the passage of erosion time. When the erosion flow rate is 3 L/min and the erosion angle is 15°, there is no breach in the dam specimen. However, with the increase in erosion angle and flow rate, the expansion speed of the breach will accelerate at first and then tend to be stable. The reason for this phenomenon is that the overtopping dam failure belongs to the type of scouring failure, which first occurs at the contact between the dam crest and the downstream slope. In the case of a small flow rate, the scouring speed is relatively slow [41]. During the EICP mineralization process, urease-catalyzed hydrolysis of urea produces carbonate ions [6]. The calcium carbonate precipitate produced by carbonate ions and calcium ions tightly binds with the sand particles on the surface of the dam, forming a strong hard shell that prevents the shear stress generated by water flow from damaging the top and downstream of the dam. However, with the increase in water flow rate and erosion angle, the shear stress will increase accordingly. Therefore, under the action of erosion time, the breach first appeared at the weak surface cementation site, where the shear stress was higher than the surrounding area, and the breach developed rapidly. However, due to the large cementation depth of EICP, the development speed of the breach is relatively slow.

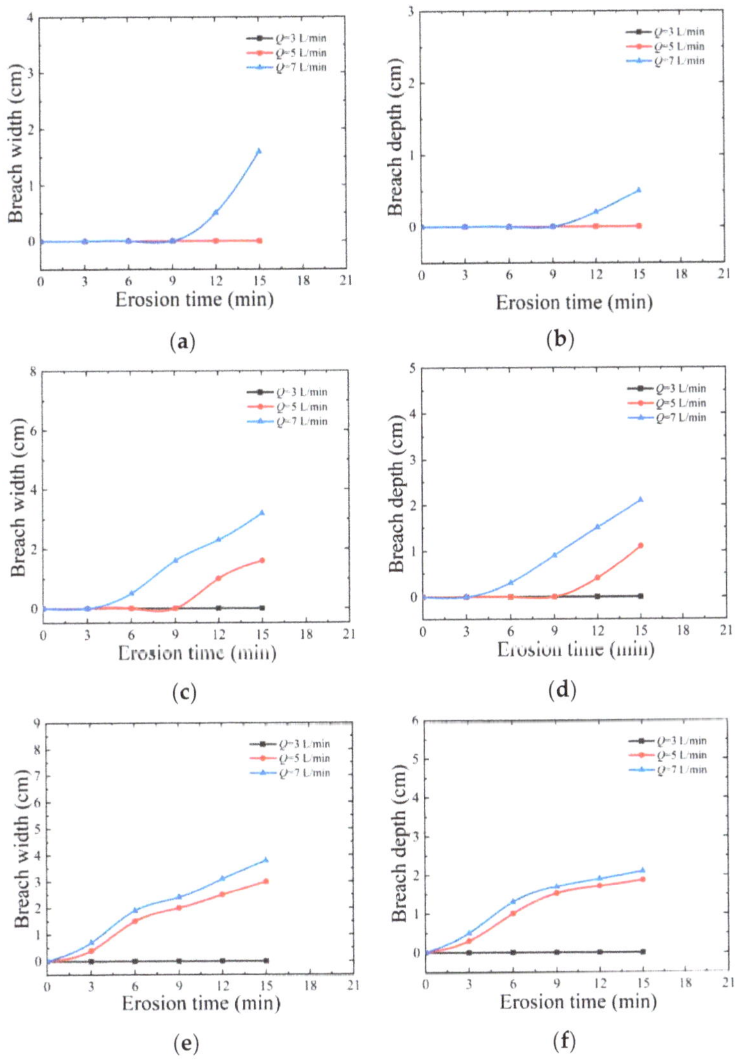

Figure 13. Influence of erosion time on the top depth and widening of dam central axis breach: (**a**) α = 15°, breach width; (**b**) α = 15°, breach depth; (**c**) α = 30°, breach width; (**d**) α = 30°, breach depth; (**e**) α = 45°, breach width; (**f**) α = 45°, breach depth.

4.5. Analysis of Erosion Failure Mode

Figure 14 shows the failure process of dam overtopping. Overtopping dam failure is a form of traceability scouring, and dams without EICP reinforcement will collapse and break under current erosion. The failure begins in the contact area between the dam crest and the downstream slope, during which the current erosion forms a gully and extends upstream. As time goes by, the gully expands and deepens to form an obvious breach. When the breach extends to the upstream edge of the dam body, the flow rate increases sharply and the erosion intensifies, which leads to the expansion and deepening of the breach and the collapse of the dam bodies on both sides. After the breach runs through, the water flow velocity increases sharply, the water level drops sharply, the dam erosion becomes more intense, the breach continues to expand and deepen, and the collapse of the dams on both sides intensifies. Due to the uneven particle distribution and structure in the dam, the breach is often asymmetric, which leads to the flow around the breach and aggravates the erosion at the foot of the slope. With the continuous discharge of water flow, the upstream water level gradually decreases, and the erosion of the upstream part weakens, finally emerging from the water surface. At this time, the main erosion focus of the breach shifts to the downstream side wall, and the water flow continues to erode the foot of the breach slope, causing the slope to collapse [42].

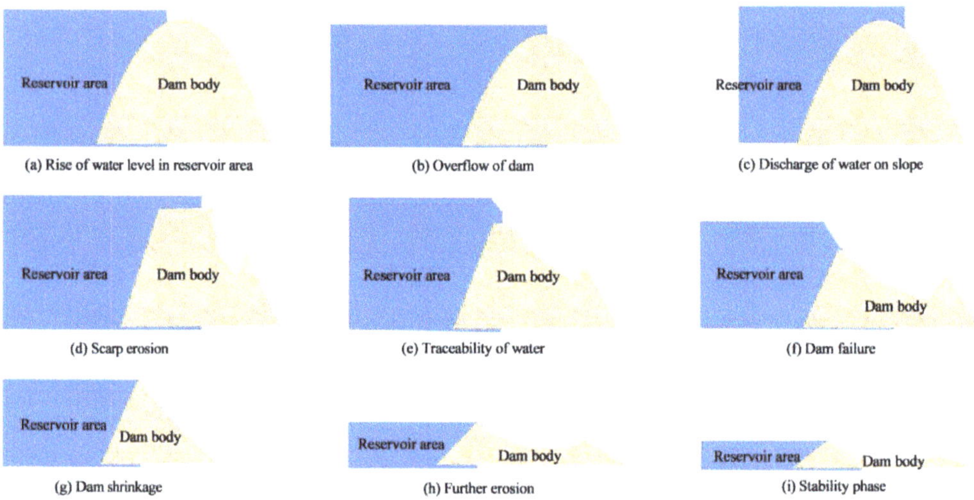

Figure 14. Failure process of dam overtopping: (**a**) rise of water level in reservoir area; (**b**) overflow of dam; (**c**) discharge of water on slope; (**d**) scarp erosion; (**e**) traceability of water; (**f**) dam failure; (**g**) dam shrinkage; (**h**) further erosion; (**i**) stability phase.

Figure 15 shows the erosion pattern of the dam strengthened by EICP after failure. As shown in the figure, there is no serious collapse or dam break under various erosion conditions. When erosion occurs, the breach of the sample is uneven in shape. Weak cementation will wash the massive particles to the foot of the slope under the erosion of water flow. As shown in Figure 15a, when the erosion angle is 15, the breach is mostly shallow and wide. As shown in Figure 15a–c, when the erosion angles are 30 and 45, the breach is mostly deep and thin. The main reason for the above phenomenon is that after EICP treatment, $CaCO_3$ crystals cover the surface of sand particles, which can wrap and cement loose sand particles into a whole, and the friction and adhesion between sand particles increase at the same time, which is consistent with the findings of reference [43]. Compared with uncemented dam samples, the anti-erosion ability is significantly improved and the integrity is stronger when subjected to water erosion. At the same time, due to the deep cementation, the weak part of the cementation is washed to the foot of the slope by water flow after the breach, while the remaining part can still resist the erosion of water flow.

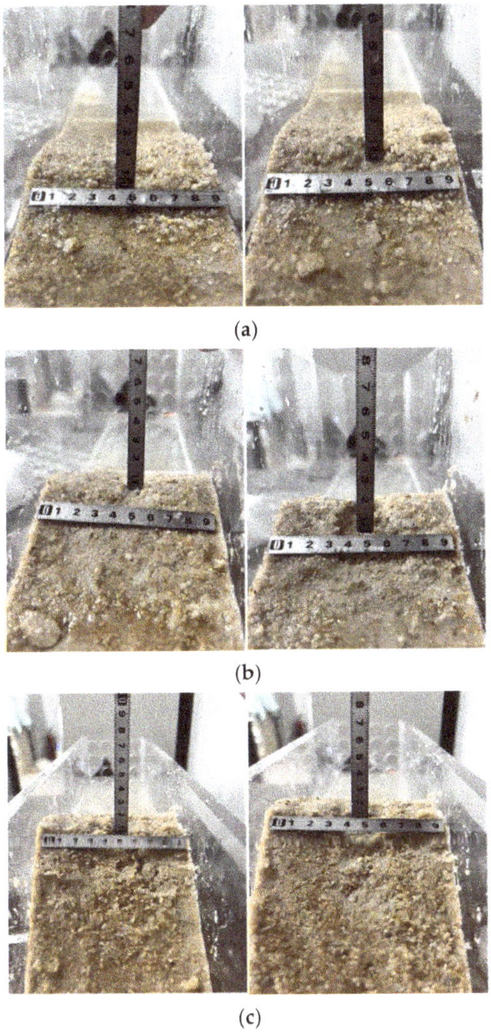

Figure 15. Erosion failure mode: (**a**) $\alpha = 15°$; (**b**) $\alpha = 30°$; (**c**) $\alpha = 45°$.

5. Conclusions

The effects of dry density, cementation times, standing time, and confining pressure on the shear strength of EICP-solidified Aeolian sand were analyzed through a triaxial consolidated undrained shear test. The quantitative relationship between $CaCO_3$ generation, cohesion, and internal friction angle was established. The effects of erosion angle, erosion flow rate, and erosion time on the erosion resistance of EICP-solidified sand were analyzed through the erosion model test. The main conclusions are as follows:

1. When the dry density of the sample is high, the smaller the distance between sand particles, the better the cementation effect of $CaCO_3$. With the increase in the cementation times, the strength of standard sand increases, and the brittle failure is obvious.

With the increase in standing time, the peak strength of the deviator stress of standard sand also increases.
2. When the sample with low $CaCO_3$ content reaches its peak strength, the internal structure changes little, the difference between residual strength and peak strength is small, and the strain softening is not obvious. When the sample with high $CaCO_3$ content reaches peak strength, the internal $CaCO_3$ crystal is destroyed, the integrity of sand particles is damaged, the shear strength is greatly reduced, the difference between residual strength and peak strength is large, and the strain softening is obvious.
3. The local deformation of the specimen under low confining pressure is earlier than that under high confining pressure. Under low confining pressure and high cementation, the deviator stress increases rapidly with the increase in strain, then decreases or stabilizes after the peak value, and the specimen appears to have strain-softening characteristics, with the peak value appearing as a shear band. The strength of EICP-solidified sand is obviously decreased, which indicates that EICP-solidified sand has poor plastic deformation ability under low confining pressure and high cementation levels and is more prone to brittle failure.
4. With the increase in $CaCO_3$ content, the cohesion and the angle of internal friction increase, showing a positive correlation. $CaCO_3$ crystals produced in the EICP process play a role in filling and cementing sand particles, which increases the cohesion and internal friction angle between sand particles and makes the loose sand particles be cemented together to bear the external load together, thus significantly improving their shear strength.
5. The smaller the erosion angle, the better the stability of the dam, and the longer the dam break time. When the erosion angle is 15° and the erosion flow rate is 3 L/min, there is no obvious damage, and the overall integrity of the dam sample is good. The larger the erosion angle, the greater the overtopping shear stress of the water flow. With the increase in erosion time, the breach appears at the weak surface cementation.
6. With the increase in erosion flow, the impact capacity of water flow is gradually enhanced. When overtopping failure occurs, the shear stress will increase, and at the same time, when the dam breaks, the released energy will become greater, which will lead to an increase in the degree of harm caused by it. In addition, in the process of overflow scouring, the increase in flow rate will accelerate the scouring speed and make the breach gradually larger.

Author Contributions: Conceptualization, G.L.; Methodology, J.L.; Validation, Q.Z. and C.L.; Writing—Original Draft, G.L. and J.Z.; Writing—Review and Editing, J.Z.; Funding Acquisition, G.L. All authors have read and agreed to the published version of the manuscript.

Funding: This study was supported by the Natural Science Basic Research Program of Shaanxi Province (2021JM-535) and the Scientific Research Program Funded by the Education Department of Shaanxi Provincial Government (23JS061).

Institutional Review Board Statement: Not applicable.

Informed Consent Statement: Not applicable.

Data Availability Statement: All data generated or analyzed during this study are included in this published article.

Conflicts of Interest: The authors declare no conflicts of interest.

References

1. Su, Z.; Zhang, K.; Liu, C. Dynamic risk assessment of slope stability of homogeneous earth-rock dam under action of multiple hazards. *Simulation* **2022**, *98*, 699–710. [CrossRef]
2. Arab, M.G.; Alsodi, R.; Almajed, A.; Yasuhara, H.; Zeiada, W.; Shahin, A.M. State-of-the-art review of enzyme-induced calcite precipitation (EICP) for ground improvement: Applications and prospects. *Geosciences* **2021**, *11*, 492. [CrossRef]

3. Almajed, A.; Lateef, M.A.; Moghal, A.A.B.; Lemboye, K. State-of-the-art review of the applicability and challenges of microbial-induced calcite precipitation (MICP) and enzyme-induced calcite precipitation (EICP) techniques for geotechnical and geoenvironmental applications. *Crystals* **2021**, *11*, 370. [CrossRef]
4. Arab, M.G.; Refaei, M.; Alotaibi, E.; Omar, M.; Almajed, A.; Haridy, S. Optimizing the Compressive Strength of Sodium Alginate-Modified EICP-Treated Sand Using Design of Experiments. *J. Mater. Civil Eng.* **2024**, *36*, 04024017. [CrossRef]
5. Zhang, J.; Yin, Y.; Shi, W.; Bian, H.; Shi, L.; Wu, L.; Han, Z.; Zheng, J.; He, X. Strength and uniformity of EICP-treated sand under multi-factor coupling effects. *Biogeotechnics* **2023**, *1*, 100007. [CrossRef]
6. Cao, G.; Ma, L.; Ngo, I.; Osemudiamhen, A.E.; Guo, Z. Experimental Investigation on the Combination of Enzyme-Induced Calcium Carbonate Precipitation and Organic Materials for Underground Backfilling Preparation. *Minerals* **2024**, *14*, 153. [CrossRef]
7. Ahenkorah, I.; Rahman, M.M.; Karim, M.R.; Beecham, S. Characteristics of MICP-and EICP-treated sands in simple shear conditions: A benchmarking with the critical state of untreated sand. *Géotechnique* **2023**. *ahead of print*. [CrossRef]
8. Cui, M.J.; Lai, H.J.; Hoang, T.; Chu, J. One-phase-low-pH enzyme induced carbonate precipitation (EICP) method for soil improvement. *Acta Geotech.* **2021**, *16*, 481–489. [CrossRef]
9. Almajed, A.; Abbas, H.; Arab, M.; Arab, M.; Alsabhan, A.; Hamid, W.; Al-Salloum, Y. Enzyme-Induced Carbonate Precipitation (EICP)-Based methods for ecofriendly stabilization of different types of natural sands. *J. Cleaner Prod.* **2020**, *274*, 122627. [CrossRef]
10. Rosa, D.; Verdirame, L.; Bavasso, I.; Bracciale, M.P.; Di-Palma, L. Soil Biocementation via Enzyme Induced Carbonate Precipitation (EICP) Method Employing Soybeans as a Source of Cheap Enzyme. *Chem. Eng. Trans.* **2023**, *99*, 157–162. [CrossRef]
11. Sun, X.; Miao, L.; Wang, H.; Yuan, J.; Fan, G. Enhanced rainfall erosion durability of enzymatically induced carbonate precipitation for dust control. *Sci. Total Environm.* **2021**, *791*, 148369. [CrossRef]
12. Hu, W.; Cheng, W.C.; Wen, S.; Yuan, K. Revealing the enhancement and degradation mechanisms affecting the performance of carbonate precipitation in EICP process. *Front. Bioeng. Biotechnol.* **2021**, *9*, 750258. [CrossRef] [PubMed]
13. Saif, A.; Cuccurullo, A.; Gallipoli, D.; Perlot, C.; Bruno, A.W. Advances in enzyme induced carbonate precipitation and application to soil improvement: A review. *Materials* **2022**, *15*, 950. [CrossRef] [PubMed]
14. Li, S.; Huang, M.; Cui, M.; Jiang, Q.; Xu, K. Thermal conductivity enhancement of backfill material and soil using enzyme-induced carbonate precipitation (EICP). *Acta Geotech.* **2023**, *18*, 6143–6158. [CrossRef]
15. Alwalan, M.; Almajed, A.; Lemboye, K.; Alnuaim, A. Direct shear characteristics of enzymatically cemented sands. *KSCE J. Civ. Eng.* **2023**, *27*, 1512–1525. [CrossRef]
16. Meng, H.; Shu, S.; Gao, Y.; Yan, B.; He, J. Multiple-phase enzyme-induced carbonate precipitation (EICP) method for soil improvement. *Eng. Geol.* **2021**, *294*, 106374. [CrossRef]
17. He, J.; Fang, C.; Mao, X.; Qi, Y.; Zhou, Y.; Kou, H.; Xiao, L. Enzyme-induced carbonate precipitation for the protection of earthen dikes and embankments under surface runoff: Laboratory investigations. *J. Ocean Univ. China* **2022**, *21*, 306–314. [CrossRef]
18. Wang, Y.; Wang, Z.; Chen, Y.; Cao, T.; Yu, X.; Rui, P. Experimental study on bio-treatment effect of the dredged Yellow River silt based on soybean urease induced calcium carbonate precipitation. *J. Build.* **2023**, *75*, 106943. [CrossRef]
19. Gao, Y.F.; Meng, H.; He, J.; Qi, Y.S.; Hang, L. Field trial on use of soybean crude extract for carbonate precipitation and wind erosion control of sandy soil. *J. Cent. South Univ.* **2020**, *27*, 3320–3333. [CrossRef]
20. Almajed, A.; Lemboye, K.; Arab, M.G.; Alnuaim, A. Mitigating wind erosion of sand using biopolymer-assisted EICP technique. *Soils Found.* **2020**, *60*, 356–371. [CrossRef]
21. Wu, L.; Miao, L.; Kawasaki, S.; Wang, H. Effects of Reaction Conditions on EICP-Treated Desert Aeolian Sand. *KSCE J. Civ. Eng.* **2022**, *26*, 2662–2674. [CrossRef]
22. Alotaibi, E.; Arab, M.G.; Abdallah, M.; Nassif, N.; Omar, M. Life cycle assessment of biocemented sands using enzyme induced carbonate precipitation (EICP) for soil stabilization applications. *Sci. Rep.* **2022**, *12*, 6032. [CrossRef]
23. Miao, L.; Wu, L.; Sun, X. Enzyme-catalysed mineralisation experiment study to solidify desert sands. *Sci. Rep.* **2020**, *10*, 10611. [CrossRef]
24. Lee, S.; Kim, J. An experimental study on enzymatic-induced carbonate precipitation using yellow soybeans for soil stabilization. *KSCE J. Civ. Eng.* **2020**, *24*, 2026–2037. [CrossRef]
25. Baruah, P.; Sharma, S. Chemico-mechanical activation of EICP-modified soil: Role of urease enzyme. *Environ. Earth Sci.* **2023**, *82*, 563. [CrossRef]
26. Xu, K.; Huang, M.; Zhen, J.; Xu, C.; Cui, M. Field implementation of enzyme-induced carbonate precipitation technology for reinforcing a bedding layer beneath an underground cable duct. *J. Rock Mech. Geotech. Eng.* **2023**, *15*, 1011–1022. [CrossRef]
27. Shen, D.; Liu, Z.; Song, Z.; Song, Z.; Wu, C. Reinforcement mechanism and erosion resistance of loess slope using enzyme induced calcite precipitation technique. *Sustainability* **2023**, *15*, 1044. [CrossRef]
28. *GB/T 17671-1999*; Method of Testing Cements-Determination of Strength. National Bureau of Quality and Technical Supervision: Beijing, China, 1999.
29. *GB/T 50123-2019*; Standard for Geotechnical Testing Method. Construction Ministry of PRC: Beijing, China, 2019.
30. Chen, Y. Soil–water retention curves derived as a function of soil dry density. *GeoHazards* **2018**, *1*, 3–19. [CrossRef]
31. Cui, M.; Zheng, J.; Zhang, R.; Lai, H.; Zhang, J. Influence of cementation level on the strength b ehaviour of bio-cemented sand. *Acta Geotech.* **2017**, *12*, 971–986. [CrossRef]

32. Yu, X.; Qian, C.; Xue, B.; Wang, X. The influence of standing time and content of the slurry on bio-sandstone cemented by biological phosphates. *Constr. Build. Mater.* **2015**, *82*, 167–172. [CrossRef]
33. Arab, M.G.; Rohy, H.; Zeiada, W.; Almajed, A.; Omar, M. One-phase EICP biotreatment of sand exposed to various environmental conditions. *J. Mater. Civil Eng.* **2021**, *33*, 04020489. [CrossRef]
34. Fang, X.; Yang, Y.; Chen, Z.; Liu, H.; Shen, C. Influence of fiber content and length on engineering properties of MICP-treated coral sand. *Geomicrobiol. J.* **2020**, *37*, 582–594. [CrossRef]
35. Feng, K.; Montoya, B.M. Influence of confinement and cementation level on the behavior of microbial-induced calcite precipitated sands under monotonic drained loading. *J. Geotech. Geoenviron.* **2016**, *142*, 04015057. [CrossRef]
36. Ahenkorah, I.; Rahman, M.M.; Karim, M.R.; Beecham, S.; Saint, C. A review of enzyme induced carbonate precipitation (EICP): The role of enzyme kinetics. *Sust. Chem.* **2021**, *2*, 92–114. [CrossRef]
37. Muhammed, A.S.; Kassim, K.A.; Ahmad, K.; Zango, M.U.; Makinda, J. Influence of multiple treatment cycles on the strength and microstructure of biocemented sandy soil. *Int. J. Environ. Sci. Technol.* **2021**, *18*, 3427–3440. [CrossRef]
38. Wu, Z.; Xu, J.; Chen, H.; Shao, L.; Zhou, X.; Wang, S. Shear strength and mesoscopic characteristics of basalt fiber–reinforced loess after dry–wet cycles. *Mater. Civ. Eng.* **2022**, *34*, 04022083. [CrossRef]
39. Kang, W.; Ko, D.; Kang, J. Erosion resistance performance of surface-reinforced levees using novel biopolymers investigated via real-scale overtopping experiments. *Water* **2021**, *13*, 2482. [CrossRef]
40. Xiao, Y. Study on slope soil erosion at different anglesof Guizhou Province in karst area. *Ground Water* **2024**, *46*, 234–235. [CrossRef]
41. Zhou, J.; Liu, J.; Wang, W. The surface velocity distribution measurement based on ls-pivtechnology on in the downstream slop face of a dam. *Pearl River* **2017**, *38*, 58–61.
42. Shi, Z.; Wang, Y.; Peng, M.; Xu, Q. Large scale shaking table testson failure and breach process of landslide dams under aftershocks. *J. Eng. Geol.* **2014**, *22*, 71–77. [CrossRef]
43. Tang, C.; Pan, X.; Lu, C.; Dong, Z.; Liu, B.; Wang, D.; Li, H.; Cheng, Y.; Shi, B. Bio-geoengineering technology and the applications. *Geol. J. China Univ.* **2021**, *27*, 625. [CrossRef]

materials

Article

Experimental Study on the Wind Erosion Resistance of Aeolian Sand Solidified by Microbially Induced Calcite Precipitation (MICP)

Jing Qu [1], Gang Li [1,*], Bin Ma [1], Jia Liu [1], Jinli Zhang [2], Xing Liu [1] and Yijia Zhang [1]

1 Shaanxi Key Laboratory of Safety and Durability of Concrete Structures, Xijing University, Xi'an 710123, China
2 State Key Laboratory of Coastal and Offshore Engineering, Dalian University of Technology, Dalian 116024, China
* Correspondence: t_bag945@126.com

Abstract: Microbially induced calcite precipitation (MICP) is an emerging solidification method characterized by high economic efficiency, environmental friendliness, and durability. This study validated the reliability of the MICP sand solidification method by conducting a small-scale wind tunnel model test using aeolian sand solidified by MICP and analyzing the effects of wind velocity (7 m/s, 10 m/s, and 13 m/s), deflation angle (0°, 15°, 30°, and 45°), wind erosion cycle (1, 3, and 5), and other related factors on the mass loss rate of solidified aeolian sand. The microstructure of aeolian sand was constructed by performing mesoscopic and microscopic testing based on X-ray diffraction analysis (XRD), Fourier-transform infrared spectroscopy (FTIR), and scanning electron microscopy (SEM). According to the test results, the mass loss rate of solidified aeolian sand gradually increases with the increase in wind velocity, deflation angle, and wind erosion cycle. When the wind velocity was 13 m/s, the mass loss rate of the aeolian sand was only 63.6%, indicating that aeolian sand has excellent wind erosion resistance. $CaCO_3$ crystals generated by MICP were mostly distributed on sand particle surfaces, in sand particle pores, and between sand particles to realize the covering, filling, and cementing effects.

Keywords: MICP; aeolian sand; wind erosion resistance; model test; solidified mechanism

Citation: Qu, J.; Li, G.; Ma, B.; Liu, J.; Zhang, J.; Liu, X.; Zhang, Y. Experimental Study on the Wind Erosion Resistance of Aeolian Sand Solidified by Microbially Induced Calcite Precipitation (MICP). *Materials* **2024**, *17*, 1270. https://doi.org/10.3390/ma17061270

Academic Editors: Madhav Baral and Charles Lu

Received: 6 February 2024
Revised: 4 March 2024
Accepted: 7 March 2024
Published: 9 March 2024

1. Introduction

Arid and semi-arid regions account for 40% of the world's total land area. Under the action of wind erosion, land desertification and sandstorm events have happened more frequently in these regions, further deteriorating the ecological environment. Therefore, it is urgent to implement wind prevention and sand solidification. Traditional sand fixation methods include mechanical sand fixation [1], plant sand fixation [2], and chemical sand fixation [3]. However, these methods have disadvantages such as a long construction period, high cost, and environmental pollution. The microbially induced calcite precipitation (MICP) technique is an emerging sand solidification method characterized by high efficiency, environmental protection, and durability [4]. With respect to the principle of MICP mineralization, the microorganisms that produce urease are used to hydrolyze urea into CO_3^{2-}, which generates $CaCO_3$ deposits together with Ca^{2+} in the environment for solidification [5]. MICP has been applied to improve soils' properties [6], crack repair [7], treatment of pollution soil [8], and waste [9]. Ghalandarzadeh et al. [10] used MICP to improve the unconfined compressive strength of kaolinite clay. Behzadipour et al. [11] explored the application of MICP technology in improving the shear strength of gold mine tailings. The experimental results showed that compared with the untreated samples, the cohesion intercept and friction angle of the treated tailings samples increased by about 19 kPa and 5°, respectively. Prongmanee et al. [12] stabilized clayey soil by producing ammonium carbonate supernatant to generate calcite precipitation. The results showed that the soil stabilized with calcite had higher compressive strength than the untreated soil. The microcosmic test showed that calcite filled the voids between soil particles and resulted in the denser package of soil. Wang et al. [13] studied the effect of temperature on the cemented structure of sand treated by MICP. The results showed that the generation of $CaCO_3$ crystals by temperature would lead to changes in the internal friction angle, cohesion, stiffness, peak strength, residual strength, and expansion of sand samples treated by MICP. When carbonate crystals produced at 4 °C and 50 °C were fewer and smaller, they had lower strength reinforcement. In contrast, more larger crystal clusters were produced at 20 °C and 35 °C, which have effectively reinforced the sand particles. Jiang et al. [14] prepared cementing solution in hydrochloric acid solution to promote the solidification rate in the MICP reaction and evaluate its effectiveness. Research had shown that this method promoted the rapid bonding of calcareous sand particles, resulting in an unconfined compressive strength (UCS) of 1312.6 kPa for the sand column after five treatments. Compared with the conventional test group, the UCS of the test group containing HCl increased by about 1357%. Naskar et al. [15] investigated the effect of MICP on the mechanical properties of coal fly ash (CFA). After research, it was found that the specimens cured by MICP had higher strength, stiffness, and cohesion. Within a 28-day processing period, the permeability of CFA decreased by 78%, and the precipitation rate of calcite increased by 8%. Dubey et al. [16] conducted biological cementation treatment on desert sand through MICP and carried out soil erosion tests in indoor wind tunnels. The test results showed that single doses of 0.5 M and 1 M cementing solutions could continuously produce crusts with depths of 2 cm and 3.5 cm, thus effectively reducing erosion under the maximum velocity of 55 km/h. Devrani et al. [17] treated sand with MICP and biopolymer. Wind tunnel tests showed that the threshold friction velocity (TFV) increased from 20 km/h of the untreated sand to 45 km/h of the sand treated by MICP and biopolymer. They also observed that the mass loss rate of sand decreased from 75.23% of the untreated sand to 0%. Hang et al. [18] formed a cemented layer on the surface of desert sand through MICP to resist erosion and studied various factors relating to the erosion resistance of desert sand. The results showed that the erosion resistance of desert sand and the penetration resistance of sand surfaces were improved with the increase in treatment temperature and cementing solution concentration.

MICP technology has high bonding strength. It is not only suitable for soil reinforcement but also has a broad application prospect in controlling pollution elements, repairing cracks, repairing cultural relics, and anti-seepage treatment. However, aeolian sand has small particles, low water content, poor permeability, and low shear strength. Further validation is necessary for aeolian sand solidification using MICP. By taking aeolian sand solidified by MICP as the research object and conducting a small-scale wind tunnel model test, we analyzed the effects of wind velocity, deflation angle, wind erosion cycle, and other related factors on the mass loss rate of solidified aeolian sand and evaluated the erosion resistance of solidified aeolian sand. All the findings of this study provide an important reference value and scientific basis for the practice of wind prevention and sand solidification in desert regions.

2. Materials and Methods

2.1. Test Materials

In order to reduce the negative effects of sandstorm brought by Mu Us Desert to the northwest of China, aeolian sand was taken from the desert (Figure 1). The physical properties of aeolian sand were tested in accordance with the standard for the geotechnical testing method (GB/T 50123-2019) [19]. The main physical properties of aeolian sand are shown in Table 1. Aeolian sand can be identified as poorly graded fine sand according to the coefficient of curvature and coefficient of nonuniformity.

Figure 1. Aeolian sand taken from Mu Us Desert in Yulin, Shaanxi, China.

Table 1. Basic physical index of aeolian sand.

G_s	ρ_{dmax} (g/cm^3)	ρ_{dmin} (g/cm^3)	d_{10} (mm)	d_{30} (mm)	d_{60} (mm)	C_u	C_c
2.65	1.85	1.47	0.086	0.120	0.198	2.30	0.85

The test strain was *Sporosarcina pasteurii* (ATCC 11859), purchased from Shanghai Bioresource Collection Center. The bacterial culture medium contained 5 g of NaCl, 20 g of agar powder, 900 mL of purified water, and 100 mL of 20% urea. During the test, the pH value of the culture solution was adjusted to 9.0, and the solution was sterilized under a high-pressure steam at 125 °C for 20 min. After the solution was cooled to 60 °C, 100 mL of sterilized 20% urea was added. Then, 1 mL of bacterial solution (OD_{600} = 1.0) was inoculated into the culture solution, and the mixed solution was placed onto a shaking table at 30 °C and 200 rpm for culture. The culture was stopped when the bacterial solution concentration of OD_{600} was 1.5, and the mixed solution was then placed into a 4 °C refrigerator for storage. The mixed solution of urea and calcium chloride with an equal concentration and volume was prepared as the cementing solution with a concentration of 1.25 mol/L and pH of 9.0.

2.2. Sample Preparation

During the test, an appropriate amount of aeolian sand was filtered by a 0.5 mm sieve and stored for further use. A small transparent plastic tray was used as a sand table for sample preparation. The top side of the sand table was 23.5 cm long, the bottom side was 15.4 cm long, the width was 16.5 cm, and the height was 3.0 cm. In order to ensure the discharge of bacterial and cementing liquid during the solidification process of the sample, 9 small holes with a diameter of 0.5 cm were drilled at the bottom of the sand tray at equal intervals. To prevent sand from accumulating at the bottom and flowing away, a layer of filter paper was laid at the bottom of the tray before loading. Subsequently, 40 sand tables were prepared under the condition of 1.45 g/cm^3 dry density for the wind tunnel model test. Among them, 36 were treated with MICP curing and 4 were loose sand. The test set the loading height of aeolian sand at 2.5 cm and calculated that the sand required for each sand table was 1050.9 g, and that the volume was 724.78 cm^3. Seventy milliliters of bacterial solution was evenly sprayed to the surface of aeolian sand using a handheld sprayer. Seventy milliliters of cementing solution with the same concentration was sprayed 1 h later. The steps above were repeated every 24 h to spray the bacterial solution and cementing solution 5 times, respectively. The wind tunnel test was conducted 10 to 15 days after the solidification samples were dried. The mass of the solidified aeolian sand samples was 1489.7 g (Figure 2).

Figure 2. Sand table device.

2.3. Test Method

The wind tunnel test was carried out by means of the SNDY-4000 wind tunnel testing machine (manufactured by Nanjing Meiwen Science and Education Instrument Co., Ltd., Nanjing, China) (Figure 3a). The test section of the wind tunnel testing machine was 1.0 m long, 0.3 m wide, and 0.3 m high, with a wind rate of 11,000 m^3/h and a rotating speed of 2800 r/min. During the test, the HT-9829 anemometer (manufactured by Shanghai Shouni Electric Technology Co., Ltd., Shanghai, China) was used to measure the wind velocity, and the anemometer was calibrated before each test to ensure the accuracy of the velocity, as shown in Figure 3b. Taking into account the effect of wind velocity (v), deflation angle (α), wind erosion cycle (n), and other related factors on the quality loss rate of aeolian sand (φ) considered during the test, we set 5 min as the deflation time. The next blow was carried out at an interval of 5 min, thus completing a cycle (Table 2). After the completion of all tests in each group, the mass loss rate of aeolian sand was obtained by measuring the mass of aeolian sand samples before and after the test. A lower mass loss rate of aeolian sand samples indicates stronger resistance of the samples against wind erosion.

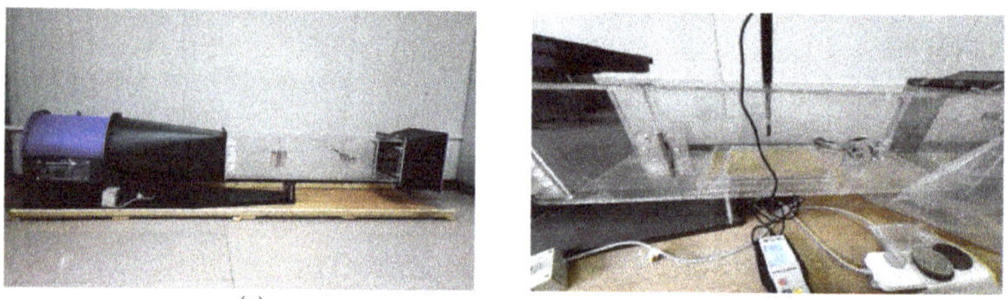

(a) (b)

Figure 3. Wind tunnel testing equipment: (**a**) wind tunnel testing machine; (**b**) anemometer.

Table 2. Wind tunnel testing scheme.

Deflation Angle (°)	Wind Velocity (m/s)	Wind Erosion Cycle
0	7	1, 3, 5
	10	1, 3, 5
	13	1, 3, 5
15	7	1, 3, 5
	10	1, 3, 5
	13	1, 3, 5
30	7	1, 3, 5
	10	1, 3, 5
	13	1, 3, 5
45	7	1, 3, 5
	10	1, 3, 5
	13	1, 3, 5

The action mechanism of aeolian sand solidification by MICP was unveiled by selecting MICP solidified aeolian sand as samples, conducting XRD (BrukerAXS D8 X-ray diffractometer, Bruker Corporation, Billerica, MA, USA), FTIR (Thermo Scientific Nicolet iS20 Fourier-transform infrared spectrometer, Thermo Fisher Scientific, Waltham, MA,

USA), and SEM (ZEISS Sigma 300 high-resolution scanning electron microscope, Carl ZEISS, Jena, Germany) testsand comprising and analyzing the crystal phase, characteristic functional group, and microstructure of aeolian sand before and after solidification.

3. Results and Discussion

3.1. *Analysis of the Effect Caused by Wind Velocity*

Wind velocity has a significant effect on the threshold friction velocity (TFV), flying height, and migration distance of aeolian sand. Figure 4 presents the curve of mass loss rate variation in aeolian sand with wind velocity. As can be seen from the figure, the mass loss rate of aeolian sand is positively correlated with wind velocity and keeps increasing with the increase in the wind erosion cycle. When the deflation angle and number of wind erosion cycles are constant and the wind velocity increases from 7 m/s to 13 m/s, the maximum mass loss rate of aeolian sand reaches 92.68%. When the deflation angle is 15° and the number of wind erosion cycles is three, the mass loss rate of aeolian sand increases from 3.34% to 56.57% with the increase in wind velocity. When the deflation angle is 30° and the number of wind erosion cycles is fiveand the wind velocity increases from 10 m/s to 13 m/s, the mass loss rate of aeolian sand shows a downward trend compared with that in the wind velocity range of 7 m/s to 10 m/s. The fundamental reason for the above situation is that after cementation through MICP, a cemented layer with a certain thickness is formed on the surface of aeolian sand, thereby increasing TFV.A higher TFV usually indicates stronger resistance of samples against wind erosion [17]. The surface of aeolian sand is broken with the continuous increase in wind velocity. When the wind velocity reaches a certain level, the surface cemented layer is destroyed. As a result, the mass loss rate of aeolian sand in sand tables increases rapidly. After the loose aeolian sand is blown away, the mass loss rate tends to be stable. Nikseresht et al. [20] conducted similar wind tunnel tests. The samples were blown for 5 min at a wind velocity of 10 m/s, 20 m/s, and 30 m/s, and the soil loss was analyzed after repeated operation. The conclusion is similar to that in this paper, in that it can have a strong soil stabilization effect against wind erosion, which is primarily related to the formation of $CaCO_3$ content. However, it is difficult to make a comprehensive comparison for different additives, wind speeds, and soil types.

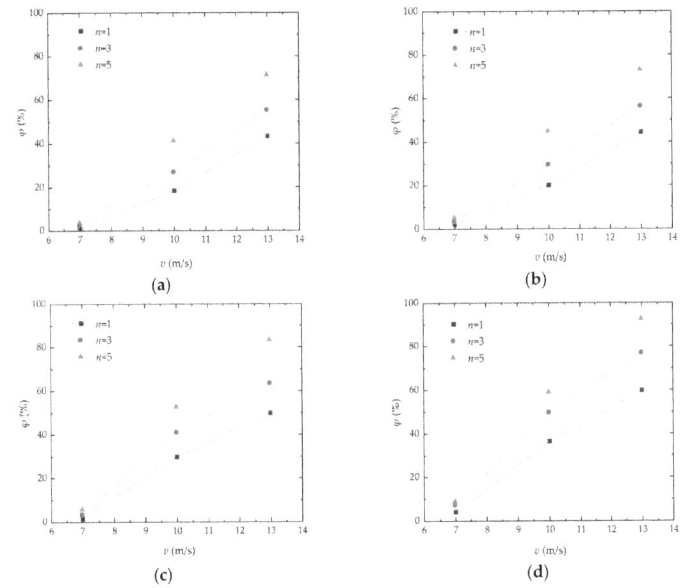

Figure 4. Curve of mass loss rate variation in aeolian sand with wind velocity: (**a**) $\alpha = 0°$; (**b**) $\alpha = 15°$; (**c**) $\alpha = 30°$; (**d**) $\alpha = 45°$.

3.2. Analysis of the Effect Caused by Deflation Angle

Deflation angle has a significant effect on the mass loss rate of aeolian sand. Figure 5 presents the curve of the mass loss rate variation in aeolian sand with the deflation angle. As can be seen from the figure, the mass loss rate of aeolian sand increases with the increase in the deflation angle under different wind velocities. The mass loss rate of aeolian sand also increases with the increasing number of wind erosion cycles. When the wind velocity and number of wind erosion cycles are constant and the deflation angle increases from 0° to 45°, the mass loss rate of aeolian sand shows a stepwise upward trend. When the wind velocity is 13 m/s and the number of wind erosion cycles is one, the mass loss rate of aeolian sand increases from 43.28% to 44.46%, it then increases from 44.46% to 49.99%, and it finally increases from 49.99% to 59.66%. The fundamental reason for this is that with the continuous increase in the deflation angle, the wind erosion experienced by the samples increase and the corresponding TFV decreases, so that sand particles are blown away by wind more easily. As a result, the mass loss rate of aeolian sand changes steadily when the deflation angle increases from 0° to 30°, but it changes sharply when the deflation angle increases from 30° to 45°. Low wind velocity cannot destroy the solid surface of aeolian sand [21]. When the deflation angle increases, the surface area of wind acting on aeolian sand increases gradually, which will blow the solid sand lumps on the outer surface of aeolian sand away from sand tables, leading to a steady increase in the deflation mass of internal loose sand. Tominaga et al. [22] conducted a wind tunnel experiment of sand erosion/deposition around the cube. The conclusion is that as the wind velocity increased, the mass transport rate increased sharply. And, most importantly, considerable erosion occurred at the windward corners of the cube. All these findings can be echoed in this paper.

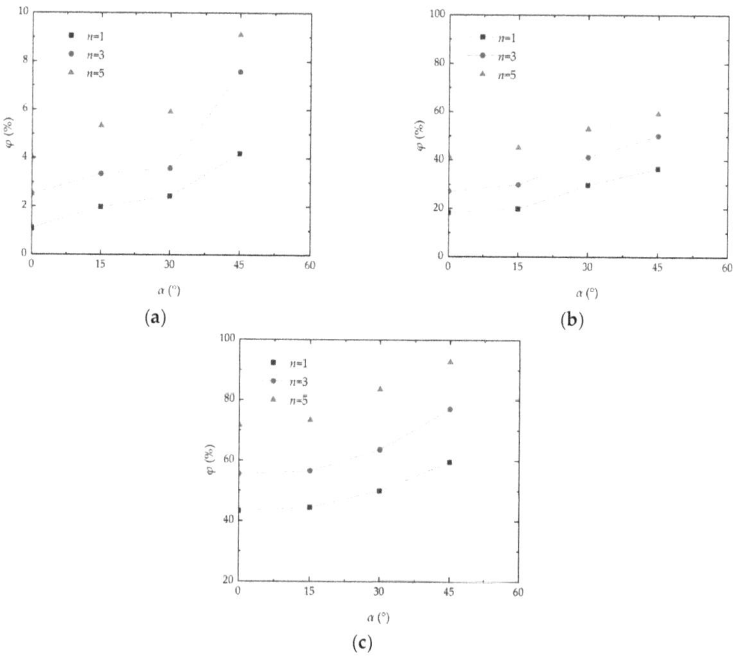

Figure 5. Curve of mass loss rate variation inaeolian sand with wind velocity: (**a**) v = 7 m/s; (**b**) v = 10 m/s; (**c**) v = 13 m/s.

3.3. Analysis of the Effect Caused by Wind Erosion Cycles

The wind erosion cycle sfactor is one of the key factors affecting the mass loss rate of aeolian sand. Figure 6 presents the curve of the mass loss rate variation in aeolian sand with wind erosion cycles. As illustrated, under the action of different wind speeds, the mass loss rate of aeolian sand increases with the increase in wind erosion cycles, and the increase in amplitude becomes larger with the increase in the deflation angle. When the wind velocity is 7 m/s and the deflation angle is 45°, the mass loss rates of aeolian sand after 1, 3, and 5 wind erosion cycles are 4.19%, 7.57%, and 9.11%, respectively. When the wind velocity is 13 m/s and the deflation angle is 45°, the mass loss rates of aeolian sand after 1, 3, and 5 wind erosion cycles are 59.66%, 77.04%, and 92.68%, respectively. The fundamental reason for the above situation is that when there are fewer wind erosion cycles, the surface of aeolian sand only suffers weak erosion and the crust of aeolian sand is not completely destroyed, resulting in a lower mass loss rate of aeolian sand. When the number of wind erosion cycles increases, continuous wind erosion destroys the solidified layer of aeolian sand and the internal loosened aeolian sand will be further eroded, thereby increasing the mass loss rate of aeolian sand. The increase in the number of wind erosion cycles is equivalent to the increase in the total time of blowing erosion. Desert environments are harsh, and the ability of aeolian sand to withstand prolonged erosion is one of the major challenges.

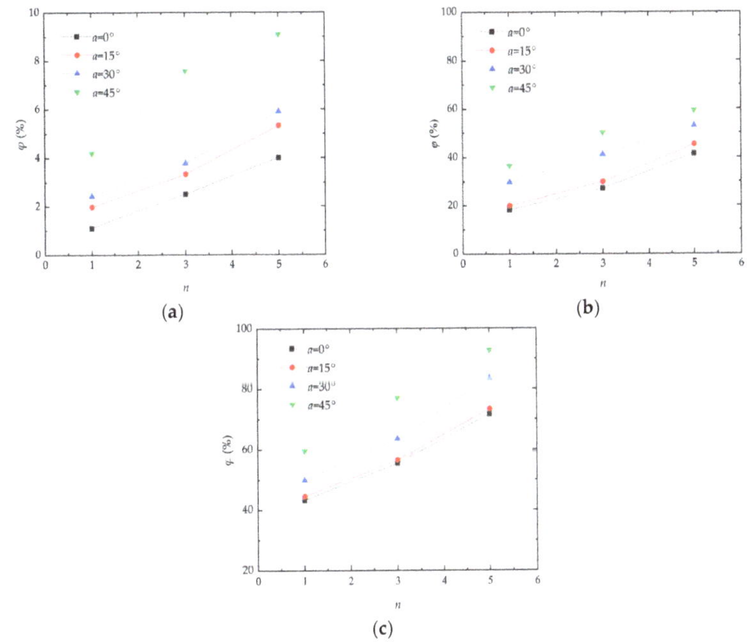

Figure 6. Curve of mass loss rate variation in aeolian sand with the number of wind erosion cycles: (**a**) v = 7 m/s; (**b**) v = 10 m/s; (**c**) v = 13 m/s.

3.4. Analysis of Morphological Characteristics of Aeolian Sand

Figure 7 presents the wind erosion morphology distributions of loose aeolian sand samples under different wind velocities, deflation angles, and wind erosion cycles; the deflation direction is from left to right. As can be seen from Figure 7a, when the wind velocity is 7 m/s, the deflation angle is 15°, and the number of wind erosion cycles is one; both the wind force and deflation angle are small, and the duration of wind action is short. Therefore, much aeolian sand is distributed in sand tables, and the aeolian sand moves from the middle to both sides. As can be seen from Figure 7b,c, the aeolian sand in the middle of the sand table is gradually blown away when the wind velocity is 10 m/s and the deflation angle is 30°, and the central blank area gradually increases with the increase in wind erosion cycles. According to Figure 7d, most of the aeolian sand in the sand table is blown and eroded, and only a small amount of aeolian sand is distributed at the end nearby wind source and the end far from the wind source when the wind velocity is 13 m/s, the deflation angle is 15°, and the number of wind erosion cycles is one. The above fact suggests that loose aeolian sand has poor resistance against wind erosion. Therefore, sandstorms and other natural disasters occur frequently in desert areas.

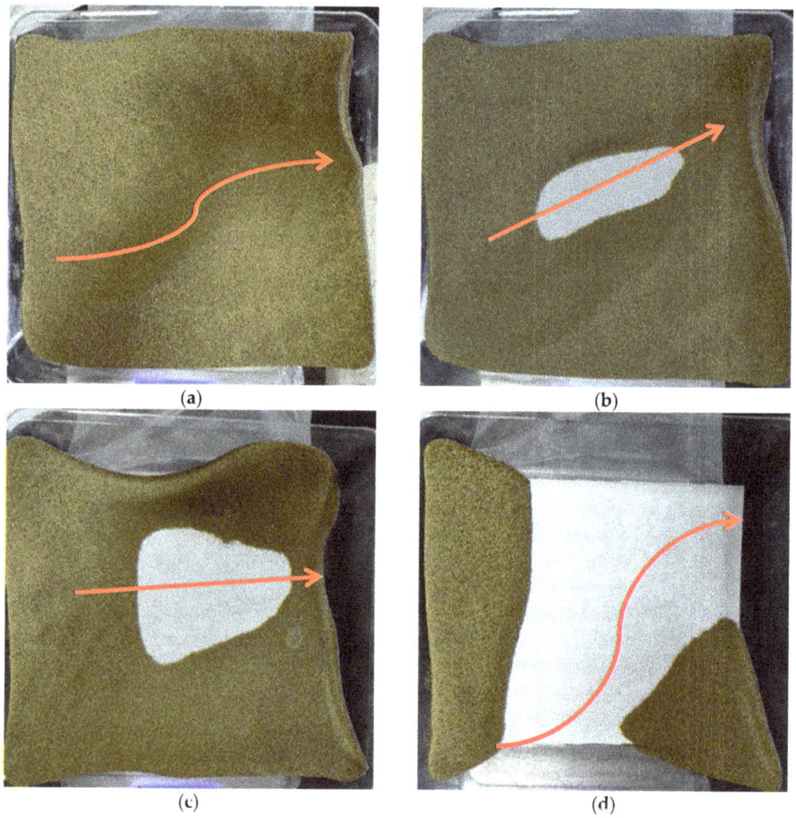

Figure 7. Wind erosion morphology of loose aeolian sand (The arrow indicate the dispersion direction of the aeolian sand): (**a**) $v = 7$ m/s, $\alpha = 15°$, $n = 1$; (**b**) $v = 10$ m/s, $\alpha = 30°$, $n = 1$; (**c**) $v = 10$ m/s, $\alpha = 30°$, $n = 3$; (**d**) $v = 13$ m/s, $\alpha = 15°$, $n = 1$.

Figure 8 presents the wind erosion morphology distributions of solidified aeolian sand samples under different wind velocities, deflation angles, and wind erosion cycles; the deflation direction is from left to right. As figured, the deflation of solidified aeolian samples is significantly less than that of loose aeolian sand when the wind velocity is 7 m/s, the deflation angle is 30°, and the number of wind erosion cycles is three [23] because solid crusts can prevent aeolian sand from deflation to some extent. Small solidified aeolian sand samples are blown away when the wind velocity is 10 m/s, and some sand particles abrade sample surfaces during movement, which will further deteriorate the destruction caused by wind erosion. When the deflation angle is 45°, the area under the action of wind erosion increases, and the middle part of aeolian sand suffers more serious erosion than the two sides. We found that $CaCO_3$ crystals mainly precipitated in the shallow layer because MICP only consolidated the surface layer of aeolian sand, and the experimental results of Wang et al. [24] are very consistent with this. When the wind velocity is 13 m/s, aeolian sand lumps are eroded and loose sand particles are blown away, resulting in a mass loss rate of 63.6%. Therefore, different from loose aeolian sand, a solid layer is formed on the surface of the aeolian sand solidified by MICP and has a certain ability to resist wind erosion. This will be of great help to reduce the occurrence of natural disasters in desert areas and reduce the current situation of air pollution.

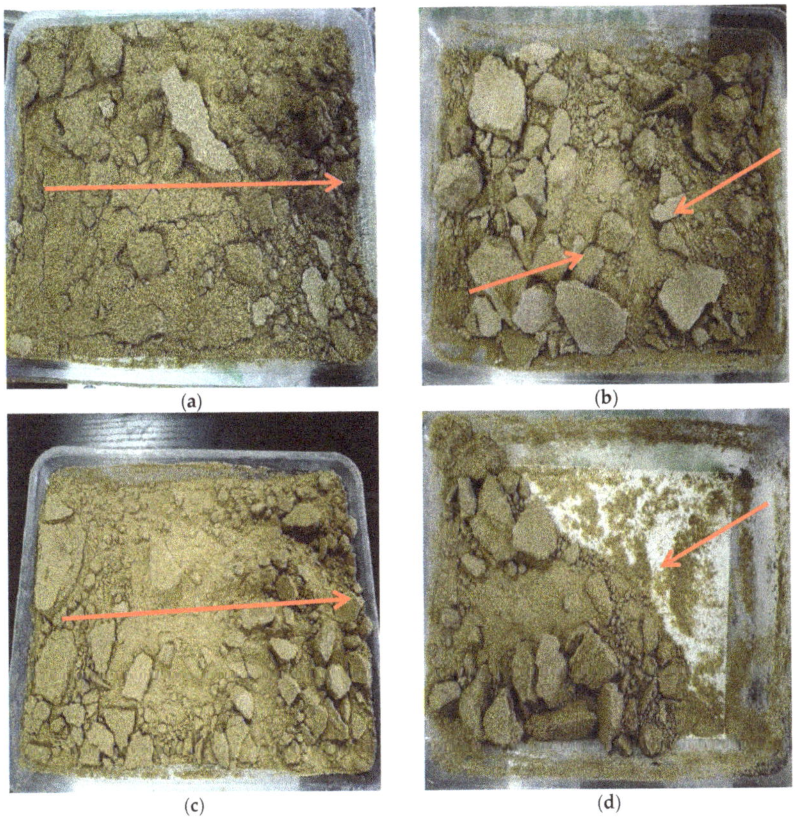

Figure 8. Wind erosion morphology of aeolian sand solidified by MICP (The arrow indicate the dispersion direction of the aeolian sand): (**a**) v = 7 m/s, α = 30°, n = 3; (**b**) v = 10 m/s, α = 30°, n = 3; (**c**) v = 10 m/s, α = 45°, n = 1; (**d**) v = 13 m/s, α = 30°, n = 3.

3.5. Analysis of the MICP Solidification Mechanism

In order to compare the mineral composition of aeolian sand before and after solidification and evaluate the effect of the MICP solidification reaction on aeolian sand, the crystal phase of the samples was determined by carrying out an XRD test. Figure 9 presents the XRD spectrum of both loose aeolian sand and MICP-solidified aeolian sand. As shown, quartz is the main mineral component of loose aeolian sand, showing fewer diffraction peaks, but MICP-solidified aeolian sand shows more characteristic diffraction peaks. According to the analysis conducted using the software Jade 6 (TiLab, Beijing, China), they are the characteristic diffraction peaks of calcite, indicating that calcite is the main crystal settling in the samples treated by MICP.

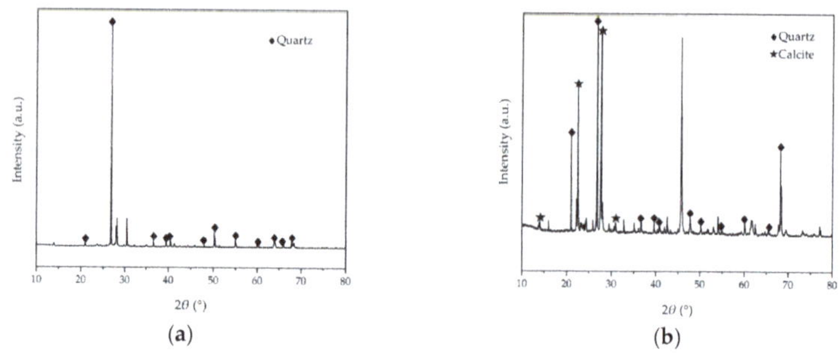

Figure 9. XRD spectrum: (**a**) loose aeolian sand; (**b**) MICP-solidified aeolian sand.

Figure 10 shows the FTIR testing results of both loose aeolian sand and MICP-solidified aeolian sand. As shown in the figure, the absorption peak of the two samples is mainly a narrow peak, and the position and quantity of absorption peaks vary slightly. The peaks at the location of 3621.02 cm^{-1} are O-H and N-H stretching vibration peaks; the peaks at the locations of 777.17 cm^{-1}, 694.09 cm^{-1}, and 462.09 cm^{-1} are Si-O symmetric stretching vibration peaks. The peak position at the location of 3233.70 cm^{-1} is the O-H stretching vibration peak, the peak at the location of 2512.77 cm^{-1} is the asymmetric vibration peak of CO_3^{2-}, and the absorption peak at the location of 1408.14 cm^{-1} is the C-O antisymmetric stretching vibration peak of CO_3^{2-}, as well as the characteristic peak of vaterite. The peak at the location of 1033.58 cm^{-1} is the C-O symmetric stretching vibration peak of CO_3^{2-}, indicating the presence of carbonate. The peaks at the locations of 777.77 cm^{-1}, 694.21 cm^{-1}, and 457.97 cm^{-1} are Si-O symmetric stretching vibration peaks, and SiO_2 is the main component of sand particles, which is consistent with the XRD test results. It can be concluded that the carbonate generated in the process of MICP mineralization is $CaCO_3$.

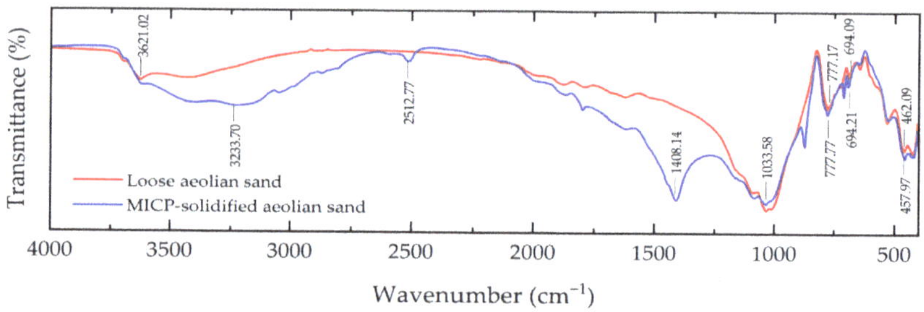

Figure 10. FTIR spectrum.

Figure 11 presents the microstructure of the aeolian sand solidified by MICP in order to further analyze the action mechanism of aeolian sand solidification by MICP. As illustrated, $CaCO_3$ crystals generated by MICP mineralization are mostly distributed on sand particle surfaces, in sand particle pores, and between sand particles to achieve covering, filling, and cementing effects [25]. In the process of MICP mineralization, bacteria are first adsorbed on the surface of aeolian sand particles, providing nucleation sites for the formation and superposition of $CaCO_3$ crystals [26]. The $CaCO_3$ crystals deposit, accumulate, and grow between adjacent particles to cement the adjacent particles into a whole, turning the point contact between particles into surface contact and improving the overall stability of the samples. $CaCO_3$ crystals can increase the surface roughness and cohesive force of sand particles, thereby improving the mechanical strength and erosion resistance of solidified aeolian sand. In Figure 11d, it can be seen that the $CaCO_3$ crystals generated by MICP are diamond-shaped, indicating that the generated $CaCO_3$ crystals are mainly calcite [27]. Calcite minerals are formed in the pores of particles to bind them. And there is a favorable relationship between its content and the strength of the sample [15]. These are consistent with the results of the XRD and FTIR experiments.

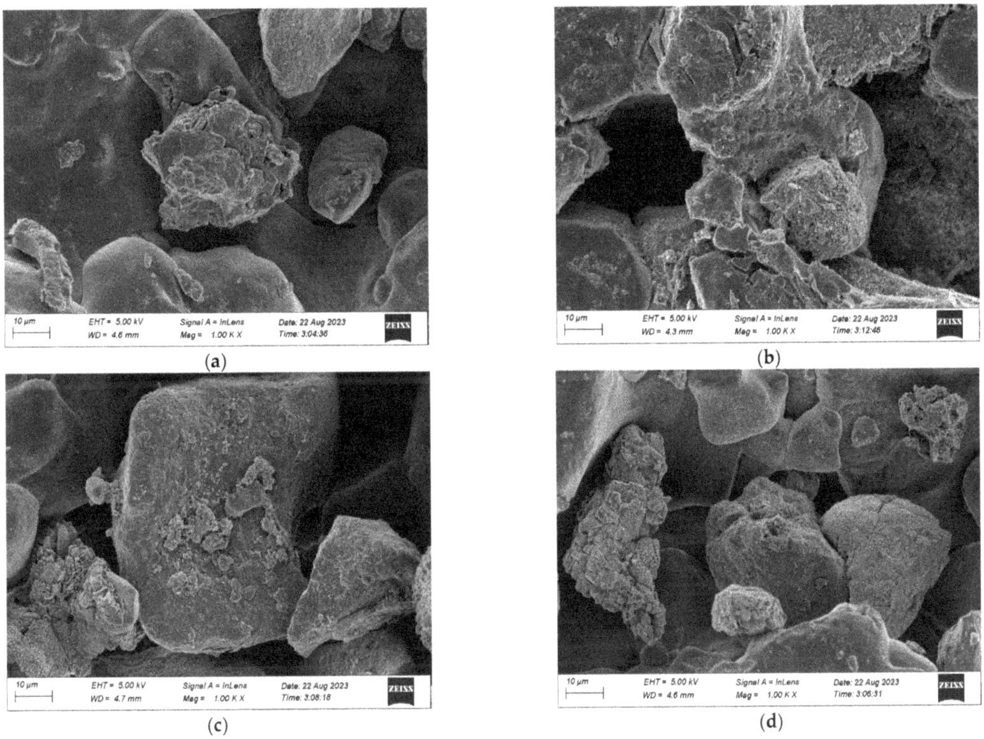

Figure 11. SEM image: (**a**) covering effect; (**b**) filling effect; (**c**) cementing effect; (**d**) calcite.

4. Conclusions

This study validated the reliability of the MICP sand solidification method by conducting a small-scale wind tunnel model test using aeolian sand solidified by MICP and analyzing the effects of wind velocity, deflation angle, wind erosion cycle, and other related factors on the mass loss rate of solidified aeolian sand. The microstructure of aeolian sand was constructed by performing mesoscopic and microscopic testing (XRD, FTIR, and SEM), thus revealing the mechanism of aeolian sand solidification via MICP. The main conclusions are as follows:

1. The mass loss rate of aeolian sand is positively correlated with wind velocity and keeps increasing with the increase in the wind erosion cycle. When the deflation angle and number of wind erosion cycle are constant and the wind velocity increases from 7 m/s to 13 m/s, the maximum mass loss rate of aeolian sand reaches 92.68%.
2. The mass loss rate of aeolian sand increases with the increase in the deflation angle. The mass loss rate of aeolian sand also increases with the increasing number of wind erosion cycles. When the wind velocity and number of wind erosion cycle are constant and the deflation angle increases from 0° to 45°, the mass loss rate of aeolian sand shows a stepwise upward trend.
3. Under the action of different wind speeds, the mass loss rate of aeolian sand increases with the increase in wind erosion cycles, and the increase amplitude becomes larger with the increase in the deflation angle. When the wind velocity is 7 m/s and the deflation angle is 45°, the mass loss rate of aeolian sand after 1, 3, and 5 wind erosion cycles is 4.19%, 7.57%, and 9.11%, respectively.

4. Loose aeolian sand has poor resistance against wind erosion. With the increase in wind velocity, the aeolian sand in the middle of the sand table is gradually blown away. Only a small amount of aeolian sand is distributed at the end near the wind source and the end far from the wind source. The solid layer formed on the surface of MICP-solidified aeolian sand has a certain ability to resist wind erosion. The mass loss rate of such aeolian sand is only 63.6% when the wind velocity is 13 m/s.
5. Quartz is the main mineral component of loose aeolian sand, while new calcite is the main mineral component of MICP-solidified aeolian sand. $CaCO_3$ crystals generated by MICP mineralization were mostly distributed on sand particle surfaces, in sand particle pores, and between sand particles to achieve covering, filling, and cementing effects.
6. The solidification of aeolian sand will be particularly important for subgrade filling, soil anti-seepage, and erosion resistance, as well as slope protection. However, the environment in the desert area is complex and harsh, and factors such as freeze–thaw and ultraviolet light have not been effectively solved, which is expected to be supplemented and improved on in future related research.

Author Contributions: Conceptualization, J.Q. and B.M.; methodology, J.L., J.Z. and X.L.; validation, G.L.; writing—original draft preparation, J.Q. and G.L.; writing—review and editing, J.Q. and Y.Z.; funding acquisition, G.L. All authors have read and agreed to the published version of the manuscript.

Funding: This study was supported by the Natural Science Basic Research Program of Shaanxi Province (2021JM-535) and Scientific Research Program Funded by Education Department of Shaanxi Provincial Government (23JS061).

Institutional Review Board Statement: Not applicable.

Informed Consent Statement: Not applicable.

Data Availability Statement: Data are contained within the article.

Conflicts of Interest: The authors declare no conflicts of interest.

References

1. Pinho-Lopes, M. Sand reinforced with recycled cotton textiles from waste blue-jeans: Stress–strain response. *Int. J. Geosynth. Ground Eng.* **2022**, *8*, 59. [CrossRef]
2. Fattahi, S.M.; Soroush, A.; Huang, N. Wind erosion control using inoculation of aeolian sand with cyanobacteria. *Land Degrad. Dev.* **2020**, *31*, 2104–2116. [CrossRef]
3. Arias-Trujillo, J.; Matías-Sanchez, A.; Cantero, B.; López-Querol, S. Effect of polymer emulsion on the bearing capacity of aeolian sand under extreme confinement conditions. *Constr. Build. Mater.* **2020**, *236*, 117473. [CrossRef]
4. Almajed, A.; Lateef, M.A.; Moghal, A.A.B.; Lemboye, K. State-of-the-art review of the applicability and challenges of microbial-induced calcite precipitation (MICP) and enzyme-induced calcite precipitation (EICP) techniques for geotechnical and geoenvironmental applications. *Crystals* **2021**, *11*, 370. [CrossRef]
5. Stocks-Fischer, S.; Galinat, J.K.; Bang, S.S. Microbiological precipitation of $CaCO_3$. *Soil Biol. Biochem.* **1999**, *31*, 1563–1571. [CrossRef]
6. Rahman, M.M.; Hora, R.N.; Ahenkorah, I.; Beecham, S.; Karim, M.R.; Iqbal, A. State-of-the-art review of microbial-induced calcite precipitation and its sustainability in engineering applications. *Sustainability* **2020**, *12*, 6281. [CrossRef]
7. Mengistu, D.M.; Mamo, A.N.; Gemeda, M.T. Isolation and characterization of calcite precipitating bacteria from soda lakes that have the capability to produce biocement for self-healing concretes. *Constr. Build. Mater.* **2023**, *408*, 133510. [CrossRef]
8. Wilcox, S.M.; Mulligan, C.N.; Neculita, C.M. Microbially Induced Calcium Carbonate Precipitation as a Bioremediation Technique for Mining Waste. *Toxics* **2024**, *12*, 107. [CrossRef] [PubMed]
9. Fouladi, A.S.; Arulrajah, A.; Chu, J.; Horpibulsuk, S. Application of Microbially Induced Calcite Precipitation (MICP) technology in construction materials: A comprehensive review of waste stream contributions. *Constr. Build. Mater.* **2023**, *388*, 131546. [CrossRef]
10. Ghalandarzadeh, S.; Maghoul, P.; Ghalandarzadeh, A.; Courcelles, B. Effect of nanoparticle-enhanced biocementation in kaolinite clay by microbially induced calcium carbonate precipitation. *Constr. Build. Mater.* **2024**, *414*, 134939. [CrossRef]
11. Behzadipour, H.; Sadrekarimi, A. Effect of microbial-induced calcite precipitation on shear strength of gold mine tailings. *Bull. Eng. Geol. Environ.* **2023**, *82*, 331. [CrossRef]
12. Prongmanee, N.; Horpibulsuk, S.; Dulyasucharit, R.; Noulmanee, A.; Boueroy, P.; Chancharoonpong, C. Novel and simplified method of producing microbial calcite powder for clayey soil stabilization. *Geomech. Energy Environ.* **2023**, *35*, 100480. [CrossRef]

13. Wang, Y.; Wang, Y.; Konstantinou, C. Strength Behavior of Temperature-Dependent MICP-Treated Soil. *J. Geotech. Geoenviron. Eng.* **2023**, *149*, 04023116. [CrossRef]
14. Jiang, Z.; Wei, R.; Dai, D.; Li, L.; Shang, Z.; Tang, J.; Peng, J.; Li, P. Experimental Study on Bio-Reinforcement of Calcareous Sand through Hydrochloric Acid Solution Precipitation into Cementing Solution. *Materials* **2023**, *16*, 6348. [CrossRef] [PubMed]
15. Naskar, J.; Sharma, A.K. Assessment of enhanced strength and stiffness properties of bio-engineered coal fly ash. *Constr. Build. Mater.* **2024**, *413*, 134793. [CrossRef]
16. Dubey, A.A.; Devrani, R.; Ravi, K.; Dhami, N.K.; Mukherjee, A.; Sahoo, L. Experimental investigation to mitigate aeolian erosion via biocementation employed with a novel ureolytic soil isolate. *Aeolian Res.* **2021**, *52*, 100727. [CrossRef]
17. Devrani, R.; Dubey, A.A.; Ravi, K.; Sahoo, L. Applications of bio-cementation and bio-polymerization for aeolian erosion control. *J. Arid Environ.* **2021**, *187*, 104433. [CrossRef]
18. Hang, L.; Yang, E.; Zhou, Y.; Song, W.; He, J. Microbially induced calcite precipitation (MICP) for stabilization of desert sand against the wind-induced erosion: A parametric study. *Sustainability* **2022**, *14*, 11409. [CrossRef]
19. *GB/T 50123-2019*; Standard for Geotechnical Testing Method. Construction Ministry of PRC: Beijing, China, 2019.
20. Nikseresht, F.; Landi, A.; Sayyad, G.; Ghezelbash, G.R.; Schulin, R. Sugarecane molasse and vinasse added as microbial growth substrates increase calcium carbonate content, surface stability and resistance against wind erosion of desert soils. *J. Environ. Manag.* **2020**, *268*, 110639. [CrossRef]
21. Meng, H.; Shu, S.; Gao, Y.; He, J.; Wan, Y. Kitchen waste for *Sporosarcina pasteurii* cultivation and its application in wind erosion control of desert soil via microbially induced carbonate precipitation. *Acta Geotech.* **2021**, *16*, 4045–4059. [CrossRef]
22. Tominaga, Y.; Okaze, T.; Mochida, A. Wind tunnel experiment and CFD analysis of sand erosion/deposition due to wind around an obstacle. *J. Wind Eng. Ind. Aerod.* **2018**, *182*, 262–271. [CrossRef]
23. Naeimi, M.; Chu, J.; Khosroshahi, M.; Zenouzi, L.K. Soil stabilization for dunes fixation using microbially induced calcium carbonate precipitation. *Geoderma* **2023**, *429*, 116183. [CrossRef]
24. Wang, Z.; Zhang, N.; Ding, J.; Lu, C.; Jin, Y. Experimental study on wind erosion resistance and strength of sands treated with microbial-induced calcium carbonate precipitation. *Adv. Mater. Sci. Eng.* **2018**, *2018*, 3463298. [CrossRef]
25. An, R.; Gao, H.; Zhang, X.; Chen, X.; Wang, Y.; Xu, H. Mechanical behaviour and microstructure of granite residual bio-cemented soil by microbially induced calcite precipitation with different cementation–solution concentrations. *Environ. Earth Sci.* **2024**, *83*, 31. [CrossRef]
26. Li, J.; Ma, J.; Tong, Y.; Fei, E.; Zhang, Z. Study on fractal characteristics of pores of NAS reinforced by MICP under the control of electric field. *Constr. Build. Mater.* **2021**, *271*, 121540. [CrossRef]
27. Srinivas, M.K.; Alengaram, U.J.; Ibrahim, S.; Vello, V.; Phang, S.M. Feasibility study on the use of microalgae as an external crack healing agent for cement mortar rehabilitation. *J. Sustain. Cem.-Based Mater.* **2024**, *13*, 17–32. [CrossRef]

MDPI AG
Grosspeteranlage 5
4052 Basel
Switzerland
Tel.: +41 61 683 77 34

Materials Editorial Office
E-mail: materials@mdpi.com
www.mdpi.com/journal/materials

www.ingramcontent.com/pod-product-compliance
Lightning Source LLC
Chambersburg PA
CBHW061807210926
51850CB00006B/635
* 9 7 8 3 7 2 5 8 1 9 1 6 4 *